Forest Ecology

Forest Ecology

Editor: Aduardo Hapke

Callisto Reference
www.callistoreference.com

Callisto Reference,
118-35 Queens Blvd., Suite 400,
Forest Hills, NY 11375, USA

Visit us on the World Wide Web at:
www.callistoreference.com

ISBN: 978-1-63239-833-8 (Hardback)

The publisher's policy is to use permanent paper from mills that operate a sustainable forestry policy. Furthermore, the publisher ensures that the text paper and cover boards used have met acceptable environmental accreditation standards.

Printed in China.

Cataloging-in-publication Data

Forest ecology / edited by Aduardo Hapke.
p. cm.
Includes bibliographical references and index.
ISBN 978-1-63239-833-8
1. Forest ecology. 2. Forest biodiversity. 3. Jungle ecology. I. Hapke, Aduardo.
QH541.5.J8 F67 2017
577.3--dc23

Table of Contents

Preface

Forest ecology can be defined as the interaction of flora and fauna in forests. Forests can be of various types according to location and climate, and the patterns of conservation, preservation and analysis of forest ecosystems change according to these factors. The objective of this book is to give a general view of the different areas of forest ecology. It explores all the important aspects of this field in the present day scenario and provides significant information of this discipline to help develop a comprehensive understanding. It includes contributions of experts and scientists which will provide innovative insights into this field. This book on forest ecology will prove to be thought-provoking for students and researchers in the fields of ecological risk assessment, wildlife conservation and ecosystem engineering.

After months of intensive research and writing, this book is the end result of all who devoted their time and efforts in the initiation and progress of this book. It will surely be a source of reference in enhancing the required knowledge of the new developments in the area. During the course of developing this book, certain measures such as accuracy, authenticity and research focused analytical studies were given preference in order to produce a comprehensive book in the area of study.

This book would not have been possible without the efforts of the authors and the publisher. I extend my sincere thanks to them. Secondly, I express my gratitude to my family and well-wishers. And most importantly, I thank my students for constantly expressing their willingness and curiosity in enhancing their knowledge in the field, which encourages me to take up further research projects for the advancement of the area.

Editor

1

A New Tool for Exploring Climate Change Induced Range Shifts of Conifer Species in China

Xiaojun Kou[1,2], Qin Li[3], Carl Beierkuhnlein[4], Yiheng Zhao[2], Shirong Liu[5]*

1 State Key Laboratory of Earth Surface Processes and Resource Ecology, Beijing Normal University, Beijing, PR China, 2 College of Life Sciences, Beijing Normal University, Beijing, PR China, 3 Department of Botany, University of British Columbia, Vancouver, B.C., Canada, 4 Department of Biogeography, University of Bayreuth, Bayreuth, Germany, 5 Institute of Forest Environment and Ecology, Chinese Academy of Forestry, Beijing, PR China

Abstract

It is inevitable that tree species will undergo considerable range shifts in response to anthropogenic induced climate change, even in the near future. Species Distribution Models (SDMs) are valuable tools in exploring general temporal trends and spatial patterns of potential range shifts. Understanding projections to future climate for tree species will facilitate policy making in forestry. Comparative studies for a large number of tree species require the availability of suitable and standardized indices. A crucial limitation when deriving such indices is the threshold problem in defining ranges, which has made interspecies comparison problematic until now. Here we propose a set of threshold-free indices, which measure range explosion (I), overlapping (O), and range center movement in three dimensions (Dx, Dy, Dz), based on fuzzy set theory (Fuzzy Set based Potential Range Shift Index, F-PRS Index). A graphical tool (PRS_Chart) was developed to visualize these indices. This technique was then applied to 46 Pinaceae species that are widely distributed and partly common in China. The spatial patterns of the modeling results were then statistically tested for significance. Results showed that range overlap was generally low; no trends in range size changes and longitudinal movements could be found, but northward and poleward movement trends were highly significant. Although range shifts seemed to exhibit huge interspecies variation, they were very consistent for certain climate change scenarios. Comparing the IPCC scenarios, we found that scenario A1B would lead to a larger extent of range shifts (less overlapping and more latitudinal movement) than the A2 and the B1 scenarios. It is expected that the newly developed standardized indices and the respective graphical tool will facilitate studies on PRS's for other tree species groups that are important in forestry as well, and thus support climate adaptive forest management.

Editor: Keping Ma, Institute of Botany, Chinese Academy of Sciences, China

Funding: This research was carried out with the support of the Special Research Program for Public-Welfare Forestry (Grant No. 201404201) and NSFC's Project (Grant No. 31370486). The funders had no role in study design, data collection and analysis, decision to publish, or preparation of the manuscript.

Competing Interests: The authors have declared that no competing interests exist.

* Email: Liusr@caf.ac.cn

Introduction

The need to predict potential changes in species distribution in response to climate change has stimulated the development of Species Distribution Models (SDMs), or more generally named Ecological Niche Models (ENMs), which combine current species occurrence data with recorded environmental data layers to detect the distribution of and fundamental niche for a given species [1,2,3]. SDMs are reported to have moderate to high accuracy in reproducing current distributions based on training data sets [4]. Although still controversial regarding their transferability to future climates [5,6,7,8], SDMs are the most effective and widely used approach to forecast future ranges of species [2].

In addition to the current progress in theoretical or methodological studies [9,10,11,12,13], SDMs have been increasingly applied in practice-oriented evaluations and assessments for climate adaptive management. This includes the conservation of endangered species [14,15,16], natural reserve networks [3,10,17], biodiversity changes [18,19,20], and climate change driven invasion processes [21]. In ecological or biodiversity studies, there is the trend that a large number of species are modeled to make inferences or test hypotheses. These species are chosen either from eminent taxonomical groups, or from major life forms of some regional species pools [20,22,23].

A systemic exploration of potential range shifts in response to future climates for large numbers of tree species is still not feasible (but see [4,24,25]). Studies on individual tree species range shifts are invaluable for climate adaptive forest management, but systematic studies on large groups of species or at least on a set of key species in a focal area are even more important. Most species are expected to experience range shifts to some extent; forest policy should understand the likelihood of changes by considering the sensitivity of species to a modified climate, the intensity of response, and the future availability of niche space. Identifying potential range changes, or ideally the patterns (spatially or taxonomically) of changes within a region, will guide forest policy in management [25].

Unlike in biodiversity studies, which are based on established indices for diversity and species turnover [23], species range studies lack standardized measures and quantitative indices. Approaches that are comparative, cumulative, or aggregated across species are less developed, however this is exactly what is needed in practice when a large set of species is concerned (as in forestry) and arbitrary preferences must be avoided.

SDM's can be realized for hundreds or thousands of species, but it is difficult to generalize results that subsume a large number of species projections. In addition, the projections are usually not illustrated in traditional binary map format but rather in continuous value format, which is suitable for representing potential habitat suitability or probability of occurrence [1]. Thresholds are adopted to convert these range maps to binary ones [23,26,27,28].

The currently used threshold method makes interspecies comparisons of range shifts problematic. Firstly, how to best choose a threshold is still an unsolved problem. Furthermore, even if some standard species-based threshold selection procedure were widely accepted, the resultant thresholds would still be species specific, making comparisons among different species unreasonable. The development of taxonomical group level threshold selection procedure could be more complicated and is not foreseeable. Thus, a set of standard indices, which are threshold-free, are essential for both the researchers' and policy makers' perspectives. Fuzzy set theory is providing a promising theoretical concept for this purpose [29].

The adaptation of management directives to novel or changing conditions depends on successful communication between researchers and policy makers. Graphical representation is one of the most effective ways of facilitating interdisciplinary communication [30]. In the case of range shift studies, maps are a common method of graphical representation. However, preparing hundreds to thousands of individual maps for policy makers is not practical. Critical characteristics of range shifts and generality cannot be identified and the construction of simplified figures is challenging as well. It is difficult to summarize key patterns of range change and still maintain the information on individual species. Combining range shift indices with intuitive and easy-to-understand diagrams, based on precise data, would support the implementation of knowledge in climate adaptive forest management.

Due to the longevity of trees, forest management and adaptation must be proactively implemented as soon as possible if climatic features are changing. Conifers constitute perhaps the most important taxonomic group of trees in the world as much for their spatial extent as for their role in economy. This is also true for China, where ambitious afforestation programs are aiming at the maintenance and improvement of ecosystem services such as erosion control, water management, and the generation of biotic resources.

Pinaceae are representing the largest extant conifer family, including more than 230 species and with an extensive distribution [31,32,33,34]. Most members of this family are trees that are usually dominant or co-dominant in their ecosystems [35]. The world's largest biome and ecosystem type, the boreal forests, is characterized by Pinaceae trees. Over a large scale, these conifers form the northern tree line and are thus of major importance when discussing temperature-dependent biome shifts. Pinaceae species also dominate mountain forests at lower latitudes in the Northern hemisphere, and are major components of temperate mixed forests [31,33,35]. Needle-leaved trees represent an important Plant Functional Type in global vegetation models [36,37]. As mostly evergreen species (with the exception of the genus *Larix*), conifers are closely linked to climatic conditions throughout the year. Changing climate can affect conifers throughout all seasons.

Forests that are dominated by conifers are of global relevance in terms of economic resources, habitats for many species, and for the global carbon pool [31,33,35]. Thus, Pinaceae constitute one of the most important taxonomical groups in global forest ecosystems and in world forestry. Case studies on the future potential distribution of these trees are urgently needed.

In this study, we first project future ranges of the most abundant Chinese Pinaceae species using SDM's. We then develop and apply a quantitative index and a novel graphical tool to illustrate species potential range shifts. In a next step, we explore generality in the response of the whole group of Pinaceae species to changing climatic features. Finally, we discuss potential implications of this new approach in climate adaptive forest management.

Materials and Methods

Ethics statement

We confirm that no specific permissions were required for these study sites, the location is not privately-owned or protected in any way, and the field studies did not involve endangered or protected species.

Diversity of the Pinaceae family in China

Eastern Asia is one of the most important origins for Pinaceae, and one of the major refuges during the Pleistocene glaciation periods for this family that is concentrated to the Holarctic realm. However, during their evolution, the Pinaceae family has experienced considerabe shifts in latitudinal and altitudinal distribution [38]. Today, China is a hotspot of species diversity for Pinaceae. Among the 11 genera of Pinaceae, the following genera are represented in China: *Abies*, *Keteleeria*, *Nothotsuga*, *Tsuga*, *Pseudolarix*, *Larix*, *Cathaya*, *Pseudotsuga*, *Picea*, and *Pinus*. Only the genus *Cedrus* is not represented. A great variety of Pinaceae species are distributed throughout the Chinese mainland, Taiwan and the Hainan islands, spanning tropical to subarctic climate zones [33,35]. More than half (about 130 of 230) of the world's Pinaceae species occur in China [35].

Forests between the Great Xing'an Mountains and the Himalayas, and between Taiwan and the Altaic mountains, are dominated or co-dominated by Pinaceae species. In particular, four genera (*Abies*, *Picea*, *Larix* and *Pinus*) are common and species-rich in China, and constitute the major coniferous forests. The species of these four genera exhibit broad ranges in both geographical and environmental space. Single species are limited in their distribution areas, however, the taxonomic group is represented in across a wide range of climatic conditions. A clear geographical substitution sequence exists along the temperature gradient. For example, in the genus *Pinus*, *P. latteri*, *P. massoniana*, *P. tabulaeformis*, *P. koraiesis*, and *P. sylvestris* var. *mongolica* range from tropical to subarctic zones and exhibit a geographical substitution pattern [35]. The genus *Larix* is adapted to extremely cold temperatures during winter, as the species of this genus are deciduous and summer-green, avoiding frost damage especially in continental regions. Precipitation gradients are especially important at the margins of distribution, both as limiting factors of tree growth and as drivers of competition with other tree species.

Species and environmental data

Species distribution data. We extracted distribution maps of the species from the digital version of the *Atlas of Chinese Vegetation 1:1,000,000* [39] using ArcInfo Workstation version 8.3 (ESRI, Redlands, CA, USA). Within the four genera under focus, species were chosen that dominated or co-dominated forests based on a minimum threshold of thirty locations of presence. Rare and less abundant species were excluded in this study that is aiming towards ecosystem functions and forest management. A location of presence was defined as a $5' \times 5'$ (geographical coordinates) cell that contained at least 1% forest area dominated or co-dominated by the target species. This treatment ensures that sample points are at least 5′ apart, without missing small patches (above 1% of the cell area) of species distribution within the cell. We modeled the distribution ranges where the species were dominant and disregarded the non-dominant distribution or scattered distribution areas of individual species. Whether a species dominated or co-dominated a forest was based on whether the vegetation type was named after the target species at the unit of 'formation', as defined in the *Vegetation of China* [35].

Based on these criteria, twelve *Abies* species, twelve *Picea* species, six *Larix* species and sixteen *Pinus* species (summing up to 46 tree species) were selected for modeling as quantitatively relevant and representative species out of the total of 130 Chinese Pinaceae species. General information on the prevalence and centroid locations for all species is shown in Figure 1 and listed in Table S1.

Environmental data. We extracted the predictive dataset from BioPlant (download from http://www.zoology.ubc.ca/~qinli/BioKou/), a world plant bioclimatic dataset with a 10′ (latitude/longitude) sample resolution [40] and downscaled it to 5′ using change factor downscaling techniques. The BioPlant dataset calculates 15 layers of variables derived from monthly temperature and precipitation data, which were downscaled from general circulation model (GCM) predictions. Change factor downscaling techniques were adopted to obtain a fine scale dataset [41], which fully considered the effects of elevation on climatic variables.

Climatic data for calculating bioclimatic variables representing the present were downloaded from the WorldClim dataset [42], which has a resolution and interpolating schedule consistent with those used for our future scenarios. The same procedure was applied to calculate the 15 plant biological variables. Aridity was defined as the natural logarithm of the quotient of total potential evapotranspiration to total precipitation in the growing season (monthly temperature greater than 0°C in this case). Because the original BioPlant dataset did not contain this layer, we calculated it using the GRID module in ArcInfo Workstation v8.3. The time period in this study was set to the end of this century, and the future environmental layers were the 20-year means (from 2081 to 2100).

Species Distribution Model (SDM) implementation

There are various methods of creating an ensemble of single models to reduce prediction uncertainties. In this paper, we emphasize General Circulation Model (GCM) induced and predictor selection-caused uncertainties. We used MaxEnt to build the ensembles without taking into account uncertainties related to modeling techniques.

Each ensemble was composed of 25 combinations from five GCM's and five predictor sets. The five GCM's were extracted from twenty-three GCM-derived climatic predictions from the BioPlant data set. The five GCMs were EH5, HAD, IM3, MER and PC1, of which MER and PC1 predicted the highest and lowest temperature rise, EH5 and IM3 predicted the most divergent local patterns of annual precipitation, and HAD predicted intermediate temperature and precipitation [40].

For predictor selection, we choose the most ecologically relevant bioclimatic variables based on general plant distribution theory and our understanding of the possible factors limiting Pinaceae distribution [34,43,44]. The five predictor sets were 1) GDD0, and Aridity; 2) T_mean, P_total, and Aridity; 3) T_cold, GDD0, and Aridity; 4) T_mean, P_total, P_season, TP_syn, and Aridity; and 5) T_cold, GDD0, P_season, TP_syn, and Aridity. In the predictor sets, GDD0, Aridity, T_mean, P_total, T_cold, P_season, TP_syn represent growing degree days above 0°C, aridity of growing season, annual temperature mean, annual total precipitation, mean temperature of the coldest month, precipitation seasonality, and synchronicity between temperature and precipitation, respectively [40].

There are considerations on the spans of predictors: the number of predictors (2, 3, and 5), whether the variables represent annual mean or seasonal extremes (Tmean vs T_cold), temperature or precipitation constraints, and whether seasonal patterns are important (P_season, TP_syn). It is evident that aridity is a major factor limiting Pinaceae distribution, thus this variable is always included in the predictor sets.

Ensembles were built for all species and scenarios (integrating 5 GCMs plus 5 predictor sets for each species). MaxEnt was used to build individual SDM's for the current distribution of each species. Then the predicted future occurrence values based on the different climate models was summarized by the median. This provided the consensus prediction of the ensemble on a cell-by-cell basis [10]. We adopted logistic output format, and the default settings for all model parameters in MaxEnt (version 3.31). Scripts were compiled to perform a batch run with MaxEnt [45,46].

Measurements of model performances

The model performances were measured by the area under the receiver operating characteristic curve (AUC), maximum Kappa value (*Max* κ), and the maximum true skill score (*Max TSS*) for each species. All κ and *TSS* values were calculated with a series of thresholds ranging from 0.05 to 0.95, and having an interval of 0.05. The *Max* κ and *Max TSS* were selected from those values for each species. Arc Macro language (AML) (in ArcInfo Workstation version 8.3) scripts were adopted to implement the calculations. Means and standard deviations of AUC, *Max* κ and *Max TSS* indices were aggregated to the genera level (*Abies, Picea, Larix*, and *Pinus*).

Definitions for the Fuzzy Set based Potential Range Shift (F-PRS) set of indices and graphical representation

Threshold-free range shift indices. To quantify trends in range changes, two indices were used to measure the fundamental aspects of range shifts. The range increment index (I) measures the change in potential range area under a future climate scenario compared with the current climate. A negative value indicates range contraction and a positive value indicates range expansion. The range overlap index (O) measures the proportion of range overlap under a future climate scenario with the range overlap under the current climate. A value of 0 indicates that no overlap exists and a value of 1 indicates complete congruence. The translocation of the species range center is measured by three indices: Dx, Dy, and Dz, with the first two representing horizontal movements along longitudinal and latitudinal directions, and Dz representing the vertical movement.

The definition of a specific species range can be ambiguous when predictions are made in terms of continuous values, such as suitability or probability of occurrence [1]. Arbitrary threshold selection or selecting the 'best threshold' may cause incongruence in what constitutes a range, resulting in deviating predictions [23,26,27,47].A threshold-free range definition based on fuzzy set theory avoids this problem. Fuzzy set theory does not perform a discrete classification among members of a data set. Actually, it numerically estimates how well a member fits into certain categories. Thus, a member can belong to several categories to some degree as measured by its memberships [48].

This study focuses on how many cells could be classified as having a specific species listed as "present." A cell is not defined as

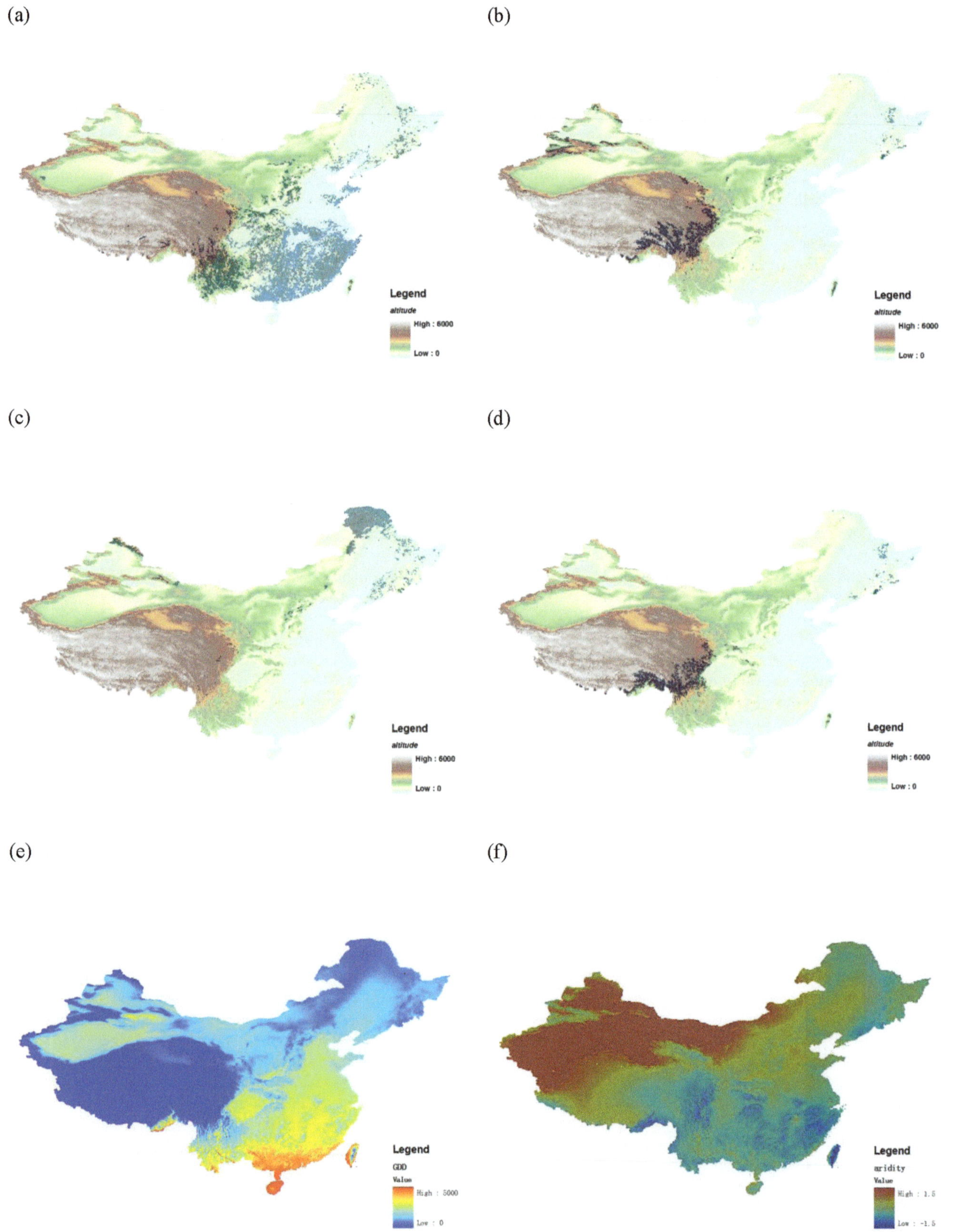

Figure 1. Species distributions and environmental layers. Figure (a–d) are distribution maps of species pooled to genera *Pinus*, *Abies*, *Larix* and *Picea*, respectively, superimposed over background digital elevation model (DEM) maps with 50% transparence. Figure (e) and (f) are maps of the environmental factors gross degree days (GDD) and aridity respectively.

'belonging' or 'not belonging' to the range but rather partially belongs to the range with a degree of membership [8,29]. Membership is a function of suitability or probability of occurrence, respectively. Here, I and O indices based on the membership function were applied [29]. The equations to calculate these indices are:

$$I = (c(F_f) - c(F_p))/c(F_p).$$

$$O = c(F_f \cap F_p)/c(F_f \cup F_p).$$

where, c represents fuzzy set cardinality, F_p represents the fuzzy set of the current distribution, and F_f is the fuzzy set of the future distribution. The indices were calculated using AML scripts in ArcInfo Workstation (Appendix File S1).

The calculated species range movements were based on the movement of the centers of each species range between time periods, where the range center was defined as the geometrical centroid of the entire potential range. To avoid the threshold selection problem, a fuzzy set style centroid was defined, instead of a traditional centroid, which did not depend on partitioning between "present" or "absent". All cells in a grid map that contributed to the centroid were weighted by the membership of those cells that pertain to a specific species. Cells with a membership of zero do not actually contribute to the centroid but can be dealt with in the same way as any other cell.

The three dimensions (*x*, *y*, *z*) of the centroid can be calculated by:

$$Cx = \sum (M_{ij} \times X_{ij}) / \sum X_{ij}$$

$$Cy = \sum (M_{ij} \times Y_{ij}) / \sum Y_{ij}.$$

$$Cz = \sum (M_{ij} \times Z_{ij}) / \sum Z_{ij}.$$

where *Cx*, *Cy*, and *Cz* are the *x*, *y*, *z* coordinates of the centroid. The *i* and *j* are the column number and row number of a cell in a grid map. X_{ij}, Y_{ij}, and Z_{ij} are the coordinates of the $cell_{ij}$, and M_{ij} is a binary variable indicating whether a focal species is present or absent in $cell_{ij}$. $\sum$ are the sums of all cells within a map.

Below, *Dx*, *Dy*, and *Dz* represent east/west, north/south, and higher/lower elevation movements, respectively. The range movement indices were defined as:

$$Dx = Cx_{end} - Cx_{start}$$

$$Dy = Cy_{end} - Cy_{start}$$

$$Dz = Cz_{end} - Cz_{start}$$

where the subscripts *end* and *start* were attached to the coordinates of centroids to indicate time periods.

In this study, membership is set equal to the logistic output of MaxEnt. Here, the units of *Dx* and *Dy* were angular degrees (longitude and latitude, respectively), and the unit of *Dz* was meters a.s.l.. All calculations were completed using ArcInfo Workstation version 8.3 (Appendix File S2).

Potential Range Shift chart (PRS_Chart) tool. A diagram tool named Potential Range Shift chart (PRS_Chart) (Figure 2) was developed to illustrate the five indices of Potential Range Shifts (I, O, Dx, Dy, and Dz). A unit circle was used to represent current distribution area for the focal species, and another circle with a radius proportional to the ratio of distribution areas between current and projected future distributions was used to indicate the future distribution area. Thus, I value was represented by $(Area_{black} - Area_{blue})/Area_{black}$. A sector of the unit circle with proportional area equal to the O value was used to demonstrate range overlap.

Dx and Dy combined were represented by a directional arrow line. The direction of the arrow represents the actual movement direction of distribution centers between current and future climates. The length of the arrow tail represents the distance of the movement. In order to depict a larger range of movement distance in the chart, logarithm transformation can be applied. In this study, a base 10 logarithm transformation was used, and the unit length of the arrow represents 1.0 degree (latitude and longitude) of movement. In this manner, a 10 degree movement was drawn 2 units long. Similarly, the distance and direction of vertical movement of the range center (Dz) was represented by a vertical arrow. It is also recommended to conduct a logarithm transformation to movement distance. A base 10 logarithm transformation was used with a unit length representing 100 meters of vertical movement.

An R script file was developed to draw the PRS_Chart (Appendix 2). Range shift charts of all combinations of species and climate scenarios were drawn using this script file (Appendix File S3).

Statistical analyses

The mean and standard deviation of I and O indices were calculated for each genus and for the three climate change scenarios under consideration. The mean and standard deviation were also calculated at the family level for each scenario. Grand means and standard deviations were finally calculated by pooling all species and scenarios.

Double-sided student distributions were used for all means of I, Dx, Dy, Dz values to test if these indices differed significantly from zero ($\alpha = 0.05$). Analysis of Variance (ANOVA) testing was conducted to determine whether the difference of the five indices among scenarios was significant ($\alpha = 0.05$). If any of the ANOVA analyses were significant, the least significant difference t-test (LSD-t) test was conducted for scenario to scenario comparisons ($\alpha = 0.10$). This test was applied to all four genera and to the whole Pinaceae family by pooling the four genera together. All calculations were conducted in Microsoft Visual Fortran (with IMSL Stat Library).

Results

Performances of the SDM modeling

The high AUC values illustrate very good discriminating power of the applied models. The lowest mean AUC value among the four genera, found in the genus *Pinus*, reached 0.975. Standard deviation was generally low (with highest standard deviation of 0.018, also in *Pinus*) (Table 1). The MaxTSS values were slightly lower than those of AUC (with lowest mean MaxTSS of 0.897 in *Pinus*) and were more variable (with highest standard deviation of 0.066 in *Pinus*) (Table 1). *Max* κ, however, indicated only fair performances (with the lowest mean of 0.371 in *Abies*, and highest

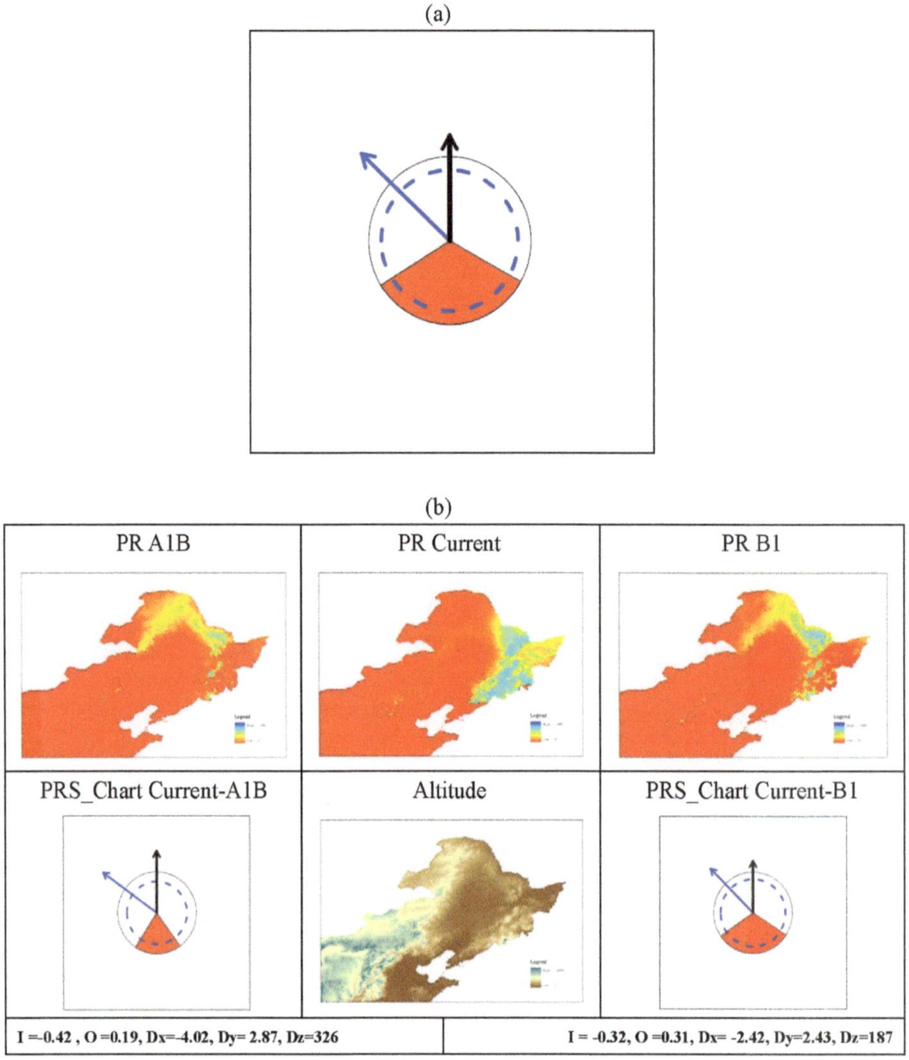

Figure 2. PRS_Chart and an example for *Pinus koraiensis*. Figure (a) is an example potential range shift chart (PRS_Chart) to visualize PRS indices. A unit circle (with thin black continuous line) was used to represent current distribution area for the focal species, and another circle (with broad blue broken line) with radium proportional to the ratio of distribution areas between current and projected future distributions was used to represent future distribution area. A sector of the unit circle (in red) with proportional area equal to the **O** value was used to represent range overlapping. **Dx** and **Dy** combined were represented by a directional arrow line (in blue). The distance and direction of vertical movement of range center (**Dz**) was represented by a vertical arrow (in black). Figure (b) are maps and PRS_Charts that served as an example for spices *Pinus koraiensis*. Maps in first row are potential range maps for climate scenario A1B (in the end of this century), current climate, and B1, respectively. The map in the middle of second row is digital elevation map, and charts in the left and right are PRS_Charts for A1B and A2 respectively. The values listed in bottom row are values of the PRS indices for A1B and B1.

standard deviation of 0.180 in *Pinus*) (Table 1), although it should be noted that the scales for AUC and *MaxTSS* are from 0.5 to 1.0, representing completely random prediction to perfect prediction, while the scale for *Max* κ is from 0.0 to 1.0.

General patterns of Potential Range Shifts (PRS's)

According to the PRS_Charts (Figure 3, 4, 5, 6), the following general traits of spatial patterns can be detected:

(1) Considerable interspecies variation exists in PRS's for each of the five indices;

(2) PRS indices were very consistent among the three climate change scenarios for all species.

(3) No clear inter-genera differences appeared in all five indices.

(4) No clear range expansion or contraction trends could be detected in the I index.

(5) The values of O index were generally low, mostly below 0.5 and peaking around 0.3 to 0.4. The A1B scenario was related to smaller or similar O values compared to those of A2 and B1.

(6) As to horizontal movements (Dx, Dy), there was a general trend for northward spreading, but no clear trend in eastward or northward direction. Thus, the prevalent moving directions were northeast, north, and northwest.

(7) As to vertical movement (Dz), almost all species moved upward with very few exceptions.

(8) Species that moved south were always accompanied by exceptionally strong upward shift (large Dz), indicating a

Table 1. Performance of SDM modeling measured by model evaluation indices.

	Abies	*Picea*	*Larix*	*Pinus*
AUC	0.984 (0.009)	0.983 (0.012)	0.982 (0.005)	0.975 (0.018)
Max TSS	0.935 (0.036)	0.913 (0.053)	0.920 (0.029)	0.897(0.066)
Max κ	0.371 (0.096)	0.375 (0.140)	0.372 (0.150)	0.380(0.180)

Notes: Table content is means and standard deviations of model evaluation indices of species distribution model (SDM) aggregated from all combinations species within the genra and climate scenarios (A1B, A2, and B1). **AUC** is area under the receiver operating characteristic curve. **Max TSS** is maximum true skill score. **Max κ** is maximum Kappa value.

Figure 3. PRS_Chart for *Abies*. Each chart is a Potential range shift chart (PRS_Chart) that is explained in Figure 2. Charts in the tabular style figure representing combinations of species (SP1-12) in *Abies* and climate scenarios (A1B, A2, and B1). General species information can be found in Table S1.

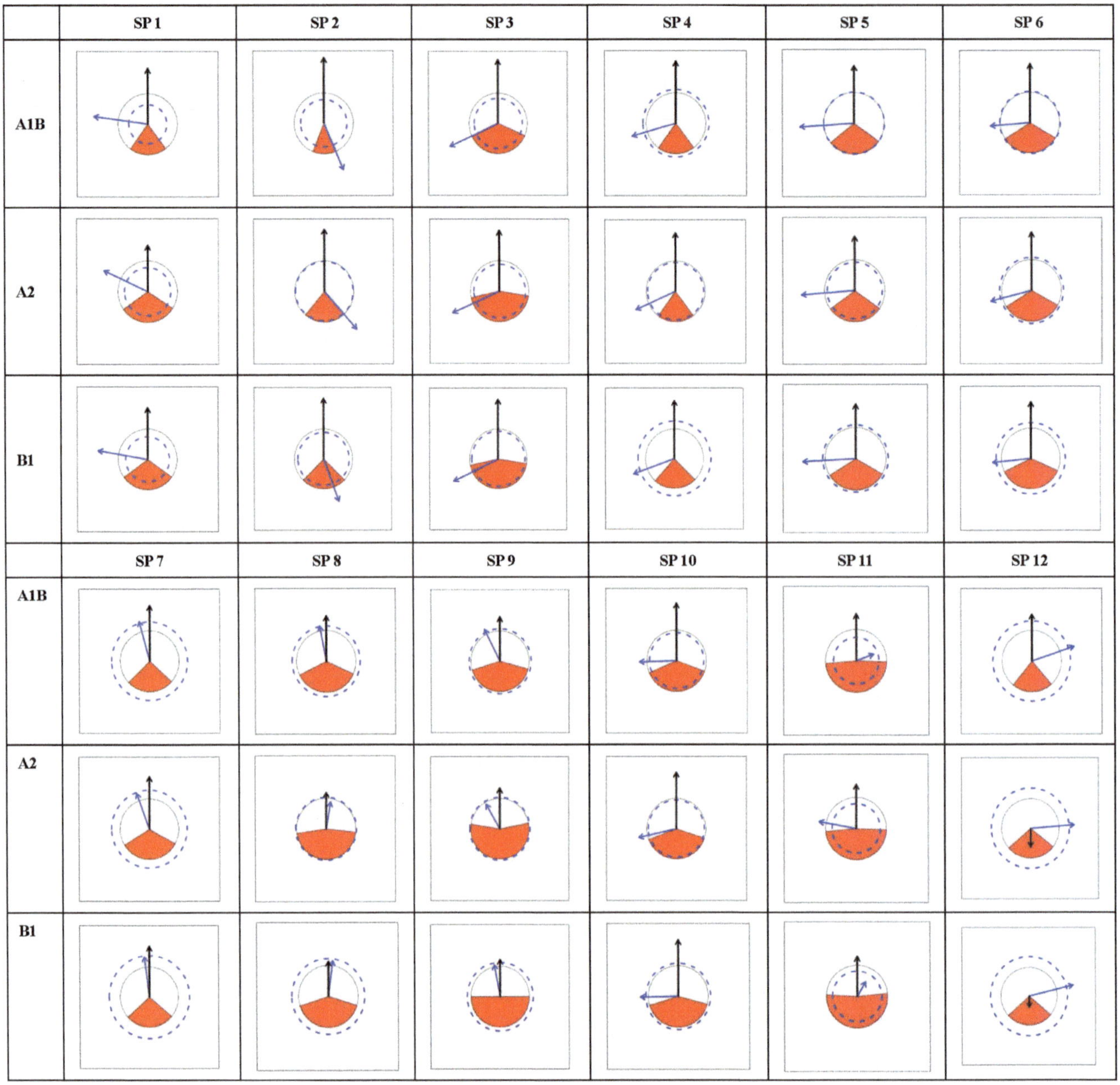

Figure 4. PRS_Chart for *Picea*. The same as legend of Figure 3 except the species number is for *Picea*.

compensatory effect between upward and northward movement trends.

Quantitative summarization of Potential Range Shifts

The mean range size changes (I index in table 2) for Pinaceae species were relatively small (0.09, 0.12, and 0.22 for A1B, A2, and B1 respectively) in comparison with large standard deviations (0.59, 0.59, and 0.63 for A1B, A2, and B1). No significant ($\alpha = 0.05$) range size changes were detected both at family level and genus level for all three climate scenarios.

The range overlapping (O index in Table 2) showed low values (with means 0.29, 0.36 and 0.35 for A1B, A2, and B1) and relatively small interspecific variations (with standard deviations 0.14, 0.13 and 0.13 for A1B, A2, and B1) at family level and a similar pattern at the genus level.

The longitudinal location of the range center (Dx index in Table 2) was found to move slightly westward (with mean −1.15, −0.60 and −0.70 degrees for A1B, A2, and B1) but this trend was not statistically significant and was connected with large interspecific variation (standard deviation 4.12, 3.52, and 3.72 degrees for A1B, A2, and B1). Comparable results were found for all genera.

The latitudinal displacement of the range center at the family level (Dy in Table 2) was slightly larger (with mean1.27, 1.03, and 1.13 degrees for A1B, A2, and B1) than the longitudinal trend, but had much smaller interspecific variations (with standard deviation 2.18, 1.86, and 1.84 degrees for A1B, A2, and B1). The northward movement was highly significant ($\alpha = 0.01$) at the family level. Comparable trends were found in *Pinus* and *Abies* but did not

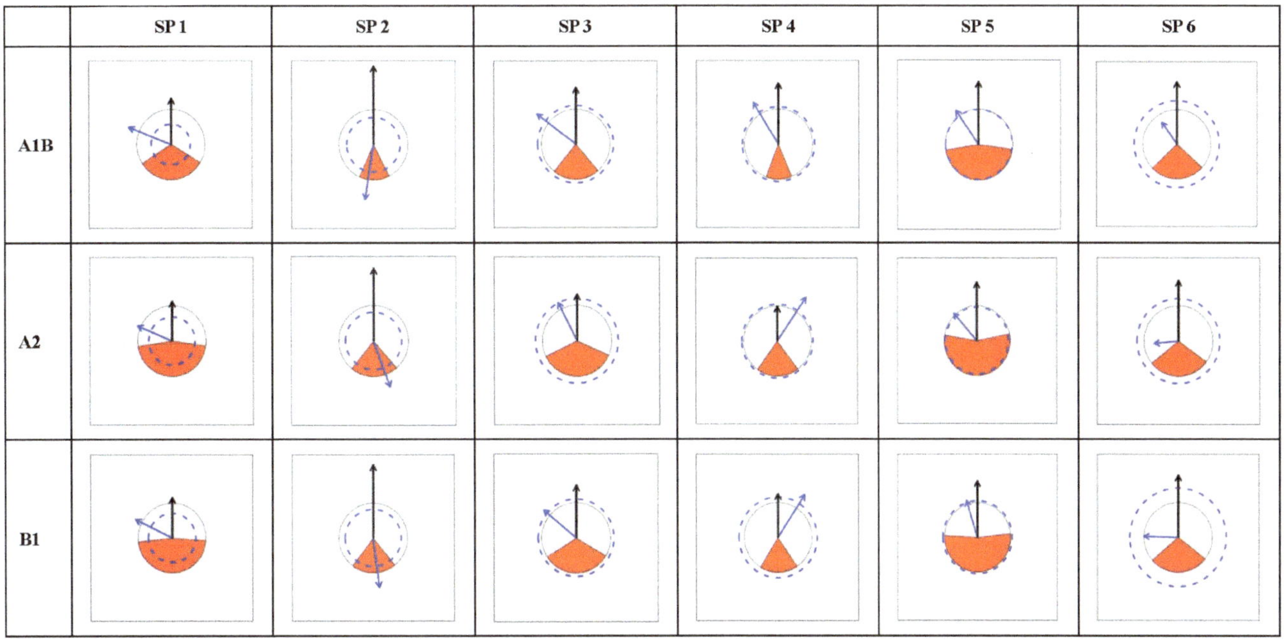

Figure 5. PRS_Chart for *Larix*. The same as legend of Figure 3 except the species number is for *Larix*.

emerge in *Picea* and *Larix*. This indicates that the northward trend at the family level can be attributed mainly to pine and fir species.

The vertical movement (Dz in Table 2) trends were highly significant at the family level and were also highly significant in *Pinus* and *Picea*, but the significance of *Abies* and *Larix* depends on climate scenarios. Mean movement distances of the family were 727, 339, and 364 meters, with standard deviations 392, 345, and 342 meters for A1B, A2, and B1, respectively.

Inter-scenario comparisons

The ANOVA (Table 3) indicated no significant difference ($\alpha = 0.05$) among scenarios in horizontal directions (Dx and Dy) both at the genus and at the family level. There was also no significance among scenario differences in range size changes (O index) at family and genus level with the exception of *Picea*, which responded to scenario B1 with a significantly ($\alpha = 0.10$ by LSD-t test) larger range expansion compared to A1B or A2.

Generally, there were highly significant differences among scenarios ($\alpha = 0.10$ by ANOVA in Table 3) in range overlap (O index) and vertical movement (Dz) at family level. O values in A1B scenario were found to be significantly smaller than those in A2 and B1, and Dz values in A1B were significantly larger than those in A2 and B1. A similar pattern was also found at the genus level but with a few cases of ANOVA results being significant instead of highly significant. An exception was the genus *Larix*, which did not differ in vertical movement among scenarios. Consistent lower O values and larger Dz values indicated that a realization of scenario A1B would lead to the strongest range shifts.

Discussion

Performance of the novel set of indices for tree species range shift

Modeling results on the future potential optimum distribution of tree species are urgently needed for forest ecosystem management and silviculture in order to avoid the loss of resources and ecosystem services. Climate change projections are still vague to a certain degree mainly due to uncertainties in global economic development, mitigation policy, and technological progress. However, general trends emerge especially when scenarios are compared.

Here, a new quantitative index is presented as a tool for forestry, ecology, and biogeography that circumvents restrictions and disadvantages of former species distribution modeling approaches according to their focus on selected species, scenarios and directions, and intensity of shift. This research contributes to the rapid development and progress of biogeographical modeling in response to the ongoing global changes [49,50,51,52,53,54]. In this study, models are based on data for grid cells where the species were common or even dominant in order to avoid bias from small species populations on special sites with extraordinary microclimatic conditions that were not representative for the climate at the landscape scale of the modeled grid cells. Distributional data of this type did not differ substantially from total species ranges inferred from other map sources [39,55].

As a novel and robust approach, the Fuzzy Set based Potential Range Shift Index (F-PRS Index) is introduced in this study. Here, the application is focused on dominant, widespread, and economically important conifer species of the Pinaceae family in China. However, the method can be implemented to other species groups and regions in the future.

The F-PRS Index: (1) helps detect patterns that hide behind large amounts of information, and thus deepens understanding of the studied phenomena; (2) delivers a powerful tool for the building of statistically testable hypotheses; and (3) supports the efficient implementation of meta-analyses. Here, the novel F-PRS Set of Indices performance is demonstrated according to the first two aspects.

One important piece of information on the upward movement trends becomes evident by the Dz index, which integrates three levels of spatial information (current distribution, future distribution, and altitude). It is expected that this will contribute to the

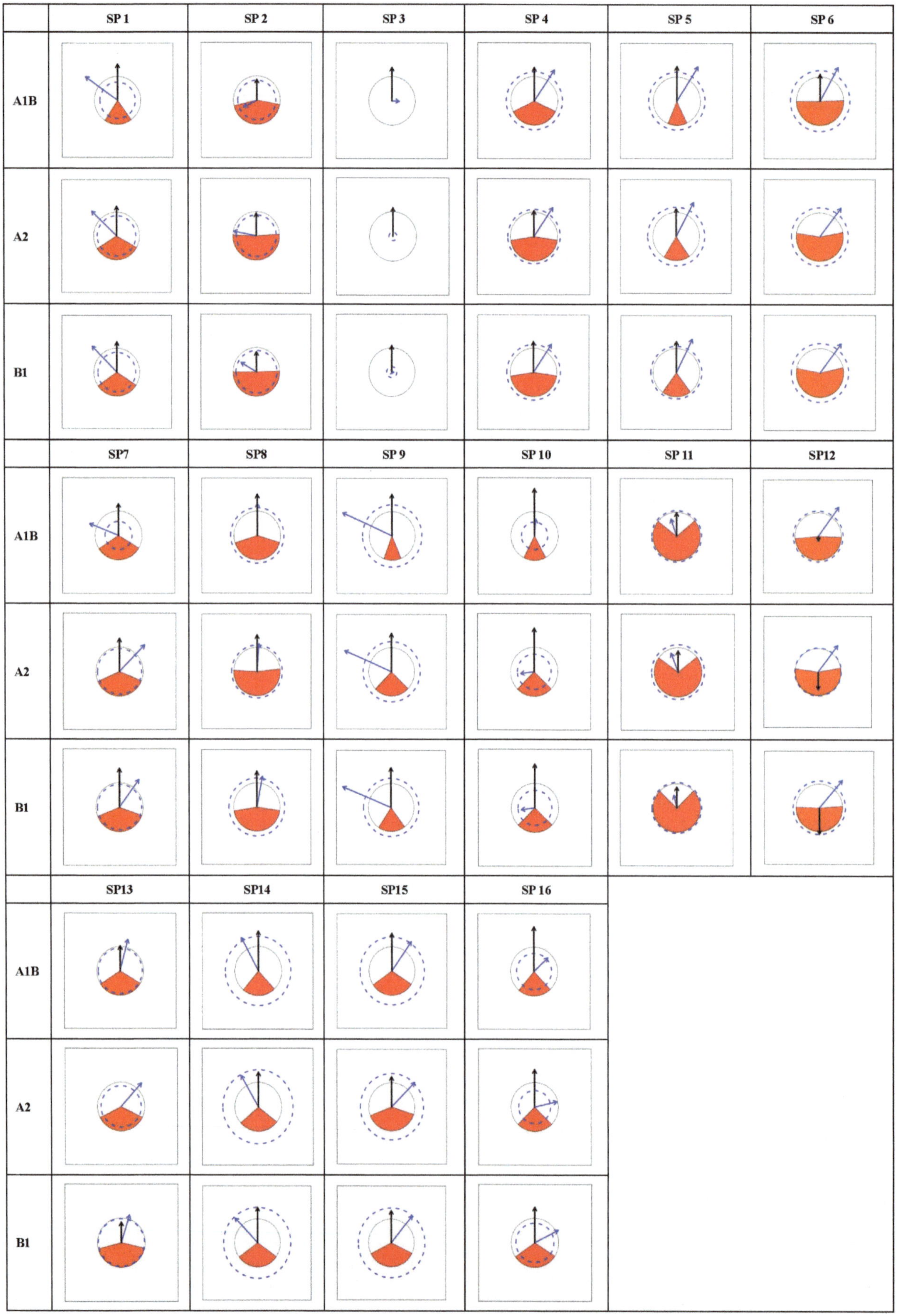
SP 1
SP 2
SP 3
SP 4
SP 5
SP 6
A1B
A2
B1
SP7
SP8
SP 9
SP 10
SP 11
SP12
A1B
A2
B1
SP13
SP14
SP15
SP 16
A1B
A2
B1

Figure 6. PRS_Chart for *Pinus*. The same as legend of Figure 3 except the species number is for *Pinus*.

third F-PRS Index aspect, and delivers an excellent basis for meta-analyses after wider applications.

General requirements for indices are: (1) The ability to extract key attributes that are essential to the studied phenomena out of complex interactions and processes; (2) The capacity to reach standardization and comparability among studies; (3) The stability against random perturbations; and (4) The independence of the members of the indices set.

The F-PRS indices set introduced in this study definitively fulfills the first criterion because the five aspects represented by the five member indices are all critical attributes for range studies. Species range size change is one of most concerning topics in climate change biology [56,57]. Range size conservation, and range size being identified as species level trait to test species level selection have instigated a large number studies in evolutionary ecology [58,59]. Range overlapping that measures the ratio of area shared by the current range and future range has a profound application in conservation biology [60]. Range position change is a hot topic in biogeography.

The drivers of northward and upward movements for species in response to climate warming are intensely studied both empirically and theoretically [61,62,63,64,65,66]. The temperature related "altitude for latitude" movement hypothesis is intensely debated in biogeography [67]. However, longitudinal movement, for instance in response to changes in precipitation regimes, is a little touched topic due to a lack of methodological concepts. Here, a tool that enables the detection of anisotropy in temporal trends in all directions is presented.

One preeminent advantage of the F-PRS indices set is that they are standardized and can be adopted to make comparisons among different studies. Although it requires that the SDM's predicted suitability maps are consistent, which is not granted in all cases, it can at least be used to make comparisons among studies using the same SDM. Ensemble models that integrate various species distribution modeling algorithms such as BIOMOD can serve as a standard pre-analysis in order to avoid deviations between specific SDM's [13].

The fuzzy set style indices presented here are more stable than former threshold defined measurements of range shifts [29]. Two sources of mechanisms may produce unstable results for threshold measurements. The first one is that small perturbations (such as random processes during SDM modeling and different strategies in selecting pseudo absent data) may cause large differences in selecting "best" threshold values, thus resulting in deviating range maps. The second source of uncertainty in traditional approaches comes from the fact that even small perturbations in threshold can cause large differences in range maps in case that the predicted suitability map contains a large area of very low values. The fuzzy set indices are certainly free from the first source of instability, because they are threshold free, and performed much better in dealing with the second source [29].

Lastly, it is evident that all the five indices in the F-PRS indices set have clearly differentiated meanings. Thus, we believe the new F-PRS indices set is good based on the above criterions.

Applicability in range change studies and forest managements

In practice, it is critical to achieve mutual understanding between researchers and policy makers via readily available communication techniques. Visualization is one of the most emphasized strategies to facilitate communication, the exchange of information, and its implementation [30]. In this case study, a graphical representation of range shifts (Figure 3, 4, 5, 6) is introduced, which will contribute to an efficient climate adapted management policy for coniferous forest in China. The effects that climate change is likely to bring about to Chinese forests can be instantly recognized. Consequences for climate adapted management becomes obvious.

Visualization is not only beneficial for forest management but also for research. Although it seems that all the information presented in the graphs has already been contained in the calculated indices, all the spatial information that is emerging in the patterns may not be revealed from numerical indices. Graphs are more appropriate than single values or numbers to demonstrate large-scale spatial patterns, which is exemplified by the eight general patterns identified in this study (section3.2). Based on these spatial findings, more specific hypotheses can be raised to be tested in the future.

Unlike theoretical studies, which usually emphasize summarizing general patterns and further reveal underlying mechanisms that form a large amount of case studies [68,69,70,71], management sectors such as forestry and conservation biology need to keep balance between generality and detail [72,73]. At one end, overall policy to maintain forest ecosystems is highly emphasized, at the other end, not a single species can be neglected in forestry or nature conservation. As the former has been fully emphasized, it is necessary to draw some attention on details, especially a few exceptional cases that can be clearly seen from the graphs (Figure 3, 4, 5, 6). For example, AbiesT05 (*Abies spectabilis*) had exceptionally large range expansion (several times the current range) and showed a strong downward movement trend in each scenario. PinusT03 (*Pinus pumila*) seemed to lose almost its complete potential range in each of the scenarios. PinusT11 (*Pinus massoniana*) had very large overlap between scenarios and little movement, indicating a stable range position. *Picea* speciesT11 (*Picea morrisonicola*) had very similar horizontal movement trends between scenarios A1B and B1, but was very different in A2, and PiceaT12 (*Picea spinulosa*) displayed large upward movement in A1B but moved slightly downward in A2 and B1. Apart from the fact that knowing the reasons that brought about those exceptions might have theoretic significance, knowing the facts themselves also has conservation and management importance.

Assisted migration of tree species in climate adaptive forest management

The results of this study show that for various species with considerable range displacement, assisted migration becomes more or less inevitable in order to maintain the functioning of ecosystems and to preserve the biological diversity of tree species. Intended translocation of endangered species has been proposed to avoid climate change caused extinction [74]. Immediate concerns about unintended side effects were expressed [75,76] and debates were aroused [77,78,79,80,81].

Here, extending the idea of assisted migration and endangered species translocation in biological conservation is proposed for the rearrangement of a larger amount of, as well rare, economically important tree species. However, the design of "novel ecosystems" is currently under debate and ecological and practical consequences are not solved, yet. Based on this case study on Pinaceae species in China, there are only around one third of overlapping

Table 2. Statistics of potential range shift indices.

Shift Indices	Emission Scenarios	*Abies*		*Picea*		*Larix*		*Pinus*		Family pooled	
		Mean	**Std**	**Mean**	**Std**	**Mean**	**Std**	**Mean**	**Std**	**Mean**	**Std**
I	**A1B**	0.17	0.82	0.02	0.42	−0.03	0.45	0.09	0.58	0.09	0.59
	A2	0.26	0.85	0.03	0.37	0.00	0.40	0.11	0.57	0.12	0.59
	B1	0.37	0.91	0.23	0.45	0.13	0.53	0.15	0.56	0.22	0.63
O	**A1B**	0.29	0.12	0.30	0.10	0.25	0.12	0.29	0.18	0.29	0.14
	A2	0.35	0.11	0.36	0.11	0.35	0.14	0.38	0.17	0.36	0.13
	B1	0.34	0.11	0.36	0.10	0.33	0.14	0.37	0.16	0.35	0.13
Dx (Degree)	**A1B**	−0.97	4.53	−1.87	3.14	−0.19	0.71	−0.72	5.28	−1.15	4.12
	A2	−0.17	4.14	−1.51	2.84	−0.08	1.14	−0.44	4.15	−0.60	3.52
	B1	−0.30	4.11	−1.41	2.80	−0.36	1.14	−0.61	4.72	−0.70	3.72
Dy (Degree)	**A1B**	1.56*	1.85	−0.13	1.82	0.42	2.09	2.42**	2.14	1.27**	2.18
	A2	1.45*	1.94	−0.25	1.52	0.52	1.64	1.87**	1.65	1.03**	1.86
	B1	1.51*	1.83	0.00	1.52	0.46	1.71	1.94**	1.72	1.13**	1.84
Dz (Meters)	**A1B**	526**	424	720**	380	669*	510	329**	238	527**	392
	A2	295*	408	542**	364	419*	379	188**	168	339**	345
	B1	357*	412	528**	351	472*	398	205**	173	364**	342

Notes: Table contents are mean(**mean**) and standard deviation (**Std**) of potential range shift (PRS) indices representing by combinations of **I**, **O**, **Dx**, **Dy**, **Dz** and CO_2 emission scenarios **A1B**, **A2**, and **B1** shown in first two columns. Superscripts ** after mean values indicating that the mean is very significant from zero ($\alpha = 0.01$), and * is significant different from zero ($\alpha = 0.05$) by using double sided student distribution. No superscript after mean value means it is not significant from zero, except the mean of O index which did not test for deviation from zero.

Table 3. Comparisons of Potential range shift (PRS) indices among climate scenarios.

Shift Indices	Tested Item	*Abies*	*Picea*	*Larix*	*Pinus*	Family pooled
I	ANOVA	X	P = 0.032	X	X	X
	A1B	/	0.02[a]	/	/	/
	A2	/	0.03[a]	/	/	/
	B1	/	0.23[b]	/	/	/
O	ANOVA	P = 0.034	P = 0.007	P = 0.018	P = 0.010	P = 0.000
	A1B	0.29[a]	0.30[a]	0.25[a]	0.29[a]	0.29[a]
	A2	0.35[b]	0.36[b]	0.35[b]	0.38[b]	0.36[b]
	B1	0.34[b]	0.36[b]	0.33[b]	0.37[b]	0.35[b]
Dx	ANOVA	X	X	X	X	X
	A1B	/	/	/	/	/
	A2	/	/	/	/	/
	B1	/	/	/	/	/
Dy	ANOVA	X	X	X	X	X
	A1B	/	/	/	/	/
	A2	/	/	/	/	/
	B1	/	/	/	/	/
Dz	ANOVA	P = 0.030	P = 0.026	X	P = 0.000	P = 0.000
	A1B	526[a]	720[a]	/	329[a]	527[a]
	A2	295[b]	542[b]	/	188[b]	339[b]
	B1	357[b]	528[b]	/	205[b]	364[b]

otes: The table contents after row **ANOVA** are results of analysis of variance for each genus comparing among climate scenarios, in which X representing not significant ($\alpha = 0.05$) or P-values is provided significant. Contents after row **A1B**, **A2**, and **B1** are results of multiple comparisons if the ANOVA result is significant by using least significant difference t test (LSD-t). the/means test is not carried out. Number and superscript represent mean value and multiple comparison result, with same superscript characters meaning not significant and different character meaning significant ($\alpha = 0.10$).

areas that are both suitable for the species at current climate and under climatic conditions in the end of this century, not accounting for species adaptive ability and other sources of uncertainty.

The huge overall changes in suitable ranges require that foresters track the potential ranges of most of the Pinaceae species, otherwise a tremendous decline of coniferous forest cannot be excluded. It is unfeasible and unrealistic to let natural processes (i.e. nature dispersal or local adaptation) fulfill this task for two reasons: firstly, current speed of anthropogenic caused climate change is much faster than natural climate variability in history [82], thus this will leave no chance for tree species to adapt in a short time period [83]. Secondly, the high fragmentation of natural habitats segregated by agricultural land uses, urban areas and road systems is suppressing dispersal and colonization of new habitats [84,85,86]. More broadly, forest management needs to identify and prepare in advance habitats for the establishment of larger numbers of the tree species close to but perhaps outside their previous "native" range, or even novel forest communities because they were formerly considered to be potentially natural communities, not representing future climate anymore [87].

Fundamental changes in forestry will become unavoidable. Although there are many risks connected to the introduction of species to new localities outside of their previous distribution, such as biological invasion or unwanted species interactions, keeping all species in their current locations will not be unfeasible either. In this study it isshown that about two-thirds of the coniferous forests in China may experience species loss and become restricted in their functioning.

We support the point made by Thomas [81] that the question is not on whether or not to make changes, but on how to manage forests in a way that promotes advantages and avoids disadvantages. This involves a broad and deep understanding of the processes and the functioning of ecosystems. Thus, the idea of climate adaptive management requires not only best practice and reliable projections for species range shifts, but also the integration of knowledge of all basic branches of biological science, such as ecology, biogeography, genetics and evolution.

Limits and perspectives

Species distribution models and climate envelope modeling features some intrinsic limitations, such the disregard of evolutionary processes and adaptation, or the ignorance of within-species genotypic and ecological variability. Experimental research must be designed to specifically address these issues that cannot be tackled by the actual nature of species distribution models. However, in the case of key tree species, rapid directives are needed in order to adapt species composition and structure of forests at an early stage to the environmental conditions that are expected in the near future during the life-time of the planted saplings. Evolutionary processes are irrelevant at this time scale. However, it is important to know genetic and ecological variability, especially for tree species with a large spatial distribution [88].

The introduced set of indices delivers a powerful and practical approach for adapting the Chinese forests to future developments that are forecasted in climate scenarios and global climate models. Uncertainty according to realized scenarios and climate model output must be considered in forest management [89]. Especially the role of climatic extremes that are expected to become more and more frequent and intense (IPCC 2012) is an issue that is important in forest practice but not satisfactory in its reflection in climate models. Climatic extremes are much more important for long-lived species such as trees than long-term trends in average values [90].

In consequence, flexible approaches are asked for which allow management to be adapted for the developing state of knowledge, to further develop societal and emission scenarios, and to improved climatic projections. The introduced set of indices, which can be applied for larger groups of tree taxa, offers these options and will allow the implementation of new findings, in order to identify their relevance in forestry.

In perspective, it is not only the application of such models but also the assignment and upscaling of species distribution models to global vegetation models is required. Key species for important biomes, such as the members of the Pinaceae family, are neglected in the current state-of-the-art approaches for global vegetation models that are the basis for the calculation of future carbon fluxes and budgets. The introduced approach in this study brings these modelling philosophies one step closer and may contribute to make vegetation models that integrate plant functional types such as needle-leaved trees more reliable.

Conclusions

As a consequence of the long-term life cycles of trees, there is an urgent need in forestry for reliable forecasts on future ranges of tree species. Species Distribution Models are providing such projections but at continental scales, many species must be taken into account. The Pinaceae family represents a high number of ecologically and economically important tree species in China. For this reason, this family was chosen to develop and implement a new set of indices that enable the detection of directions and intensity of range shifts at the species, genus and family level.

This new threshold free set of indices for tree range shifts features several advantages compared to traditional approaches. Besides its capacity to integrate across a large number of species and to identify general patterns of range shifts for large species groups, the PRS_Chart graphical tool offers an intuitive visualization tool for the science-policy interface which will support the translation of research into practice.

Particularly, the predicted low range overlap between current and future tree species ranges combined with a significant northward as well as upward movement trend for the whole group of species implies that assisted migration can become increasingly important in climate adaptive forest management.

Supporting Information

File S1 AML file for calculating PRS indices (I, O).An AML script file under ARCGIS 8.3 Workstation for calculating PRS indices I and O.

File S2 AML file for calculating PRS indices (Dx, Dy, Dz).An AML script file under ARCGIS 8.3 Workstation for calculating PRS indices Dx, Dy, and Dz.

File S3 R file for plotting PRS_Chart. A R script file under R version 2.15.1 for plotting PRS_Chart.

Table S1 General information of species studied.Notes: SP.No.: species number; Prev.: species prevalence indicated by ratios*10^{-4} of cells occupied by that species to total study area; LG and LT: longitude and latitude of the species centroid.

Acknowledgments

This research was carried out with the support of the Special Research Program for Public-Welfare Forestry (Grant No. 200804001) and NSFC's Project (Grant No.31370486).

Author Contributions

Conceived and designed the experiments: XK SL. Performed the experiments: XK QL YZ. Analyzed the data: XK QL YZ. Contributed reagents/materials/analysis tools: XK QL. Wrote the paper: XK CB SL.

References

1. Elith J, Graham CH, Anderson RP, Dudik M, Ferrier S, et al. (2006) Novel methods improve prediction of species' distributions from occurrence data. Ecography 29: 129–151.
2. Peterson AT, Soberon J, Pearson RG, Anderson RP, Martinez-Meyer E, et al. (2011) Ecological Niches and Geogrpahical Distributions Princeton, Oxford: Princeton University Press.
3. Araujo MB, Cabeza M, Thuiller W, Hannah L, Williams PH (2004) Would climate change drive species out of reserves? An assessment of existing reserve-selection methods. Global Change Biology 10: 1618–1626.
4. McKenney DW, Pedlar JH, Rood RB, Price D (2011) Revisiting projected shifts in the climate envelopes of North American trees using updated general circulation models. Global Change Biology 17: 2720–2730.
5. Guisan A, Thuiller W (2005) Predicting species distribution: offering more than simple habitat models. Ecology Letters 8: 993–1009.
6. Hampe A (2004) Bioclimate envelope models: what they detect and what they hide. Global Ecology and Biogeography 13: 469–471.
7. Ibanez I, Clark JS, Dietze MC, Feeley K, Hersh M, et al. (2006) Predicting biodiversity change: Outside the climate envelope, beyond the species-area curve. Ecology 87: 1896–1906.
8. Real R, Marquez AL, Olivero J, Estrada A (2010) Species distribution models in climate change scenarios are still not useful for informing policy planning: an uncertainty assessment using fuzzy logic. Ecography 33: 304–314.
9. Araujo MB, Guisan A (2006) Five (or so) challenges for species distribution modelling. Journal of Biogeography 33: 1677–1688.
10. Araújo MB, New M (2007) Ensemble forecasting of species distributions. Trends in Ecology and Evolution 22: 42–47.
11. Austin MP, Van Niel KP (2011) Improving species distribution models for climate change studies: variable selection and scale. Journal of Biogeography 38: 1–8.
12. Elith J, Graham CH (2009) Do they? How do they? WHY do they differ? On finding reasons for differing performances of species distribution models. Ecography 32: 66–77.
13. Thuiller W, Lafourcade B, Engler R, Araujo MB (2009) BIOMOD - a platform for ensemble forecasting of species distributions. Ecography 32: 369–373.
14. Engler R, Guisan A, Rechsteiner L (2004) An improved approach for predicting the distribution of rare and endangered species from occurrence and pseudo-absence data. Journal of Applied Ecology 41: 263–274.
15. Jaeschke A, Bittner T, Reineking B, Beierkuhnlein C (2013) Can they keep up with climate change? - Integrating specific dispersal abilities of protected Odonata in species distribution modelling. Insect Conservation and Diversity 6: 93–103.
16. Wilson CD, Roberts D (2011) Modelling distributional trends to inform conservation strategies for an endangered species. Diversity and Distributions 17: 182–189.
17. Bittner T, Jaeschke A, Reineking B, Beierkuhnlein C (2011) Comparing modelling approaches at two levels of biological organisation - Climate change impacts on selected Natura 2000 habitats. Journal of Vegetation Science 22: 699–710.
18. Araujo MB (2002) Biodiversity hotspots and zones of ecological transition. Conservation Biology 16: 1662–1663.
19. Botkin DB, Saxe H, Araujo MB, Betts R, Bradshaw RHW, et al. (2007) Forecasting the effects of global warming on biodiversity. Bioscience 57: 227–236.

20. Thuiller W, Lavorel S, Araujo MB, Sykes MT, Prentice IC (2005) Climate change threats to plant diversity in Europe. Proceedings of the National Academy of Sciences of the United States of America 102: 8245–8250.
21. Bradley BA, Oppenheimer M, Wilcove DS (2009) Climate change and plant invasions: restoration opportunities ahead? Global Change Biology 15: 1511–1521.
22. Hickling R, Roy DB, Hill JK, Fox R, Thomas CD (2006) The distributions of a wide range of taxonomic groups are expanding polewards. Global Change Biology 12: 450–455.
23. Thuiller W (2004) Patterns and uncertainties of species' range shifts under climate change. Global Change Biology 10: 2020–2027.
24. Chambers D, Perie C, Casajus N, de Blois S (2013) Challenges in modelling the abundance of 105 tree species in eastern North America using climate, edaphic, and topographic variables. Forest Ecology and Management 291: 20–29.
25. Iverson LR, Prasad AM, Matthews SN, Peters M (2008) Estimating potential habitat for 134 eastern US tree species under six climate scenarios. Forest Ecology and Management 254: 390–406.
26. Jimenez-Valverde A, Lobo JM (2007) Threshold criteria for conversion of probability of species presence to either-or presence-absence. Acta oecologica 31: 361–369.
27. Liu CR, Berry PM, Dawson TP, Pearson RG (2005) Selecting thresholds of occurrence in the prediction of species distributions. Ecography 28: 385–393.
28. Liu CR, White M, Newell G (2013) Selecting thresholds for the prediction of species ocurrence with presence-only data. Journal of Biogeography 40: 778–789.
29. Kou XJ, Li Q, Liu SR (2011) Quantifying species' range shifts in relation to climate change: A case study of Abies spp. in China. Plos ONE 6(8): e23115. doi:23110.21371/journal.pone.0023115.
30. Ficarra FVC, Nicol E, Ficarra VM (2011) Research and Development: Business into Transfer Information and Communication Technology. In: Ficarra FVC, Lozano CD, Jimenez MP, Kratky ENA, CipollaFicarra M, editors. Advances in New Technologies, Interactive Interfaces, and Communicability: First Iinternational Conference. pp.44–61.
31. Eckenwalder JE (2009) Conifers of the World. Portland: Timber Press.
32. Farjon A (2008) A Natural history of Conifers. Portland: Timber Press.
33. Farjon A (2010) A handbook of the World's Conifers. Boston: Brill.
34. Smith WK, Hinckley TM, editors (1995) Ecophysiology of coniferous forests. San Diego Academic Press.
35. Wu ZY (1980) Vegetation of China. Beijing: Science Press.
36. Kattge J, Diaz S, Lavorel S, Prentice C, Leadley P, et al. (2011) TRY - a global database of plant traits. Global Change Biology 17: 2905–2935.
37. Prentice IC, Cramer W, Harrison SP, Leemans R, Monserud RA, et al. (1992) A GLOBAL BIOME MODEL BASED ON PLANT PHYSIOLOGY AND DOMINANCE, SOIL PROPERTIES AND CLIMATE. Journal of Biogeography 19: 117–134.
38. LePage BA (2003) The evolution, biogeography and palaeoecology of the Pinaceae based on fossil and extant representatives. In: Mill RR, editor. Proceedings of the Fourth International Conifer Conference: Conifers for the Future? pp.29–52.
39. Hou XY (2001) The Vegetation Atlas of China(1:1000 000). Beijing: Science Press.
40. Kou XJ, Li Q, Liu SR (2011) High-resolution bioclimatic dataset derived from future climate projections for plant species distribution modeling. Ecological Informatics 6: 196–204.
41. Tabor K, Williams JW (2010) Globally downscaled climate projections for assessing the conservation impacts of climate change. Ecological Applications 20: 554–565.
42. Hijmans RJ, Cameron SE, Parra JL, Jones PG, Jarvis A (2005) Very high resolution interpolated climate surfaces for global land areas. International Journal of Climatology 25: 1965–1978.
43. Harrison SP, Prentice IC, Barboni D, Kohfeld KE, Ni J, et al. (2010) Ecophysiological and bioclimatic foundations for a global plant functional classification. Journal of Vegetation Science 21: 300–317.
44. Woodward FI (1987) Climate and plant distribution. Cambridge UK: Cambridge University Press.
45. Elith J, Phillips SJ, Hastie T, Dudik M, Chee YE, et al. (2011) A statistical explanation of MaxEnt for ecologists. Diversity and Distributions 17: 43–57.
46. Phillips SJ, Dudik M (2008) Modeling of species distributions with Maxent: new extensions and a comprehensive evaluation. Ecography 31: 161–175.
47. Allouche O, Tsoar A, Kadmon R (2006) Assessing the accuracy of species distribution models: prevalence, kappa and the true skill statistic (TSS). Journal of Applied Ecology 43: 1223–1232.
48. Zadeh LA (1965) Fuzzy sets. Information and Control 8: 338–353.
49. Ives AR, Helmus MR (2010) Phylogenetic Metrics of Community Similarity. American Naturalist 176: E128–E142.
50. Kupfer JA (2012) Landscape ecology and biogeography: Rethinking landscape metrics in a post-FRAGSTATS landscape. Progress in Physical Geography 36: 400–420.
51. Muenkemueller T, de Bello F, Meynard CN, Gravel D, Lavergne S, et al. (2012) From diversity indices to community assembly processes: a test with simulated data. Ecography 35: 468–480.
52. Swenson NG (2011) Phylogenetic Beta Diversity Metrics, Trait Evolution and Inferring the Functional Beta Diversity of Communities. Plos One 6.
53. Tuomisto H (2010) A diversity of beta diversities: straightening up a concept gone awry. Part 1. Defining beta diversity as a function of alpha and gamma diversity. Ecography 33: 2–22.
54. Uuemaa E, Mander U, Marja R (2013) Trends in the use of landscape spatial metrics as landscape indicators: A review. Ecological Indicators 28: 100–106.
55. Fang JY, Wang ZH, Tang ZY (2009) Atlas of Woody Plants in China. Beijing: Higher Education Press.
56. Thomas CD, Cameron A, Green RE, Bakkenes M, Beaumont LJ, et al. (2004) Extinction risk from climate change. Nature 427: 145–148.
57. Thomas CD, Franco AMA, Hill JK (2006) Range retractions and extinction in the face of climate warming. Trends in Ecology & Evolution 21: 415–416.
58. Jablonski D (2008) Species Selection: Theory and Data. Annual Review of Ecology Evolution and Systematics. pp. 501–524.
59. Webb TJ, Gaston KJ (2003) On the heritability of geographic range sizes. American Naturalist 161: 553–566.
60. Mouillot D, Gaston K (2009) Spatial overlap enhances geographic range size conservatism. Ecography 32: 671–675.
61. Colwell RK, Brehm G, Cardelus CL, Gilman AC, Longino JT (2008) Global warming, elevational range shifts, and lowland biotic attrition in the wet tropics. Science 322: 258–261.
62. Crimmins SM, Dobrowski SZ, Greenberg JA, Abatzoglou JT, Mynsberge AR (2011) Changes in Climatic Water Balance Drive Downhill Shifts in Plant Species' Optimum Elevations. Science 331: 324–327.
63. Dobrowski SZ, Crimmins SM, Greenberg JA, Abatzoglou JT, Mynsberge AR (2011) Response to Comments on "Changes in Climatic Water Balance Drive Downhill Shifts in Plant Species' Optimum Elevations". Science 334.
64. Feeley KJ, Silman MR, Bush MB, Farfan W, Cabrera KG, et al. (2011) Upslope migration of Andean trees. Journal of Biogeography 38: 783–791.
65. Frei E, Bodin J, Walther G-R (2010) Plant species' range shifts in mountainous areas-all uphill from here? Botanica Helvetica 120: 117–128.
66. Lenoir J, Gegout JC, Marquet PA, de Ruffray P, Brisse H (2008) A significant upward shift in plant species optimum elevation during the 20th century. Science 320: 1768–1771.
67. Jump AS, Matyas C, Penuelas J (2009) The altitude-for-latitude disparity in the range retractions of woody species. Trends in Ecology & Evolution 24: 694–701.
68. Engler R, Randin CF, Thuiller W, Dullinger S, Zimmermann NE, et al. (2011) 21st century climate change threatens mountain flora unequally across Europe. Global Change Biology 17: 2330–2341.
69. Reyer CPO, Leuzinger S, Rammig A, Wolf A, Bartholomeus RP, et al. (2013) A plant's perspective of extremes: terrestrial plant responses to changing climatic variability. Global Change Biology 19: 75–89.
70. Bellard C, Bertelsmeier C, Leadley P, Thuiller W, Courchamp F (2012) Impacts of climate change on the future of biodiversity. Ecology Letters 15: 365–377.
71. Thuiller W, Lavergne S, Roquet C, Boulangeat I, Lafourcade B, et al. (2011) Consequences of climate change on the tree of life in Europe. Nature 470: 531–534.
72. Curtis JMR, Vincent ACJ (2008) Use of Population Viability Analysis to Evaluate CITES Trade-Management Options for Threatened Marine Fishes. Conservation Biology 22: 1225–1232.
73. Kery M, Royle JA, Schmid H, Schaub M, Volet B, et al. (2010) Site-Occupancy Distribution Modeling to Correct Population-Trend Estimates Derived from Opportunistic Observations. Conservation Biology 24: 1388–1397.
74. Thomas CD (2011) Translocation of species, climate change, and the end of trying to recreate past ecological communities. Trends in Ecology & Evolution 26: 216–221.
75. Vila M, Hulme PE (2011) Jurassic Park? No thanks. Trends in Ecology & Evolution 26: 496–497.
76. Webber BL, Scott JK, Didham RK (2011) Translocation or bust! A new acclimatization agenda for the 21(st) century? Trends in Ecology & Evolution 26: 495–496.
77. Carrete M, Tella JL (2012) Is assisted colonization feasible? Lessons from past introductions. Frontiers in Ecology and the Environment 10: 12–13.
78. Corlett RT (2012) Climate change in the tropics: The end of the world as we know it? Biological Conservation 151: 22–25.
79. Kreyling J, Bittner T, Jaeschke A, Jentsch A, Steinbauer MJ, et al. (2011) Assisted Colonization: A Question of Focal Units and Recipient Localities. Restoration Ecology 19: 433–440.
80. Schwartz MW, Hellmann JJ, McLachlan JM, Sax DF, Borevitz JO, et al. (2012) Managed Relocation: Integrating the Scientific, Regulatory, and Ethical Challenges. Bioscience 62: 732–743.
81. Thomas CD (2011) Anthropocene Park? No alternative. Trends in Ecology & Evolution 26: 497–498.
82. Dobrowski SZ, Abatzoglou J, Swanson AK, Greenberg JA, Mynsberge AR, et al. (2013) The climate velocity of the contiguous United States during the 20th century. Global Change Biology 19: 241–251.
83. Etterson JR, Shaw RG (2001) Constraint to adaptive evolution in response to global warming. Science 294: 151–154.
84. Pearson RG, Dawson TP (2005) Long-distance plant dispersal and habitat fragmentation: identifying conservation targets for spatial landscape planning under climate change. Biological Conservation 123: 389–401.
85. Summers DM, Bryan BA, Crossman ND, Meyer WS (2012) Species vulnerability to climate change: impacts on spatial conservation priorities and species representation. Global Change Biology 18: 2335–2348.

86. Feeley KJ, Rehm EM (2012) Amazon's vulnerability to climate change heightened by deforestation and man-made dispersal barriers. Global Change Biology 18: 3606–3614.
87. Chiarucci A, Araujo MB, Decocq G, Beierkuhnlein C, Maria Fernandez-Palacios J (2010) The concept of potential natural vegetation: an epitaph? Journal of Vegetation Science 21: 1172–1178.
88. Oney B, Reineking B, O'Neill G, Kreyling J (2013) Intraspecific variation buffers projected climate change impacts on Pinus contorta. Ecology and Evolution 3: 437–449.
89. Millar CI, Stephenson NL, Stephens SL (2007) Climate change and forests of the future: Managing in the face of uncertainty. Ecological Applications 17: 2145–2151.
90. Jentsch A, Beierkuhnlein C (2008) Research frontiers in climate change: Effects of extreme meteorological events on ecosystems. Comptes Rendus Geoscience 340: 621–628.

2

Spatial and Temporal Patterns of Carbon Storage in Forest Ecosystems on Hainan Island, Southern China

Hai Ren[1*], Linjun Li[1], Qiang Liu[2], Xu Wang[3], Yide Li[4], Dafeng Hui[5], Shuguang Jian[1], Jun Wang[1], Huai Yang[4], Hongfang Lu[1], Guoyi Zhou[1], Xuli Tang[1], Qianmei Zhang[1], Dong Wang[6], Lianlian Yuan[1], Xubing Chen[1]

1 Key Laboratory of Vegetation Restoration and Management of Degraded Ecosystems, South China Botanical Garden, Chinese Academy of Sciences, Guangzhou, China, **2** College of Life Science, Hainan Normal University, Haikou, China, **3** College of Environment and Plant Protection, Hainan University, Haikou, China, **4** Research Institute of Tropical Forestry, Chinese Academy of Forestry, Guangzhou, China, **5** Department of Biological Sciences, Tennessee State University, Nashville, TN, United States of America, **6** School of Life Sciences, Central China Normal University, Wuhan, China

Abstract

Spatial and temporal patterns of carbon (C) storage in forest ecosystems significantly affect the terrestrial C budget, but such patterns are unclear in the forests in Hainan Province, the largest tropical island in China. Here, we estimated the spatial and temporal patterns of C storage from 1993–2008 in Hainan's forest ecosystems by combining our measured data with four consecutive national forest inventories data. Forest coverage increased from 20.7% in the 1950s to 56.4% in the 2010s. The average C density of 163.7 Mg C/ha in Hainan's forest ecosystems in this study was slightly higher than that of China's mainland forests, but was remarkably lower than that in the tropical forests worldwide. Total forest ecosystem C storage in Hainan increased from 109.51 Tg in 1993 to 279.17 Tg in 2008. Soil C accounted for more than 70% of total forest ecosystem C. The spatial distribution of forest C storage in Hainan was uneven, reflecting differences in land use change and forest management. The potential carbon sequestration of forest ecosystems was 77.3 Tg C if all forested lands were restored to natural tropical forests. To increase the C sequestration potential on Hainan Island, future forest management should focus on the conservation of natural forests, selection of tree species, planting of understory species, and implementation of sustainable practices.

Editor: Ting Wang, Wuhan Botanical Garden, Chinese Academy of Sciences, Wuhan, China, China

Funding: This research was supported by the "Strategic Priority Research Program" of the Chinese Academy of Sciences, grant No. XDA05050206. The funders had no role in study design, data collection and analysis, decision to publish, or preparation of the manuscript.

Competing Interests: The authors have declared that no competing interests exist.

* Email: renhai@scbg.ac.cn

These authors contributed equally to this work.

Introduction

Carbon (C) storage in forest ecosystems is one of the largest and most active components of C cycling in terrestrial ecosystems and plays an important role in global C cycling and climate change [1,2]. Information on the spatial distribution of C sources and sinks and their temporal changes is critical for understanding C cycle mechanisms and is essential for formulating climate change policies [3]. As a result, estimation of C budgets at large spatial scales has received increasing attention in recent years [4].

While occupying only 6% of land area, tropical forests contain about 40% of the stored C in the terrestrial biosphere, with vegetation accounting for 58% and soil accounting for 41% [5]. However, there is substantial uncertainty about the estimates of C storage. Conflicting results on tropical forest C storage have been reported. Houghton et al. (1992), for example, indicated that tropical forests are a C source (from 1.2 to 2.2 Pg C/yr) because of deforestation and forest degradation [6]. Malhi and Grace (2000), in contrast, reported that tropical forests are C sinks (1–3 Pg C/yr) while northern forests are C sources [7]. Further studies on C storage in tropical forests at large scales are still needed.

Hainan, the largest tropical island and the second largest island province in China, is part of the Indo-Burma biodiversity hotspot and harbors large areas of tropical forests. Several studies have been conducted on forest resources and C storage on Hainan Island, but produced remarkably varying results. For example, Fang et al. (1996) reported that the total biomass of forests on Hainan Island was 59.79 Tg during 1984–1988 [8]. Zhao and Zhou (2004) found that the forest C storage on the island was 30.92 Tg during 1989–1993 [9]. After considering forest age and vegetation types, Wang (2001) reported that the forest C storage was only 23.21 Tg [10]. Cao et al. (2002) reported that forest C stored in vegetation increased from 30.45 Tg in 1979 to 37.74 Tg in 1993 [11]. Li and Lei (2010) estimated that the total C storage was as high as 50.83 Tg in 2004–2008, while Guo et al. (2013) recently reported the total forest C storage was 37.3 Tg [12,13].

The large discrepancies among those studies are probably due to differences in the methods used to calculate C storage. While all studies used the data from national forestry inventories (seven inventories have been conducted since 1973) conducted by the Sate Forest Agency on Hainan Island, the studies used different inventory datasets, different components of C storage, C concentration coefficients (i.e. the proportion of carbon contained in dry mass of plant organs), or age structures. For example, Cao et al. (2002), used a C concentration coefficient of 0.50 while Wang et al. (2001) used a coefficient of 0.45 [11,14]. Although C storage in ecosystems includes both biomass C and soil C, all of the previous studies considered only the C stored in tree vegetation and failed to consider that stored in the understory or soil. In addition, the spatial distribution of C storage on Hainan Island has not been reported. Thus, it remains unclear how the spatial and temporal patterns of C storage have changed in forest ecosystems during 1993–2008 on Hainan Island, Southern China.

The goal of this study was to examine the spatial and temporal patterns of C storage in forest ecosystems on Hainan Island, China. The specific objectives were to determine: 1) changes in C density of forest vegetation on Hainan Island from 1993–2008; 2) the temporal and spatial patterns of C storage in forest ecosystems on Hainan Island during this period; and 3) how the potential for C storage can be increased.

Materials and Methods

Ethics Statement

This study was based on forest inventory data and our field measurements. For the field study, all necessary permits were obtained from Hainan Bureau of Forestry. The field study did not involve endangered or protected species.

Description of Hainan Island

Hainan Island has a land area of 33,920 km^2 and is located at the northern edge of the tropics (latitude 18°10′–20°10′N, longitude 108°37′–111°03′E). Its tropical monsoon climate includes distinct dry and wet seasons and typhoons. Average annual rainfall is 1500–2500 mm, and average annual temperature is 22–26°C. The soil type is mainly laterite. The main zonal vegetation types include tropical rain forest and tropical mountain rain forest. The island contains more than 4200 plant species (259 families, 100 genera) including about 2000 tropical species [15].

Vegetation classification based on remote sensing and image processing

We collected the Landsat TM satellite images (November 2008), 1:250,000 Digital Elevation Model (DEM), Hainan forest maps (1:500,000), and administrative maps. The images were processed using ERDAS IMAGINE 8.31 [4]. This included geometric correction processing, unsupervised classification method, vegetation information extraction, image classification, and determination of area statistics [16]. The image contained a total of 17 spectral clusters of land cover of which nine were vegetation. These nine spectral clusters were merged into six vegetation types based on the Chinese vegetation taxonomy system [17]: tropical natural rain forest, *Eucalyptus* plantation, rubber plantation, *Casuarina* plantation, coniferous plantation, and orchard. The spatial location of the six vegetation types was overlaid with the Hainan forest maps to show the actual geographical distribution of the studied vegetation types and created the distribution map of forests on Hainan Island in 2008. Finally, we selected the field control points to verify and correct the distribution map, and overlaid the digital map of the administrative boundary onto the processed TM image to estimate the area of each forest type [4,18,19].

Forest inventory data

Forest area and timber volume for each age class and forest type have been inventoried in China once every five years since 1973 [20]. The systematic inventorying of forests on Hainan Island began in 1989 after the island became a province split from Guangdong Province. The forest inventory database used in this study included four inventories, each of which covered a 5-year period: 1989–1993, 1994–1998, 1999–2003, and 2004–2008. The inventory data included statistical report data, a plot database, and a sample trees database. The plot database contained more than 60 factors including plot number, name of dominant species, average tree diameter at breast height (DBH), average tree height, stand volume, number of standing trees (or bamboo), and litter thickness. The sample trees database contained 11 factors including the number of sampled trees, stand type, plot number, DBH, and volume. For the plot database, plots were established using a systematic sampling method. The southwest crossing point of each grid was used as a reference point to establish a 25.82-m×25.82-m plot within a 4-km×6-km grid in 1989 (1421 plots in total). Grid size was changed to 4-km×3-km in 1994 (2829 plots in total) [21].

Field survey plots in 2012 (field sampling data)

To verify the accuracy of the forest inventory data and to estimate C storage in the understory, litter, and soil layers, we established 100 field survey plots in 2012. The plots were selected based on forest type, spatial distribution, forest area, stand volume, and age class on the island. The number of plots for each forest type was as follows: 50 for natural forest (tropical rain forest), 24 for rubber plantation, 8 for eucalyptus plantation, 3 for *Acacia* plantation, 3 for *Pinus* plantation, 2 for *Casuarina* plantation, 1 for mixed coniferous and broad-leaved species forest, 3 for mango orchard, 3 for betel nut orchard, 2 for lychee orchard, 1 for longan orchard, and 1 for other hardwood forest. There were three replicated quadrats in each plot. The area per quadrat was 3600 m^2 for natural forest, 800 m^2 for plantation, and 400 m^2 for orchard. The measured variables were the same as in forest inventory. In addition, for each quadrat, we sampled plant tissue in the tree and understory layer, litter, and soil for laboratory analysis.

Estimation of C storage in forest ecosystems

The C in forest ecosystems includes C stored in the tree layer (tree C, including tree root C), shrub layer, herb layer, litter layer, and soil layer. C storage in the tree layer was estimated by forest inventory data and validated by our field sampling data in 2012. C storage in the shrub layer, herb layer, litter layer, and soil layer in 2012 was calculated using our field sampling data. The methods for estimating C storage in these layers were described below. Since the data of C storage in the shrub layer, herb layer, litter layer, and soil layer were not included in the forest inventories, we estimated these data using the relationships between measurements of shrub, herb litter, soil layer C and tree layer C biomass measurements developed using the measurements in 2012. While C storage in shrub, herb, litter and soil layer, and tree layer C biomass varied among years, we assumed that the relationships did not change.

Estimation of C storage in the tree layer based on forest inventory

The biomass of trees was calculated using the Biomass Expansion Factor (f_{BEF}) method [4].

$$f_{BEF} = a + b/V \tag{1}$$

where V is forest stand volume (V, m^3 ha^{-1}, for the measurement method see reference [21]), and a and b are parameters of the conversion factor of a specific tree species from volume to biomass. The conversion factor values for different dominant tree species were obtained from previous studies on Hainan Island (Table 1).

The tree biomass at the forest stand scale (B, Mg ha^{-1}) was calculated using the following formula:

$$B = \sum_{i=1}^{k} A_i \times f_{BEF_i} \times V_i \tag{2}$$

where i is the dominant species of the forest type, A_i is the forest stand area, V_i is the average storage volume, and f_{BEFi} is the corresponding conversion factor of the i dominant species in the forest type.

The data of C storage in trees were also calculated at the city scale (Hainan Island has 18 cities. Each city represents an administration area). The biomass of the j-th plot in the i-th city (Bij) can be calculated using the following formula:

$$B_{ij} = aV_{ij} + b \tag{3}$$

where the units of B_{ij} and V_{ij} are Mg ha^{-1} and m^3 ha^{-1}, respectively, and a and b are conversion factors of the dominant species (Table 1).

The formula for determining the average biomass of trees in the i-th city (B_i, Mg ha^{-1}) was:

$$B_i = \frac{1}{n} \sum_{j=1}^{n} B_{ij} \tag{4}$$

where n is the total number of plots in the i-th city. The formula for determining the total biomass of trees in the i-th city (Ti) was:

$$T_i = 100 * A_i * C_i * B_i \tag{5}$$

where A_i is the land area (unit: km^2) in the i-th city, C_i is the percentage of forest coverage in the i-th city, B_i is the average biomass of tree in the i-th city (Mg ha^{-1}), and 100 is the unit conversion factor.

The total tree biomass in Hainan Province (T) was summed for all 18 county/city-level cities as:

$$T = \sum_{i=1}^{18} T_i \tag{6}$$

Tree C storage on Hainan Island was calculated by multiplying forest biomass (T) by the C concentration. The C concentration was measured in 2012.

Estimation of C storage in the understory layer based on field sampling and laboratory analysis

The understory layer included a shrub layer (0.5 to 1.5 m tall) and a herb layer (<0.5 m tall). To estimate C storage in the understory, we collected all plant individuals including seedlings from three 5-m×5-m subquadrats in each quadrat. The collected material was dried and weighed, and 30% of the dried material per subquadrat was used for determination of C concentration by the potassium dichromate oxidation method [28]. C storage in understory layers was estimated by multiplying the dry mass of the ground layer collected from each plot and the corresponding ground layer C concentration [29].

Estimation of C storage in the litter layer based on field sampling and laboratory analysis

To determine litter layer C, we collected all litter from three 1-m×1-m subquadrats is each quadrat. The methods used for

Table 1. The conversion formulas used in previous studies for estimating the biomass of dominant tree species on Hainan Island.

Dominant species	Biomass expansion factor (f_{BEF}) formula	n*	R^2**	Reference
Eucalyptus	$f_{BEF} = 0.8873 + 4.5539/V$	20	0.80	Han et al., 2010 [22]
Rubber	$f_{BEF} = 0.7975V + 0.4204$	18	0.87	Cao et al., 2009 [23]
Pinus	$f_{BEF} = 0.5101V + 1.0451$	12	0.92	Fang et al., 2001 [24]
Cunninghamia lanceolata	$f_{BEF} = 0.3999V + 22.541$	56	0.95	Fang et al., 1996, 2001 [8,24]
Native broad-leaved species plantation (soft wood#)	$f_{BEF} = 0.7564V + 8.3103$	12	0.91	Cao et al., 2009 [23]
Native broad-leaved species plantation (hard wood#)	$f_{BEF} = 0.6255V + 91.0013$	19	0.86	Li, 1993 [25]
Tropical rain forest species	$f_{BEF} = 1.0357V + 8.0591$	17	0.89	Li et al., 1995 [26]
Acacia	$f_{BEF} = 0.6255V + 91.0013$	19	0.86	Zhou et al., 2008 [27]
Fruit species	$f_{BEF} = 0.3154V + 3.4171$	6	0.76	Cao et al., 2009 [23]
Casuarina equisetifolia	$f_{BEF} = 0.7441V + 3.2377$	10	0.95	Fang et al., 2001; Zhou et al., 2008 [24,27]
Mixed coniferous and broad-leaved tree species	$f_{BEF} = 0.8136V + 18.4660$	10	0.99	Fang et al., 2001; Zhou et al., 2008 [24,27]

*: n is the number of trees used in developing the regression model.
**: R^2 is the coefficient of determination. All the regression models are significant ($P<0.05$).
#: hard wood (wood density >0.7); soft wood (wood density <0.7).

collection, analysis, and calculation were the same as those used for the understory.

Estimation of C storage in the soil layer based on field sampling and laboratory analysis

For determination of C in the forest soil, we collected three soil cores (4 cm diameter and 100 cm deep) per subquadrat with a soil auger. We separated each 10-cm layer for the top 70 cm (seven layers), while the 70–100 cm depth was sampled as one layer because of its relatively constant C concentration. Soil depth varied among subquadrats, and we collected cores to the maximum depth in each case. The soil bulk density was measured in accordance with the soil layers of every 1 meter soil profile [30]. The samples were processed by the potassium dichromate oxidation method for determination of the organic matter [28].

C storage in the soil of the *j*-th plot of the *i*-th city (SOC_{ij}) was calculated as:

$$SOC_{ij} = 0.58 * 100 * W_{ij} * D_{ij} * R_{ij} \tag{7}$$

where the units for SOC_{ij} are Mg ha^{-1}; W_{ij} is soil bulk density (g cm^{-3}); D_{ij} is soil depth (cm, soil depth ranged from 60 to 100 cm for different soil types, which depended on the soil layer depth in the field); R_{ij} is the average soil organic matter content (%) of the *j*-th plot in the *i*-th city; 0.58 is the conversion coefficient from organic matter to organic C [4]; and 100 is the unit conversion factor. The mean $SOCi$ of the *i*-th city was calculated as:

$$SOC_i = \frac{1}{18}\sum_{j=1}^{18} SOC_{ij} \tag{8}$$

where 18 represent that there are 18 cities in Hainan.

The total ecosystem C storage of *i*-th city (Total C_i, Mg ha^{-1}) was summed by vegetation C and SOC. Therefore, we used the same calculation method as above to obtain C storage data for different cities on Hainan Island.

Mapping methods

Based on the estimation of total C storage (Vegetation C and SOC) in each city, we produced the spatial distribution map of C storage on the administration map in Hainan. The spatial distribution maps of forest ecosystem C storage on Hainan Island in 2008 were created by overlaying the spatial distribution map of tree biomass C storage in 2008 and the spatial distribution map of C storage in the shrub, herb, litter, and soil layers in 2012.

Uncertainty analysis

There were three major sources of uncertainty in C storage estimation in forest ecosystems on Hainan Island: the uncertainty in estimation of C storage in tree layer, uncertainty in relationships used to estimate C storage in the shrub, herb, litter, and soil layers from C storage in tree layer and uncertainty in forest area estimation in our research. The uncertainty of estimations was conducted by analysis of the different error sources. The error of estimation on C storage in tree layer mainly came from the input data such as inventory of forest area and volume and model parameters associated with regression coefficients used for estimation of dominant tree biomass. The Monte-Carlo method [4] was used to calculate the uncertainty in estimation of C storage in tree layer and uncertainty in forest area estimation. The uncertainty in relationships used to calculate C storage in the shrub, herb, litter, and soil layers from C storage in tree layer in 2012 could come from two sources. One was the modeling fitting of C storage in tree layer with C storage in other layer. Another source was the application of these relationships developed in 2012 to other years. We estimated these uncertainty using error propagation method following the Guide to the Expression of Uncertainty in Measurement [31,32]. The law of the propagation of uncertainty or the Taylor method [32] was also used in this analysis.

Results

Change in forest coverage from the 1940s to the 2010s and the spatial distribution of forests on Hainan in 2008

Forest coverage (defined as the percentage of total land area in a region that is covered by any kind of natural or artificial forest) on Hainan Island increased from 20.7% in the 1950s to 56.4% in the 2010s. However, the natural forest coverage (defined as the percentage of total land area in a region that is covered by natural forests) decreased from 49.9% in the 1940s to 6.9% in the 2010s (Fig. 1). According to the forest inventory reports in 2008, Hainan Island had six types of forest ecosystems in 1993 and 11 types in 2008 (Table 2. Five new forest types were counted). Among them, natural tropical rain forests occupied the largest area, and followed with *Eucalyptus* (Table 2).

The area occupied by plantations was much larger than that occupied by natural forests in 2008 (Fig. 2). The natural tropical rain forests mainly grew in the mountainous areas of the central south of Hainan, while the plantations were distributed in the northern hilly land and the surrounding coastal plateau.

Change in C density of the forest ecosystems from 1993 to 2008

The average C density across all forest types in Hainan in 2008 was 163.7 Mg C/ha. Among the layers of tree, shrub, herb, litter, and soil, C density in the soil layer was the largest and accounted for most of the C in each forest ecosystem (Table 2). The C density in the soil layer was 121.4 Mg C/ha, which accounted for about 74% of the total C density. The vegetation C density was about

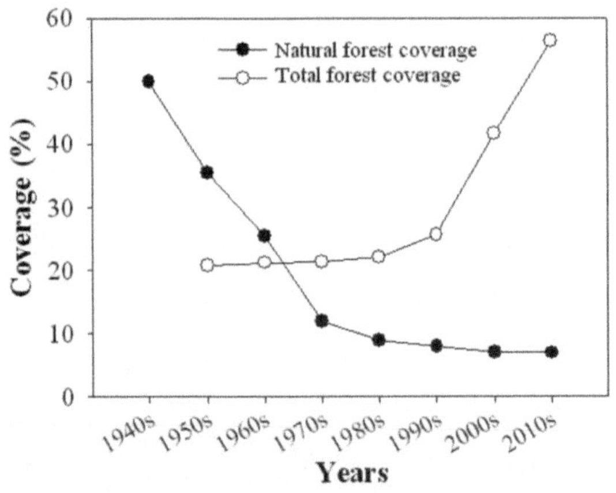

Figure 1. Total forest coverage and natural forest coverage on Hainan Island from 1940s to 2010s. Data are from the National Forest Resources Inventory.

Table 2. Carbon (C) density, forest area, and C storage of forest ecosystems on Hainan Island.

Year and forest type	Area (100 ha)	C density (t/ha)						Total forest ecosystem C storage (Tg)
		Tree layer	Soil layer	Shrub layer	Herb layer	Litter layer	Forest ecosystem	
1993								
Casuarina plantation	551	12.89	116.44	2.12	0.76	1.03	133.25	7.34
Native broad-leaved species plantation (soft wood**)	179	49.59	124.06	3.30	0.52	1.13	178.60	3.20
Eucalyptus plantation	1596	9.51	114.79	2.04	0.82	1.01	128.16	20.45
Tropical rain forest (natural+secondary)	3539	81.58	126.99	4.84	0.43	1.17	215.01	76.09
Pinus plantation	144	7.56	113.56	1.99	0.86	0.99	124.96	1.80
Cunninghamia lanceolata plantation	48	10.83	115.49	2.07	0.80	1.02	130.21	0.62
Total	6057							109.51
1998								
Casuarina plantation	588	12.73	116.38	2.12	0.77	1.03	133.02	7.82
Native broad-leaved species plantation (soft wood**)	204	67.31	125.85	4.08	0.47	1.15	198.86	4.06
Eucalyptus plantation	1703	13.88	116.85	2.15	0.75	1.04	134.66	22.93
Tropical rain forest (natural+secondary)	5169	71.19	126.18	4.27	0.46	1.16	203.26	105.06
Pinus plantation	216	8.00	113.86	2.00	0.85	1.00	125.72	2.72
Cunninghamia lanceolata plantation	48	19.60	118.76	2.30	0.69	1.06	142.41	0.68
Acacia plantation	240	49.20	124.01	3.28	0.52	1.13	178.14	4.28
Total	8168							147.55
2003								
Casuarina plantation	516	15.06	117.30	2.18	0.74	1.04	136.31	7.03
Native broad-leaved species plantation (soft wood**)	239	57.73	124.95	3.64	0.49	1.14	187.95	4.49
Eucalyptus plantation	1667	14.96	117.26	2.18	0.74	1.04	136.17	22.70
Tropical rain forest (natural+secondary)	5756	69.87	126.07	4.21	0.46	1.16	201.76	116.13
Pinus plantation	264	15.03	117.29	2.18	0.74	1.04	136.28	3.60
Cunninghamia lanceolata plantation	71	20.94	119.13	2.34	0.68	1.06	144.15	1.02
Native broad-leaved species plantation (hard wood**)	23	43.97	123.36	3.08	0.54	1.12	172.07	0.40
Acacia plantation	384	52.77	124.42	3.43	0.51	1.13	182.26	7.00
Total	8920							162.37
2008								
Casuarina plantation	312	16.30	117.73	2.21	0.72	1.05	138.01	4.31
Native broad-leaved species plantation (soft wood**)	48	46.06	123.63	3.16	0.53	1.12	174.51	0.84

Table 2. Cont.

Year and forest type	Area (100 ha)	C density (t/ha)						Total forest ecosystem C storage (Tg)
		Tree layer	Soil layer	Shrub layer	Herb layer	Litter layer	Forest ecosystem	
Eucalyptus plantation	1930	12.55	116.30	2.11	0.77	1.03	132.75	25.62
Tropical rain forest (natural+secondary)	5133	73.82	126.40	4.41	0.45	1.16	206.23	105.86
Pinus plantation	324	15.08	117.30	2.18	0.74	1.04	136.34	4.42
Cunninghamia lanceolata plantation	11	33.73	121.83	2.73	0.59	1.10	159.98	0.18
Native broad-leaved species plantation (hard wood**)	107	79.21	126.82	4.71	0.44	1.17	212.33	2.27
Acacia plantation	480	51.28	124.25	3.37	0.52	1.13	180.54	8.67
Mixed coniferous and broad-leaved tree species plantation.	72	50.71	124.19	3.34	0.52	1.13	179.89	1.30
Rubber plantation	5754	18.17	118.34	2.26	0.70	1.05	140.52	80.86
Orchard	3203	17.82	118.23	2.25	0.71	1.05	140.05	44.86
Total	17374*							279.17

*In 2008, rubber plantations, orchards, and mixed coniferous and broad-leaved tree species plantations were included in the statistics. The rubber plantation and orchard were first accounted, and some *Pinus* plantation developed into mixed coniferous and broad-leaved tree species plantation.
**hard wood (wood density >0.7); soft wood (wood density <0.7).

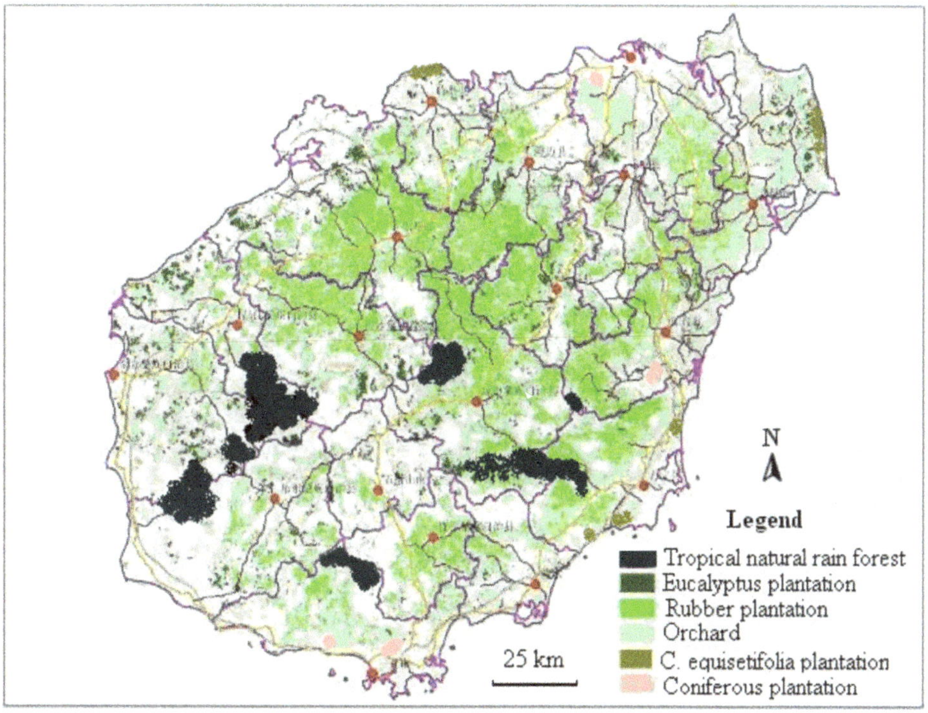

Figure 2. The distribution of forests on Hainan Island in 2008.

41.2 Mg C/ha, and the C density in litter layer was only 1.1 Mg C/ha (Table 2).

The C density of most forest ecosystems on Hainan Island gradually increased from 1993 to 2008 (Table 2). The C density of forest ecosystems in 1993 varied with forest type and ranged from 125.0 Mg C/ha in *Pinus* plantations to 215.0 Mg C/ha in natural tropical rain forests. The C density of native broad-leaved species plantations (hard wood) was the highest and followed by natural tropical rain forests. In 2008, the lowest C density was in *Eucalyptus* plantations. Overall, C density was higher in natural tropical rain forests and native broad-leaved species plantations (hard wood) than in more artificial systems such as rubber and *Eucalyptus* plantations. The C storage was higher in forest types with natural regeneration (e.g., mixed coniferous and broad-leaved species plantation) than in plantations.

The average carbon density of forest ecosystems on Hainan Island increased about 2.11 Mg C/ha from 1993 to 2008 (Excluding the statistics on rubber plantations and orchards in

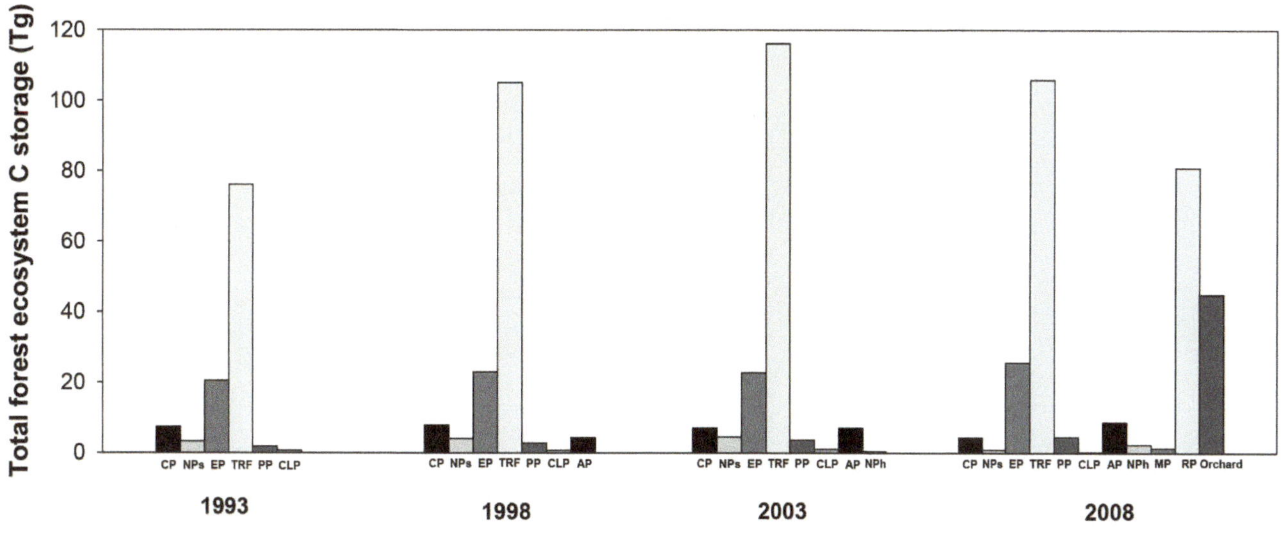

Figure 3. Total C storage of different forest ecosystems on Hainan Island during 1993–2008. CP: *Casuarina* plantation; NP (soft wood): Native broad-leaved species plantation (soft wood); EP: *Eucalyptus* plantation; TRF: Tropical rain forest (natural+secondary); PP: *Pinus* plantation; CLP: *Cunninghamia lanceolata* plantation; AP: *Acacia* plantation; NP (hard wood): Native broad-leaved species plantation (hard wood); MP: Mixed coniferous and broad-leaved tree species plantation; RP: Rubber plantation.

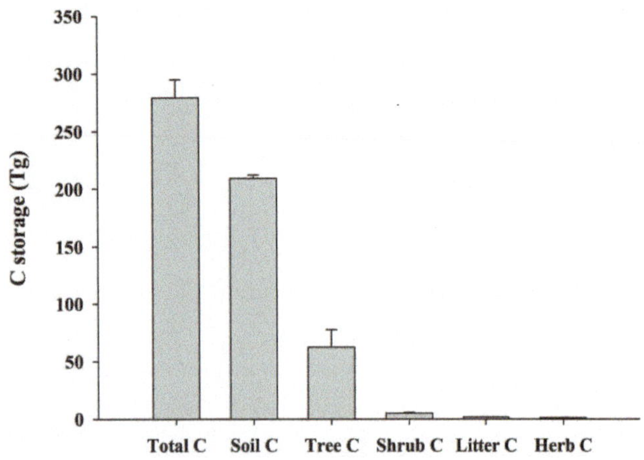

Figure 4. The C storage in different layers of forest ecosystems on Hainan Island in 2008. Tree C includes C in above- and below-ground biomass. Values are means ± SE.

2008). Although the coverage of natural forest with higher carbon density decreased, the average carbon density across all forest types increased. Since other types of forests accounted for a large area and also continued to accumulating C. The results meant that the average carbon density was strongly dependent on the spatial extent of the region and types of land uses included. The average carbon density had changed along with the shifts in forest type and forested area (Table 2; Fig. 1).

Change in C storage in forest ecosystems from 1993 to 2008

Over the past 15 years, the total forest C storage on Hainan Island gradually increased 1.55 times from 109.51 in 1993 to 279.17 in 2008. The C storage of most forest ecosystems increased from 1993 to 2009. Among them, the C storage in *Pinus* and *Eucalyptus* plantations increased 35% from 1993 to 2008. In *Casuarina* plantations and natural tropical rain forests, however, C storage increased from 1993 to 2003 but decreased from 2003 to 2008 (Fig. 3).

The forest ecosystems on Hainan island in 2008 stored about 279.17 Tg C, with 209.07 Tg in the soil layer, 62.19 Tg in the tree

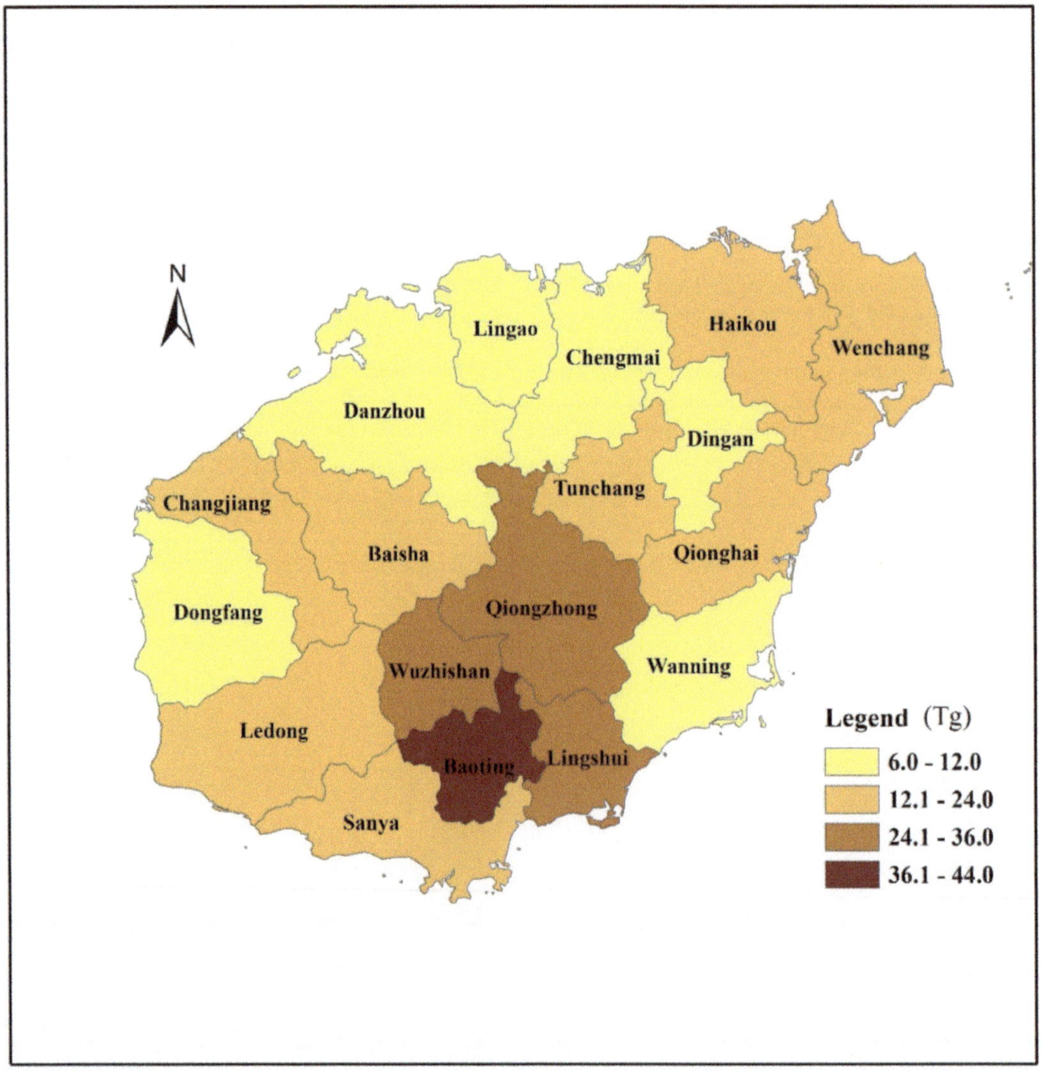

Figure 5. The spatial distribution of forest ecosystem C storage on Hainan Island in 2008.

layer, 5.06 Tg in the shrub layer, 1.06 Tg in the herb layer, and 1.79 Tg in the litter layer. Soil C accounted for 74.9% of the total C storage (Fig. 4).

The spatial distribution of forest ecosystem C storage on Hainan Island in 2008

The spatial distribution of forest ecosystem C storage (combination of tree biomass C storage and C storage in the understory, litter, and soil layers) in 2008 was not homogenous across the province (Fig. 5). The forest ecosystem C storage was highest in the south central region (24.1–44.0 Tg C), lowest in the north (6.0–12.0 Tg C), and intermediate in other regions (12.1–24.0 Tg C).

Uncertainty analysis of total forest ecosystems C density

The results of uncertainty analysis of forest ecosystem C density indicated that forest ecosystem C density errors were partly came from the uncertainty in relationships used to calculate from C storage in tree layer to that in shrub, herb, litter and soil layers among three sources, it accounted for an average 3.2% (±5.05 Mg C/ha). The uncertainty on the estimation of C storage in tree layer and uncertainty in forest area estimation accounted for an average 6.1% (±10.13 Mg C/ha) of the total error (Table 3). The uncertainty of different forest types varied remarkably (Table 3).

Discussion

The C storage in forest ecosystems is closely related to ecosystem area and forest health. Both forest ecosystem area and health have declined on Hainan Island since 1900s [15]. The tropical forest area has decreased at a rate of 2.02% per year since 1950 [15]. The main causes of the tropical forest loss were excessive lumbering, planting of rubber trees, slash-and-burn cultivation, and unrestricted deforestation for fuels and other usages. Change from natural forests to artificial plantations has caused an obvious decrease in forest quality. Fortunately, the government realized the importance of protecting natural forests in the 1990s and prohibited further deforestation on Hainan Island. Rubber plantations, *Eucalyptus* plantations, and orchards, however, remain abundant. Because of rapid population growth and economic development, natural tropical forests on the plains and hilly land and along the coast have mostly been destroyed, and only remain in the mountain areas. The current status reflects a long history of human disturbance and of persistent conflict between development and conservation.

The average C density of 163.7 Mg C/ha in Hainan's forest ecosystems as estimated in this study was slightly higher than the average in China's mainland forests. For example, Wang et al. (2001) and Ren et al. (2013) reported that China's mainland forests contain a total of 141.3–147.5 Mg C/ha, with an average of 36–42 Mg C/ha in the vegetation and 105.3 Mg C/ha in the soil [4,14]. Among those forest types in Hainan in 2008, the average C density of natural forest was the highest and was 206.23 Mg/ha in total, 78.68 Mg/ha in vegetation layer, 1.16 Mg/ha in litter layer and 126.40 Mg/ha in soil layer. The average C density of other plantation forests varied from 132.75 to 180.54 Mg C/ha. Hainan could not provide more land for planting trees [21], therefore, the potential carbon sequestration scenario was that all forested lands were restored to natural tropical forests. The potential carbon sequestration would be 77.3 Tg C. In addition, our estimation was lower than the C density in tropical forests worldwide (279 Mg/ha in total, 157 Mg/ha in vegetation and 122 Mg/ha in soil) without considering the influence of climate, fertility and other limiting factors to the forest growth [2]. The C density in forest soils on Hainan Island was close to the average C density in soil of tropical forests worldwide, but the C density in the vegetation layer in forests on the island was far less than that in tropical forests worldwide. The C storage on Hainan Island could be increased by selecting tree species with high C densities and by improving community structure. The C density in the forest ecosystems on Hainan Island was high in the soil and vegetation layers and was low in the litter layer. The low C density in the litter was reasonable because litter decomposition and nutrient cycling should occur at rapid rates under the high temperate and moisture conditions on Hainan Island.

The total forest ecosystem C storage on Hainan Island increased from 109.51 Tg in 1993 to 279.17 Tg C in 2008, with a total increase of 169.66 Tg. The increase was partially due to a 30% net increase in forest coverage during this period (Fig. 1) and partially due to the shifts in forest types. This increase was similar to the average increase in forest ecosystems in China [24]. It is worth noting that, if the C stored in rubber plantations and orchards was removed from the calculation, the total forest C storage on Hainan Island would increase by only 43.94 Tg. Another finding was that

Table 3. Estimations of carbon density in different layers of major forest ecosystems in 2008 with uncertainty analysis (Mean±SE).

Forest type	C density (t/ha)					
	Tree layer	Soil layer	Shrub layer	Herb layer	Litter layer	Forest ecosystem
Tropical rain forest (natural+ secondary)	67.67±20.35	125.88±1.78	4.10±1.00	0.47±0.05	1.15±0.02	199.27±20.45
Native broad-leaved species plantation (hard wood**)	66.34±20.19	125.76±1.80	4.03±0.98	0.47±0.05	1.15±0.02	197.75±20.30
Acacia plantation	54.56±8.13	124.61±0.87	3.50±0.34	0.50±0.03	1.14±0.01	184.32±8.19
Native broad-leaved species plantation (soft wood**)	47.32±4.80	123.78±0.59	3.21±0.18	0.53±0.02	1.13±0.01	175.96±4.84
Cunninghamia lanceolata plantation	24.89±0.00	120.10±0.00	2.45±0.00	0.65±0.00	1.08±0.00	149.17±0.00
Pinus plantation	24.20±8.65	119.94±2.02	2.43±0.25	0.65±0.06	1.08±0.03	148.30±8.89
Eucalyptus plantation	18.92±9.48	118.56±2.79	2.28±0.26	0.69±0.09	1.06±0.04	141.51±9.89
Casuarina plantation	13.07±7.80	116.52±3.27	2.13±0.20	0.76±0.11	1.03±0.04	133.51±8.46

**hard wood (wood density >0.7); soft wood (wood density <0.7).

C storage from 2003 to 2008 did not increase much or even decreased slightly. This occurred because local farmers planted large areas with rubber plantations, *Eucalyptus* plantations, and orchards for economic reasons. The local government encouraged farmers to convert the existing commercial forest stand such as rubber plantations, *Eucalyptus* plantations, and orchards into ecological forest (i.e. the forests or plantations to provide ecosystem services and social services in important eco-regions or fragile regions). However, the annual compensation fee of ecological forests was only about 25% of the commodity value of plantation such as timber, rubber, and fruits [34,35].

Rubber plantations, pulp plantations (*Eucalyptus* and *Acacia*), and orchards represent a serious threat to Hainan's natural tropical forests and C storage. The regrowth of tropical secondary forests and plantations cannot offset the C that is released as a consequence of forest deforestation, resulting in an overall net C loss on tropical lands. The C density varied among different forest types, and the C density of natural tropical rain forest was higher than that of other forest types on Hainan Island. Although Song et al. (2014) hypothesized that rubber plantations in tropical China may act as a large C sink, they were not a C sink when the deforestation of pre-existing tropical forests was considered during the establishment of rubber plantations [33]. Our previous study of C storage in *Eucalyptus* plantations and orchard on mainland China showed similar results as the rubber plantations [4]. Those studies indicated that the conversion from natural forest to plantation would result in decreasing C storage. However, farmers preferred to cut natural forests, grow the fast growing commercial trees, and sell timber to obtain the immediate economic benefits. They seldom considered the tradeoffs between conservation and agriculture [15]. Conversion of remaining natural forests to plantations would result in a loss of 105 Tg C, thus preservation of remaining natural tropical forests could make an important contribution to carbon sequestration and other ecosystem services on the island. Therefore, we provide the following recommendations to increase C sequestration in forest ecosystems: protection of all natural forests, afforestation in barren lands or waste lands, planting hard wood native species with high C fixation abilities, and restoration of forests from croplands in low productivity areas.

The estimation of forest C storage on Hainan Island varied among studies due to that various methods were used by investigators and the forest ecosystems are complex in nature. To guide climate change studies, Intergovernmental Panel on Climate Change (IPCC) published a methodological and technological guideline [36]. We applied a method that similar to the method recommended by the IPCC. Compared to other methods in previous studies [8–12], our estimation accounted for additional C storage in the understory layer, litter layer and soil layer and directly measured C concentration coefficient of plant organs. In addition, the uncertainty analysis of ecosystem C density, and the temporal and spatial heterogeneity of C storage were first studied, which provided more useful information for forest management in Hainan.

Conclusions

By combining field measurements with data from forest inventories, we quantified the C storage in tropical forest ecosystems on Hainan Island between 1993 and 2008. The average C density in Hainan's forests in 2008 was 163.7 Mg C/ha, with 121.4 Mg C/ha in the soil, 1.1 Mg C/ha in the litter, and 41.2 Mg C/ha in the vegetation (trees, shrubs, herbs, including their roots). The C density of Hainan's forests was higher than the average C density of terrestrial forest ecosystems in China but lower than the worldwide average for such ecosystems. Hainan's tropical forest ecosystems stored 109.51, 147.55, 162.37, and 279.17 Tg C in total in 1993, 1998, 2003, and 2008, respectively. The total C storage in the above- and below-ground portions of forest ecosystems increased over time because of the increase in forest area and the forest type change. The spatial distribution of forest C storage on Hainan Island has been and remains uneven, and the spatial heterogeneity is related to land use, forest type, soil type and climate factors. With the increase in forest area and forest development on Hainan Island, C storage is expected to continuously increase. The potential carbon sequestration was 77.3 Tg C if all the forest stands were still natural tropical forests in 2008. From C sequestration point of view, future forest management should focus on the selection of tree species, the rational planting of understory vegetation at plantations, and implementation of sustainable practices.

Acknowledgments

The authors are indebted to Prof. Bruce Jaffee for English editing, and the anonymous reviewers for their constructive comments.

Author Contributions

Conceived and designed the experiments: HR QL XW YL GZ. Performed the experiments: HR QL XW SJ JW HY HL XT QZ LY XC. Analyzed the data: HR LL DH DW. Contributed to the writing of the manuscript: HR LL DH JW HL.

References

1. Ciais P (1995) A large northern hemisphere terrestrial CO_2 sink indicated by the $^{13}C/^{12}C$ ratio of atmospheric CO_2. Science 269: 1098–1102.
2. Lal R (2005) Forest soils and carbon sequestration. For Ecol Manag 220: 242–258.
3. Houghton RA (2005) Aboveground forest biomass and the global C balance. Glob Change Biol 11:945–958.
4. Ren H, Chen H, Li L, Li P, Hou C, et al. (2013) Spatial and temporal patterns of carbon storage from 1992 to 2002 in forest ecosystems in Guangdong, Southern China. Plant Soil 63: 123–138.
5. Ashton MS, Tyrrell ML, Spalding D, Gentry B (2012) Managing forest carbon in a changing climate. London: Springer. 397p.
6. Houghton RA (1992) Tropical forests and climate. Paper presented at the International workshop on ecology, conservation and management of Southeast Asian rainforests, October 12–14, Kuching, Sarawak.
7. Malhi Y, Grace J (2000) Tropical forests and atmospheric carbon dioxide. Trends Ecol Evol 15: 332–337.
8. Fang JY, Liu GH, Xu SL (1996) Biomass and net production of forest vegetation in China. Acta Ecol Sin 16: 497–508.
9. Zhao M, Zhou GS (2004) Carbon storage of forest vegetation and its relationship with climatic factors. Sci Geogr Sin 4: 50–54.
10. Wang XK, Feng ZW, Ouyang ZY (2001) Vegetation carbon storage and density of forest ecosystems in China. Chinese J Appl Ecol 12: 13–16.
11. Cao J, Zhang YL, Liu YH (2002) Changes in forest biomass carbon storage in Hainan Island over the past 20 years. Geogr Res 21: 551–560.
12. Li HQ, Lei YC (2010) Estimation and evaluation of forest biomass carbon storage in China. Beijing: China Forestry Publishing House. 60p.
13. Guo ZD, Hu HF, Li P, Li NY, Fang JY (2013) Spatial-temporal changes in biomass carbon sinks in China's forests during 1977–2008. Sci China Life Sci 43: 421–431.
14. Wang XK, Feng ZW, Ouyang ZY (2001) Vegetation carbon storage and density of forest ecosystems in China. Chinese J Appl Ecol 12: 13–16.
15. Zhou G (1995) Influences of tropical forest changes on environmental quality in Hainan province, P.R. of China. Ecol Eng 4: 223–229.
16. Jobin B, Beaulieu J, Grenier M, Bélanger L, Maisonneuve C, et al. (2003) Landscape changes and ecological studies in agricultural regions, Québec, Canada. Landscape Ecol 18: 575–590.
17. Hou XY (2001) The vegetation atlas of China. Beijing: Science Press. 280p.
18. Achard F, Eva H, Mayaux P (2001) Tropical forest mapping from coarse spatial resolution satellite data: production and accuracy assessment issues. Int J Remote Sens 22: 2741–2762.

19. Lee TM, Yeh HC (2009) Applying remote sensing techniques to monitor shifting estuary mangrove communities, Taiwan. Ecol Eng 35: 487–496.
20. Guo Z, Fang J, Pan Y, Birdsey R (2010) Inventory-based estimates of forest biomass carbon stocks in China: A comparison of three methods. For Ecol Manag 259: 1225–1231.
21. Hainan Bureau of Forestry (1999) Forest resource statistics of China (1994–1998). Department of Forest Resource and Management, Hainan Bureau of Forestry, Haikou, China.
22. Han FY, Zhou QY, Chen SX, Chen WP, Li TH, et al. (2010) Study on biomass and energy of two different-aged Eucalyptus stands. Forest Res 23: 690–696.
23. Cao JH, Jiang JS, Lin WF, Xie GS, Tao ZL (2009) Biomass of *Hevea* clone PR107. Chinese J Trop Agr 29: 1–8.
24. Fang J, Chen A, Peng C, Zhao S, Ci L (2001) Changes in forest biomass carbon storage in China between 1949 and 1998. Science 292: 2320–2322.
25. Li YD (1993) Comparative analysis for biomass measurement of tropical mountain rain forest in Hainan Island, China. Acta Ecol Sin 13: 314–320.
26. Li YD, Wu ZM, Zeng QB, Zhou GY, Chen BF, et al. (1998) Carbon pool and carbon dioxide dynamics of tropical mountain rain forest ecosystem at Jianfengling, Hainan Island. Acta Ecol Sin 18: 371–378.
27. Zhou C, Wei X, Zhou G, Yan J, Wang X, et al. (2008) Impacts of a large-scale reforestation program on C storage dynamics in Guangdong, China. For Ecol Manag 255:847–854.
28. Liu GS, Jiang NH, Zhang LD, Liu ZL (1996) Soil physical and chemical analysis and description of soil profiles. Beijing: Standards Press of China. 266p.
29. Liu H, Ren H, Hui D, Wang W, Liao B, et al. (2014) Carbon stocks and potential carbon storage in the mangrove forests of China. J Environ Manag 133: 86–93.
30. Zhang JP, Shen CD, Ren H, Wang J, Han WD (2011) Estimating change in sedimentary organic carbon content during mangrove restoration in Southern China using carbon isotopic measurements. Pedosphere 22: 58–66.
31. Cox M, Harris P, Siebert BPL (2003) Evaluation of measurement uncertainty based on the propagation of distributions using Monte Carlo simulation. Meas Tech 46: 824–833.
32. Krouwer JS (2003) Critique of the guide to the expression of uncertainty in measurement method of estimating and reporting uncertainty in diagnostic assays. Clin Chem 49: 1818–1821.
33. Song QH, Tan ZH, Zhang YP, Sha LQ, Deng XB, et al. (2014) Do the rubber plantations in tropical China act as large carbon sinks? Forest 7: 42–47.
34. Deng F, Chen Q, Chen X (2007) Comparison of ecological service among natural forest, rubber and Eucalyptus plantations. J South China Univer of Trop Agr 13:19–23.
35. Ren H, Shen W, Lu H, Wen X, Jian S (2007) Degraded ecosystems in China: Status, causes, and restoration efforts. Landscap Eco Engine 3:1–13.
36. IPCC (2000) Land use, land-use change, and forestry. Cambridge: Cambridge University Press. 375p.

3

Forest Structure in Low-Diversity Tropical Forests: A Study of Hawaiian Wet and Dry Forests

Rebecca Ostertag[1]*, Faith Inman-Narahari[2], Susan Cordell[3], Christian P. Giardina[3], Lawren Sack[4]

1 Department of Biology, University of Hawai'i at Hilo, Hilo, Hawai'i, United States of America, **2** Department of Natural Resources and Environmental Management, University of Hawai'i at Mānoa, Honolulu, Hawai'i, United States of America, **3** Institute of Pacific Islands Forestry, Pacific Southwest Research Station, USDA Forest Service, Hilo, Hawai'i, United States of America, **4** Department of Ecology and Evolutionary Biology, University of California Los Angeles, Los Angeles, California, United States of America

Abstract

The potential influence of diversity on ecosystem structure and function remains a topic of significant debate, especially for tropical forests where diversity can range widely. We used Center for Tropical Forest Science (CTFS) methodology to establish forest dynamics plots in montane wet forest and lowland dry forest on Hawai'i Island. We compared the species diversity, tree density, basal area, biomass, and size class distributions between the two forest types. We then examined these variables across tropical forests within the CTFS network. Consistent with other island forests, the Hawai'i forests were characterized by low species richness and very high relative dominance. The two Hawai'i forests were floristically distinct, yet similar in species richness (15 vs. 21 species) and stem density (3078 vs. 3486/ha). While these forests were selected for their low invasive species cover relative to surrounding forests, both forests averaged 5–>50% invasive species cover; ongoing removal will be necessary to reduce or prevent competitive impacts, especially from woody species. The montane wet forest had much larger trees, resulting in eightfold higher basal area and above-ground biomass. Across the CTFS network, the Hawaiian montane wet forest was similar to other tropical forests with respect to diameter distributions, density, and aboveground biomass, while the Hawai'i lowland dry forest was similar in density to tropical forests with much higher diversity. These findings suggest that forest structural variables can be similar across tropical forests independently of species richness. The inclusion of low-diversity Pacific Island forests in the CTFS network provides an ~80-fold range in species richness (15–1182 species), six-fold variation in mean annual rainfall (835–5272 mm yr^{-1}) and 1.8-fold variation in mean annual temperature (16.0–28.4°C). Thus, the Hawaiian forest plots expand the global forest plot network to enable testing of ecological theory for links among species diversity, environmental variation and ecosystem function.

Editor: Bruno Hérault, Cirad, France

Funding: The major funding for this research came from National Science Foundation's EPSCoR Grants No. 0554657 and No. 0903833 to the University of Hawai 'i. Major in-kind support was provided by the Pacific Southwest Research Station of the United States Forest Service. Logistical or financial support that was supplemental was provided for by the Smithsonian Tropical Research Institute Center for Tropical Forest Science, and the University of California, Los Angeles. The funders had no role in study design, data collection and analysis, decision to publish, or preparation of the manuscript.

Competing Interests: The authors have declared that no competing interests exist.

* Email: ostertag@hawaii.edu

Introduction

High species richness is a hallmark of many tropical forests [1,2]. Indeed, the latitudinal gradient and equatorial peak in plant diversity has attracted attention for centuries e.g., [1,3,4,5]. Numerous studies have focused on the causes of high diversity in tropical forests [1,6,7,8,9,10], and theories have been formulated to explain how species or functional diversity in turn affects ecosystem function [11,12]. However, these linkages have rarely been tested, and not all tropical forests are diverse. For example, legume-dominated swamp forests, peat forests, pine savannas, and oceanic islands that are geographically isolated can have low to very low diversity [13,14,15,16]. Such low-diversity forests are understudied, and there is no clear answer to the simple question of whether the structure of a low-diversity tropical forest would be expected to be similar to or different from that of a high-diversity tropical forest with comparable climate.

Indeed, the question of how forest structure—i.e., physiognomy, basal area, density, diameter size class distributions, biomass, and evenness—varies with species diversity is itself understudied, likely an effect of the paucity of studies of the structure of low-diversity tropical forests. Some have hypothesized that forest structure and species-richness might be related, if structure acts as a habitat scaffold or template that precedes and enables species assembly and diversity by providing an increased variety of habitat niches (e.g., nurse logs for seedlings, perches for birds that disperse seeds, climbing structures for vines [17,18]. Alternatively, higher diversity may enhance forest structure, if more species correspond to a wider variety of size classes, strata, and crown architectures [17]. Both processes are not mutually exclusive and may operate simultaneously, creating a positive feedback cycle that would enhance diversity and influence various forest structural attributes. Recent efforts have examined some structural variables, such as latitudinal trends in height across forests e.g., [19,20] and the effects of diversity and spatial scale on standing forest biomass [12], but very low-diversity tropical forests were not considered in these analyses. The tropical forests in the Hawaiian Islands represent a low-diversity extreme, as a result of its young geological origins [21] and extreme isolation from continental land masses: at approximately 4000 km from the nearest continent, Hawai'i is the world's most isolated archipelago. The resulting native flora in Hawai'i is disharmonic (i.e., missing many functional groups) and is about 90% endemic [22]. While long-term plot-based ecological measurements across the tropics have focused on high-diversity forests, there have been surprisingly few data from low-diversity tropical forests [23,24,25,26]. Such low-diversity forests present many interesting contrasts to other tropical forests, and within Hawai'i they also fall across striking environmental gradients (Table 1).

The aim of this study was to: 1) characterize and compare two extremely low-diversity Hawaiian forests, montane wet and lowland dry forest, and 2) compare the structural attributes of these two forests to more diverse tropical forests within the Center for Tropical Forest Science (CTFS) permanent plot network. Including the Hawaiian plots as part of a cross-plot analysis allows, for the first time, examination of forest structure along a diversity gradient that varies almost 80-fold across large-scale plots with consistent measurement protocols.

We used the initial census of large-scale permanent plots in Hawai'i to examine structural and floristic characteristics of two forests that are geographically close but located in widely contrasting environments. The two Hawaiian forest types examined in this first census were montane wet forest (MWF) and lowland dry forest (LDF). Many studies have shown that forests established in areas with higher rainfall or temperature have higher diversity [1,27,28], and also greater basal area, tree height, and above-ground biomass [28,29,30,31]. Further, forests in higher rainfall areas tend to have a greater representation of larger trees, but lower tree densities [32]. We therefore ask: 1) How do the two Hawaiian forests compare in terms floristic and life form composition, stand structure, species diversity, and non-native species cover? Our study was not designed to specifically examine the effects of climate on forest structure and composition, but we used this study design, to test a prediction based on the previous literature that Hawaiian dry forest would have greater stem density, lower diversity, and smaller diameter trees than wet forest [32]. To place our findings in a broader context, we also asked: 2) Can the extremely low forests of Hawai'i have similar structural attributes to more diverse tropical forests? To examine this question, we compared Hawaiian forests with others in the CTFS network enabling the comparison of forest structural variables across a range of environments and diversity levels [1,28,31,33,34,35]. If Hawaiian forests converge with other tropical forests, the importance of climate in determining forest structure is highlighted.

Materials and Methods

Study Sites

In 2008 and 2009, we established two forest dynamics plots (FDPs) on Hawai'i Island – one within montane wet forest (MWF) and one within lowland dry forest (LDF), to initiate the Hawai'i Permanent Plot Network (HIPPNET; Fig. 1). We focused our study on Hawai'i Island, because it has the greatest area of intact forests, a complete map of lava flow ages, and excellent infrastructure for ecological studies. As the youngest island in the archipelago (<700,000 years), it has had the least time for plant colonization and subsequent speciation, and thus has lower species richness relative to its size than the older islands [36]. We selected areas in excellent ecological condition that are representative of a given forest type, with high native species cover, and a commitment by ownership to long-term conservation objectives. Notably, all forests in Hawai'i are affected to some degree by altered trophic interactions due to invasion of non-native species or extinction of the native species [37], but this is not unique to Hawai'i [38]. Non-native stems that were encountered were measured for percent cover, and then controlled mechanically or

Table 1. Distinctive structural and demographic features of Hawaiian forests.

Environmental Conditions
Large variation in elevation, rainfall, temperature and soils among forests that are geographically close [1–2]
High light levels in intact wet, mesic, and dry forest (1.9–40% diffuse light transmission) [3–9]
Species Composition and Diversity Patterns
A global biodiversity hotspot due to high endemism and number of endangered species [10–11]
Same species distributed in many habitats differing in environmental conditions, demonstrating exceptional phenotypic plasticity [10–14]
Tree ferns common and often the understory dominant in wet forests at all elevations, whereas outside of Hawai'i they tend to be more restricted [10]
Monodominance by a few canopy species [15]
Autecology of Plant Species
Metrosideros polymorpha dominant in wet forests throughout succession (as pioneer and late successional species) [15–17]
Extremely slow growth of primary pioneer species, *M. polymorpha* (1–2 mm/year diameter) [18–20]
Nurse logs serve as a substrate for seedling regeneration [21]
Dieback and regeneration of canopy dominant *M. polymorpha* in cohorts contribute strongly to gap dynamics [16, 22]
Trophic Interactions
Evolution without land mammals [23, 24]
Documented extinctions of plants, pollinators and dispersers may influence present day evenness and rarity measures [23]
Animal dispersal of seeds conducted entirely by birds before human contact [24, 25]
Apparently low rates of insect herbivory [26] and seed predation [27]
Presence of invasive weeds, ungulates, and birds may alter present-day plant-animal interactions [25–26, 28]

Superscripts refer to references listed in Table S5 in File S2.

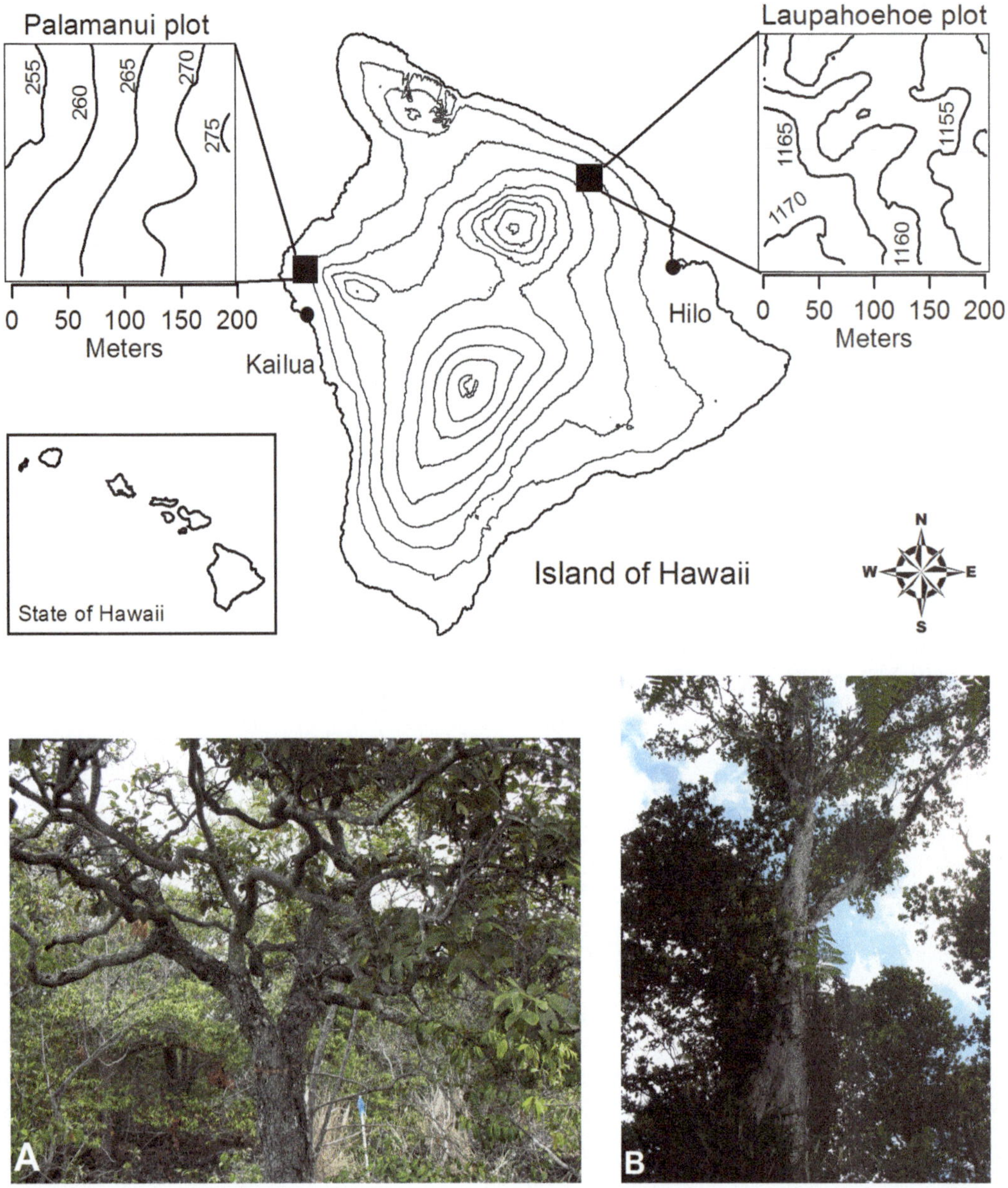

Figure 1. Contour map of the two 4-ha forest plots on Hawai'i Island. Pālamanui site in west Hawai'i is lowland dry forest (LDF; left panel showing the dominant canopy tree *Diospyros sandwicensis* and the open canopy and understory structure of small trees and shrubs); Laupāhoehoe plot in east Hawai'i is montane wet forest (MWF; right panel showing *Metrosideros polymorpha* tree and *Cibotium* spp. tree fern understory).

chemically (see "*Plot Establishment and Vegetation Measurements*" below) and were not considered in the census of stems.

Montane wet forest (MWF). The 4-ha Laupāhoehoe FDP (19°55' N, 155°17' W) is located within the state-owned Laupāhoehoe Natural Area Reserve section of the Hawai'i Experimental Tropical Forest (HETF) on the northeast slope of Mauna Kea volcano. Permits were obtained for work in the HETF through the Institute of Pacific Islands Forestry and the Hawai'i Division of Forestry and Wildlife/Department of Land and Natural Resources. The mean elevation of the plot is 1120 m.a.s.l. with slopes of 0–20%, and the overall direction of downslope is northwards towards the Pacific Ocean. The substrate within the plot is 4000-14,000 years old [39]. Soils were formed from weathered volcanic material, and are deep, rocky, and moderately well-drained silty clay loam in the Akaka series, and classified as hydrous, ferrihydritic, isothermic Acrudoxic Hydrudands (websoilsurvey.nrcs.usda.gov). Rainfall at the MWF is dominated by tradewind-driven precipitation [40]. Interpolated mean annual precipitation, based on analysis of climate station data over 30 years, is 3440 mm with no distinct dry season [41] and mean annual air temperature is 16°C [42]. The forest consists of evergreen broad-leaved trees, and the ~25–28 m canopy is dominated by *Metrosideros polymorpha* (Myrtaceae; Fig. 1) and to a lesser extent, *Acacia koa* (Fabaceae). Vegetation at the MWF is highly representative of this forest type in Hawai'i [43] (see references in Table 1).

The dominant pre-human contact disturbance regime in this forest type was single-to multiple-tree falls, with the maximum gap size averaging 21.5 m^2 [44]. Larger openings coincide with dieback due to cohort senescence of older *M. polymorpha* stands

[45]. Following contact, large *A. koa* trees were occasionally harvested for traditional canoe building. In modern times, limited *A. koa* logging occurred in the HETF but was restricted to < 100 m of an unimproved road that traverses areas. There is no evidence of logging within the MWF [46], which >500 m from the road. Non-native wild pigs disturb soils while rooting, as well as tree ferns [47], with damage over a large area.

Lowland dry forest (LDF). The 4-ha Pālamanui FDP is an example of one of the world's most endangered forest types, and is located on a privately-owned tract of dry forest on the northwest slope of Hualālai Volcano in the district of North Kona (240 m elevation, 19°44' N, 155°59' W). A memorandum of understanding was established with the land owners and managers, the Palāmani Group, for permission to conduct research in the lowland dry forest site. The mean elevation of the plot is 240 m.a.s.l. Geological substrate in the Pālamanui area consists of 'a'ā lava with scattered pāhoehoe flows dating to 1,500–3,000 years old [48]. Soils developing at this site are shallow, rocky, highly organic, and classified as euic, isothermic, shallow Lithic Ustifolist (websoilsurvey.nrcs.usda.gov). Interpolated mean annual precipitation at the LDF site is 835 mm [41,49], with large within- and between-year variability [50]. For the LDF, major rainfall events typically occur in the winter as low pressure storms ("Kona lows") while summers tend to be dry and characterized by small convective storms. Mean daily air temperature is approximately 20°C (wrcc.dri.edu). Native vegetation consists of evergreen broad-leaved trees and shrubs that form an open-canopy forest that reaches heights of ~7–8 m dominated by *Diospyros sandwicensis* (Ebenaceae) and *Psydrax odorata* (Rubiaceae; Fig. 1). One species (*Erythrina sandwicensis*) is drought deciduous and is only represented by a few individuals.

Pre-contact disturbance regimes likely included tree falls. Following contact, selective harvesting of valuable woods (e.g., sandalwood) occurred throughout the area but we do not know of any logging that occurred within the plot. In the last 200 years, much of the lowland dry forest in Hawai'i has been subjected to grazing and browsing by exotic ungulates, with remnants impacted by wildfire carried by non-native grasses [51]. These factors have reduced the native forest to a fraction of its original extent [52]. While the area containing the FDP has not been burned or significantly browsed by ungulates, the surrounding area is a matrix of degraded LDF and open grassland, and in 2009, a fence and firebreak were installed around the area to protect it from ungulates and fire.

Plot Establishment and Vegetation Measurements

We applied field methodology developed by the Center for Tropical Forest Science global FDP network [53]. Both of our 4-ha FDPs (200×200 m) were oriented north-south and located at the center of a 16 ha buffer area, with all edges at least 100 m from any road or major trail where possible. From 2008 to 2009, we tagged all live, native woody plants ≥1 cm diameter at breast height (DBH, at 130 cm), and mapped tagged plants relative to 5 m×5 m grids installed throughout the plots. Each tagged plant was identified to species and measured for DBH. More detailed methods are in Methods S1 in File S1.

Finally, we estimated and mapped cover of abundant non-native herbaceous, shrub and tree species, which will be important for understanding long-term vegetation change. At each site, we chose six abundant focal species or life forms that were considered "invasive pests" according to their Hawai'i Weed Risk Assessment scores (Daehler 2004; www.botany.hawaii.edu/faculty/daehler/wra/full_table.asp). Percent cover within each 5×5 m subquadrat was estimated in the following categories: 0: absent, 1: <5%, 2: 5–25%, 3: 25–50%, 4: 50–100%. Non-native trees with stems ≥ 1 cm at 130 cm were individually mapped. The DBH of the largest stem of non-native trees <5 cm was estimated to the nearest centimeter and measured to the nearest centimeter if > 5 cm. For trees with multiple stems, we counted the total number of stems ≥1 cm at 130 cm. After the non-native trees were mapped, they were girdled and sprayed with herbicide. We did not spray herbicide on the grasses in the LDF, nor the vine *Passiflora tarminiana* in the MWF.

Data Analyses

Stand structure. We determined stand structural characteristics based on DBH measurements. We considered multiple-stemmed plants as single individuals for the calculation of stem density, and summed the basal area of all stems for the calculation of basal area (m^2/ha). For each species, we calculated relative abundance (RA, %) as the number of individuals of that species/total number of individuals, relative dominance (RD, %) as the basal area of that species/total basal area, and relative frequency (RF, %) as the number of quadrats with that species/total number of quadrats.

Above-ground biomass. To estimate above-ground biomass (AGB) for the two plots, we used site-specific and species-specific information whenever possible for wood specific gravity, tree height, and DBH (equations derived from 52,54,55,56; see Table S1 in File S2). When these were not available, we compiled data from global databases, utilizing equations based on other sites, and in some cases for other species from within the same genera [56,57]. Previous studies have reported that genus means are reasonable proxies for species values for specific gravity (r^2>0.70; [58,59]).

To determine tree height, we applied species-specific equations of [54] giving the relationship of tree height vs. DBH, to each individual tree for 12 of the MWF species and 4 of the LDF species (Table S1 in File S2). For the other species, we used the general wet and dry forest equations [55] to determine tree height. We used these tree height estimates to calculate AGB for each tree using published equations that also included DBH and wood specific gravity. Hawai'i-specific equations for AGB were available for 5 MWF species and 4 LDF species. For another two species of the LDF, *D. sandwicensis* and *P. odorata*, equations were available that were developed specifically from our study site [52] (Table S1 in File S2).

Species richness and diversity. Species area curves were generated by plotting cumulative number of species against area for the 20 m×20 m quadrats. Rarefaction analyses were based on 999 permutations (PRIMER-E v. 6, PRIMER-E Ltd, Plymouth, UK), which randomized the sampling order and resulted in a robust average curve. We present several indices: Sobs (the observed number of species), Chao 1 based on rare species (non-parametric), and Michaelis-Menten (parametric), given uncertainty in the ideal estimator [60,61,62]. We used the program EstimateS 9.1.0 to calculate species diversity indices and an estimate of error. We report Fisher's alpha, Shannon diversity index, and Simpson's index (inverse form) following standard formulas [63]. Overlap in species composition between the two sites was determined using the Sørenson similarity index (SI):

$$SI = \frac{number\ of\ species\ shared\ in\ both\ sites}{(no.\ of\ species\ in\ MWF + no.\ of\ species\ in\ LDF)}$$

Table 2. Diversity and forest structure characteristics of plots in the Center of Tropical Forest Science global plot network, including the Hawaiian plots, arranged in order of descending species richness.

CTFS plot location	Plot code	Latitude	Mean annual rain (mm)	Dry season months	Mean elevation (m)	Mean annual temp (°C)	Land type	Plot size (ha)	No. of species	No. of families	Mean species/ family	Fisher's α/ha	H'/ha	Trees/ha	Basal area (m^2/ha)	Dom. of most common family	Above-ground biomass (Mg/ha)[7]
Lambir, Malaysia	LAM	4.19	2664	0	170	26.6	I	52	1182	83	14.2	165	2.40	6915	43.5	41.0	497.2
Yasuní, Ecuador	YAS	−0.69	3081	0	230	28.4	M	50	1114	81	13.8	187	2.44	3026	33.0	14.9	282.4
Pasoh, Malaysia	PAS	2.98	1788	1	80	28.0	M	50	814	82	9.93	124	2.31	6708	31.0	28.2	339.8
Khao Chong, Thailand	KHA	7.54	19851	32	140	27.3	M	24	593	na	na	na	Na	5063	na	na	na
Korup, Cameroon	KOR	5.07	5272	3	200	26.7	M	50	494	62	7.97	48.0	1.75	6580	32.0	16.1	na
Ituri, Dem. Rep. of Congo3	ITU	1.44	1700	3–4	780	23.1	M	404	445	na	na	na	Na	7200	na	na	na
Palanan, Philippines	PAL	17.04	3379	4–5	110	23.5	I	16	335	60	5.58	43.4	1.92	4125	39.8	52.8	290.1
Bukit Timan, Singapore	BUK	1.25	2473	0	150	26.9	I	2	329	62	5.31	60.0	1.90	5950	34.5	38.4	na
BCI, Panama	BCI	9.15	2551	3	140	26.9	M	50	299	58	5.16	34.6	1.62	4168	32.1	11.4	306.5
Mo Singto, Thailand	MOS	14.43	2200	5	770	23.0	M	30.5	262	na	na	na	Na	Na	na	na	na
Huai Kha Khaeng, Thailand	HKK	15.63	1474	6	590	24.1	M	50	251	58	4.33	23.3	1.50	1450	31.2	21.2	211.2
La Planada, Colombia	LPL	1.16	4084	0	1840	19.0	M	25	240	54	4.44	30.6	1.72	4216	29.8	14.9	177.6
Dinghushan, China	DIN	23.16	1985	0	350	20.9	M	20	210	na	na	na	Na	3581	na	na	na
Sinharaja, Sri Lanka	SIN	6.4	5016	0	500	22.6	I	25	204	46	4.43	24.4	1.53	7736	45.6	26.7	357.9
Doi Inthanon, Thailand	DOI	18.52	1908	6	1700	21.4	M	15	192	na	na	na	Na	4913	na	na	na
Luquillo, Puerto Rico	LUQ	18.33	3548	0	380	22.8	I	16	138	47	2.94	13.5	1.45	4194	38.3	17.3	276.1
Nanjenshan, Taiwan	NAN	22.06	3582	0	320	23.5	I	3	125	41	3.05	15.6	1.64	12133	36.3	32.3	na
Ilha do Cardoso, Brazil5	ILH	−25.1	2100	0	5	22.4	M	10	106	na	na	na	Na	Na	na	na	na
Mudumalai, India	MUD	11.6	1250	6	1050	22.8	M	50	72	29	2.48	6.20	0.944	510	25.5	28.2	174.2
Laupāhoehoe, USA	LAU	19.93	3440	1	1150	16.0	I	4	21	15	1.40	2.58	1.82	3078	67.36	37.4	247.9
Pālamanui, USA	PLN	19.74	835	12	240	20.0	I	4	15	15	1.00	1.18	1.00	3487	8.6	74.2	29.4

Southern hemisphere latitudes are negative; land types are island (I) and mainland (M); dry season months are as those with <100 mm precipitation (Richards 1996). Data are from [104] and ctfs.si.edu unless indicated by footnotes.

[1]Mean annual rainfall data for the nearby city of Songkhla, Thailand (www.world-climates.com)

[2]Kira T (1998) NPP Tropical Forest: Khao Chong, Thailand, 1962–1965. Data set. Available on-line [http://www.daac.ornl.gov] from Oak Ridge National Laboratory Distributed Active Archive Center, Oak Ridge, Tennessee, U.S.A.

[3]Average of 4 plots (2 monodominant forest, 2 mixed forest)

[4]Divided into four 10-ha plots

[5]Data from Ferreira de Lima RA, Oliveira AAD, Martini AMZ, Sampaio D, Souza VC, Rodrigues RR (2011) Structure, diversity, and spatial patterns in a permanent plot of a high restinga forest in Southeastern Brazil. Acta Botanica Brasilica 25: 633–645.

[6]Basal area including tree ferns; 36.1m^2/ha without tree ferns

[7]Chave J and 37 others (2008) Assessing evidence for a pervasive alteration in tropical tree communities. PLoS Biology 6: 455–462.

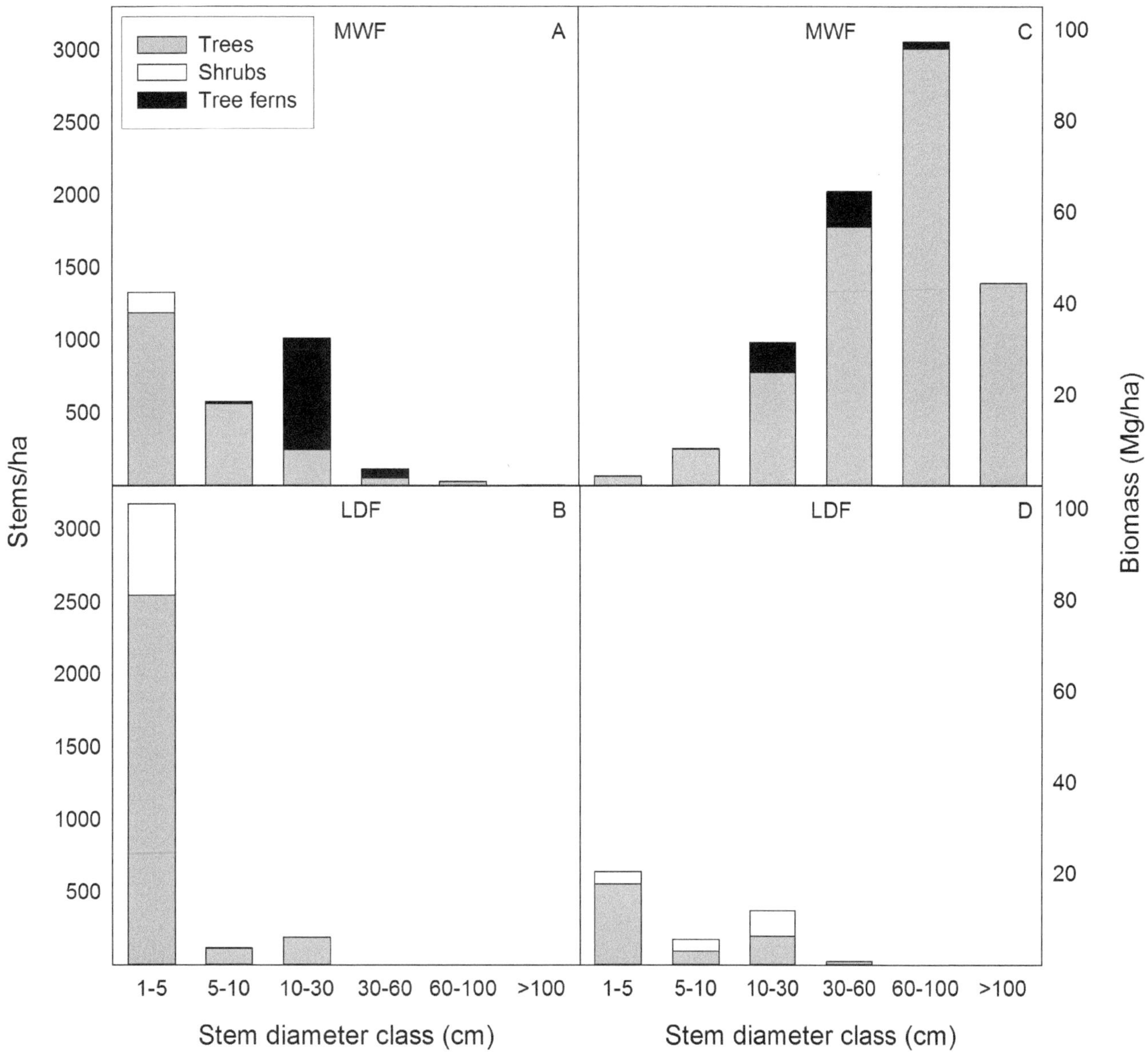

Figure 2. Life form distribution of stems and biomass by diameter size intervals. In (A) Hawaiian montane wet forest (MWF) and (B) lowland dry forest (LDF), stems represent the number of main stems (i.e., one per individual, not including other multiple stems). In (C) MWF and (D) (LDF), biomass calculations were made for all stems (including multiple stemmed individuals). Diameter classes are 1–4.99 cm, 5 - <9.99 cm, 10–29.99 cm, 30–59.99 cm, 60–99.99 cm, and ≥100 cm.

Forest type comparisons. We compiled data for 19 additional mainland and island CTFS tropical plots for which climate and structure data were available (Table 2). Differences between the Hawai'i plots and other CTFS plots were assessed using one sample *t*-tests. Differences in the characteristics of island and mainland plots were assessed using Wilcoxon signed-rank tests [64]. These statistics were analyzed with JMP v. 6 [29,65].

Results

Comparison of Floristics and Life Forms in Hawaiian Forests

The two Hawai'i forests were distinct in floristic composition (Table S2 in File S2). The plots had a very low Sørenson similarity index of 0.06 (a value of 1 would indicate complete overlap). Only *M. polymorpha* occurred in both forests; it was the second most common species in MWF but was represented by only 5 individuals in the 4-ha LDF plot. Species richness was 21 in the MWF and 15 in the LDF. Fifteen families were represented at each site, and the canopy trees at the two sites were from different families, though four families were represented in the understory or the midstory at both sites (Euphorbiaceae, Fabaceae, Myrtaceae, and Rubiaceae).

The plots differed in their distribution of plant life forms. In the MWF, 68% of stems were trees, 4.5% were shrubs, and 28% were tree ferns, accounting for 45%, 8.3% and 46% of the basal area respectively. In the LDF, 82% of stems were trees and 18% were shrubs, accounting for 95% and 5% of basal area respectively (Fig. 2). In the MWF, a large proportion of stems (31%) were growing on non-soil substrates, primarily tree ferns, logs or rocks,

Table 3. Statistics on abundance, basal area, and frequency of the species in the Laupāhoehoe (montane wet forest) plot, with data displayed on an absolute and a relative basis.

Laupāhoehoe montane wet forest

Species	No. individuals	Basal area (m^2/ha)	Presence (no. of quadrats)	Relative abundance (%)	Relative dominance (%)	Relative frequency (%)	IV (%)
METPOL	2631	25.2	100	21.4	37.5	10.2	69.1
CIBGLA	2274	17.7	100	18.5	26.4	10.2	55.1
CHETRI	3320	4.17	100	27.0	6.20	10.2	43.4
CIBMEN	1076	13.2	100	8.74	19.6	10.2	38.6
COPRHY	972	0.585	99	7.90	0.870	10.1	18.9
ILEANO	965	0.466	99	7.84	0.692	10.1	18.7
ACAKOA	141	5.49	57	1.15	8.16	5.84	15.1
BROARG	271	0.0454	74	2.20	0.067	7.58	9.85
MYRLES	237	0.0571	70	1.93	0.085	7.17	9.18
VACCAL	255	0.0328	51	2.07	0.0488	5.23	7.35
HEDHIL	43	0.0200	29	0.349	0.0297	2.97	3.35
PERSAN	35	0.0084	28	0.284	0.0125	2.87	3.17
CIBCHA	34	0.232	17	0.276	0.345	1.74	2.36
CLEPAR	19	0.00230	17	0.154	0.00342	1.74	1.90
MELCLU	13	0.00235	11	0.106	0.00349	1.13	1.24
PSYHAW	10	0.00219	9	0.0812	0.00326	0.922	1.01
MYRSAN	6	0.00166	6	0.0487	0.00246	0.615	0.666
PIPALB	4	0.00226	4	0.0325	0.00335	0.410	0.446
TREGRA	2	0.000107	2	0.0162	0.000159	0.205	0.221
LEPTAM	2	0.0000682	2	0.0162	0.000101	0.205	0.221
ANTPLA	1	0.000154	1	0.0081	0.000229	0.102	0.111
Total	*12311*	*67.3*					

Table 4. Statistics on abundance, basal area, and frequency of the species in the Pālamanui (lowland dry forest) plot, with data displayed on an absolute and a relative basis.

Pālamanui lowland dry forest

Species	No. individuals	Basal area (m^2/ha)	Presence (no. of quadrats)	Relative abundance (%)	Relative dominance (%)	Relative frequency (%)	IV (%)
DIOSAN	2208	6.41	99	15.8	74.2	18.3	108.3
PSYODO	8640	1.27	100	62.0	14.7	18.5	95.2
DODVIS	2301	0.359	94	16.5	4.15	17.4	38.1
SOPCHR	5	0.21	4	0.0359	2.38	0.741	3.20
SANPAN	275	0.156	32	1.97	1.81	5.93	9.70
OSTANT	147	0.0900	40	1.05	1.04	7.41	9.50
WIKSAN	88	0.0890	44	0.631	1.03	8.15	9.81
EUPMUL	134	0.023800	54	0.961	0.275	10.0	11.2
SENGAU	70	0.013300	27	0.502	0.154	5.00	5.66
METPOL	12	0.013	9	0.0860	0.152	1.67	1.90
MYOSAN	54	0.007480	26	0.387	0.087	4.81	5.29
SIDFAL	1	0.0008	1	0.0072	0.00947	0.185	0.202
PLEHAW	1	0.0007	1	0.0072	0.00843	0.185	0.201
ERYSAN	2	0.000513	2	0.0143	0.00594	0.370	0.391
PITTER	8	0.00048	7	0.0574	0.00551	1.30	1.36
Total	*13946*	*8.64*					

Presence based on 100 20×20 m quadrats per plot; species sorted by importance value (IV), which is the sum of the three relative measures (max 300%); species abbreviations as in Table S2 in File S2.

Table 5. Aboveground biomass listed by species for the two Hawai'i forest plots; species abbreviations as in Table S2 in File S2.

Laupāhoehoe montane wet forest			Pālamanui lowland dry forest		
Species	**Biomass (Mg/ha)**	**Relative biomass (%)**	**Species**	**Biomass (Mg/ha)**	**Relative biomass (%)**
METPOL	186	74.9	PSYODO	15.3	51.9
ACAKOA	31.1	12.5	DIOSAN	10.5	35.8
CHETRI	12.4	4.99	METPOL	1.40	4.78
CIBMEN	10.9	4.39	DODVIS	0.921	3.14
CIBGLA	4.55	1.83	OSTANT	0.525	1.79
COPRHY	1.59	0.64	SANPAN	0.359	1.22
ILEANO	1.27	0.51	WIKSAN	0.181	0.615
CIBCHA	0.184	0.07	MYOSAN	0.109	0.372
MYRLES	0.109	0.04	SOPCHR	0.0446	0.152
VACCAL	0.0947	0.04	SENGAU	0.0398	0.135
HEDHIL	0.0456	0.02	EUPMUL	0.0199	0.068
BROARG	0.0394	0.02	PITTER	0.00435	0.0148
PERSAN	0.0124	0.00499	PLEHAW	0.00217	0.00740
PSYHAW	0.00384	0.00155	SIDFAL	0.00187	0.00637
MELCLU	0.00343	0.00138	ERYSAN	0.000871	0.00297
CLEPAR	0.00327	0.00132			
MYRSAN	0.00319	0.00128			
PIPALB	0.00274	0.00110	*Total*	*29.4*	
ANTPLA	0.000287	0.000116			
LEPTAM	0.00017	0.0000685			
TREGRA	0.0000703	0.0000283			
Total	*247.9*				

with 17% of all individuals growing on dead tree ferns (Table S3 in File S2). In contrast, in the LDF, all trees were growing on soil or broken lava, and tree ferns were absent.

Comparison of Stand Structure in Hawaiian Forests

The MWF had larger trees and lower stem density than the LDF (3078±1.21 and 3487±1.40 stems/ha respectively; Table 3). The tree size class distributions differed between the two forests as expected based on their contrasting climates: the LDF had mainly small stems and the MWF had a much more even spread of size classes (Fig. 2). Because the stems in the LDF were small, total basal area and biomass values were low. Thus, the MWF had a nearly eight-fold higher basal area than the LDF (67.3 vs. 8.6 m^2/ha respectively; Tables 3–4), and tree ferns accounted for 31.2 m^2/ha basal area. Above-ground biomass in the MWF was also more than eight times higher than the LDF (248 Mg/ha vs. 29.4 Mg/ha respectively; with 15.6 Mg/ha in the MWF accounted for by tree ferns; Table 5). The above-ground biomass value for the MWF was consistent with that previously estimated for surrounding forest in the same reserve [66]. In both forest types, the two most common canopy species represented 87–88% of biomass (Table 5). In the MWF, the very large trees (≥60 cm) made up the greatest proportion of the biomass, but in the LDF the majority of the biomass was in the 1–5 cm size class (Fig. 2). More multi-stemmed individuals make up the LDF, a mean of 3.2 stems/individual, compared to 1.4 stems/individual in the MWF plot) (Table S4 in File S2).

Comparison of Community Structure in Hawaiian Forests

For both the Hawaiian MWF and LDF, rarefaction curves indicated that a 1 ha sample was sufficient to capture 90% of the species present in the larger 4-ha area (Fig. 3). In the MWF, diversity values for the plot were 2.46 (Fisher's alpha), 1.98 (Shannon), and 5.74 (Simpson); in the LDF values were 1.66 (Fisher's alpha), 1.15 (Shannon), and 2.29 (Simpson). When viewed graphically, there was no overlap in any index value between the two forests (Fig. 4): the Hawaiian MWF was more diverse than the LDF. In the MWF, species evenness was higher than in the LDF, primarily because *P. odorata* in the LDF had a relative abundance of over 60%. In contrast the forests were similar in the abundance of uncommon species, and ~20% of species were rare, i.e., having ≤1 stem/ha (Tables 3–4).

Non-native Species in the Plots

Invasive species made up a larger presence in LDF than MWF (Figs.5–6). The grass *Pennisetum setaceum* was most widespread in the LDF and the herbaceous weed *Persicaria punctata* was most common in the MWF, where it tended to dominate low-lying boggy areas. In the LDF there was a greater overall weed cover, particularly of woody weeds (Fig. 7; see also Methods S1 in File S1). In the MWF there were only a few stems that qualified for DBH measurements (>5 cm). *F. uhdei* averaged 32.3 cm (n = 1) and *P. cattleianum* averaged 6.2 cm (n = 1). In the LDF, average DBH for *Grevillia robusta* was 20.0±8.2 cm SE (n = 27), for *Leucaena leucocephalum* it was 5.9±0.8 cm SE (n = 7), and for *Schinus terebinthius* it was 11.0±1.5 cm SE (n = 20).

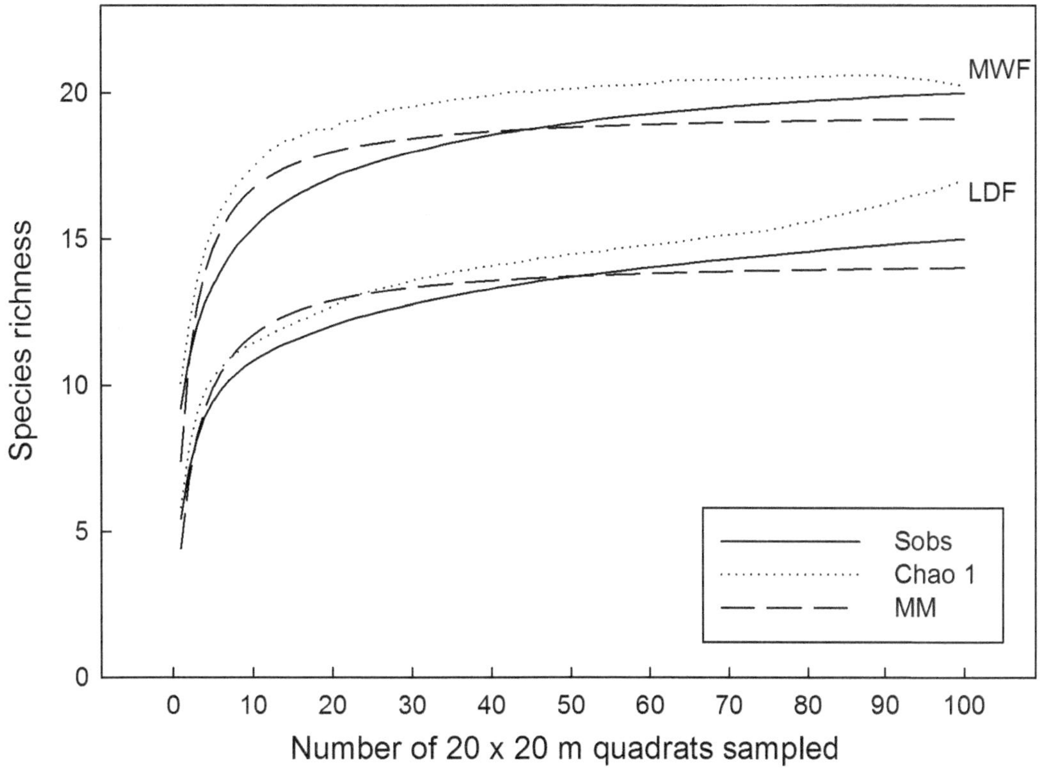

Figure 3. Species accumulation curves. Species number is shown cumulatively, as additional 20 m×20 m quadrats are sampled, until the entire 4-ha plot is represented (100 quadrats), for Hawaiian montane wet forest (MWF) and lowland dry forest (LDF). Three rarefaction techniques are used: Sobs (observed species number), Chao 1, and MM (Michaelis-Menten).

Comparison of Hawai'i to Other CTFS Plots

Diversity of both Hawai'i forests was very low relative to other CTFS forests, including those on islands and with dry climates (Fig. 8). Across the CTFS network, the mean Fisher's alpha per ha ± SE was 59.7±16.6 (n = 13), and the two Hawaiian forests were statistical outliers, with diversity values more than 2 SD lower (t = 3.45 and 3.53, P<0.005). The MWF had approximately 15% as many species as the most comparable island site with tropical wet forest (Luquillo, Puerto Rico). Compared with the next two driest CTFS sites, the Hawai'i LDF had 21% of the number of species found in the Mudumalai, India plot and just 6% of the number of species found at the Huai Kha Khaeng, Thailand plot (Table 2).

In contrast to biodiversity, the structural comparisons across the CTFS network revealed complicated patterns. The MWF was similar to other CTFS sites with respect to tree size class distribution (Fig. 9), and was not significantly different from other CTFS plots with respect to standing above-ground biomass/ha. However, the MWF had 35% lower stem density than the all-forest mean of 4733±722 SE (t = 2.29, P = 0.039, n = 14). Further, the MWF had a 92% higher basal area than the mean of other tropical FDPs due to its high tree fern abundance (t = −21.6, P< 0.0001). When tree ferns were excluded, the basal area of the Hawaiian MWF was within the range of that for other FDPs. For the LDF, stem density was not significantly different than the all-forest mean (t = 1.79, P = 0.097, n = 16, Table 2), but the LDF was an outlier with its very low basal area (t = 10.40, P<0.001, n = 19) and above-ground biomass (t = 7.12, P<0.001, n = 12). The LDF was especially distinctive in having virtually all small stems (Fig. 9).

Discussion

Comparing Hawaiian Wet and Dry Forests

Examination of how the two Hawaiian forests compared in terms of composition and structure matched and extended the paradigm for differences between mature wet and dry forests described on other tropical islands [32]. As predicted, our results for Hawai'i are in agreement with comparisons of mature wet and dry forests in Puerto Rico [32]: wet forest had larger diameter trees, greater basal area, and higher biomass than dry forest, and differences between wet and dry forest in tree density, dominance, and species richness were minor (Tables 3–4). Indeed, across many forests, biomass and basal area are typically correlated with climatic variables such as MAT, MAP, and water deficit within and across sites [31,70,71,72,73,74,75]. However, other variables such as substrate age and type [76] are likely to have contributed to structural and floristic differences between LDF and MWF and we cannot ascribe results solely to climate. While both forests occur on young lava, the much higher rainfall of the MWF and variation in substrate type and texture contributed to greater soil development. Lava flow age and substrate type are important determinants of successional stage in Hawai'i [39]. In addition, differences in disturbance regimes between the two sites may have influenced their forest structure, and their invasive species cover. In the MWF, almost a third of stems were found growing on a substrate other than soil, such as nurse logs and living tree ferns, which likely reflects preferential survivorship on those substrates. Further, canopy dieback of *Metrosideros polymorpha* [45], wind storms [44], and invasive animals [77] may be important factors influencing forest structure. Canopy gaps are larger in the MWF due to the much larger and taller trees that make up the canopy.

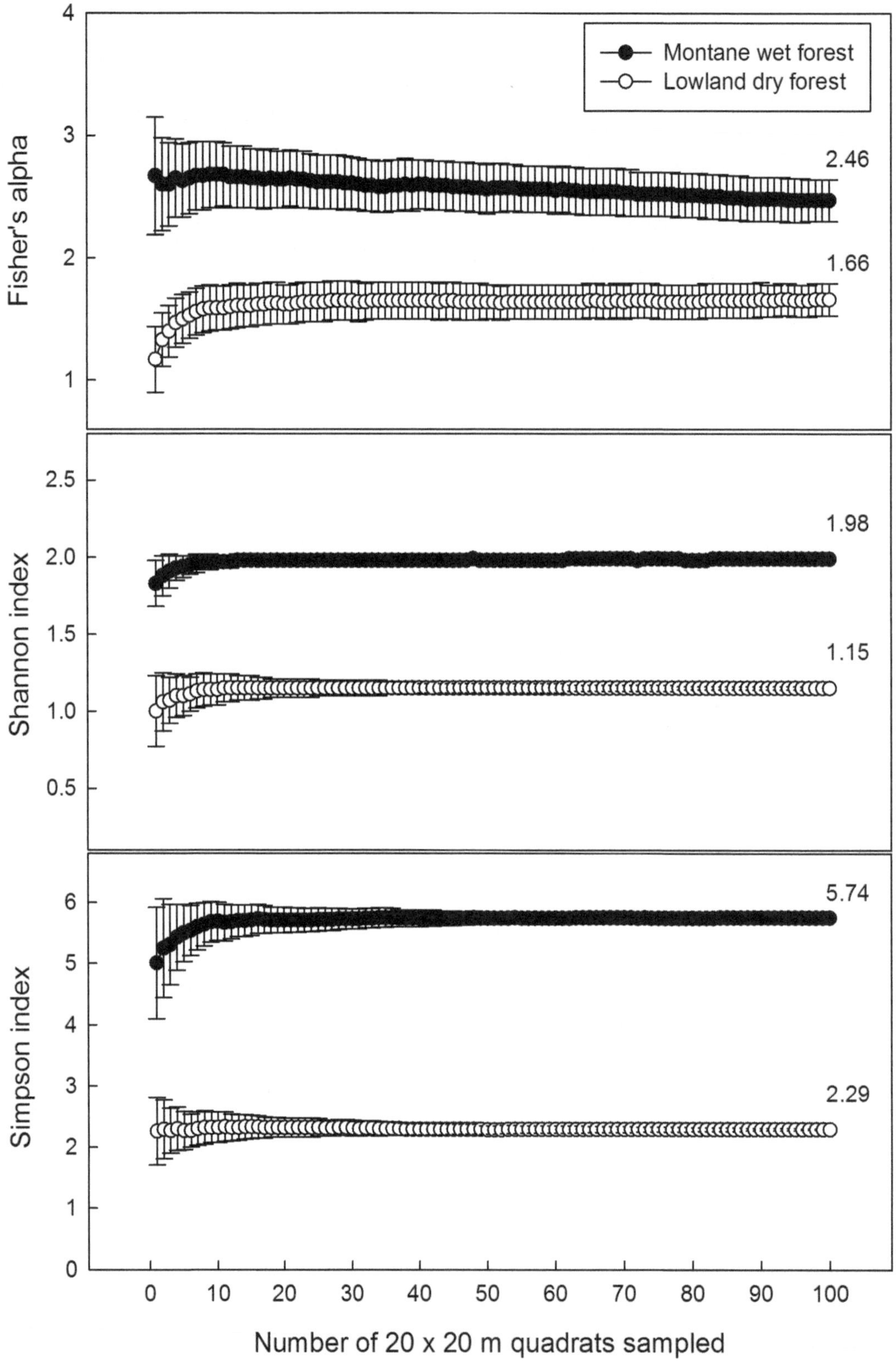

Figure 4. Species diversity indices. Fisher's alpha, Shannon index, and Simpson index for the Hawaiian montane wet forest (MWF) and lowland dry forest (LDF). Each 20×20 m subplot is shown, with the values being cumulative and number above each line representing the entire plot area (4-ha). Values are the diversity index and standard deviation, as estimated by the program EstimateS.

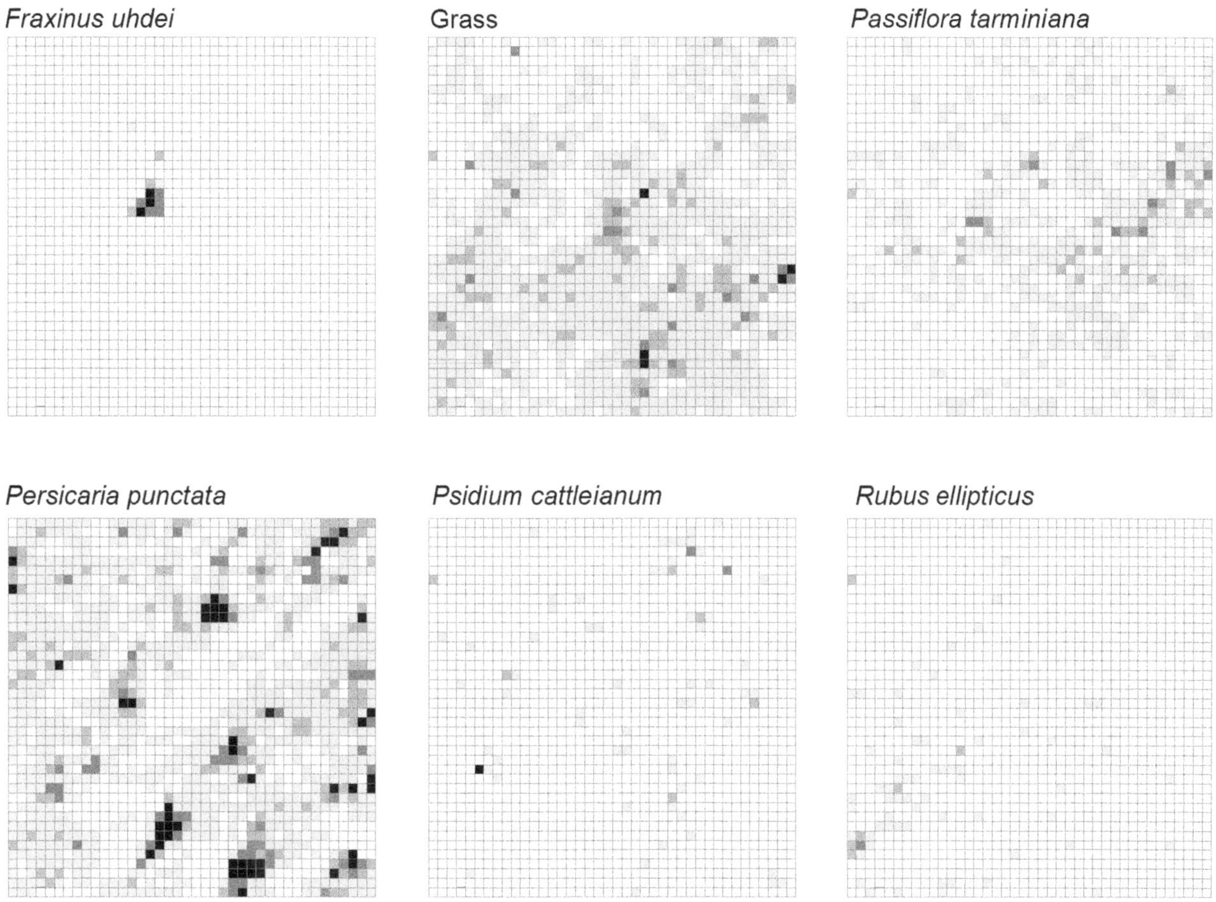

Figure 5. Invasive species cover distribution. Map showing percent cover and locations of invasive species in the MWF. Each grid square represents one 5×5-m subquadrat white: absent, light grey: present to <5%, medium grey: 5–25%, dark grey: 25–50%, black: >50% cover).

There is also evidence of pig rooting that may affect seedling regeneration preferences [47,78], explaining why many stems grow on substrates other than soil. The LDF site is currently fenced from ungulates but the large proportion of multi-stemmed trees and the higher prevalence of regeneration by sprouting suggest adaptation to disturbance [79,80,81].

The MWF site distinguishes itself in its abundance and dominance of tree ferns, which form a distinct mid-canopy layer approximately 5 m above the ground. Notably, tree ferns also make up a large proportion of stand basal area or stem density in some temperate rain forests [82,83,84] and tropical cloud forests [85,86,87,88] but not in other CTFS sites. In Hawai'i, tree ferns are common in wet forests at all elevations, and are particularly abundant in areas with more well-developed organic soils as opposed to young lava flows. While the dynamics of tree ferns have not been well studied in tropical environments [89], in Hawaiian forests tree ferns undoubtedly influence forest function, due to their long lifespans, high frond area, slow growth [90], and slow decomposition rates [91]. They also play a critical role as a substrate for tree seedlings [47,78].

Unlike other CTFS plots that are not heavily impacted by non-native plant species, all forests in Hawai'i have been invaded to some degree. We purposefully chose sites with low non-native species abundance, but cataloged cover before removal for future long-term studies. Because we are removing the invasive species after data collection, we are not examining the consequences of invasion, but previous work in Hawai'i has shown that invaders can significantly alter forest functioning [50,51,52,92,94]. A debate in invasion biology is whether invaders owe their success to their introducing a new function to the community (e.g., N-fixing species) or are simply better competitors [92] and we argue that it is the latter case at our sites. At our sites, invasive grasses were widespread, but, woody invaders are a greater competitive threat (Figs. 7). In the LDF, *Pennisetum setaceum* is widespread, but the vegetation is still dominated by woody species with moderate canopy closure, and reduction of grass cover and fire prevention will reduce its competitive effect in the future. While non-native grasses and herbs are more common than non-native trees across the MWF, their abundance is strongly related to boggy areas, canopy openings, and pig disturbance, and these patches are not likely to expand, but rather to be shaded out in the long term. In MWF, woody invaders such as *Psidium cattleianum* represent much greater threats based on their extreme abundance elsewhere, and traits such as shade tolerance, vegetative reproduction, and animal-dispersed fruits [93,94]. At present, the MWF has limited cover of woody invaders (Fig. 7), and in that respect is in better condition that the LDF.

Clearly, site-specific properties influence the structure and species composition between the two sites, but our study also highlights that at the island scale (1 million ha), climate likely exerts a strong influence, both directly and indirectly [67]. These differences matched patterns found in continental forests, where diversity measures as well as structural measures correlate negatively with the length or severity of the dry season [68,69].

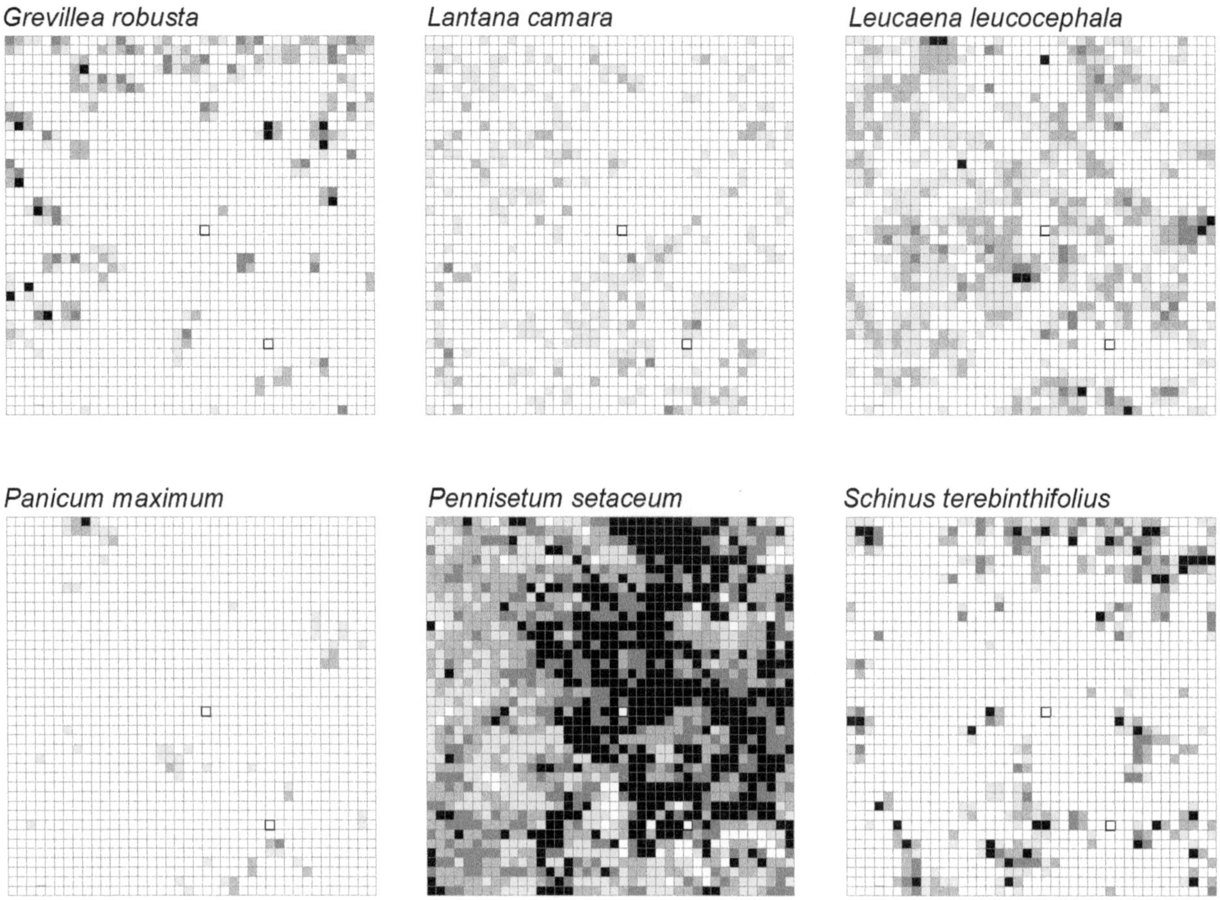

Figure 6. Map showing percent cover and locations of invasive species in the LDF. Each grid square represents one 5×5-m subquadrat (white: absent, light grey: present to <5%, medium grey: 5–25%, dark grey: 25–50%, black: >50% cover).

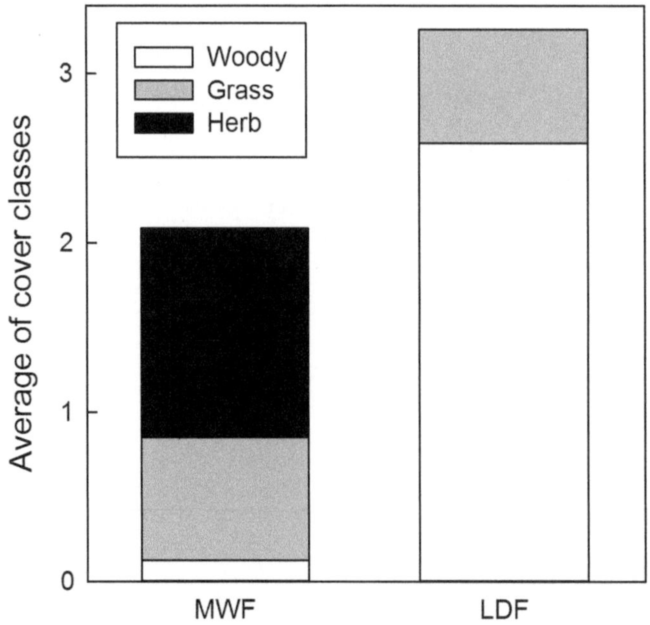

Figure 7. Combined invasive species cover. In each 5×5 m subquadrat a cover score from 0–4 was given based on cover classes (see Methods). The y axis represents the average cover class across the 400 subquadrats, separated by life form: grasses, herbaceous, or woody (shrubs and trees). The combined cover represents the species shown in Figure 5.

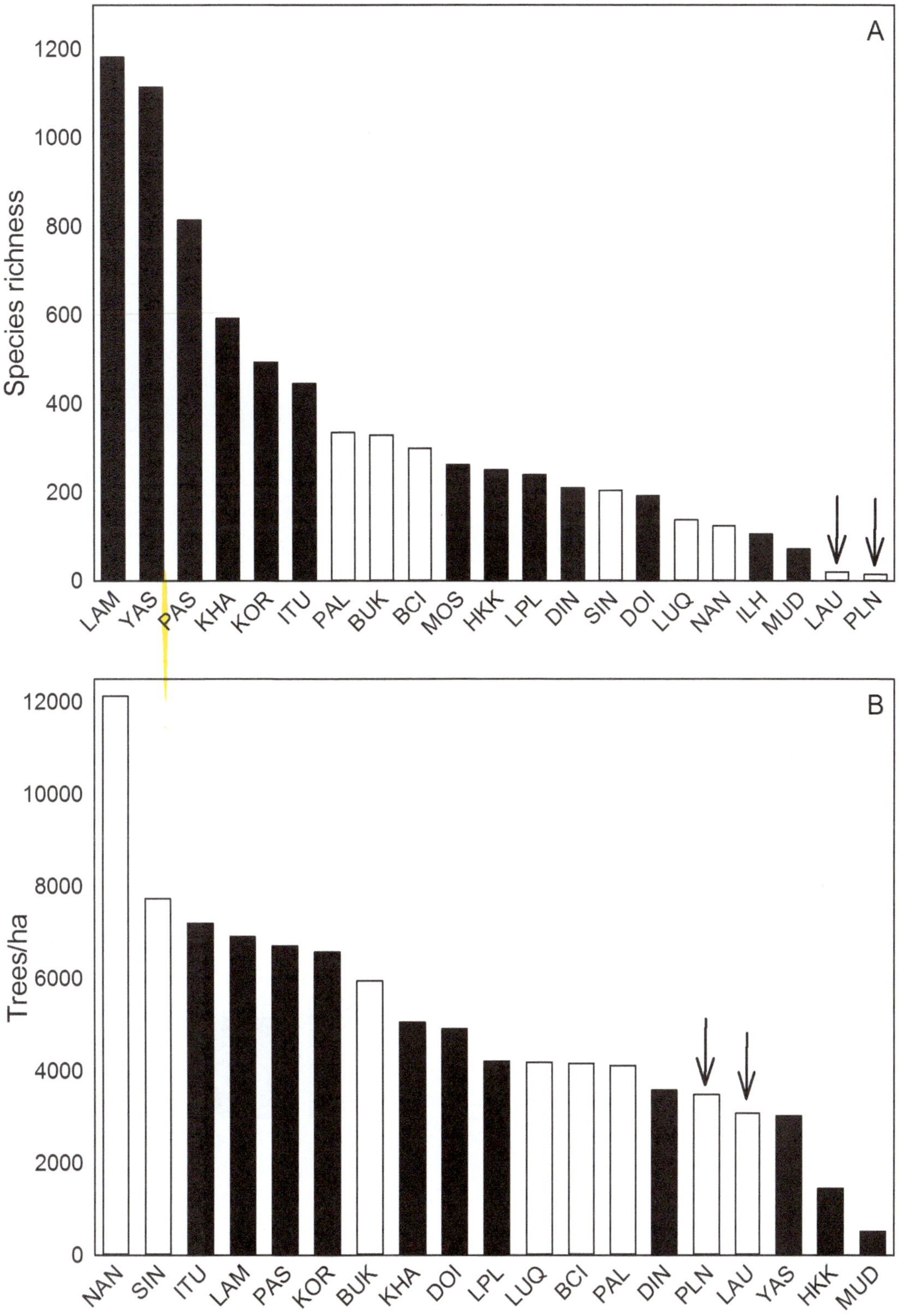

Figure 8. Comparisons of species richness and stem density across a series of CTFS plots. Black bars represent continents and open bars represent islands. Abbreviations as in Table 2. Data from Losos and Leigh, Jr. (2004) and www.ctfs.si.edu.

Structure and Diversity across Tropical Forests Globally

The Hawaiian forest data allowed for the examination of the question of how forest structure varies across species diversity gradients across a much wider range of tree species diversity than was previously available. One of the most striking conclusions of our study is that, despite the extremely low species richness of the Hawai'i FDPs, some structural variables, particularly those for the wet forest, were well within the range of values for the world's most

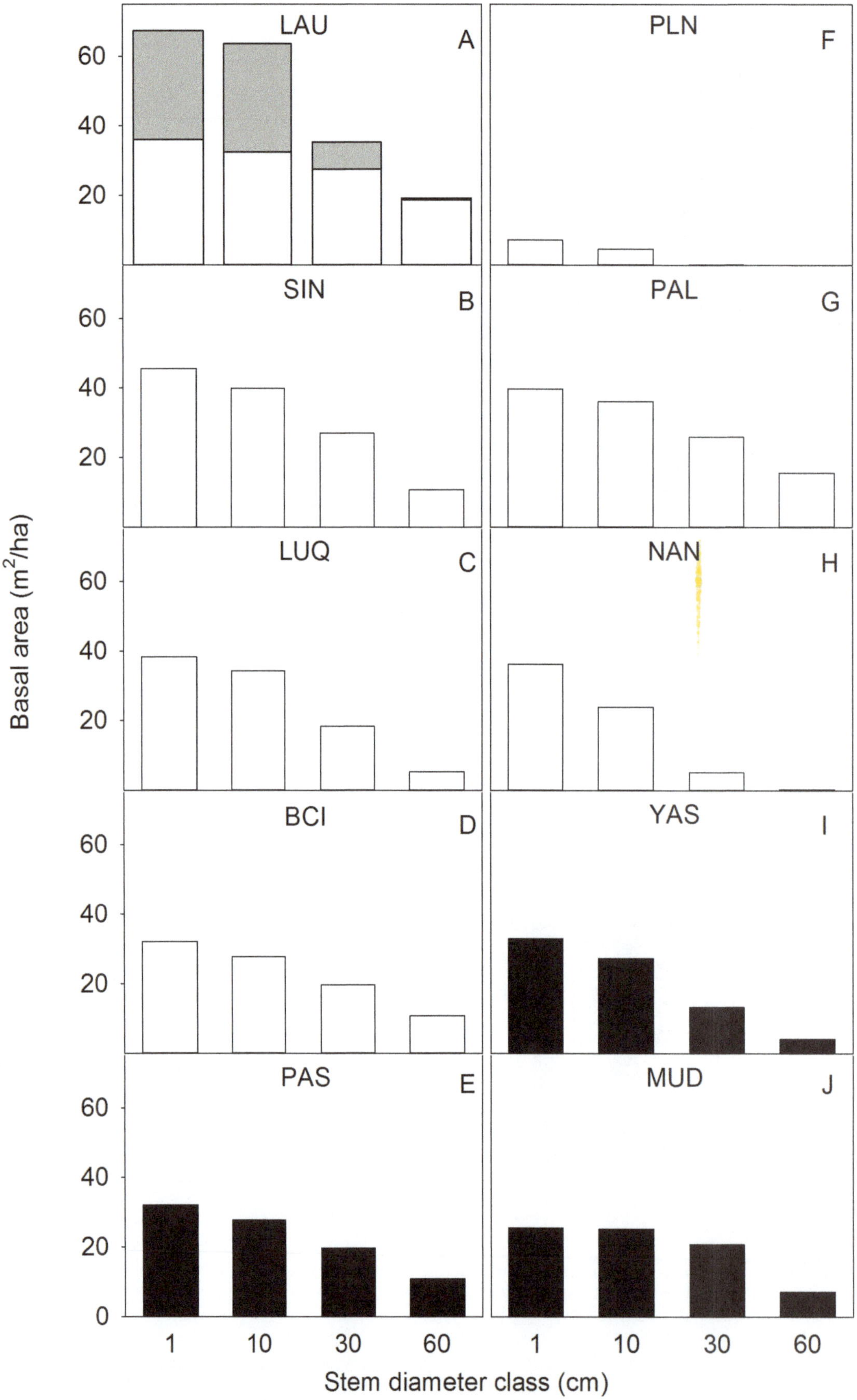
LAU
A
PLN
F
SIN
B
PAL
G
LUQ
C
NAN
H
BCI
D
YAS
I
PAS
E
MUD
J
Basal area (m²/ha)
0
20
40
60
1
10
30
60
Stem diameter class (cm)

Figure 9. A reverse-cumulative distribution of basal area by size class. Size classes are: ≥1 cm, ≥10 cm, ≥30 cm, and ≥60 cm. Data shown for the Hawaiian montane wet forest (LAU) and lowland dry forest (PLN) (top row) and for selected other CTFS plots. Tree ferns (found only at LAU) are symbolized by the gray bars. Island sites are open bars and continental sites are filled bars. Abbreviations as Table 2. Data from Losos and Leigh, Jr. (2004) and www.ctfs.si.edu.

diverse tropical forests (Table 2). For example, stem densities for Hawaiian MWF and LDF were similar to those of the hyper-species-rich Yasun??? FDP (Fig. 8), while biomass and basal area of the Hawai'i MWF (excluding tree ferns) were similar to those of the higher diversity forests in the CTFS network (Fig. 9). The inclusion of tree ferns increased basal area values by 52%, but only increased biomass by 11% (Tables 3 and 4). Notably, the LDF had among the lowest basal area and biomass in the CTFS network, consistent with this site having the lowest precipitation of all FDPs (Table 2). It should also be noted that the LDF is dry year-round, while other dry sites in the CTFS network are seasonally dry (Table 2).

The low floristic richness and population structure of the Hawaiian forest plots represented strong convergence with other island forests. Hawaiian forests had fewer species per family and greater average population densities for each species, as seen in other very isolated sites [95]. High relative dominance values were consistent with island forests having greater dominance by the most common family than mainland tropical forests (Table 2). In the MWF and LDF, 37% and 74% of basal area respectively were accounted for by a single canopy dominant species. On average 20% of species were rare in Hawaiian forests (defined as ≤1 tree/ha), by contrast with 42% on average across other high-diversity forests [96]. It is likely that the patterns of high basal area dominance in Hawaiian forests arose due to the biogeographic consequences of isolation, but we cannot rule out species loss due to human disturbance and invasive species of multiple trophic levels [97,98,99,100,101].

In conclusion, Hawaiian forests have among the lowest species richness and highest endemism rates globally, but in a number of key structural variables both of these forests were similar to even the highest diversity tropical forests in the CTFS network. Future work could examine the evolutionary consequences of such a limited species pool. Biodiversity theory developed in high-diversity tropical forests emphasizes that competitive interactions among species are unlikely on evolutionary time scales because any given two species are rarely consistent neighbors [9]. However, in low-diversity forest any two given species have far greater potential for competitive interactions than in high-diversity tropical forests [102,103]. The addition of Hawai'i to the global plot network enables investigations of the consequences of such differences across a very wide range in species diversity and environmental gradients.

Supporting Information

File S1 Methods S1. Detailed methods and description of situations where field site conditions dictated a different or entirely new methodology by adopted than standardized CTFS protocol in [1].

File S2 Supporting tables. Table S1. Values and equations used for estimating aboveground biomass (AGB) in the montane wet forest (MWF) and lowland dry forest sites (LDF). **Table S2**. Species ≥1 cm diameter at breast height recorded in Laupāhoehoe (montane wet forest) plot with canopy dominants in bold. **Table S3**. Percentage of individuals in the Laupāhoehoe (montane wet forest) plot growing on each substrate type. **Table S4**. Size and multiple stem characteristics of the species species in Laupāhoehoe (montane wet forest) and Pālamanui (lowland dry forest) plots; species abbreviations as in Table S4. **Table S5**. References from Table 1.

Acknowledgments

The Hawai'i Permanent Plot Network thanks the USFS Institute of Pacific Islands Forestry (IPIF) and the Hawai'i Division of Forestry and Wildlife/Department of Land and Natural Resources for permission to conduct research within the Hawai'i Experimental Tropical Forest; the Pālamanui Group, especially Roger Harris, for access to the lowland dry forest site. We thank N. DiManno, L. Ellsworth, B. Hwang, R. Moseley, M. Murphy, K. Nelson-Kaula, M. Nullet, C. Perry, J. Schulten, M. Snyder, and J. VanDeMark for logistical assistance, among the many others who served as project interns and volunteers (see hippnet.hawaii.edu); G. Asner, S. Davies, T. Giambelluca, J. Mascaro, D. Metcalfe, J. Michaud, and J. Thompson for technical advice and/or comments on the manuscript

Author Contributions

Conceived and designed the experiments: RO FIN SC CPG LS. Performed the experiments: RO FIN SC CPG LS. Analyzed the data: RO FIN LS. Contributed reagents/materials/analysis tools: RO FIN SC CPG LS. Wrote the paper: RO FIN SC CPG LS.

References

1. Gentry AH (1988) Changes in plant community diversity and floristic composition on environmental and geographical gradients. Annals of the Missouri Botanical Garden 75: 1–34.
2. Givnish TJ (1999) On the causes of gradients in tropical tree diversity. Journal of Ecology 87: 193–210.
3. Phillips O, Gentry AH, Hall P, Sawyer S, Vasquez R (1994) Dynamics and species richness of tropical rain forests. Proceedings of the National Academy of Sciences 91: 2805–2809.
4. Mittelbach GG, Schemske DW, Cornell HV, Allen AP, Brown JM, et al. (2007) Evolution and the latitudinal diversity gradient: speciation, extinction and biogeography. Ecology Letters 10: 315–331.
5. Hawkins BA, Rodríguez MÁ, Weller SG (2011) Global angiosperm family richness revisited: linking ecology and evolution to climate. Journal of Biogeography 38: 1253–1266.
6. Condit R (1998) Tropical forest densus plots: Methods and results from Barro Colorado Island, Panama and a comparison with other plots. Berlin, Germany, and Georgetown, Texas, USA: Springer-Verlag and R. G. Landes Company. 211 p.
7. Rosenzweig ML (1995) Species diversity in space and time. Cambridge, UK: Cambridge University Press.
8. Givnish TJ (1999) On the causes of gradients in tropical tree diversity. Journal of Ecology 87: 193–210.
9. Hubbell SP (2001) The unified neutral theory of biodiversity and biogeography. Princeton, New Jersey, USA: Princeton University Press.
10. Zimmerman JK, Thompson J, Brokaw N (2008) Large tropical forest dynamics plots: testing explanations for the maintenance of species diversity. Chichester, UK: Wiley-Blackwell.
11. Hooper DU, Chapin FS, Ewel JJ, Hector A, Inchausti P, et al. (2005) Effects of biodiversity on ecosystem functioning: A consensus of current knowledge. Ecological Monographs 75: 3–35.
12. Chisholm RA, Muller-Landau HC, Abdul Rahman K, Bebber DP, Bin Y, et al. (2013) Scale-dependent relationships between tree species richness and ecosystem function in forests. Journal of Ecology 101: 1214–1224.
13. Torti SD, Coley PD, Kursar TA (2001) Causes and consequences of monodominance in tropical lowland forests. American Naturalist 157: 141–153.

14. Ghazoul J, Sheil D (2010) Tropical rain forest ecology, diversity, and conservation. Oxford: Oxford University Press.
15. Whittaker RJ, Triantis KA, Ladle RJ (2008) A general dynamic theory of oceanic island biogeography. Journal of Biogeography 35: 977–994.
16. Kier G, Kreft H, Lee TM, Jetz W, Ibisch PL, et al. (2009) A global assessment of endemism and species richness across island and mainland regions. Proceedings of the National Academy of Sciences 106: 9322–9327.
17. Terborgh J (1985) The vertical component of plant species diversity in temperate and tropical forests. The American Naturalist 126: 760–776.
18. Spies TA (1998) Forest structure: A key to the ecosystem. Northwest Science 72: 34–39.
19. Whitmore TC (1998) An introduction to tropical rain forests. Oxford, UK: Oxford Press.
20. Moles AT, Bonser SP, Poore AGB, Wallis IR, Foley WJ (2011) Assessing the evidence for latitudinal gradients in plant defence and herbivory. Functional Ecology 25: 380–388.
21. Price JP, Clague DA (2002) How old is the Hawaiian biota? Geology and phylogeny suggest recent divergence. Proceedings of the Royal Society of London, Series B Biological Sciences 269: 2429–2435.
22. Wagner WL, Herbst DR, Sohmer S (1999) Manual of the flowering plants of Hawaii. 2nd ed. Honolulu, HI: Bishop Museum.
23. Condit R (1995) Research in large, long-term tropical forest plots. Trends in Ecology and Evolution 10: 18–21.
24. Connell JH, Green PT (2000) Seedling dynamics over thirty-two years in a tropical rain forest. Ecology 81: 568–584.
25. Hobbie JE, Carpenter SR, Grimm NB, Gosz JR, Seastedt TR (2003) The US long term ecological research program. BioScience 53: 21–32.
26. Hubbell SP (2004) Two decades of research on the BCI Forest Dynamics Plot. In: Losos EC, Leigh EG Jr, editors. Tropical forest diversity and dynamism: Findings from a large-scale plot network. Chicago: University of Chicago Press. pp. 8–30.
27. Clinebell RR, Phillips OL, Gentry AH, Stark N, Zuuring H (1995) Prediction of neotropical tree and liana species richness from soil and climatic data. Biodiversity and Conservation 4: 56–90.
28. Malhi Y, Wood D, Baker TR, Wright J, Phillips OL, et al. (2006) The regional variation of aboveground live biomass in old-growth Amazonian forests. Global Change Biology 12: 1107–1138.
29. Saatchi SS, Houghton RA, Dos Santos Alvalá RC, Soares JV, Yu Y (2007) Distribution of aboveground live biomass in the Amazon basin. Global Change Biology 13: 816–837.
30. Slik JWF, Aiba S-I, Brearley FQ, Cannon CH, Forshed O, et al. (2010) Environmental correlates of tree biomass, basal area, wood specific gravity and stem density gradients in Borneo's tropical forests. Global Ecology and Biogeography 19: 50–60.
31. Toledo M, Poorter L, Peña-Claros M, Alarcón A, Balcázar J, et al. (2011) Climate and soil drive forest structure in Bolivian lowland forests. Journal of Tropical Ecology 27: 333–345.
32. Lugo AE, Scatena FN, Silver WL, Molina Colón S, Murphy PG (2002) Resilience of tropical wet and dry forests in Puerto Rico. In: Gunderson LH, Pritchard L Jr, editors. Resilience and the behavior of large-scale systems. Washington, DC: Island Press. pp. 195–225.
33. Bellingham PJ, Stewart GH, Allen RB (1999) Tree species richness and turnover throughout New Zealand forests. Journal of Vegetation Science 10: 825–832.
34. Sagar R, Singh JS (2006) Tree density, basal area and species diversity in a disturbed dry tropical forest of northern India: implications for conservation. Environmental Conservation 33: 256–262.
35. Stegen JC, Swenson NG, Enquist BJ, White EP, Phillips OL, et al. (2011) Variation in above-ground forest biomass across broad climatic gradients. Global Ecology and Biogeography: 744–754.
36. Price JP, Wagner WL (2004) Speciation in Hawaiian angiosperm lineages: cause, consequence, and mode. Evolution 58: 2185–2200.
37. Ziegler AC (2002) Hawaiian natural history, ecology, and evolution. Honolulu, HI: University of Hawaii Press.
38. Harrison RD (2011) Emptying the forest: Hunting and the extirpation of wildlife from tropical nature reserves. BioScience 61: 919–924.
39. Vitousek PM, Farrington H (1997) Nutrient limitation and soil development: experimental test of a biogeochemical theory. Biogeochemistry 37: 63–75.
40. Juvik SP, Juvik JO, Paradise TR (1998) Atlas of Hawaii, 3rd edition. Honolulu, HI: University of Hawaii Press.
41. Giambelluca TW, Chen Q, Frazier AG, Price JP, Chen Y-L, et al. (2013) Online Rainfall Atlas of Hawai'i. Bulletin of the American Meteorological Society 94: 313–316.
42. Crews TE, Kitayama K, Fownes JH, Riley RH, Herbert DA, et al. (1995) Changes in soil phosphorus fractions and ecosystem dynamics across a long chronosequence in Hawaii. Ecology 76: 1407–1424.
43. Tosi JAJ, Watson V, Bolaños R (2002) Life zone maps Hawaii, Guam, American Samoa, Northern Mariana Islands, Palau, and the Federated States of Micronesia. San Jose, Costa Rica and Hilo, Hawaii, USA: Tropical Science Center and the Institute of Pacific Islands Forestry, USDA Forest Service.
44. Kellner JR, Asner GP (2009) Convergent structural responses of tropical forests to diverse disturbance regimes. Ecology Letters 12: 1–11.
45. Mueller-Dombois D (2000) Rain forest establishment and succession in the Hawaiian Islands. Landsc Urban Plann 51: 147–157.
46. Friday JB, Scowcroft PG, Ares A (2008) Responses of native and invasive plant species to selective logging in an *Acacia koa-Metrosideros polymorpha* forest in Hawai'i. Applied Vegetation Science 11: 471–482.
47. Murphy M, Inman-Narahari F, Ostertag R, Litton CM (2014) Invasive feral pigs impact native tree ferns and woody seedlings in Hawaiian forest. Biological Invasions 16: 63–71.
48. Moore RB, Claque DA, Rubin M, Bohrson WA (1987) Hualalai volcano: A preliminary summary of geologic, petrologic, and geophysical data. In: Decker RW, Wright TL, Stauffer PH, editors. Volcanism in Hawaii. Washington, DC: U.S. Geological Service Professional Paper 1350, U'. Government Printing Office. pp. 571–585.
49. Giambelluca T, Chen Q, Frazier A, Price J, Chen Y-L, et al. (2011) The Rainfall Atlas of Hawai'i. Available: http://rainfall.geography.hawaii.edu. Accessed 2014 Aug 10.
50. Thaxton JM, Cole TC, Cordell S, Cabin RJ, Sandquist DR, et al. (2010) Native species regeneration following ungulate exclusion and nonnative grass removal in a remnant Hawaiian dry forest. Pacific Science 64: 533–544.
51. Freifelder RR, Vitousek PM, D'Antonio CM (1998) Microclimate change and effect on fire following forest-grass conversion in seasonally dry tropical woodland. Biotropica 30: 286–297.
52. Litton CM, Sandquist DR, Cordell S (2006) Effects of non-native grass invasion on aboveground carbon pools and tree population structure in a tropical dry forest of Hawaii. Forest Ecology and Management 231: 105–113.
53. Condit R (1998) Tropical forest census plots. Berlin: Springer-Verlag, Berlin, Germany.
54. Asner GP, Hughes RF, Mascaro J, Uowolo AL, Knapp DE, et al. (2011) High-resolution carbon mapping on the million-hectare island of Hawai'i. Frontiers in Ecology and the Environment 9: 434–439.
55. Chave J, Andalo C, Brown S, Cairns MA, Chambers JQ, et al. (2005) Tree allometry and improved estimation of carbon stocks and balance in tropical forests. Oecologia 145: 87–99.
56. Zanne AE, Lopez-Gonzalez G, Coomes DA, Ilic J, Jansen S, et al. (2009) Towards a worldwide wood economics spectrum. Dryad Digital Repository doi: 10.5061/dryad.5234.
57. Chave J, Coomes D, Jansen S, Lewis SL, Swenson NG, et al. (2009) Towards a worldwide wood economics spectrum. Ecology Letters 12: 351–366.
58. Slik JWF (2006) Estimating species-specific wood density from the genus average in Indonesian trees. Journal of Tropical Ecology 22: 481–482.
59. Chave J, Alonso D, Etienne RS (2006) Theoretical biology: Comparing models of species abundance. Nature 441: E1.
60. Gotelli NJ, Colwell RK (2001) Quantifying biodiversity: procedures and pitfalls in the measurement and comparison of species richness. Ecology Letters 4: 379–391.
61. Chiarucci A, Enright NJ, Perry GLW, Miller BP, Lamont BB (2003) Performance of nonparametric species richness estimators in a high diversity plant community. Diversity and Distributions 9: 283–295.
62. Colwell RK, Mao CX, Chang J (2004) Interpolating, extrapolating, and comparing incidence-based species accumulation curves Ecology 85: 2717–2727.
63. Magurran A (2004) Measuring biological diversity. Oxford, UK: Blackwell Publishing.
64. Sokal RR, Rohlf FJ (1994) Biometry: The principles and practice of statistics in biological research. 3rd edition. New York, New York: W.H. Freeman.
65. SAS Institute (2005) JMP introductory guide, release 6. Cary, NC: SAS Institute.
66. Asner GP, Hughes RF, Varga TA, Knapp DE, Kennedy-Bowdoin T (2009) Environmental and biotic controls over aboveground biomass throughout a tropical rain forest. Ecosystems 12: 261–278.
67. Brenes-Arguedas T, Roddy A, Coley P, Kursar T (2010) Do differences in understory light contribute to species distributions along a tropical rainfall gradient? Oecologia: 1–14.
68. Ashton PS (2004) Floristics and vegetation of the forest dynamics plots. In: Losos E, Leigh EG Jr, editors. Tropical forest diversity and dynamism: Findings from a large-scale plot network. Chicago, Illinois, USA: University of Chicago Press.
69. Davidar P, Puyravaud JP, Leigh EG Jr (2005) Changes in rain forest tree diversity, dominance and rarity across a seasonality gradient in the Western Ghats, India. Journal of Biogeography 32: 493–501.
70. Clark JS, McLachlan JS (2003) Stability of forest diversity. Nature 423: 635–638.
71. Clark DA, Piper SC, Keeling CD, Clark DB (2003) Tropical rain forest tree growth and atmospheric carbon dynamics linked to interannual temperature variation during 1984–2000. Proceedings of the National Academy of Sciences 100: 5852–5857.
72. Feeley KJ, Wright SJ, Supardi MNN, Kassim AR, Davies SJ (2007) Decelerating growth in tropical forest trees. Ecology Letters 10: 1–9.
73. Clark DB, Clark DA, Oberbauer SF (2010) Annual wood production in a tropical rain forest in NE Costa Rica linked to climatic variation but not to increasing CO_2. Global Change Biology 16: 747–759.
74. Stegen JC, Swenson NG, Enquist BJ, White EP, Phillips OL, et al. (2011) Variation in above-ground forest biomass across broad climatic gradients. Global Ecology and Biogeography 20: 744–754.
75. Martinez-Yrizar A (1995) Biomass distribution and primary productivity of tropical dry forests. In: Bullock SH, Mooney HA, Medina E, editors. Seasonally

Dry Tropical Forests. Cambridge, UK: Cambridge University Press. pp. 326–345.
76. Vitousek P, Asner GP, Chadwick OA, Hotchkiss S (2009) Landscape-level variation in forest structure and biogeochemistry across a substrate age gradient in Hawaii. Ecology 90: 3074–3086.
77. Cole RJ, Litton CM, Koontz MJ, Loh RK (2012) Vegetation recovery 16 years after feral pig removal from a wet Hawaiian forest. Biotropica 44: 463–471.
78. Inman-Narahari F, Ostertag R, Cordell S, Giardina CP, Nelson-Kaula K, et al. (2013) Seedling recruitment factors in low-diversity Hawaiian wet forest: towards global comparisons among tropical forests. Ecosphere 4: 24. http://dx.doi.org/10.1890/ES1812-00164.00161.
79. Busby PE, Vitousek PM, Dirzo R (2010) Prevalence of tree regeneration by sprouting and seeding along a rainfall gradient in Hawai'i. Biotropica 42: 80–86.
80. Kammesheidt L (1999) Forest recovery by root suckers and above-ground sprouts after slash-and-burn agriculture, fire and logging in Paraguay and Venezuela. Journal of Tropical Ecology 15: 143–157.
81. Bellingham PJ, Sparrow AD (2009) Multi-stemmed trees in montane rain forests: their frequency and demography in relation to elevation, soil nutrients, and disturbance. Journal of Ecology 97: 472–483.
82. Lehmann A, Leathwick JR, Overton JM (2002) Assessing New Zealand fern diversity from spatial predictions of species assemblages. Biodiversity and Conservation 11: 2217–2238.
83. Coomes DA, Allen RB, Bentley WA, Burrows LE, Canham CD, et al. (2005) The hare, the tortoise and the crocodile: the ecology of angiosperm dominance, conifer persistence and fern filtering. Journal of Ecology 93: 918–935.
84. Bellingham PJ, Richardson SJ (2006) Tree seedling growth and survival over 6 years across different microsites in a temperate rain forest. Canadian Journal of Forest Research 36: 910–918.
85. Tanner EVJ (1983) Leaf demography and growth of the tree-fern *Cyathea pubescens* Mett ex Kuhn in Jamaica Bot J Linn Soc 87: 213–227.
86. Bernabe N, Williams-Linera G, Palacios-Rios M (1999) Tree ferns in the interior and at the edge of a Mexican cloud forest remnant: Spore germination and sporophyte survival and establishment. Biotropica 31: 83–88.
87. Arens NC (2001) Variation in performance of the tree fern *Cyathea caracasana* (Cyatheaceae) across a successional mosaic in an Andean cloud forest. American Journal of Botany 88: 545–551.
88. Williams-Linera G, Palacios-Ríos M, Hernández-Gómez R (2005) Fern richness, tree species surrogacy and fragments complementarity in a Mexican tropical montane cloud forest. Biodiversity and Conservation 14: 119–133.
89. Jones MM, Olivas Rojas P, Tuomisto H, Clark DB (2007) Environmental and neighbourhood effects on tree fern distributions in a neotropical lowland rain forest. Journal of Vegetation Science 18: 13–24.
90. Durand LZ, Goldstein G (2001) Growth, leaf characteristics, and spore production in native and invasive tree ferns in Hawaii. American Fern Journal 91: 25–35.
91. Amatangelo KL, Vitousek PM (2009) Contrasting predictors of fern versus angiosperm decomposition in a common garden. Biotropica 41: 154–161.
92. Mack MC, D'Antonio CM, Ley RE (2001) Alteration of ecosystem nitrogen dynamics by exotic plants: A case study of C_4 grasses in Hawaii. Ecological Applications 11: 1323–1335.
93. Huenneke LF, Vitousek PM (1989) Seedling and clonal recruitment of the invasive tree, *Psidium cattleianum*: implications for management of native Hawaiian forests. Biological Conservation 53: 199–211.
94. Hughes RF, Denslow JS (2005) Invasion by an N2-fixing tree, *Falcataria moluccana*, alters function, composition, and structure of wet lowland forests of Hawai'i. Ecological Applications 15: 1615–1628.
95. Gravel D, Canham CD, Beaudet M, Messier C (2006) Reconciling niche and neutrality: The continuum hypothesis. Ecology Letters 9: 399–409.
96. Hubbell SP, Foster RB (1986) Commonness and rarity in a neotropical forest: Implications for tropical tree conservation. In: Soule ME, editor. Conservation biology: The science of scarcity and diversity: Sunderland, Mass, USA Illus Paper Maps. Sunderland, Massachussetts: Sinauer Associates, Inc. pp. 205–231.
97. Lach L (2003) Invasive ants: Unwanted partners in ant-plant interactions? Annals of the Missouri Botanical Garden 90: 91–108.
98. Nogueira SSdC, Nogueira-Filho SLG, Bassford M, Silvius K, Fragoso JMV (2007) Feral pigs in Hawai'i: Using behavior and ecology to refine control techniques. Appl Anim Behav Sci 108: 1–11.
99. Nogueira-Filho S, Nogueira S, Fragoso J (2009) Ecological impacts of feral pigs in the Hawaiian Islands. Biodiversity and Conservation 18: 3677–3683.
100. Shiels A, Drake D (2010) Are introduced rats (*Rattus rattus*) both seed predators and dispersers in Hawaii? Biological Invasions: 1–12.
101. Shiels AB (2011) Frugivory by introduced black rats (*Rattus rattus*) promotes dispersal of invasive plant seeds. Biological Invasions 13: 781–792.
102. Gilbert B, Lechowicz MJ (2004) Neutrality, niches, and dispersal in a temperate forest understory. Proceedings of the National Academy of Sciences of the United States of America 101: 7651–7656.
103. Hubbell SP (2005) Neutral theory in community ecology and the hypothesis of functional equivalence. Functional Ecology 19: 166–172.
104. Losos E, Leigh EG Jr, editors (2004) Tropical forest diversity and dynamism: Findings from a large-scale plot network. Chicago, Illinois: University of Chicago Press.

4

Dynamic Analysis and Pattern Visualization of Forest Fires

António M. Lopes[1]*, J. A. Tenreiro Machado[2]

1 Institute of Mechanical Engineering, Faculty of Engineering, University of Porto, Porto, Portugal, **2** Institute of Engineering, Polytechnic of Porto, Porto, Portugal

Abstract

This paper analyses forest fires in the perspective of dynamical systems. Forest fires exhibit complex correlations in size, space and time, revealing features often present in complex systems, such as the absence of a characteristic length-scale, or the emergence of long range correlations and persistent memory. This study addresses a public domain forest fires catalogue, containing information of events for Portugal, during the period from 1980 up to 2012. The data is analysed in an annual basis, modelling the occurrences as sequences of Dirac impulses with amplitude proportional to the burnt area. First, we consider mutual information to correlate annual patterns. We use visualization trees, generated by hierarchical clustering algorithms, in order to compare and to extract relationships among the data. Second, we adopt the Multidimensional Scaling (MDS) visualization tool. MDS generates maps where each object corresponds to a point. Objects that are perceived to be similar to each other are placed on the map forming clusters. The results are analysed in order to extract relationships among the data and to identify forest fire patterns.

Editor: Jesus Gomez-Gardenes, Universidad de Zarazoga, Spain

Funding: The authors have no support or funding to report.

Competing Interests: The authors have declared that no competing interests exist.

* Email: aml@fe.up.pt

Introduction

Forest fires are a major concern in many countries, like United States, Australia, Russia, Brazil, China and Mediterranean Basin European regions [1–3]. Every year forest fires consume vast areas of vegetation, compromising ecosystems and contributing to the carbon dioxide emissions that are changing Earth's climate [4–5]. Besides the long-term economic implications associated to the climate change, forest fires have direct impact upon economy due to the destruction of public and private property and infrastructures [6]. Fires are mainly caused by natural factors, human negligence, or even human intent. Fire propagation and burnt area depend on many natural factors and conditions, not only on the terrain orography and the type of vegetation, but also on the efficacy of detection and suppression strategies. Moreover, fires caused by incendiaries contribute to increase the complexity of the phenomena. Understanding the underlying patterns of forest fires in terms of their size and spatiotemporal distributions may help the decision makers to take preventive measures beforehand, identifying possible hazards and deciding strategies for fire prevention, detection and suppression [7–8].

Forest fires have been studied using classical statistical tools. However, those methods reveal limitations, both in capturing all characteristics underneath forest fires dynamics, and the evolution along years [9]. Forest fires dynamics exhibits correlations in size, space and time. Size-frequency distributions unveil long range memory, which is typical in complex systems. Correlation between data is characterized by self-similarity and absence of characteristic length-scale, meaning that forest fires exhibit power-law (PL) behaviour [10–13].

Several studies have been published during the last years about this topic [14–17]. In references [18–19] it is shown that forest fires exhibit PL frequency-size relationship over many orders of magnitude and that such behaviour seems consistent with the self-organized criticality arising in complex systems. The most important practical implication of such results is that the frequency-size distribution of small and medium fires can be used to quantify the risk of large fires [19]. Nevertheless, some authors [15] suggest that a simple PL distribution of sizes may be too simple to describe the distributions of forest fires over their full range.

In reference [20] the time dynamics of forest fires is investigated and it is shown that forest fires exhibit time-clustering phenomena. More recently, the fractality of the forest fires was addressed in [21] using spatial and temporal fractal tools. The authors prove that these phenomena exhibit space–time clustering behaviour.

In this paper we look at forest fires from the perspective of dynamical systems. A public domain forest fires catalogue containing data of events occurred in Portugal, in the period 1980 up to 2012, is addressed. The data is analysed in an annual basis, modelling the occurrences as a sequence of Dirac impulses. Therefore, instead of modelling individual forest fires, we are describing the global dynamics along several decades. In this perspective, mutual information and visualization trees, generated by hierarchical clustering algorithms, are used. The Multidimensional Scaling (MDS) tool is adopted in order to compare and to

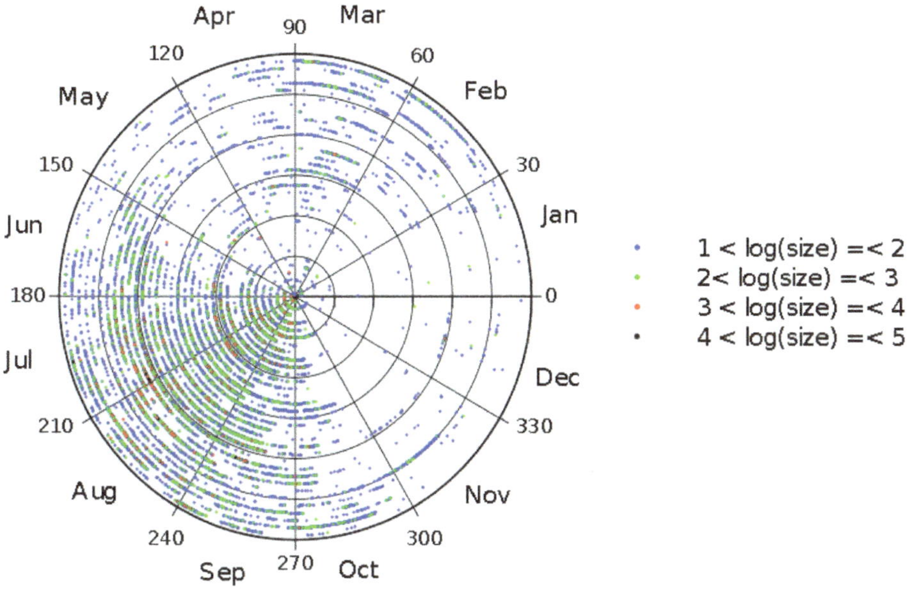

Figure 1. Temporal evolution and size (in log units) of forest fires registered in Portugal in the time period 1980–2012, with burnt area larger than A_{min} = 10 ha. Each (r, θ) point represents the time of the event and the color represents the size.

extract relationships among the data. Finally, we propose an amplitude-space embedding technique that produces a clear fire pattern classification.

Characterization of the Dataset

Data from forest fires is available online at the Portuguese Institute of Nature and Forest Conservation (INCF), http://www.icnf.pt/portal/florestas/dfci/inc/estatisticas, and the catalogue contains events since 1980 up to 2012. Ignitions might have different sources, as natural causes, human negligence or human intentionality, among others. The data analysed in this paper was retrieved in December, 2013. Each record contains information about the events date, time (with one minute resolution), geographic location and size (in terms of burnt area). We decided to discard small size events, as those are prone to measurement errors, by adopting a cutoff threshold value of A_{min} = 10 hectares.

Fig. 1 illustrates the temporal evolution and size of the events occurred in Portugal, during 1980–2012 and meeting the cutoff threshold criterion. We tackle the concept of 'circular time' (since

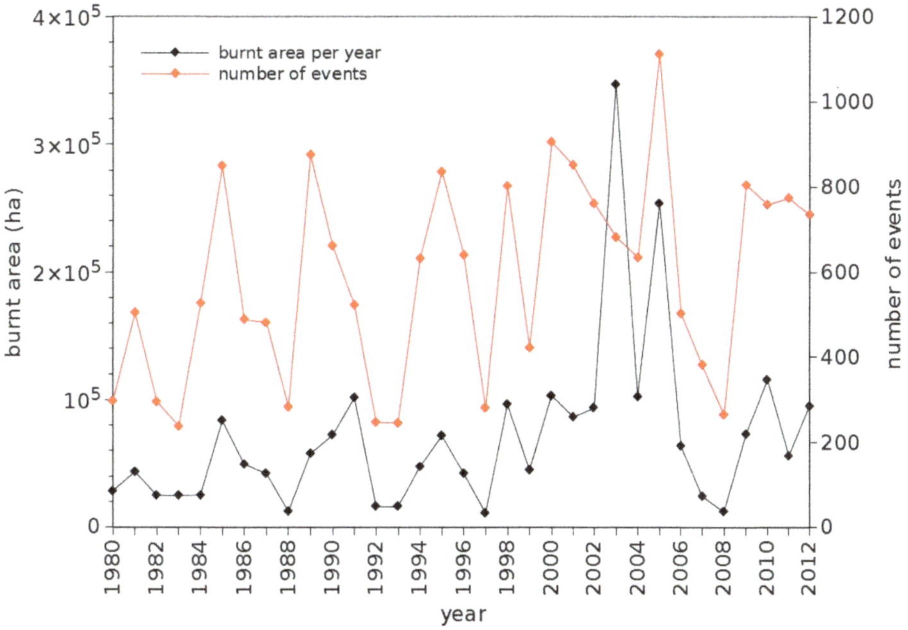

Figure 2. Evolution of burnt area per year and number of occurrences registered in Portugal in the time period 1980–2012 (are considered events with burnt area larger than A_{min} = 10 ha).

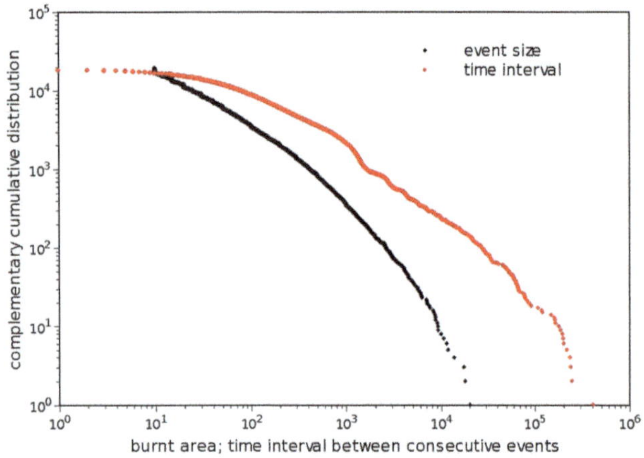

Figure 3. Complimentary cumulative distribution of events size and time interval (in minutes) between consecutive events, corresponding to occurrences registered in Portugal in the time period 1980–2012, with burnt area larger than A_{min} = 10 ha.

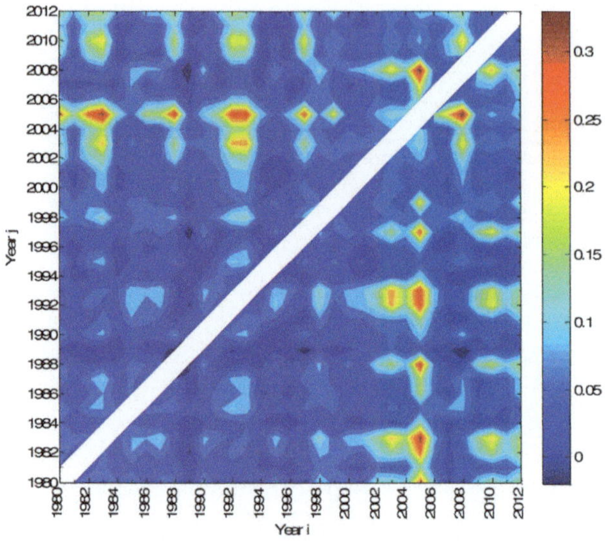

Figure 4. Contour map of the mutual information, $I_M(X_i, X_j)$, between occurrences registered in Portugal during 1980–2012. The cutoff threshold value A_{min} = 10 ha is adopted.

there is a kind of one-year periodicity, with December close to January and not the opposite, as a Cartesian scale implicitly assumes). The (circular) time scale evolves along an Archimedean spiral, with origin at the center of the circumferences, given by:

$$\theta = 2\pi \cdot \left(\frac{t}{T} + i\right) \tag{1}$$

$$r = p + q \cdot \theta \tag{2}$$

where (r, θ) denotes the radius and angle coordinates, respectively, $i = 0, \ldots, 32$, represents the year and $p = q = 1$. The burnt area is expressed in logarithmic units and is related to the color of the marks. We can note two annual cycles: the first is weaker and includes the months of February and March; the second is stronger and is due to the major incidence of fires during summer [22].

In Fig. 2 we depict the evolution of the burnt area per year and number of occurrences versus year. It is visible the increasing number of events as well as the strong activity verified around the middle of the decade 2000–2009. Nevertheless, the charts reveal a large volatility and pose difficulties to capture some trend. We observe minimal values for years 1983, 1988, …, 2008, and maximum values for 2003 and 2005, but no straightforward method to correlate data points. Fig. 3 represents the complementary cumulative distributions of the events size and the time interval between consecutive events.

The results shown above illustrate through simple statistics the increasing importance of understanding the behavior of forest fires and characterizing the spatiotemporal distributions unveiled by such a complex phenomenon. For that purpose, in the next sections we adopt several complementary mathematical tools.

Mutual Information Analysis

In this section we adopt the mutual information to correlate forest fires annual patterns. First we compute the mutual information, based on events size (i.e., burnt area), for each pair of years in the time period 1980–2012. Second, we use a hierarcHical clustering algorithm to find relationships among the data. Visualization trees are used to highlight the interpretation of the results.

Mutual Information

The mutual information is a measure of the statistical dependence between two random variables, giving the amount of information that one variable "contains" about the other. If X_i and X_j are two discrete random variables, then the mutual information, $I(X_i, X_j)$, is given by:

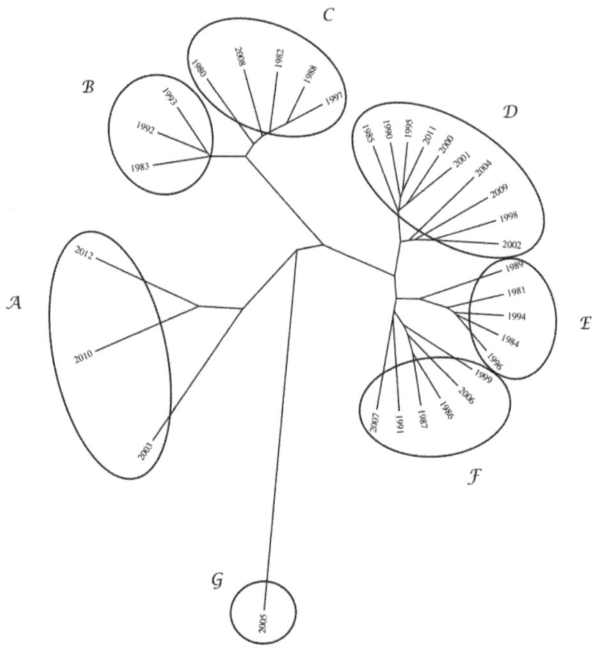

Figure 5. Tree representing mutual information, $I_M(X_i, X_j)$, between occurrences registered in Portugal during 1980–2012. The cutoff threshold value A_{min} = 10 ha is adopted.

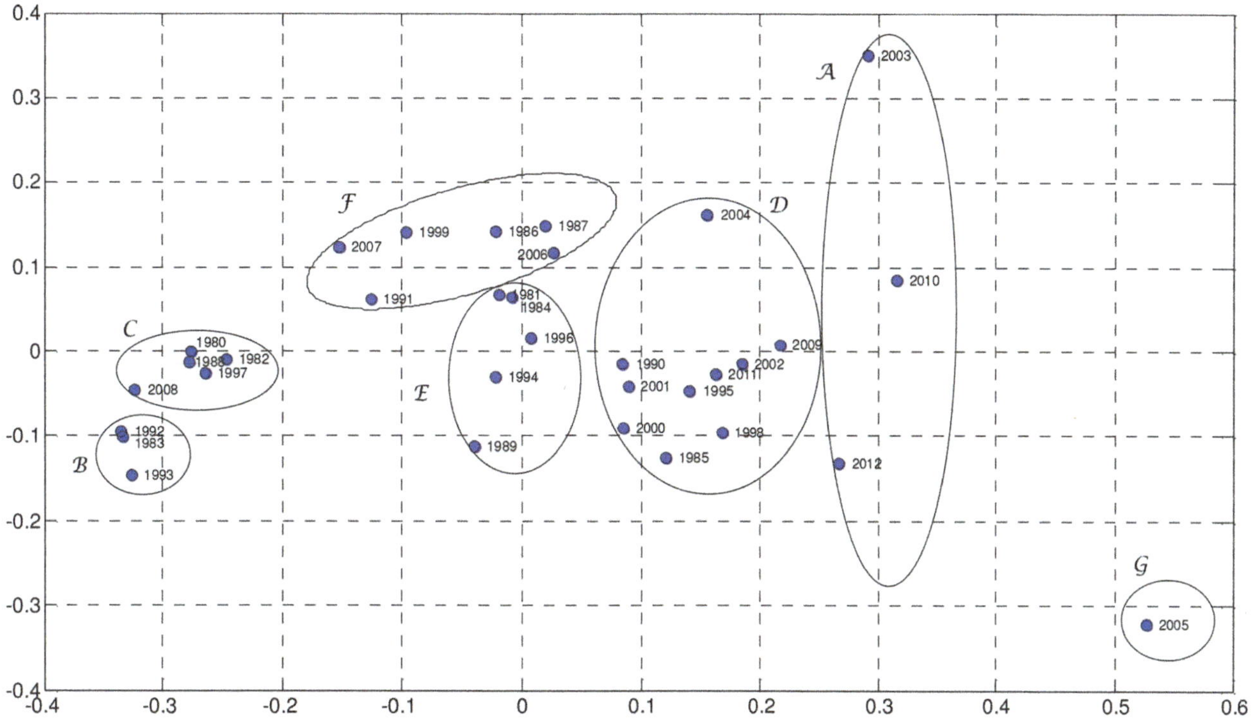

Figure 6. MDS map based on matrix D, for visualization space with dimension $m = 2$.

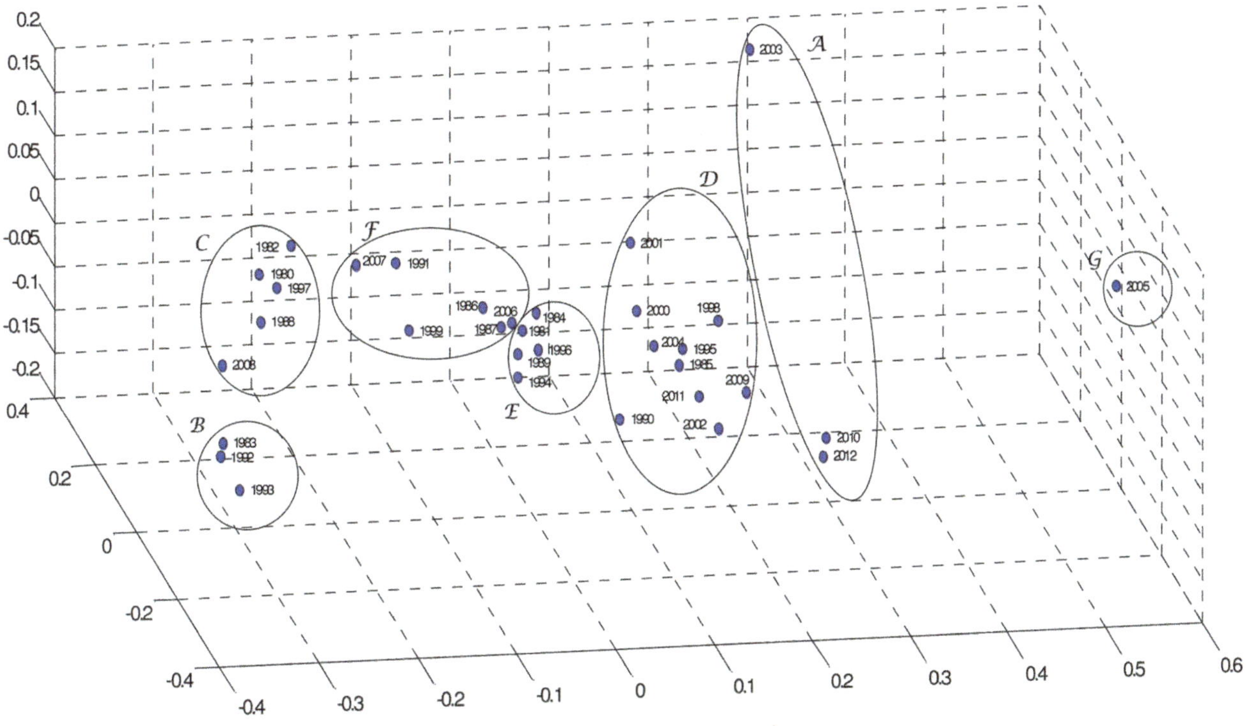

Figure 7. MDS map based on matrix D, for visualization space with dimension $m = 3$.

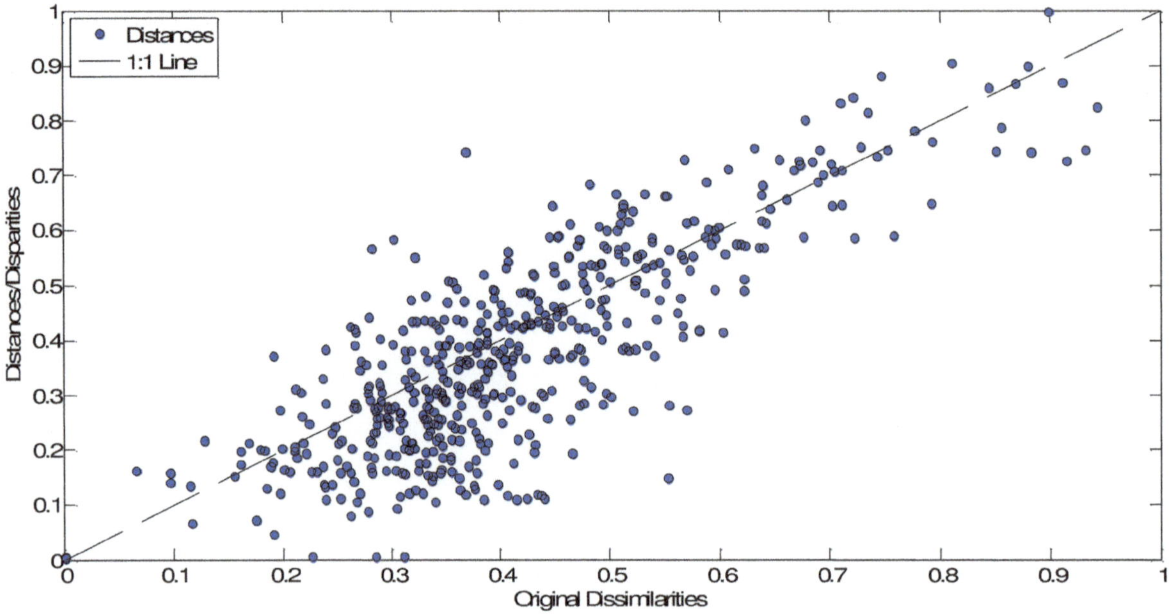

Figure 8. Shepard plot of the MDS map based on matrix D, for visualization space with dimension $m=2$.

$$I(X_i,X_j)=\sum_{x_j\in X_j}\sum_{x_i\in X_i} p(x_i,x_j)\cdot\log\left(\frac{p(x_i,x_j)}{p(x_i)\cdot p(x_j)}\right) \qquad (3)$$

where $p(x_i, x_j)$ is the joint probability distribution function of (X_i, X_j), and $p(x_i)$ and $p(x_j)$ are the marginal probability distribution functions of X_i and X_j, respectively.

The concept of mutual information comes from the information theory [23] and has been adopted in the study of complex systems from diverse fields, namely in experimental time series analysis, in DNA and symbol sequencing and in providing a theoretical basis for the notion of complexity [24–30].

In this section, instead of expression (3), we use the normalized mutual information, $I_N(X_i, X_j)$, given by [31]:

$$I_N(X_i,X_j)=\frac{I(X_i,X_j)}{H(X_i,X_j)} \qquad (4)$$

where $H(X_i, X_j)$ represents the joint entropy between X_i and X_j:

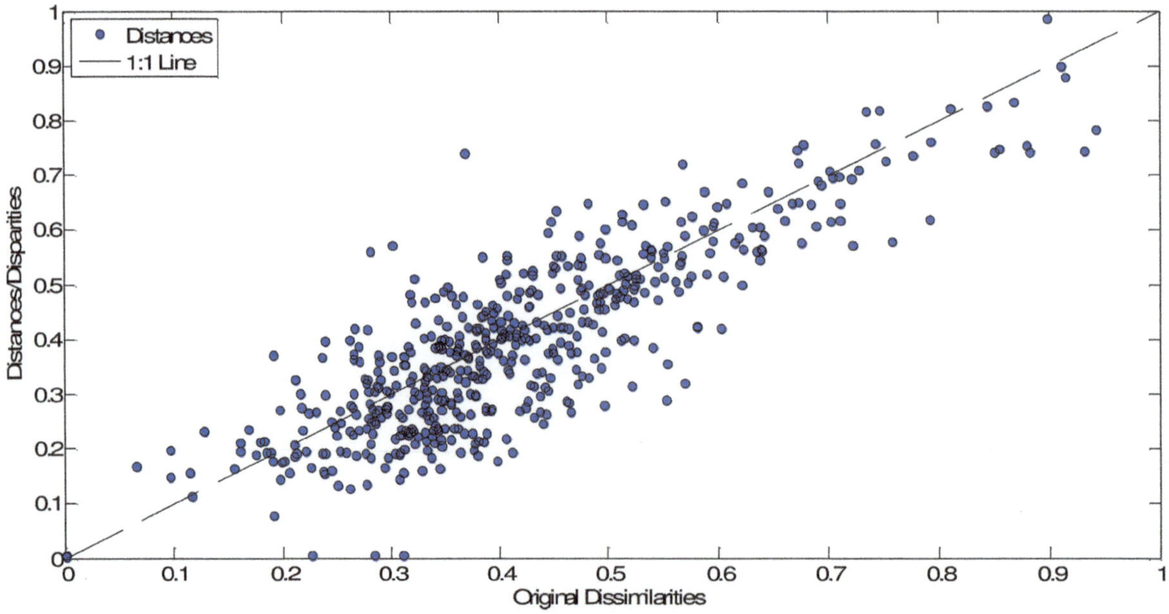

Figure 9. Shepard plot of the MDS map based on matrix D, for visualization space with dimension $m=3$.

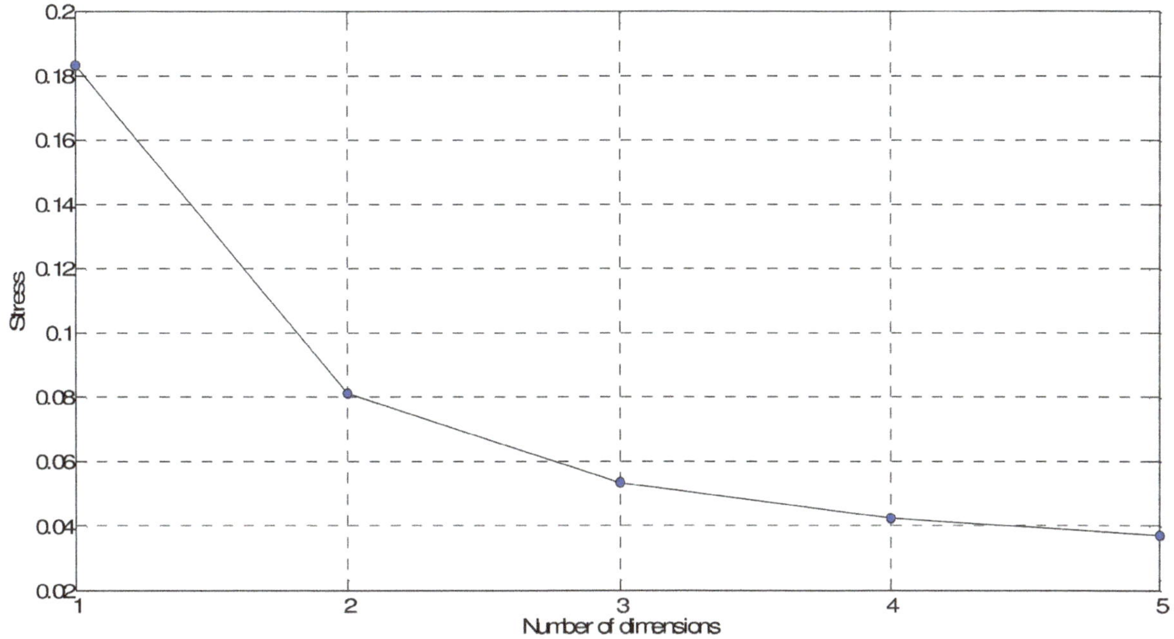

Figure 10. Stress plot for the MDS based on matrix D.

$$H(X_i,X_j) = -\sum_{x_j \in X_j}\sum_{x_i \in X_i} p(x_i,x_j)\cdot\log\left[p(x_i,x_j)\right] \quad (5)$$

The normalized mutual information $I_N(X_i, X_j) \in [0, 1]$ simplifies comparison across different conditions and improves sensitivity.

Forest fires are analysed in an annual basis. For each year, $i = 0, \ldots, 32$, in the period 1980–2012 the events are represented by:

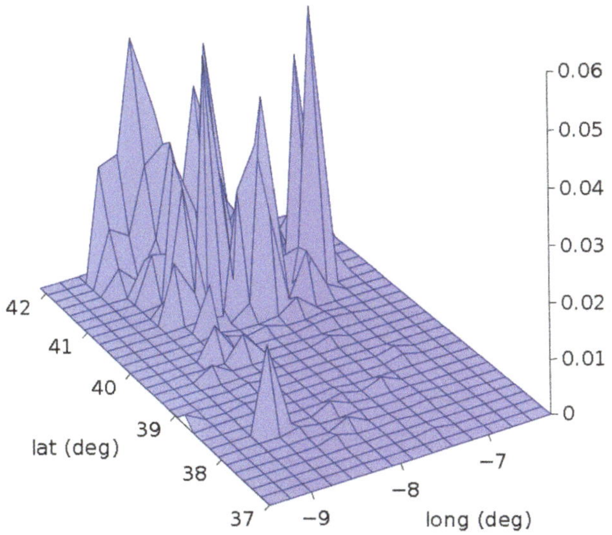

Figure 11. Bidimensional histogram of forest fires versus latitude and longitude, in Portugal, for year 2010.

$$x_i(t) = \sum_{k=1}^{T} A_k \delta(t - t_k), \quad i = 0, \cdots, 32 \quad (6)$$

leading to 33 one-year length time series. This means that the events are modelled as Dirac impulses, where A_k represents fire size (i.e., burnt area), t_k is the instant of occurrence (with one minute resolution), t represents time and T denotes the time period of one year.

The signals $x_i(t)$ are then normalized according to (7):

$$\tilde{x}_i(t) = \frac{x_i(t) - \mu}{\sigma} \quad (7)$$

where μ and σ represent the mean and standard deviation values of all events listed during 1980–2012, with magnitude larger than $A_{min} = 10$ ha. The mutual information is calculated to correlate events occurred in different years of the analysed time period.

Fig. 4 depicts in a contour map the mutual information, $I_N(X_i, X_j)$, between every pair of years $i, j = 0, \ldots, 32$. The probabilities for calculating the mutual information are estimated from the histograms of amplitudes A_k, constructed considering 476 bins, each one having width equal to 0.1 ha. To facilitate the comparison the cases $i = j$ (i.e., those with maximum value of mutual information) are removed from the graph.

The map reveals strong correlations between certain years, corresponding to higher values of mutual information. This is well noted for the years a = {2003, 1983}, b = {2003, 1993}, c = {2005, 1980}, d = {2005, 1983}, e = {2005, 1988}, f = {2005, 1993}, g = {2008, 2005} and h = {2010, 1983}. Nevertheless, the analysis is not totally assertive and requires multiple comparisons.

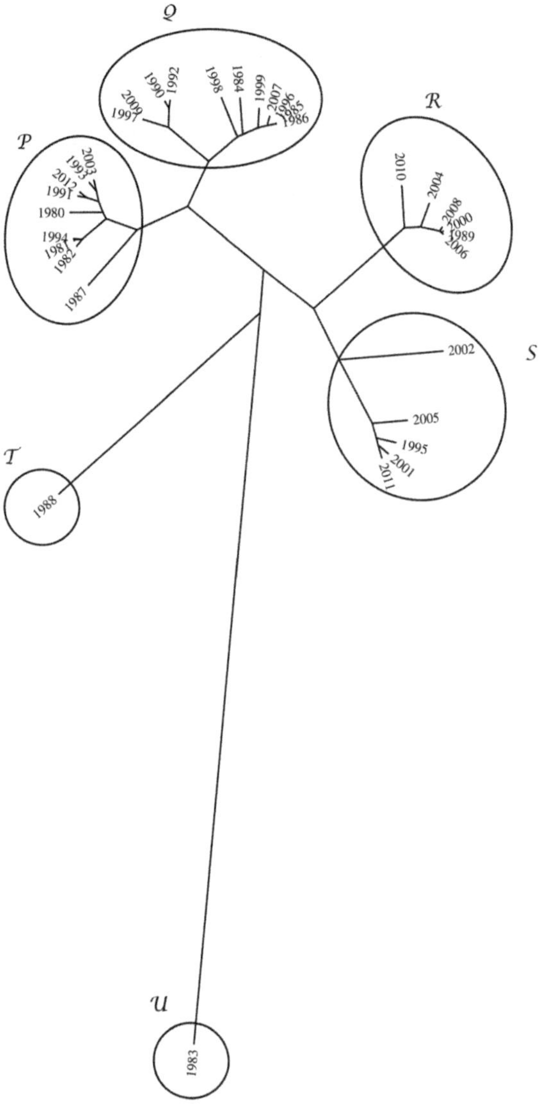

Figure 12. Tree comparing the 33 bidimensional histograms based on index d_{ij}.

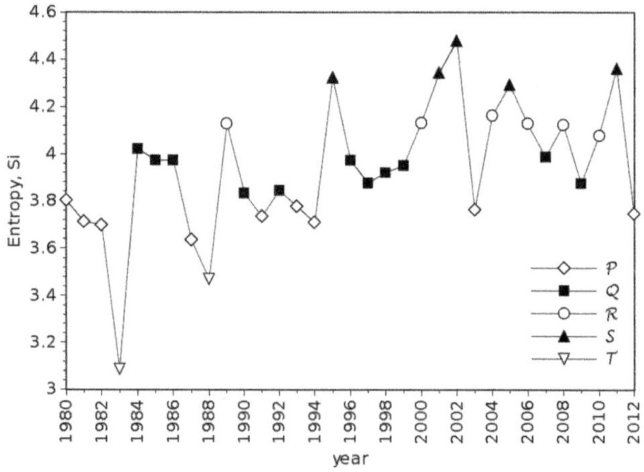

Figure 13. Entropy, S_i, versus year, during the time period 1980–2012.

Hierarchical clustering

Having in mind an efficient method to visualize and to compare results, a hierarchical clustering algorithm is adopted, based on the mutual information, $I_N(X_i, X_j)$, between pairs of objects.

The goal of hierarchical clustering is to build a hierarchy of clusters, in such a way that objects in the same cluster are, in some sense, similar to each other [30,32–33]. Based on a measure of dissimilarity between clusters, those are combined (or, alternatively, split) for agglomerative (or, alternatively, divisive) clustering. This is achieved by using an appropriate metric, quantifying the distance between pairs of objects, and a linkage criterion, defining the dissimilarity between clusters as a function of the pairwise distances between objects. The results of hierarchical clustering are presented in a phylogenetic tree adopting the successive (agglomerative) clustering and average-linkage method (Fig. 5). The software PHYLIP was used for generating both graphs (http://evolution.genetics.washington.edu/phylip.html).

Fig. 5 unveils groups of objects (years) in such a way that objects in the same group (cluster) are more similar to each other than to those in other groups. For example, we can easily identify clusters composed by years A= {2003, 2010, 2012}, B= {1983, 1992, 1993} and C= {1980, 1982, 1988, 1997, 2008,}. On the contrary, year G= {2005} is far aside, meaning that it is different from all the rest. Both representations of Fig. 4 and Fig. 5 can be used to visualise and to compare the events, in an annual basis. Fig. 5 leads to a result easier to interpret than Fig. 4, as it identifies groups of objects that are similar.

MDS Analysis and Visualization

In this section we adopt the MDS tools to handle information and the relationships embedded into the data.

MDS is a statistical technique for visualizing data that can reveal similarities between objects. The algorithm requires the definition of a similarity measure (or, inversely, of a distance) and the construction of a $s \times s$ symmetric matrix **D** of similarities (or distances) between each pair of s objects. MDS assigns a point to each object in a m-dimensional space and arranges the set in order to reproduce the observed similarities. A shorter (larger) distance between two points means that the corresponding objects are more similar (distinct). For $m=2$ or $m=3$ dimensions the resulting locations may be displayed in a "map" that can be visualized [34–39].

In our case, we obtain **D** (33×33 dimensional) by means of the mutual information (4). Fig. 6 and Fig. 7 show the MDS maps for $m=2$ and $m=3$, respectively. The Shepard and the stress plots assess the quality of the MDS maps. The Shepard diagrams (Fig. 8 and Fig. 9) show an acceptable distribution of points around the 45 degree line, which means a good fit of the distances to the dissimilarities. On the other hand, the stress plot reveals that a three dimensional space describes adequately the data (Fig. 10). This can be concluded by observing the stress line, which diminishes strongly until the dimensionality is two, moderately towards dimensionality three and weakly from then on. Often, the maximum curvature point of the stress line is adopted as the criterion for deciding the dimensionality of the MDS map.

The MDS maps of Fig. 6 and Fig. 7 confirm the groups previously identified by the hierarchical clustering and, consequently, the relationships between the corresponding yearly patterns. Comparing Fig. 5 with Fig. 6 and Fig. 7, we conclude that all allow an easy interpretation of the results. The MDS maps, in particular the 3D plot, are more intuitive than the phylogenetic

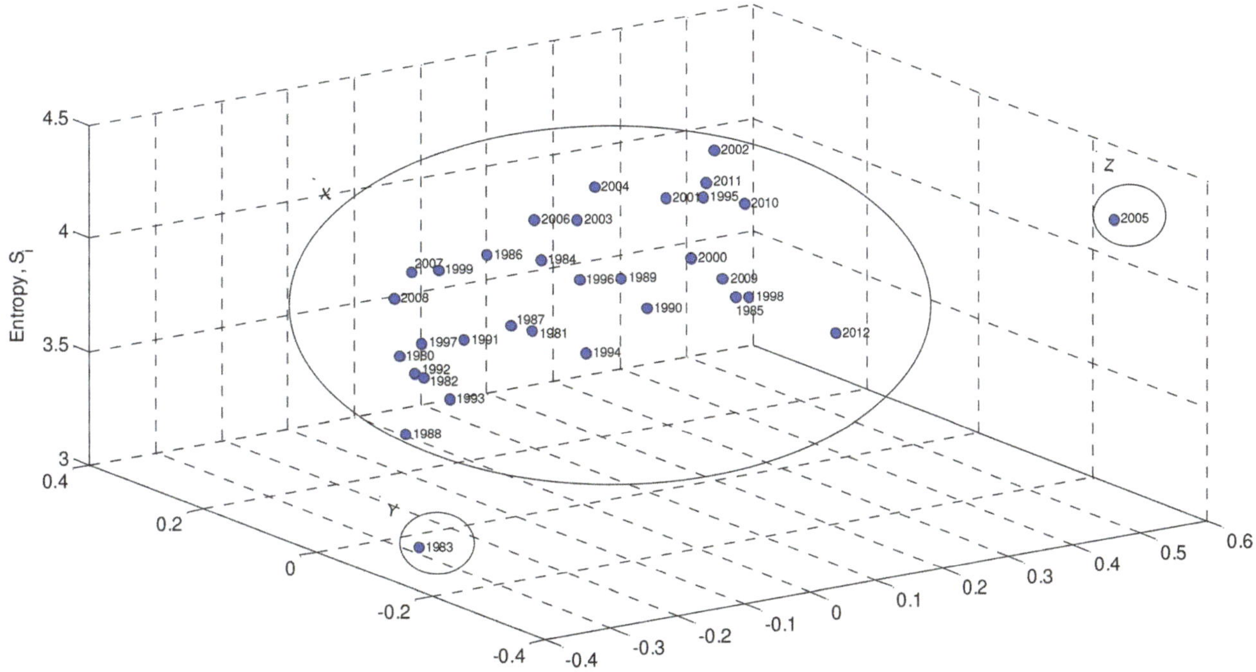

Figure 14. MDS 2D plot with the vertical axis representing Shannon entropy.

tree. Moreover, most software for MDS analysis allows the user to rotate and visualise the maps from different perspectives, easing the identification of clusters. This is useful especially when dealing with large amounts of data.

Forest Fires Spatial Patterns

In this section we study forest fires in a complementary line of thought, namely by considering spatial information. First, we divide the geographic territory under study (i.e., $36.95° \leq$ lat $\leq 42.15°$; $-9.50° \leq$ lon $\leq -6.19°$), using a $M \times N$ ($M = 30$, $N = 15$) rectangular grid, and we determine the 33 bidimensional histograms of relative frequencies for all years in the period 1980–2012. Second, for characterizing the histograms, we calculate the Shannon entropy, S_i, given by:

$$S_i = -\sum_{m=1}^{M}\sum_{n=1}^{N} p_i(m,n) \cdot \log[p_i(m,n)], \quad i = 0, \cdots, 32 \qquad (8)$$

where the probabilities $p_i(m, n)$ are approximated by the relative frequencies.

In Fig. 11, for example, we depict the bidimensional histogram for year 2010. The corresponding entropy is $S_i = 4.08$ ($i = 30$).

For histogram comparison we calculate a $M \times N$ symmetric matrix $\mathbf{D} = [d_{ij}]$, where

$$d_{ij} = |S_i - S_j|, \quad i,j = 0, \cdots, 32 \qquad (9)$$

The results are visualized in the phylogenetic tree of Fig. 12. We can observe six clusters: P = {1980, 1981, 1982, 1987, 1991, 1993, 1994, 2003, 2012}; Q = {1984, 1985, 1986, 1990, 1992, 1996, 1997, 1998, 1999, 2007, 2009}; R = {1989, 2000, 2004, 2006, 2008, 2010}; S = {1995, 2001, 2002, 2005, 2011}; T = {1983} and U = {1988}.

The evolution of S_i versus year is represented in Fig. 13, where the clusters shown in Fig. 12 are identified. In this chart is clear a large volatility and apparently some increase of entropy along time.

In a more global perspective, we verify that amplitude and space data lead to distinct observations. The conclusions are 'decoupled' and reveal that both directions must be explored, with more data, in order to include all information in a global tool of analysis.

In this line of though, we embed amplitude and space data into a single graph by adding to the bidimensional MDS plot of Fig. 6 a vertical axis representing the Shannon entropy (Fig. 14).

We note that only two years, Y = {1983} and Z = {2005}, have now a clearly distinct separation from the main cluster, X. In Fig. 13 we observed them to be located at near extreme values, but, as mentioned, it is difficult to get idea due to large volatility. The embedding of amplitude-space techniques produced a clear classification pattern.

Conclusions

We analysed forest fires from the perspective of dynamical systems. Data from a public domain forest fires catalogue, containing information of events for Portugal, during the period 1980–2012, was studied in an annual basis. Mutual information to correlate annual patterns was considered. Phylogenetic trees generated by hierarchical clustering algorithms and MDS visualization tools were used to compare to extract relationships among the data and to identify forest fire patterns. Those tools allow different perspectives over forest fires that may be used to better understand the dynamics emerging in the plethora of phenomena that occur in forest fires.

Author Contributions

Conceived and designed the experiments: AML JTM. Performed the experiments: AML. Analyzed the data: AML JTM. Contributed reagents/

materials/analysis tools: AML JTM. Contributed to the writing of the manuscript: AML JTM.

References

1. Di Bella CM, Jobbágy EG, Paruelo JM, Pinnock S (2006) Continental fire density patterns in South America. Global Ecology and Biogeography 15, 192–199.
2. Bradstock RA (2008) Effects of large fires on biodiversity in south-eastern Australia: disaster or template for diversity?. International Journal of Wildland Fire 17, 809–822.
3. Hanson CT, Odion DC (2013) Is fire severity increasing in the Sierra Nevada, California, USA?. International Journal of Wildland Fire 23, 1–8.
4. Flannigan MD, Krawchuk M, de Groot W, Wotton B, Gowman L (2009) Implications of changing climate for global wildland fire. International Journal of Wildland Fire 18, 483–507.
5. Zumbrunnen T, Pezzatti GB, Menéndez P, Bugmann H, Bürgi M, et al. (2011) Weather and human impacts on forest fires: 100 years of fire history in two climatic regions of Switzerland. Forest Ecology and Management, 261, 2188–2199.
6. Silva FR, González-Cabán A (2010) 'SINAMI': a tool for the economic evaluation of forest fire management programs in Mediterranean ecosystems. International journal of wildland fire, 19 927–936.
7. Zamora R, Molina-Martínez JR, Herrera MA, Rodríguez y Silva F (2010) A model for wildfire prevention planning in game resources. Ecological Modelling 221, 19–26.
8. Preisler HK, Westerling AL, Gebert KM, Munoz-Arriola F, Holmes TP (2011) Spatially explicit forecasts of large wildland fire probability and suppression costs for California. International Journal of Wildland Fire 20, 508–517.
9. Alvarado E, Sandberg DV, Pickford SG (1998) Modeling Large Forest Fires as Extreme Events. Northwest Science 72, 66–75.
10. Bak P, Chen K, Tang C (1990) A forest-fire model and some thoughts on turbulence. Physics Letters A 147, 297–300.
11. Barford P, Bestavros A, Bradley A, Crovella M (1999) Changes in Web client access patterns: characteristics and caching implications. World Wide Web 2, 15–28.
12. Barabási AL (2005) The origin of bursts and heavy tails in human dynamics. Nature 435, 207–211.
13. Pinto CMA, Lopes AM, Tenreiro Machado JA (2012) A review of power laws in real life phenomena. Communications in Nonlinear Science and Numerical Simulations 17, 3558–3578.
14. Ricotta C, Avena G, Marchetti M (1999) The flaming sandpile: self-organized criticality and wildfires. Ecological Modelling 119, 73–77.
15. Reed W, McKelvey K (2002) Power-law behaviour and parametric models for the size-distribution of forest fires. Ecological Modelling 150, 239–254.
16. Prentice IC, Kelley DI, Foster PN, Friedlingstein P, Harrison SP, et al. (2011) Modeling fire and the terrestrial carbon balance. Global Biogeochemical Cycles. 25, GB3005
17. Fletcher IN, Aragão LE, Lima A, Shimabukuro Y, Friedlingstein P (2013) Fractal properties of forest fires in Amazonia as a basis for modelling pan-tropical burned area. Biogeosciences Discussions 10, 14141–14167.
18. Drossel B, Schwabl F (1992) Self-organized critical forest-fire model. Physical Review Letters 69, 1629–1632.
19. Malamud B, Morein G, Turcotte D (1998) Forest Fires: An Example of Self-Organized Critical Behavior. Science 281, 1840–1842.
20. Telesca L, Amatulli G, Lasaponara R, Lovallo M, Santulli A (2005) Time-scaling properties in forest-fire sequences observed in Gargano area (southern Italy). Ecological Modelling 185, 531–544.
21. Telesca L, Amatucci G, Lasaponara R, Lovallo M, Rodrigues MJ (2007) Space–time fractal properties of the forest-fire series in central Italy. Communications in Nonlinear Science and Numerical Simulation 12, 1326–1333.
22. Marques S, Borges JG, Garcia-Gonzalo J, Moreira F, Carreiras JMB, et al. (2011) Characterization of wildfires in Portugal. European Journal of Forest Research 130, 775–784.
23. Shannon CE (1948) A Mathematical Theory of Communication. Bell System Technical Journal 27, 379–423, 623–656.
24. Herzel H, Schmitt AO, Ebeling W (1994) Finite Sample effects in Sequence Analysis. Chaos, Solitons & Fractals 4, 97–113.
25. Mori T, Kudo K, Tamagawa Y, Nakamura R, Yamakawa O, et al. (1998) Edge of chaos in rule-changing cellular automata. Physica D 116, 275–282.
26. Matsuda H (2000) Physical nature of higher-order mutual information: Intrinsic correlations and frustration. Physical Review E 62, 3096–3102.
27. Posadas A, Hirata T, Vidal F, Correig A (2000) Spatio-temporal seismicity patterns using mutual information application to southern Iberian peninsula (Spain) earthquakes. Physics of the Earth and Planetary Interiors 122, 269–276.
28. Telesca L (2011) Tsallis-Based Nonextensive Analysis of the Southern California Seismicity. Entropy 13, 1267–1280.
29. Mohajeri N, Gudmundsson A (2012) Entropies and Scaling Exponents of Street and Fracture Networks. Entropy 14, 800–833.
30. Tenreiro Machado JA, Lopes AM (2013) Analysis and Visualization of Seismic Data using Mutual Information. Entropy 15, 3892–3909.
31. Kvalseth TO (1987) Entropy and correlation: Some comments. IEEE Transactions on Systems, Man and Cybernetics 17, 517–519.
32. Jain A, Dubes R (1988) 'Algorithms for Clustering Data.' (Prentice-Hall: Englewood Cliffs, NJ).
33. Johnson SC (1996) Hierarchical clustering schemes. Psychometrika 32, 241–54.
34. Shepard R (1962) The analysis of proximities: multidimensional scaling with an unknown distance function. Psychometrika 27, 219–246.
35. Kruskal J, Wish M (1978) 'Multidimensional Scaling.' (Sage Publications: Newbury Park, CA).
36. Cox T, Cox M (2001) 'Multidimensional scaling.' (Chapman & Hall/CRC).
37. Martinez W, Martinez A (2005) 'Exploratory Data Analysis with MATLAB.' (Chapman & Hall/CRC Press: UK).
38. Tzagarakis C, Jerde TA, Lewis SM, Ugurbil K, Georgopoulos AP (2009) Cerebral cortical mechanisms of copying geometrical shapes: a multidimensional scaling analysis of FMRI patterns of activation, Experimental Brain Research 194, 369–380.
39. Costa AM, Tenreiro Machado JA, Quelhas MD (2011) Histogram-based DNA analysis for the visualization of chromosome, genome and species information. Bioinformatics 27, 1207–1214.

5

Accounting for Biomass Carbon Stock Change Due to Wildfire in Temperate Forest Landscapes in Australia

Heather Keith[1]*, David B. Lindenmayer[1], Brendan G. Mackey[2], David Blair[1], Lauren Carter[1], Lachlan McBurney[1], Sachiko Okada[1], Tomoko Konishi-Nagano[3]

1 The Fenner School of Environment and Society, Australian National University, Building 48, Canberra, ACT, Australia, 2 Griffith Climate Change Response Program, Griffith University, Queensland, Australia, 3 Fujitsu Laboratories Ltd., Kawasaki, Japan

Abstract

Carbon stock change due to forest management and disturbance must be accounted for in UNFCCC national inventory reports and for signatories to the Kyoto Protocol. Impacts of disturbance on greenhouse gas (GHG) inventories are important for many countries with large forest estates prone to wildfires. Our objective was to measure changes in carbon stocks due to short-term combustion and to simulate longer-term carbon stock dynamics resulting from redistribution among biomass components following wildfire. We studied the impacts of a wildfire in 2009 that burnt temperate forest of tall, wet eucalypts in south-eastern Australia. Biomass combusted ranged from 40 to 58 tC ha^{-1}, which represented 6–7% and 9–14% in low- and high-severity fire, respectively, of the pre-fire total biomass carbon stock. Pre-fire total stock ranged from 400 to 1040 tC ha^{-1} depending on forest age and disturbance history. An estimated 3.9 TgC was emitted from the 2009 fire within the forest region, representing 8.5% of total biomass carbon stock across the landscape. Carbon losses from combustion were large over hours to days during the wildfire, but from an ecosystem dynamics perspective, the proportion of total carbon stock combusted was relatively small. Furthermore, more than half the stock losses from combustion were derived from biomass components with short lifetimes. Most biomass remained on-site, although redistributed from living to dead components. Decomposition of these components and new regeneration constituted the greatest changes in carbon stocks over ensuing decades. A critical issue for carbon accounting policy arises because the timeframes of ecological processes of carbon stock change are longer than the periods for reporting GHG inventories for national emissions reductions targets. Carbon accounts should be comprehensive of all stock changes, but reporting against targets should be based on human-induced changes in carbon stocks to incentivise mitigation activities.

Editor: Benjamin Poulter, Montana State University, United States of America

Funding: The funder provided support in the form of salaries for the authors [HK, SO], but did not have any additional role in the study design, data collection and analysis, decision to publish, or preparation of the manuscript. The specific roles of these authors are articulated in the 'author contributions' section. The author affiliated with the funding organization contributed to the conceptual framework of the study.

Competing Interests: The authors declare that one of the authors [TKN] is affiliated with one of the funding organizations of this research study, Fujitsu Laboratories Ltd. This affiliation does not alter the authors' adherence to PLOS ONE policies on sharing data and materials.

* Email: Heather.Keith@anu.edu.au

Introduction

Stabilising the carbon stock in the atmosphere to prevent dangerous anthropogenic interference with the climate system is the objective of the United Nations Framework Convention on Climate Change (UNFCCC) [1]. Forest ecosystems naturally exchange carbon dioxide (CO_2) with the atmosphere, but losses of carbon stocks from forests due to land management activities is a significant source of greenhouse gas (GHG) emissions [2]. Losses of carbon as emissions of CO_2 from forest ecosystems to the atmosphere occur due to biological decomposition processes, natural disturbance events like fire, as well as human activities. Removals of CO_2 from the atmosphere to forest ecosystem carbon stocks occur by the uptake of carbon in growing vegetation.

Under the UNFCCC and Kyoto Protocol, each country constructs GHG inventories and reports on their net annual emissions, that is, flows of CO_2 and other GHGs, due to specific activities and sectors including Land Use, Land Use Change and Forestry. Flow-based inventories in the land sector, however, obscure important differences between ecosystems types and between the impacts of human activities and natural disturbances. A more comprehensive, stock-based approach to accounting can address these problems and provide information that complements flow-based accounts (Ajani et al. 2013). The flow-based accounts are currently used for national reporting against emission reduction targets over short time periods, such as annual or 5 to 8 year commitment periods, as determined by political negotiations. However, this short time period exposes nations to high variability in emissions due to natural disturbance events, especially wildfires, which are outside human control, yet create large GHG fluctuations. The resulting variability in national net emissions due to natural disturbance can be much greater than the changes due to land use impacts, thus confounding attempts to provide incentives for reducing emissions in the land sector through improved forest management.

Losses of carbon due to wildfire are a significant component of global annual GHG balances and so it is important that they are included in national inventories. Seasonal losses of carbon from wildfires are detected as high concentrations of trace gases in the troposphere [3,4]. Estimates of gross losses of carbon from forest fires globally range from 2.0–2.5 Pg C yr^{-1} [5,6] to 3.8–4.3 Pg C yr^{-1} [7] as annual averages. As a comparison of the magnitude of this flux, the global net losses are 1.1 PgC yr^{-1} from land-use change and 7.9 PgC yr^{-1} from fossil fuel emissions [8]. For nations with large areas of forest that are subject to periodic wildfires, such as Australia, Canada, USA, Russia and South Africa [9,10,11], the treatment of wildfires in carbon accounting for forest management is an important consideration. The rules and methodologies are complex, often controversial, and raise many issues in international accounting systems.

Carbon stock changes due to wildfires occur at different temporal scales that should be included in comprehensive stock-based accounts. Large losses occur over periods of hours to days or weeks due to combustion and are highly variable each year due to occurrence of fire. Uptake through forest regrowth and losses through decomposition of dead biomass occur over decades to centuries. These changes in carbon stocks are balanced in comprehensive accounts, and in the accumulated concentration of CO_2 in the atmosphere, when assessed over a sufficiently long time period, assuming the fire regime is stationary. The following information is required to enable inclusion of wildfire disturbance in comprehensive stock-based national carbon accounts: carbon stock loss due to combustion, redistribution of carbon stocks between living and dead biomass components; subsequent rates of decomposition; and carbon uptake by regenerating vegetation. These estimates of carbon stock change need to be calculated on a landscape-wide basis and integrated over appropriate time periods [12] with data from a range of forest types. However, methodologies for measuring carbon stock change due to wildfire are problematic for a number of reasons, including: (i) wildfires typically extend over large areas, within which fire severity varies spatially; (ii) initial carbon stocks vary across the landscape in response to differences in forest type, age, disturbance history and environmental factors; (iii) biomass components are combusted with differing efficiencies; (iv) fire occurrence is stochastic [13,14]; and (v) the dynamics of carbon stocks occur over long time periods in response to wildfire.

Of these methodological issues, the parameter with the greatest uncertainty is the amount of carbon combusted, calculated as either an amount or a proportion of the fuel load or total stock, and then upscaled from individual components to sites and landscapes [5,15]. Combustion efficiency varies due to characteristics of individual fuel materials, such as logs of varying size, wood density, state of decay and moisture content [16], and characteristics of the fire, such as intensity and residence time, which determine the oxidation conditions of flaming or smouldering combustion [7,17,18]. At the landscape scale, the proportion combusted depends on the areas burnt within the fire boundary, including small scale heterogeneity in fire occurrence and severity. The high degree of uncertainty created by the limited existing data is demonstrated when aggregated spatially in regional and global inventories. For example, estimates of carbon stock losses from wildfires in Australia derived from global analyses differ by up to six-fold [6,19,20,21,22].

The objectives of this study were to investigate the magnitude and timeframe of carbon stock changes due to wildfire and to evaluate the results in terms of implications for national carbon accounts.

(i) We assessed the magnitude of carbon stock losses due to combustion in a wildfire in terms of the amount and proportion of the total forest ecosystem carbon stock. We then considered the significance of these carbon losses compared with potential impacts from human activities.

(ii) We considered the impact of fire on the ecosystem carbon stock in terms of the timeframe of stock changes. The mean residence times of carbon stocks change in response to redistribution of carbon among biomass components post-fire and their subsequent relative rates of losses and gains. We investigated the dynamics of carbon stock change within the current disturbance regime and not predicted future regimes or climate.

(iii) We evaluated our results about the magnitude and timeframe of carbon stock changes in terms of their significance for the rules and methodologies governing national carbon accounting for forest management. Now is an important time to contribute quantitative information to the development of accounting guidelines for post-2020 agreements which include forest management and disturbance events under the Durban Platform for Enhanced Action [23].

To address these issues about accounting for carbon stock changes due to wildfire, we examined changes in carbon stocks in a temperate eucalypt forest in south-east Australia that was subjected to a wildfire in 2009. We selected a temperate forest because they contain high carbon stocks and contribute 34% to the global forest carbon sink; a proportion that has been increasing over the last two decades [24]. Changes in carbon stocks due to combustion during the fire were assessed empirically and we also investigated the long-term dynamics of biomass carbon pools post-fire using a simulation model, both at the site level and up-scaled across the landscape.

Methods

Description of study region

Access to field sites was granted by Parks Victoria and the Department of Environment and Primary Industries Victoria. Our study region (2,326 km^2) was the tall, wet sclerophyll montane ash forests (*Eucalyptus regnans, E. delegatensis* and *E. nitens)* in the Central Highlands of Victoria (Figure 1). These temperate forests are among the most carbon-dense in the world and protection of these carbon stocks offers the potential for mitigation [25]. These forests are subject to a disturbance regime of infrequent, high intensity wildfires [26,27].

Wildfire disturbance regime

The known historical fire frequency has been one in the 1700s, three in the 1800s, nine in the 1900s, and three in the 2000s [27,28], and their spatial extent was mapped across the study region (Figure 1). Occurrence of these fires does not necessarily mean that each conflagration occurred at the same locations or at the same intensity. The mean interval for fires that kill ash trees at a site has been estimated to be 75 to 150 years [29,30]. However, an estimate that approximately half the trees survive within the boundary of a wildfire [30] means that only every second fire in any one location is of sufficient intensity to kill ash trees. Hence, the mean interval of all fires is estimated to be 37 to 75 years. A maximum fire interval of 350–400 years was estimated based on maximum age of the montane eucalypt species [31].

Unlike most eucalypt species, *E. regnans* and *E. delegatensis* are obligate seeders and usually killed if their canopies are predominantly scorched by fire [32,33,34]. However, wildfires are highly

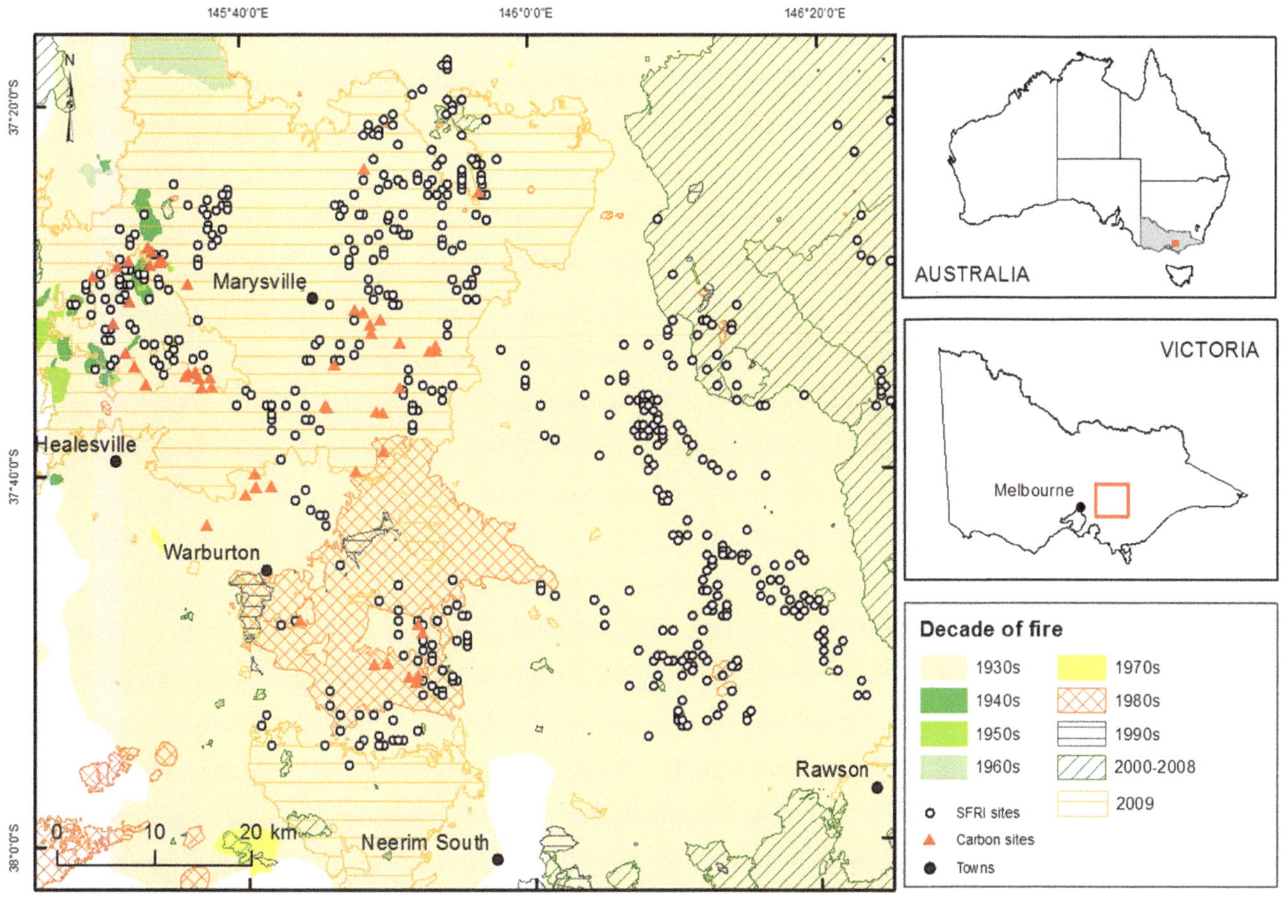

Figure 1. Location of the study region in the Central Highlands of Victoria, Australia. Location of field sites is marked, including long-term ecological monitoring sites (n = 54) where carbon stocks in biomass components were measured, and inventory sites (n = 876) where carbon stocks were estimated and used for upscaling. The spatial extent of wildfires shown represents the outer boundary of the burnt area.

variable in extent and severity, which is reflected in current stand age structures of a mosaic of even-aged and multi-aged forests [26,35,36]. Even-aged stands result from widespread mortality and regeneration following high-severity wildfire. Multi-aged stands result from partial tree death and subsequent regeneration alongside surviving trees [29,37,38]. Hence, the variability of historical fires is important for determining current stand structures. For example, our entire study area is within the boundary of the 1939 fire. However, evidence that individual trees and forest stands survived is seen by the current age structure of patches of old-growth forest where many living trees have fire scars from the 1939 fire event [39].

A wildfire started on 7th February 2009 that burnt over 450,000 ha in Victoria under extreme hot, dry weather conditions (46.4 °C in the nearby city of Melbourne), with winds up to 100 km hr^{-1} [40,41]. Within our study region of montane ash forest, 485 km^2 burnt at low-severity and 283 km^2 burnt at high-severity (Figure 2). Fire severity was mapped by the Victorian Department of Environment and Primary Industries using air photo interpretation and categorised by degree of crown scorch and subsequent tree survival or mortality.

Field sites

We selected 54 sites within the montane ash forest region of the Central Highlands of Victoria (Figure 1) from an existing network of 170 long-term ecological monitoring sites [27] to provide sites with a range of conditions of forest age and severity of the 2009 wildfire. The site categories of forest age were 1983 and 1939 regrowth after wildfire followed by salvage logging or clearcut logging, and old growth. Salvage logging was common practice after wildfire to remove the timber resource [42,43,44,45]. Selection of these regrowth sites provided reasonably even-aged stands of similar structure, although some residual older trees occur [46]. Old growth forest consisted of dominant trees greater than 120 years old but mostly 250 years old that had regenerated following a fire in approximately 1750 [39]. The site categories of severity of the 2009 wildfire were unburnt, low-severity and high-severity burns. Our distinction between low- and high-severity fire was based on site observations of full scorch of the canopy and mortality of *E. regnans and E. delegatensis* trees. The three forest age and three fire severity categories produced a matrix of nine categories, each with six replicate sites.

While the 54 sites covered a range of forest stand structures and disturbance histories, the assumption that the sites had comparable pre-fire biomass carbon stocks, disturbance history, and local biophysical conditions did not necessarily hold. Their carbon stocks likely varied because of the natural heterogeneity of the landscape and the fact that fires are less likely to burn in wet, highly productive mature and old growth forest areas [47,48,49]. Furthermore, the monitoring protocols used prior to the 2009 fire at the sites were aimed at biodiversity assessments, and did not include all the measurements required to quantify ecosystem

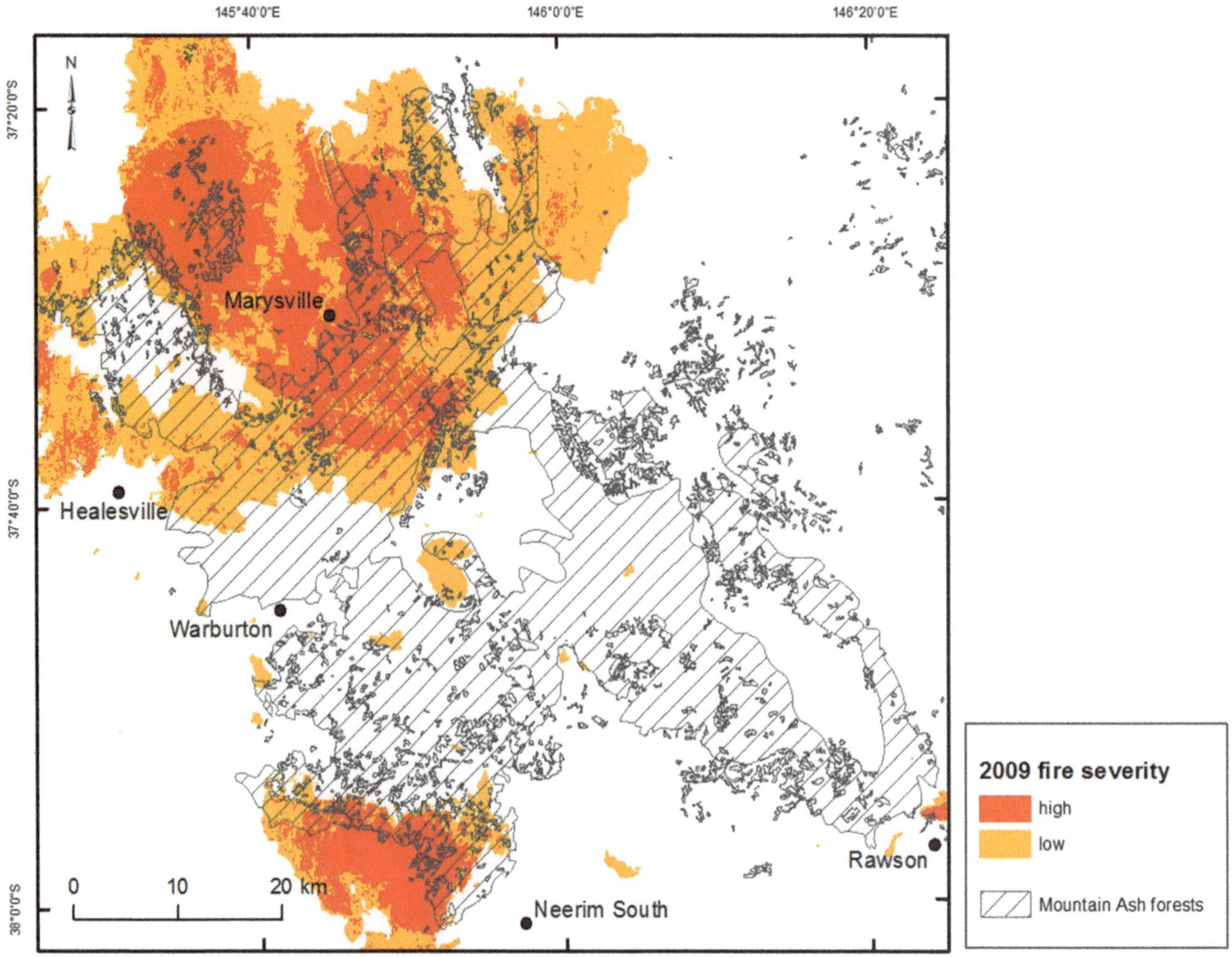

Figure 2. Spatial extent of the 2009 wildfire within the study region. The fire was categorised as low-severity where most trees survived, and high-severity where most trees were killed.

carbon stocks. Therefore, we undertook additional field survey to ensure there was a consistent set of measurements across sites.

Measurement of biomass carbon stocks and stock loss due to fire

We measured carbon stocks in all aboveground biomass components in the forest (living vegetation, dead standing trees, coarse woody debris (CWD or dead and downed woody debris), and litter) at the selected 54 sites in 2010. We defined biomass as all intact organic components of the ecosystem, both living and dead, above- and below-ground, but excluding soil organic matter. Biomass components were classified for ease of measurement and to track movement of carbon between ecosystem components. Classification was based on the status of living or dead, size and vertical location. We used different sampling strategies for each biomass component within 1 ha sites to maximise accuracy and efficiency of measurements of different sizes, densities and distributions of components (Table 1 and Figure 3).

All woody stems, both living and dead, greater than 2 m height and less than 100 cm diameter (diameter at breast height, DBH) were assessed in height and diameter categories for each species within three 10 m×10 m plots (0.03 ha). Trees greater than 100 cm DBH were measured in two perpendicular and intersecting 100 m transects by 30 m width (0.51 ha). In large trees that form buttresses with fluted stems [50], there is a cross-sectional area deficit that was accounted for by converting to a 'functional' diameter [51]. The allometric equation for *E. regnans* derived by Sillett et al. [51] was used to calculate stem and branch volume, and multiplied by wood density and carbon concentration to derive aboveground biomass carbon content. Average stem wood density was 0.520 g cm^{-3} [52,53,54,55] and branch wood density was 0.677 g cm^{-3} [51]. A carbon concentration of 0.5 gC g^{-1} was used for all biomass components [56,57]. Internal decay or hollows in stems were accounted for in the calculation of biomass derived from stem volume [58,59]. Equations to predict occurrence and volume of decay related to tree size were derived from a subset of inventory data (734 trees) [60] where dimensions of defect in sawlogs had been measured. Allometric equations for other vegetation components included *Acacia* spp as mid-storey trees [61], a general rainforest equation for other mid- and understorey species [62], and treeferns were calculated as a cylindrical volume and measured wood density. Root mass in *E. regnans* forests has not been measured, hence an average root: shoot ratio for eucalypt forests of 0.25 [63] was used to convert aboveground biomass to total tree biomass for living and dead trees.

Table 1. Methods for measuring carbon stocks in biomass components within 1 ha sites post-fire at burnt and unburnt sites.

Biomass component	Sampling strategy	Measurement
Trees (living and dead) eucalypt and rainforest species		
<100 cm diameter	Three 10 m×10 m plots (0.03 ha)	Estimated height and diameter in size categories
>100 cm diameter	Two perpendicular intersecting 100 m×30 m transects (0.51 ha)	Measured DBH and height.
Mid- and Under-storey (living and dead) non-eucalypt species		
<20 m height	Three 10 m×10 m plots (0.03 ha)	Estimated height and diameter in size categories
Coarse woody debris		
<60 cm diameter	Line intersect method along 6×10 m transects	Log diameter, decay class, hollows, charcoal and bulk density [64,65]
>60 cm diameter	2×100 m transects	
Litter layer		
<2.5 cm diameter	15 points along transects per site. 30 quadrats randomly located at sites to cover range in litter depths	Measured litter depth.
		Measured litter dry mass.

Coarse woody debris consisted of all woody material ≥25 mm diameter on the ground, and was measured using the line intersect method [64,65]. Logs less than 60 cm diameter were measured along 6×10 m transects within the site and logs greater than 60 cm diameter were measured along 2×100 m perpendicular transects. Each piece of CWD was assessed for hollows and degree of decay in three categories. Bulk density was measured for each category of decay to determine biomass. The litter layer consisted of organic material less than 25 mm diameter including leaves, twigs, insect detritus, animal scats, and comminuted material that was recognisable as organic material. Litter depth was measured at 15 points per site on a compressed litter pack to standardise the quantity of loose material. Depth was converted to biomass using a relationship derived from 30 quadrats sampled for dry weight of litter. Our results for carbon stocks refer to the total biomass of living and dead, above- and below-ground components.

We estimated biomass carbon stock loss by combustion in the 2009 wildfire from the following lines of evidence: (i) field observations and photographs taken immediately after the fire (within 2 months) [66], (ii) measurements of some components of the forest that were measured pre-fire compared with post-fire, and (iii) measurements of biomass components post-fire taken at burnt and unburnt sites. Biomass components combusted in the fire were measured specifically; including, hollow trees, decorticating and rough bark, canopy, shrub biomass, CWD and litter (Table 2 and Figure 4).

Hollow trees, both living and dead, are highly susceptible to combustion because fire is funnelled through the pipe [67]. Data for individual tree heights and diameters from 2005 and 2011 [67] were compared to assess the change pre- and post-2009 fire. The difference in carbon stock between years represented a combination of natural mortality, decay and collapse of trees plus combustion at the burnt sites. Carbon stock loss was calculated as a proportion of the initial stock and the difference between burnt and unburnt used to distinguish the effect of natural decay and combustion.

Biomass of decorticating or ribbon bark hanging from the upper stem and branches was related to tree DBH by a non-destructive sampling method using estimated standard units [68]. All decorticating bark was combusted in the fire. Biomass of rough or fibrous bark on the lower stem was estimated from measurements of bark thickness and bulk density related to tree DBH. Combustion of rough bark was determined from the difference between burnt and unburnt sites. Both decorticating and rough bark mass were calculated for all the inventory trees at the 54 sites.

Canopy leaf biomass was estimated to be approximately 7 tC ha^{-1}, derived from annual litterfall of 4.7 tC ha^{-1} yr^{-1} in ash forests [69,70] and longevity in the canopy of about 18 months [50,71]. Twig (diameter <4 mm) biomass is a similar amount to the leaves [72]. The proportion of the canopy combusted was estimated from photographs after the fire of the leaves remaining in the canopy and the scorched leaves that had fallen to the ground.

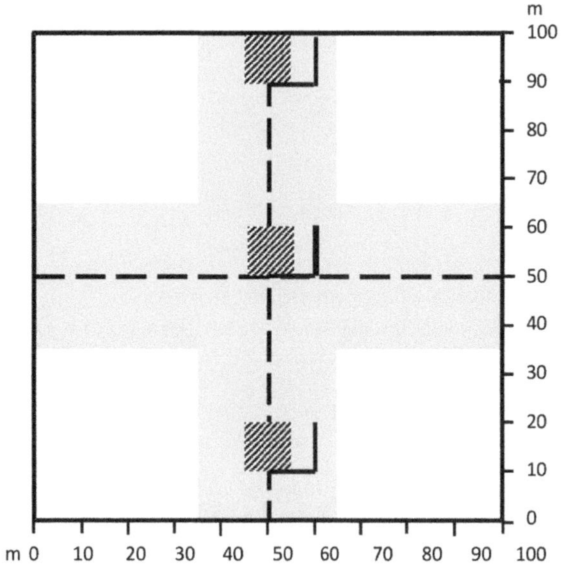

Figure 3. Design of 1 ha experimental sites. Sites of 100 m×100 m included central and perpendicular transects (dashed lines), 10 m×10 m plots (hatched), 10 m transects (black lines), and area sampled for large trees 2×30 m×100 m = 0.51 ha) (grey shaded area). Different biomass components were sampled in different parts of the site.

Table 2. Methods for measuring changes in carbon stocks of biomass components after wildfire, within 1 ha sites.

Biomass component	Sampling time	Sampling strategy	Measurement	Proportion of biomass combusted	
				Low-severity fire	High-severity fire
Hollow trees	Measured in 2005 and 2011	All trees in 1 ha	Measured DBH and height.	0.12 to 0.29	0.16 to 0.56
Decorticating bark	Post-fire, burnt and unburnt sites	Representative size range of trees.	Measured relationship between tree DBH and bark mass.	1.0	1.0
Rough bark	Post-fire, burnt and unburnt sites	Representative size range of trees.	Measured relationship between tree DBH and bark thickness plus bulk density.	0.3 (for trees >200 cm)	0.6 (for trees >200 cm)
Canopy leaves	Post-fire, burnt and unburnt sites	Observations and photos	Estimated biomass from litterfall and leaf longevity	0.5	1.0 (mostly)
Canopy twigs	Post-fire, burnt and unburnt sites	Observations and photos	Relationship between leaf and twig mass.	0.5	1.0 (mostly)
Shrubs	Post-fire, burnt sites	Representative range of shrub sizes and species.	Minimum size of tips remaining on stems after fire. Calculated biomass in each stem diameter class in plots.	0.25 to 0.5 (from branch diameter <5 cm)	0.3 to 0.75 (from branch diameter <5 cm)
Coarse woody debris	Post-fire burnt sites	Line intersect transects	Proportions of logs charred. Lines of charcoal remaining post-fire.	Highly variable dependent on log size	Highly variable dependent on log size
Litter	Photos soon after the fire	Average from all low- and high-severity burnt sites.	Proportion of litter layer combusted.	1.0	1.0

Combustion of shrub biomass in woody stems, branches and twigs was assessed using measurements of: (i) minimum sizes of tips remaining on stems after the fire, (ii) the proportion of biomass lost from a shrub when different tip sizes were removed, and (iii) the amount of biomass in each stem diameter size class of shrubs from the inventory data. A range of species and sizes of understorey shrubs was sampled.

Combustion of CWD was determined on each piece along the measurement transects by the proportion of the piece charred, or by the number of lines of charcoal remaining on the ground. The initial carbon stock of the logs that were completely converted to charcoal was assumed to be represented by the mean size of logs at the site. CWD was too variable between burnt and unburnt sites within stand age categories to allow a comparison to be indicative of the amount combusted as the mean of a category. The proportion of the litter layer combusted was estimated from photographs of quadrats taken soon after the fire in low- and high-severity burnt areas and compared with litter biomass in unburnt sites.

Spatial estimation of carbon stocks

Spatial up-scaling of the site data to the landscape level followed the statistical modelling approach of Keith et al. [73]. Calibration data for the model used the pre-2009 carbon stocks from the 54 sites described above, as well as an additional 876 sites with inventory data [60]. The statistical model regressed site carbon stocks against the following explanatory variables at a 250 m grid resolution: climate variables of precipitation, water availability index, temperature (minimum, mean, maximum) and radiation; topographic variables of topographic position, elevation, aspect and slope; substrate variables of soil parent material, lithology and soil organic carbon content; gross primary productivity (GPP, derived from remote sensing); forest type; forest management area; and disturbance history. Methods used to derive spatial estimates of the environmental and GPP variables are documented in Mackey et al. and Berry et al. [26,74,75]. Disturbance history included wildfires and logging under different silvicultural systems, both of which have complex effects on forest age structure. The statistical regression model was used to generate spatial estimates of carbon stocks across the study region.

Impacts of the wildfire on ecosystem carbon stocks were quantified by determining first, the loss in carbon stock of each biomass component combusted, and second, the redistribution of carbon among biomass components within the ecosystem. Spatial estimation of the carbon stock post-2009 fire was based on measured reductions in the stock according to the fire severity category in the burnt areas and forest age.

Modelling change in carbon stocks

Changes in carbon stocks were simulated for each forest age category after high-severity fire at the average historical return time of 112 years. Initial conditions for the model used site data of the current carbon stock in each age category and the proportion of the stock combusted. After the fire, carbon stocks were redistributed among living and dead components according to the proportions measured in each age category. For the second fire cycle, the initial condition of total biomass at the time of the fire was the stock simulated for year 112, and the proportions of the stock combusted and redistributed were taken as the average of the site data from the 1939 regrowth and old growth forests. Changes in carbon stocks over time were simulated using functions to describe the processes of regeneration, mortality, collapse of dead trees, and decomposition of CWD derived from existing data and observations for montane ash forest (Table 3).

The rate of regeneration was based on carbon stocks at sites with different times since stand-replacing disturbance, derived from the current study and inventory data [60]. Mortality due to self-thinning of the regeneration was estimated from changes in stem density with age, based on data from the literature [76,77]. Mortality of living trees that survived the fire was based on measured rates of mortality over 28 years [67] and the average longevity of *E. regnans* of at least 250 years [78], but up to 500 years [79]. The rate of collapse of dead standing trees was described by a logistic function, with a slow initial rate that

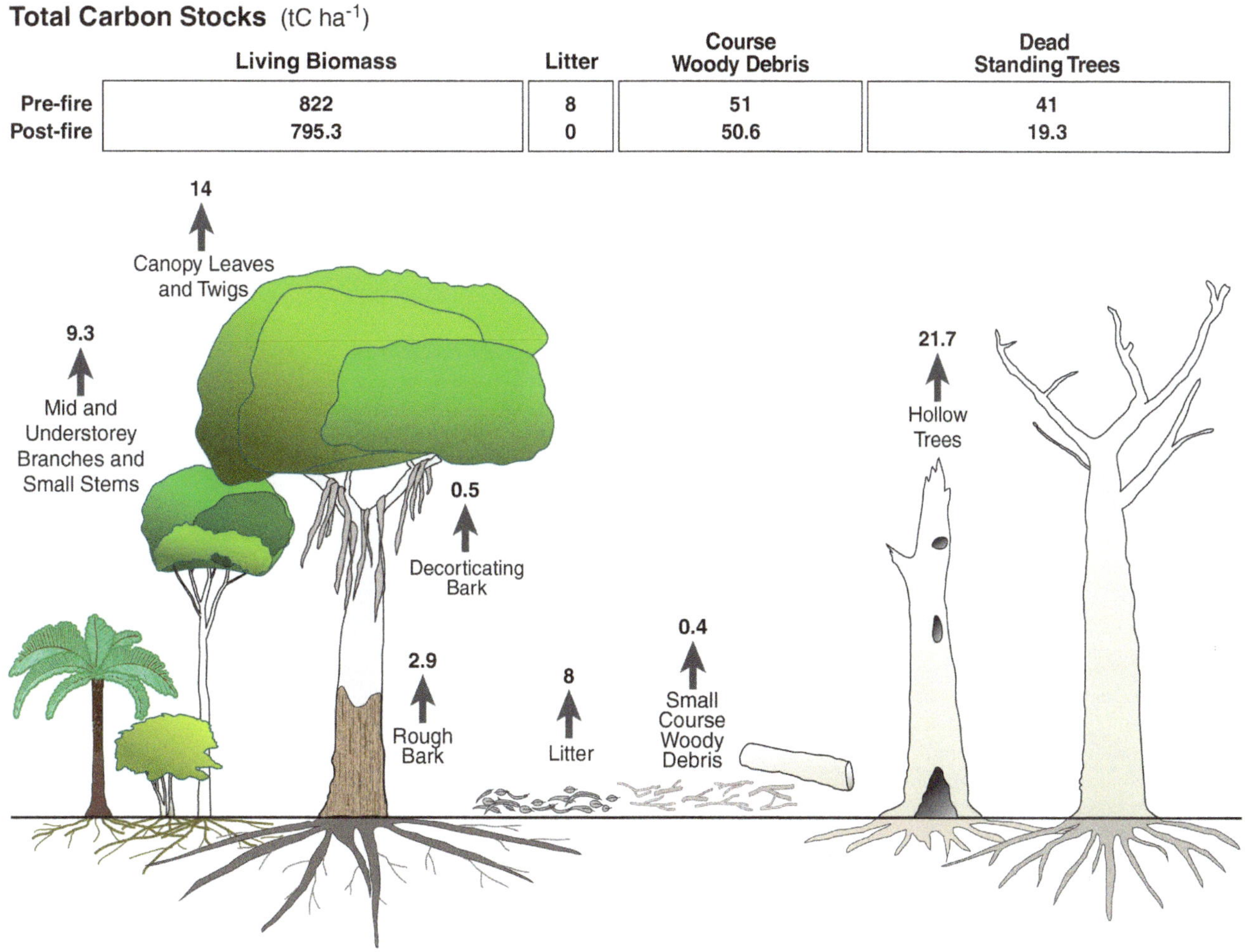

Figure 4. Schematic of carbon stocks in the forest ecosystem and stock changes resulting from fire.

increased to a maximum and then a slow final rate when few trees remain. This pattern was based on observations and evidence that dead trees remain standing for approximately 10 to 75 years [66,80]. The biomass in dead trees was then transferred to the CWD pool. The rate of decomposition of CWD and dead root biomass used in the model was 0.07 yr^{-1}, which had been derived from empirical data for *E. regnans* [55]. This rate was intermediate among published rate constants [80,81,82], and produced amounts of CWD that were consistent with observed amounts in montane ash forest.

Results

Current carbon stock

Biomass density varied substantially among sites within the montane ash forest due to the disturbance history and subsequent age distribution of the forest. Biomass density (mean ±standard error, n = 6 sites) in unburnt stands was 405±33 tC ha^{-1} in 1983 regrowth, 603±74 tC ha^{-1} in 1939 regrowth, and 1039±44 tC ha^{-1} in old growth sites. The structure of these forest stands is illustrated in terms of the distribution of their carbon stock by tree sizes (Figure 5). These tree sizes demonstrate that the regrowth stands were not necessarily even-aged and included some larger residual trees which contained high, but variable, levels of biomass.

CWD biomass was highly variable among sites in the age/fire categories, ranging from 15 to 186 tC ha^{-1}, with a mean of 67 tC ha^{-1}. There was no significant difference in CWD biomass between forest age categories, reflecting the varied disturbance history of these forests where CWD remained after logging and wildfire, and its biomass was not related to the age of the stands. Litter biomass in the unburnt sites ranged from 6 tC ha^{-1} in the 1983 regrowth to 9 tC ha^{-1} in mature and old growth stands. This litter biomass corresponded to average site depth of litter when compressed of 2 cm and 3 cm, respectively, with the majority of measured points less than 5 cm and only a few points with 5–10 cm depth.

Biomass components combusted

In areas burnt by low-severity fire, most of the over-storey trees survived as well as variable proportions of mid- and under-storey vegetation with green, scorched or combusted leaves. In areas burnt by high-severity fire, most of the leaves, twigs and decorticating bark in all canopy strata were combusted, although scorched leaves remained in patches. Some montane ash forests included a few *E. nitens* trees that regenerated by epicormic growth. Woody stems of most shrubs and mid-storey trees remained, but some were entirely combusted. The average branch size combusted from shrubs was 4 mm (maximum 19 mm) in low-severity fire, and 8 mm (maximum 42 mm) in high-severity fire.

Table 3. Input data and equations to predict biomass carbon stock (B in tC ha^{-1}) as functions of time (t) since disturbance in the model of change in carbon stocks after a high-severity wildfire in old growth montane ash forest.

Process	Function	Data	Source
Regeneration (=gain in carbon)	$B_{regen}(t) = 1200 \times (1-\exp(-0.0045 \times t))^{0.7}$	Inventory data from 99 sites	46,60
	$Log_{10}(SD_t) = 1.28+3.16 \times 0.913^t + 1.9 \times 0.99^t$ $Log_{10}(SD_t) = 11.61-1.624 \log_{10}(t)$ (average from the two functions used)	Double exponential function fitted to chronosequence site data for *E.regnans.* Function derived from site data	76,77
Mortality of regeneration (=input to dead trees)	$B_{dead}(t) = [(SD(t-1) - SD(t))/SD(t)] \times B_{regen}(t)$	Empirically derived from stem density changes	Current
Branch fall (=input to CWD)	$B_{branch}(t) = B_{regen}(t) \times 0.005$	Rate constant derived empirically to produce CWD biomass in the range observed (3–255 tC ha^{-1}).	96
Mortality of surviving trees (=input to dead trees)	$B_{dead}(t) = 135 \exp(-0.015 \times t)$	Rate of mortality estimated from site data.	67
Collapse of dead trees (=input to CWD)	$B_{CWD\_in}(t) = 687/(1+\exp(0.1 \times t - 5)$	Logistic function derived to fit observations that dead trees remain standing for 10 to 75 years.	66
Decomposition of CWD (=loss of carbon)	$B_{CWD\_loss}(t) = 124 \exp(-0.07 \times t)$	Rate constant. Modelled range in CWD consistent with site data (mean of 51 tC ha^{-1} and max. of 255 tC ha^{-1})	55,96 Current

The carbon stock losses from each biomass component and the proportion of the total biomass combusted differed with age class of the forest (Table 4). The distribution of readily combusted components within the ecosystem influenced the proportion of the stock emitted during the fire. Much of the observed change due to fire was combustion of shrubs in the understorey. However, this represented a small proportion of the total biomass in the ecosystem. Combustion of 7.8–16.2 tC ha^{-1} in the small branches on shrubs of all stem diameter categories in the understorey (Table 4) represented 1.6–4% of the total living biomass (above- and below-ground) in low-severity fire, and 1.9–6.7% in high-severity fire (range from young regrowth to old growth stands).

Some CWD was combusted during the fire, but both the absolute amount and proportion of initial biomass were highly variable spatially. Higher amounts of CWD were combusted in the 1983 and 1939 regrowth stands (range 1.37–2.41 tC ha^{-1}) than in the old growth stands (0.16–0.43 tC ha^{-1}) (Table 4). Large amounts of CWD occur in these regrowth stands due to self-thinning and residual material from previous logging and fire events. The high rates of combustion in regrowth stands occur

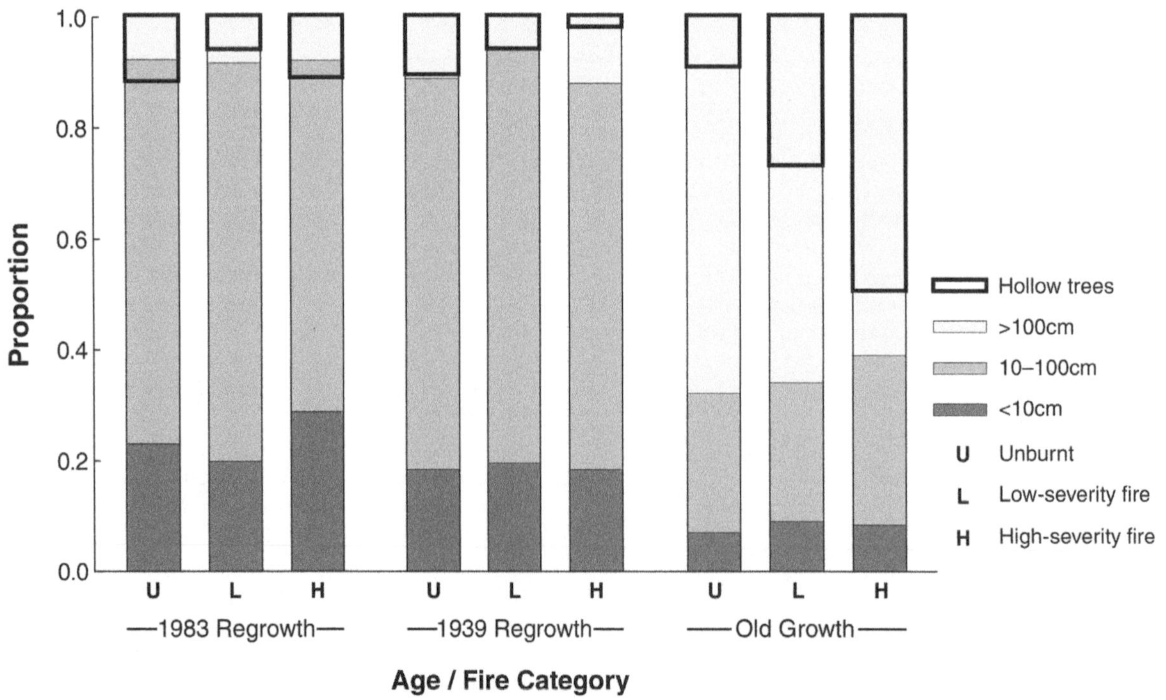

Figure 5. Distribution of carbon stocks in trees according to their size. Carbon stock is shown as the proportion in stem size classes of DHB and trees with hollows, averaged for the sites in each age/fire category (n = 6 sites per category).

Table 4. Summary of biomass components combusted by low- and high-severity fires.

	Mass of carbon[1] combusted (tC ha^{-1})					
	1983 regrowth		**1939 regrowth**		**Old growth**	
Low-severity fire	**mean**	**SE**	**mean**	**SE**	**mean**	**SE**
Litter	6	1.0	9	1.0	9	1.0
Canopy leaves	3.5		3.5		3.5	
Small stems	1.5	0.6	0.6	0.4	0.2	0.2
Small branches	11.9	1.2	11.8	2.7	7.6	3.4
Decorticating bark	0.3	0.02	0.3	0.06	0.5	0.07
Rough bark	0.6	0.2	0.6	0.1	1.3	0.3
CWD	1.4	0.6	1.8	0.7	0.2	0.2
Dead standing trees	3.8	0.9	9	1.7	17.6	5.2
Total biomass[2] combusted[2]	29	4.5	36.6	6.7	39.9	10.4
Total carbon stock[3]	433	35	475	91	564	88
Proportion combusted[4]	0.063		0.072		0.066	
High-severity fire						
Litter	6	1.0	9	1.0	9	1.0
Canopy leaves	7		7		7	
Canopy twigs	7		7		7	
Small stems	2.2	0.5	0.8	0.3	0.3	0.2
Small branches	14	3.0	15	4.0	9	2.0
Decorticating bark	0.3	0.05	0.4	0.1	0.5	0.1
Rough bark	0.8	0.1	0.7	0.2	2.9	0.4
CWD	2.4	1.7	1.7	0.5	0.4	0.3
Dead standing trees	7	1.9	14.8	1.2	21.7	6.2
Total biomass[2] combusted[2]	46.7	8.3	56.4	7.9	57.8	10.2
Total carbon stock[3]	28	46	459	102	589	63
Proportion combusted[4]	0.140		0.109		0.089	

[1]Mass of carbon (mean ±standard error, n = 6 sites).
[2]Total biomass combusted is the sum of all components combusted.
[3]Total carbon stock (above- and below-ground) pre-fire is current carbon stock before the fire.
[4]Proportion combusted is the total biomass combusted divided by the pre-fire carbon stock.

because large proportions of the CWD are small logs and drier logs (decay class 1), compared with the old growth stands. Hollow or damaged trees, either living or dead, were a major source of carbon stock loss (3.8–21.7 tC ha^{-1}) (Table 4). Where these trees had hollows in the centre of the stem, flames funnel up like a chimney and ignite exposed dry wood. In regrowth stands, hollow trees comprised 2–12% of the total tree biomass, and contained a carbon stock ranging from 7–50 tC ha^{-1}. In old-growth stands, trees with hollows generally consisted of the older cohort of trees but also included damaged younger trees. They represented a highly variable proportion of the total biomass, from 9–49%, and contained a carbon stock of 45–241 tC ha^{-1}.

The average carbon stock loss from biomass components was 40 tC ha^{-1} and 58 tC ha^{-1} in low- and high-severity fires, respectively, and this represented an estimated maximum amount (Table 4). As a proportion of the total biomass carbon stock (above- and below-ground), this loss represented 6–7% in low-severity fires and 9–14% in high-severity fires.

Distribution of carbon among components due to fire

The largest impact of fire on the ecosystem carbon stock was a shift in distribution among biomass components. The proportion of total biomass that was in living vegetation in unburnt sites was 72%, 71% and 89%, respectively in the 1983, 1939 regrowth and old growth sites. The shift from living to dead biomass was 3%, 6% and 19% due to low-severity fire, and 85%, 97% and 84% due to high-severity fire, respectively in the age categories (Figure 6). CWD increased by only a small proportion in the burnt sites; some logs were combusted and a similar amount was transferred into the CWD component.

Carbon stock dynamics

Combustion of biomass during the fire resulted in an initial loss of some carbon, but most of the biomass remained at the site. Our simulation model predicted that carbon stocks would increase for several decades after the fire due to the combination of regeneration of living biomass and persistence of dead biomass components (Figure 7). Mortality of trees during the fire produced large amounts of dead biomass, but the trees only collapsed slowly and then contributed to the CWD pool, from where carbon stock loss by decomposition occurred over many decades. The lowest carbon stock in an old growth forest occurred nearly a century after a single fire when most of the dead biomass produced by the fire had decomposed. The maximum fluctuation in carbon stock

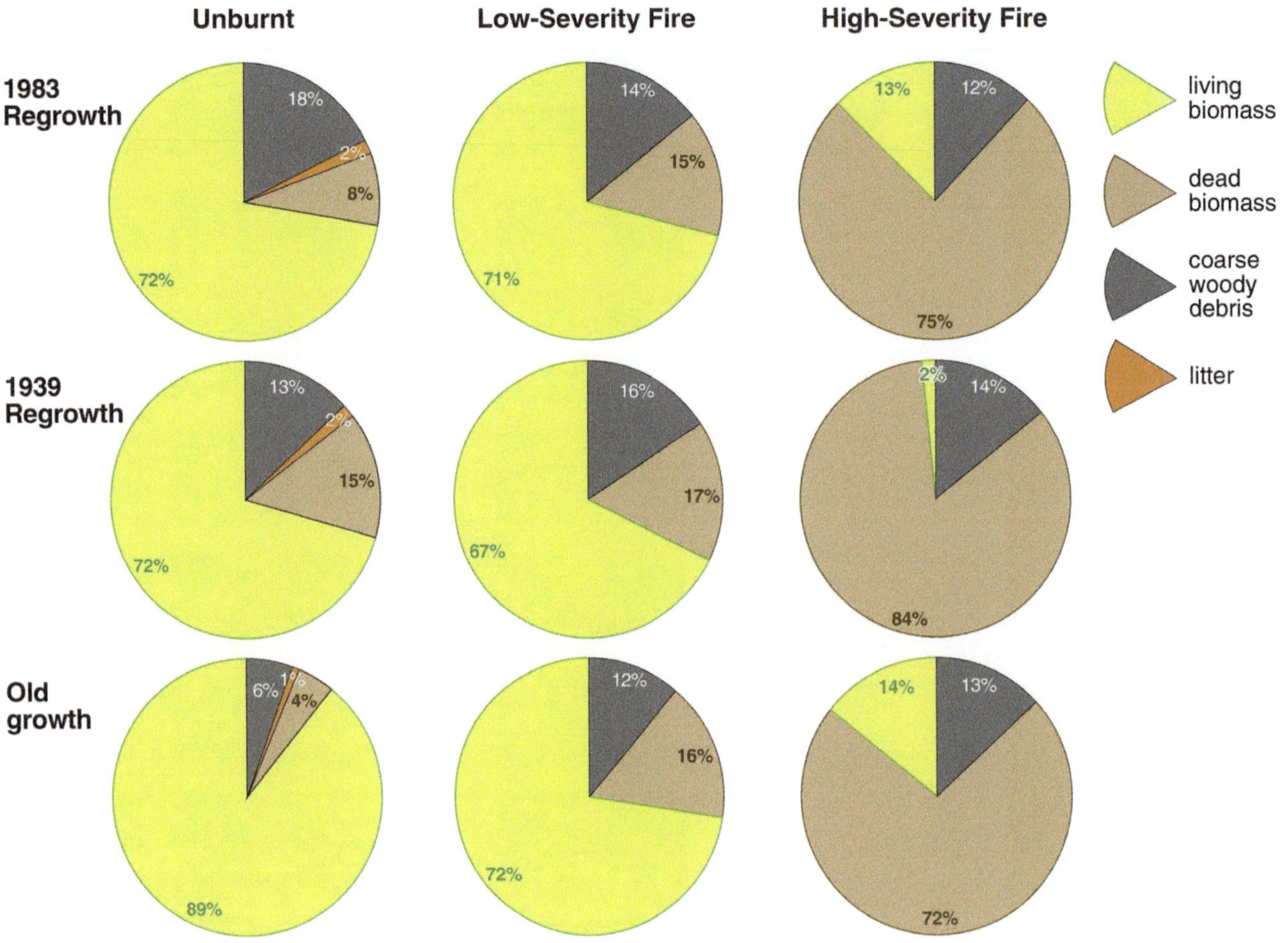

Figure 6. Distribution of carbon stocks among components in relation to stand age and fire severity. Distribution is shown as the proportion of carbon in each biomass component, including living vegetation, dead standing trees, litter and coarse woody debris.

over time as a result of fire was estimated to be 20% from the scenario of a single high-severity fire in old growth forest where there was no trend in total biomass due to changes in forest age (Figure 7a).

The carbon stock dynamics in the forests of initially different ages reflect the differences in the total biomass stock, its distribution among living and dead trees and CWD, the proportion combusted, and the proportions remaining in each of the components after the fire. The old growth forest had the largest increase in total biomass with the combined regeneration plus remaining dead trees, but subsequently the largest decline in total biomass as the dead trees collapsed and decomposed (Figure 7b). In the regrowth forests from 1939 and 1983, there was less dead biomass after the fire and the regeneration resulted in increasing total biomass carbon stock over most of the century following the fire (Figure 7c and d).

Under a regime of repeated fires on a 112-year return time, trends occurred in the total biomass carbon stock because forest age changed from that of the initial forest. The old growth forest decreased in carbon stock because of the more frequent disturbance, whereas the 1983 and 1939 regrowth forests increased in carbon stock because they remained undisturbed for a longer period than at their initial age. These scenarios of carbon stock dynamics represent an average forest stand of a given initial age subject to the average wildfire regime. A range of forest age classes occur across the landscape, including forests much older than 112 years. Forests that remain undisturbed for several hundred years have the capacity to continue accumulating carbon stocks.

Spatial estimation of carbon stocks post-2009 wildfire

The carbon stock loss post-2009 fire was estimated spatially from the distribution of biomass carbon density in the unburnt montane ash forest and the proportion of biomass combusted in low- and high-severity fire for each forest age category (Figure 8). The total carbon stock in the montane ash forest region (2,326 km^2) was 112.8 TgC pre-2009 and an estimated 3.9 TgC was emitted from biomass combustion during the 2009 wildfire (burnt area of 768 km^2 or 33% of the area of montane ash forest). The carbon emitted represented 8.5% of the total biomass carbon stock in the area of montane ash forest that was burnt.

Discussion

Although the total amount of carbon stock loss across the landscape burnt during the wildfire was large as an annual flow in the national inventory, we found that the proportion of total ecosystem carbon stock combusted was relatively small, both at site and landscape scales. The major impact of wildfire was the redistribution of carbon among biomass components, from living to dead biomass and CWD, and their longer-term rates of change due

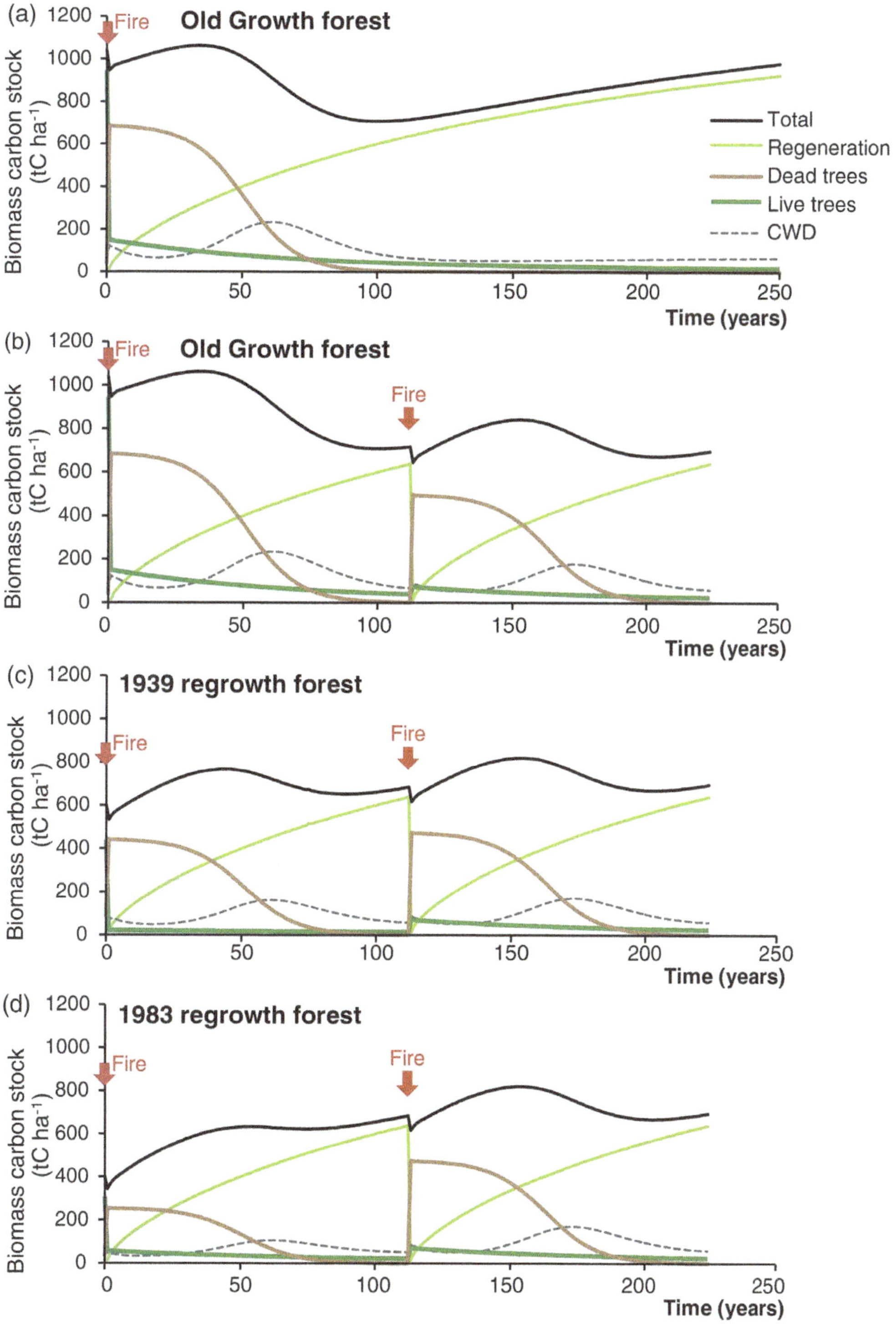

Figure 7. Simulated carbon stock change in biomass components of forests of different ages after high-severity wildfire. Initial fire occurred in year 1 and the change over time in each biomass component, plus the total biomass, is simulated; (a) old growth forest initially that is burnt in year 1 and recovery is simulated over 250 years assuming no other disturbance; (b) old growth forest initially that is burnt in years 1 and 113, according to the average fire frequency, and recovery simulated over two fire cycles; (c) 1939 regrowth forest initially that is simulated over two fire cycles; (d) 1983 regrowth forest initially that is simulated over two fire cycles.

to decomposition and regeneration. Assessing the carbon stock loss by combustion in relation to the total stock, rather than fuel load, meant that changes in stock could be tracked over time and compared between different human activities and disturbance types.

Amount and proportion of forest ecosystem carbon stock combusted

We found that the proportion of total ecosystem biomass carbon stock combusted during a wildfire in the montane ash forest was

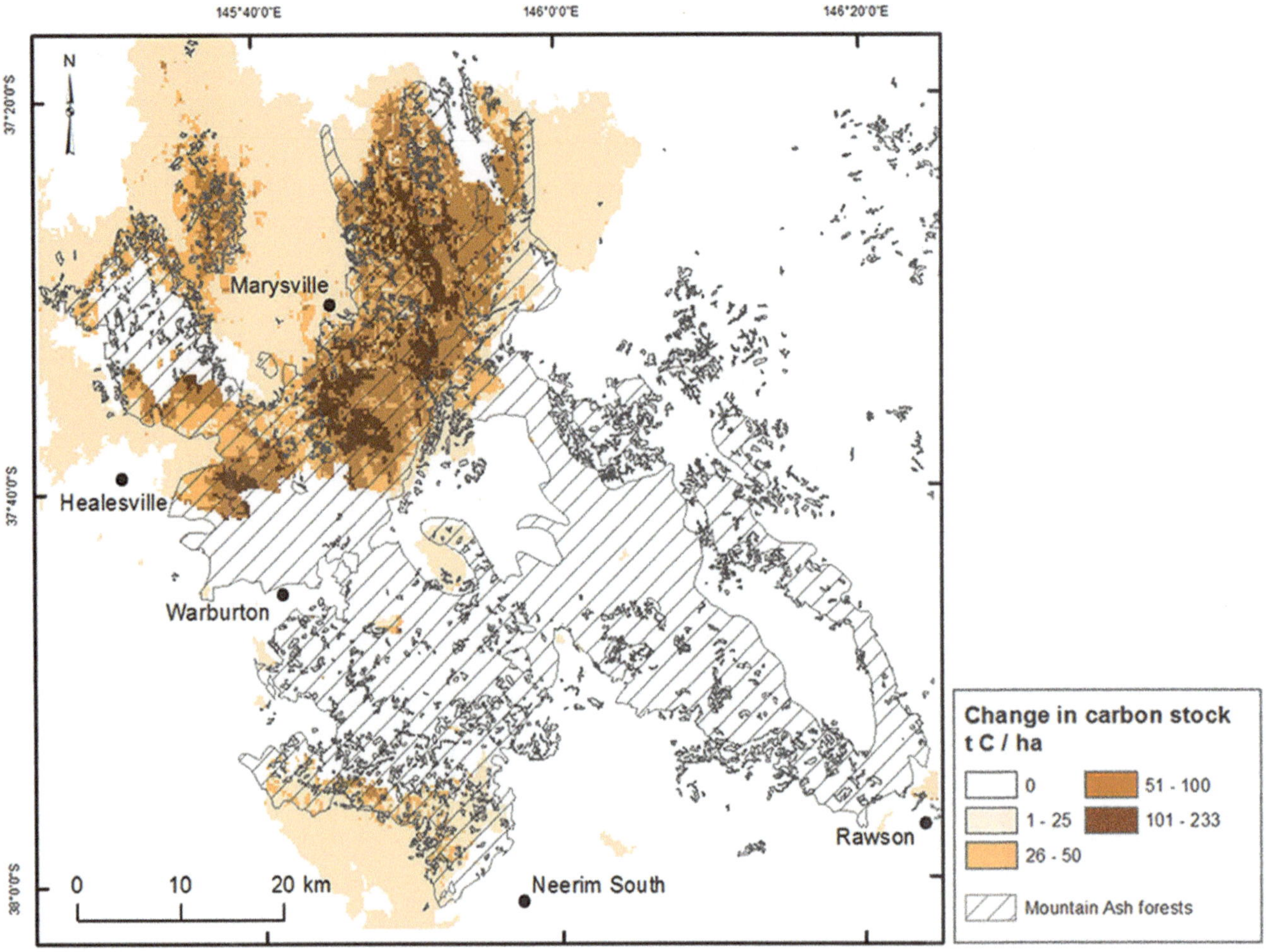

Figure 8. Spatially modelled change in carbon density (tC ha^{-1}) across the montane ash forest region. Change in carbon stock was calculated as the difference between pre- and post-2009 wildfire, including total biomass of living and dead trees, above- and below-ground, shrubs, litter and coarse woody debris.

6–14% at the site scale depending on fire severity and forest age (27 years to old growth), with an average of 8.5% across the landscape that was burnt. This is a relatively small proportion compared with impacts of human disturbance events, such as logging that removes biomass off-site [46,83,84]. Our results are similar to studies in other forest ecosystems (boreal, temperate mixed conifer, savannah woodlands, and other temperate wet sclerophyll) that estimated the relationship between carbon stock loss with total stock, but are lower than estimates of proportion of biomass combusted related to fine and coarse fuel loads (data and references in Table S1).

The estimated loss of only a small proportion of ecosystem carbon stock combusted in a fire concords with simulation results of net losses of carbon by 4 to 8% over 180 years under a fire regime scenario of increased frequency and intensity in south-east Australia [85]. Under this scenario, the loss was due to increased mortality leading to a younger age distribution of trees. Similarly, a sensitivity analysis of the carbon balance model for Canadian forests showed that a three-fold increase in area burned decreased the total forest carbon stock by 6% over 100 years [86]. These predicted losses of carbon stocks are relatively small over a century or two compared with the impact of many human activities.

Differences in the proportion of fine fuel combusted in different ecosystems are insignificant in terms of the total ecosystem carbon stock, representing 0.5 to 1% or 0.5 to 3 tC ha^{-1}. The proportion of coarse fuel combusted varies greatly depending on the size of material, which is influenced not only by forest age but also by disturbance history. Although proportions of biomass combusted were lower in montane ash forest than in other forest types, the amounts combusted in these fuel size classes were greater (Table S1). Amounts of carbon emitted during combustion in different ecosystems depend on vegetation type, past disturbances resulting in dead biomass, and fire weather conditions [87].

We conclude from our methodology that estimating carbon stock loss as a proportion of the total ecosystem carbon stock (based on tracking each biomass component) is a useful metric to compare with other impacts on ecosystem carbon stocks and other sectors of the national GHG inventory. This metric is thus more appropriate for national accounts than the current default reporting of emissions as the product of a combustion factor and fuel load in the IPCC guidelines and Australian national inventory report [6,15,82,88,89]. In our method, all biomass components are assessed without making assumptions about which components constitute fuel.

Our values of carbon stock loss in a wildfire may be small overestimates because some of the biomass combusted is not volatilised and therefore not emitted to the atmosphere. Under conditions of reduced oxygen, partial combustion forms char that remains on the ground, or particulate carbon that is transported in smoke but settles to the ground elsewhere. The proportion of char produced in wildfires has been estimated to be 1–10% of the biomass combusted [90]. Both these forms of carbon are highly resistant to decay and form a long-term stable component of soil carbon stocks [90].

Changes in forest biomass carbon dynamics

The major impact of wildfire in the montane ash forest was the redistribution of carbon among biomass components, from living to dead, and subsequent long-term stock losses from decomposition and gains by regeneration. The relative rates of these processes and subsequent mean residence times of the biomass components determine the long-term dynamics of carbon stock in the ecosystem. These long-term dynamics resulted in changes of up to 20% about the mean stock, whereas direct losses from combustion represented 6–14% of the total biomass stock.

Mean residence time of the biomass components combusted is critical for determining the impact of wildfire on longer-term carbon dynamics. More than half the stock losses from combustion were derived from biomass components with short lifetimes; in the canopy, bark, litter layer and shrubs. These components are described as fine fuel, and this amount remains reasonably constant after about 10 years post-fire in a range of forest types [91]. For example, fine litter in montane ash forest has a mean residence time of 1–4 years [50,70], and 85–95% of the litter mass has accumulated within a decade after fire [91]. Recovery of the carbon stock within a decade in the montane ash forest is within the range of up to 10–20 years for recovery of rates of carbon uptake estimated from eddy covariance measurements of net ecosystem carbon exchange at chronosequence sites across a range of forest types in North America [92].

The components combusted with longer mean residence times, such as CWD and large hollow trees, have variable stocks and proportions combusted depending on forest type, conditions for decomposition, logging and fire history. This coarse material contributed 18 to 45% of the carbon combusted in the montane ash forest, with an increasing proportion in older stands. The residence time of this coarse material is in the order of decades to centuries [67,93].

Rates of carbon accumulation in regenerating stands are difficult to estimate due to limited site data of known forest age, particularly in older aged stands. We compared carbon uptake in regenerated living biomass at a given age estimated from our site data, for example, 3.8 tC ha^{-1} yr^{-1} at 80 years, with the net ecosystem exchange (NEE) estimated from eddy covariance measurements. An average NEE over 10 years of measurements was 5.8 tC ha^{-1} yr^{-1} and the maximum was 9.3 tC ha^{-1} yr^{-1} in an 80 year old stand of *Eucalyptus delegatensis* or Alpine Ash [94]. This average NEE is comparable to the 6 tC ha^{-1} yr^{-1} in a temperate forest in Oregon, which is the highest value reported for forests of a similar age in North America [92]. Carbon accumulated in living biomass does not include uptake of carbon that is then transferred to dead biomass pools and soil organic matter.

Our simulated carbon stocks for each biomass component were within the range observed in the field. Actual magnitudes of stocks at different ages were difficult to verify because of the confounding disturbances of wildfire and salvage logging that have occurred in these forests. Among our sites, there were no younger stands that had been burnt but not logged and such stands would be rare in the landscape. Our simulated turnover times appeared to be slower than in some other ecosystems, such as boreal forests [95]. However, observations of large logs remaining after 60–90 years in the montane ash forest [50,96] suggested that even slower rates may be characteristic of large biomass components. Uncertainty about the magnitude of the changes and their timing in the long-term dynamics of carbon occurred mainly because of limited information about rates of tree fall and decomposition.

Inclusion of wildfire in national carbon accounts

Changes in carbon stocks due to natural disturbances, such as wildfire, must be included in national carbon accounts. The current UNFCCC carbon accounting guidelines for forest management (Article 3 paragraph 4 clause 33) address carbon stock change due to natural disturbances under the '*force majeure*' provision (Decision 2/CMP.7, CoP 17 Durban Climate Conference) [97]. Natural disturbances, such as wildfire, outbreaks of pests and diseases, extreme weather events and geological disturbances, are considered as extraordinary events that are outside human control. An average net stock change due to natural disturbances is calculated as a background level within the forest management reference level. These accounts for forest management and natural disturbances must be included in the 2015 inventory reports from signatory nations to the Kyoto Protocol (38 of the 43 Annex 1 nations), which are assessed for compliance against emissions reduction targets [98].

The current negotiations under the Durban Platform for Enhanced Action [23] represent a critical time to provide quantitative data on carbon stock changes due to disturbance and to interpret the implications of the inventory for maximising outcomes of mitigation activities. Nations can develop and choose methodologies, within the UNFCCC guidelines, that are appropriate for their accounts [99]. The default method is calculated as an average of net stock change from natural disturbances from 1990 to 2009, plus a margin of twice the standard deviation. Annual stock changes that are above this background level are excluded from the accounts. When this exclusion occurs, the area of land affected is removed from the inventory so that subsequent uptake by the regenerating vegetation is not counted, until pre-disturbance carbon stocks are restored [99]. The UNFCCC guidelines provide some flexibility, but the methodology should adhere to the objectives of the *force majeure* provision of minimising annual variability in the reported net stock changes due to natural disturbance events. The rationale is that this variability should not negate reductions in emissions due to human mitigation activities over the timeframe of the reporting [97].

Our results about the magnitude and timeframes of carbon stock changes due to wildfires suggest that carbon accounts need to be comprehensive and include all stocks and flows. Such accounts enable differentiation of stock changes due to natural disturbances and human activities. There are four reasons why constructing carbon accounts which differentiate carbon stock changes are beneficial.

First, our results show that in terms of national annual net change in carbon stocks, the magnitude of carbon stock loss due to combustion in a wildfire is sufficient to mask reduced emissions due to mitigation activities. Our estimated carbon stock loss was 3.9 TgC from a single (albeit large) fire in 2009 within the montane ash forest. The annual average for all fires in Australia is estimated to be 8.7 TgC yr^{-1} [100]. These single and average wildfire emissions represent about 27% and 59%, respectively, of Australia's annual emissions reduction target of 14.7 TgC yr^{-1} [100]. Similar magnitudes of emissions from individual wildfires have been quantified elsewhere, for example 2.9 TgC from the 2003 Californian fires [101]. Hence, differentiation of carbon stock changes due to natural disturbances and human activities is important when national GHG inventory reports are used for assessing progress in meeting emission reduction targets over short time periods.

Second, we demonstrated that the changes in carbon stocks in response to a disturbance event occur over the following decades. Hence, derivation of a background level of net carbon stocks over a two decade period (1990–2009) is not adequate to account for the variations in carbon dynamics within a disturbed area. However, the carbon stock at the landscape scale represents the sum of forest stands that have been burnt at different times and intensities, thus producing a mosaic of stand age classes, amounts of debris, and stages of recovery from disturbance. Under a stable disturbance regime, the carbon stock at the landscape scale remains stable when accounted for over sufficiently long time periods. Hence, carbon stock losses from combustion of biomass components do not increase the accumulated atmospheric CO_2 concentration under a stable fire regime. Carbon stock dynamics may alter at the landscape level if fire regimes change due to climate change, or due to the effects of growth from a forest age class distribution affected by historical activities [9,85,86, 102,103,104].

Third, the increases and decreases in carbon stock post-fire resulting from differential rates of regeneration, mortality and decomposition mean that excluding burnt areas from the national inventory until carbon stocks are restored to pre-fire levels, is not an adequate method for excluding the effect of natural disturbances on net changes in carbon stocks. In fact, the increase in carbon stocks for some decades post-fire would produce artificial credits, but later decreases due to decomposition would produce artificial debits in the national inventory, which are not related to human activities. Additionally, the high levels of uncertainty in these long-term dynamics could produce errors in the national inventory.

Fourth, an averaging period of net stock changes from a disturbance regime over two decades, or within the time period of adequate records, does not encompass the temporal scale of natural variability in wildfire regimes in many temperate forest ecosystems [105], such as montane ash forest with a return interval of many decades [30,31]. The frequency and intensity of wildfire is influenced by interactions of large-scale features of atmospheric circulation, such as the El Niña-Southern Oscillation, Indian Ocean Dipole, Southern Annular Mode and the Sub-Tropical Ridge, which drive multi-decadal climate variability [13,106,107]. If a background level is to be used in an accounting system, then the preferred method would be to re-calculate the actual incidence of disturbance events at the end of the reporting period [108,109,110].

Methodological and data uncertainties

Our results suggest that data from ecological monitoring sites can be used *a posteriori* to estimate carbon stock losses from wildfire, when supplemented with additional field data and ancillary data sources. Our approach used the basic framework of estimating carbon stock changes due to wildfire from the area burned, fuel load and combustion efficiency [87,90] However, our sampling also incorporated spatial variability in fire severity and initial carbon stocks at the landscape level. This provided a more representative range in carbon stock losses at the regional scale. Covering this range is important even though a balanced experimental design is difficult to obtain in studies of stochastic disturbance events where field sites may not be comparable because of uncertainties in disturbance histories and stand ages. Previous studies in Australia have used either experimental fires imposed on a planned design of sites and pre-fire measurements, where the fires were either lower severity than wildfire or regeneration burns of logging slash [89,111,112,113], or post-fire comparisons of burnt and unburnt sites albeit with limited replicate sites [41,114].

Improved assessment of fire severity as a continuous variable at the landscape scale is important to allow more accurate estimates of carbon stock changes spatially [115]. New techniques in remote sensing are providing quantitative information about fire radiative power that can be related to factors that control fire severity [116,117,118]. However, adequate field calibration of the remotely sensed metric currently limits their application [116].

The results of up-scaling the site data to estimate carbon stocks and stock losses across our study region revealed that the effect of disturbance history on tree age structure and CWD was the major source of uncertainty in spatial estimates of biomass carbon stocks. Other studies have drawn similar conclusions that the determination of forest stand age is the main factor contributing to uncertainty in assessment of regional carbon stocks [5].

Our results suggest that current national accounting that uses default values (Tier 2 methodology) for fuel combustion efficiency and fuel load can be improved by using site data from different ecosystems. Data from existing state forest inventories need to be augmented by comprehensive sampling of forest age structure, including old growth forests, and all components of the carbon stock, including dead wood. The methods used here are generic and would constitute Tier 3 methodology that is appropriate for national and project-based accounting. The sampling methods were non-destructive and feasible to apply to a large number of sites.

Conclusions

Carbon stocks in forest ecosystems change in response to wildfire in the short-term by combustion and in the long-term by redistribution among biomass components and their subsequent differential rates of decomposition and regeneration. It is important that carbon accounts include these processes because all stock changes, activities and land areas contribute to quantifying the global carbon cycle. However, the timeframes of the ecological processes that determine carbon stock changes are longer than the time periods for reporting of national inventories for compliance with emissions reductions targets. This problem of incongruous timeframes means that reporting against targets should be based on net changes in carbon stocks from human activities to incentivise mitigation activities. If the carbon accounts differentiate stock changes between the sources from natural and human-induced disturbances, then the net stock change from human activities can be extracted from the accounts for use in reporting. This distinction is currently made between the comprehensive UNFCCC accounts and the reports against targets for the Kyoto Protocol, except that the guidelines for the Protocol include net stock changes from natural disturbances that can be discounted above an averaged background level. However, our results indicate that the spatial and temporal variability in carbon dynamics is too high and the uncertainty of estimates too great to enable a realistic averaged background level that would prevent distortions of the inventory by stock changes that were outside human control.

Acknowledgments

We wish to thank Clive Hilliker for expert assistance with the figures and Steve Meacher for assistance with field measurements of shrub biomass. Helpful discussions about fire ecology with Malcolm Gill, Geoff Cary and Michael Doherty are gratefully acknowledged. The inventory plot data and spatial data for the Central Highlands were obtained from the Victorian Department of Sustainability and Environment under data access licence agreement DALA_2009_082. We thank anonymous reviewers for comments that have improved the manuscript.

Author Contributions

Conceived and designed the experiments: HK DL BM. Performed the experiments: HK DB LM SO. Analyzed the data: HK LC. Contributed to the writing of the manuscript: HK DL BM. Developed the conceptual framework and global context: HK DL BM TKN.

References

1. United Nations Framework Convention on Climate Change 1992. Available: http://unfccc.int/resource/docs/convkp/conveng.pdf. Accessed 2013 Oct 9.
2. Cubasch U, Wuebbles D, Chen D, Facchini MC, Frame D, et al. (2013) Introduction. In: Stocker TF, Qin D, Plattner G-K, Tignor M, Allen SK, et al., editors. Climate Change 2013: The Physical Science Basis. Contribution of Working Group I to the Fifth Assessment Report of the Intergovernmental Panel on Climate Change. Cambridge University Press, Cambridge, United Kingdom and New York, NY, USA. pp. 119–158.
3. Watson CE, Fishman J, Reichle HG (1990) The significance of biomass burning as a source of carbon monoxide and ozone in the southern hemisphere tropics: a satellite analysis. J Geophys Res 95: 16444–16450.
4. Cahoon DR, Stocks BJ, Levine JS, Cofer WR, O'Neill KP (1992) Seasonal distribution of African savanna fires. Nature 359: 812–815.
5. Andreae MO, Merlet P (2001) Emission of trace gases and aerosols from biomass burning. Global Biogeochem Cycles 15(4): 955–966.
6. van der Werf GR, Randerson JT, Giglio L, Collatz GJ, Kasibhatla PS, et al. (2006) Interannual variability in global biomass burning emissions from 1997 to 2004. Atmosph Chem Phys 6: 3423–3441.
7. Kasischke ES, Bruhwiler LP (2003) Emissions of carbon dioxide, carbon monoxide, and methane from boreal forest fires in 1998. J Geophys Res 108: D1,8146.
8. Le Quéré C, Raupach MR, Canadell JG, Marland G, Bopp L, et al. (2009) Trends in the sources and sinks of carbon dioxide. Nature Geosci doi: 10.1038/ngeo689.
9. Stocks BJ, Mason JA, Todd JB, Bosch EM, Wotton BM, et al. (2003) Large forest fires in Canada, 1959–1997. J Geophys Res 108:(D1), 8149.
10. Kurz WA, Stinson G, Rampley GJ, Dymond CC, Neilson ET (2008) Risk of natural disturbances makes future contribution of Canada's forests to the global carbon cycle highly uncertain. Proc Natl Acad Sci USA 105: 15551–1555.
11. Gill AM, Stephens SL, Cary GJ (2013) The worldwide "wildfire" problem. Ecol Appl 23(2): 438–454.
12. Hurst DF, Griffith DWT, Cook GD (1994) Trace gas emissions from biomass burning in tropical Australian savannas. J Geophys Res 99(D8): 16,441–6,456.
13. van der Werf GR, Randerson JT, Collatz GJ (2004) Continental-scale partitioning of fire emissions during the 1997 to 2001 El Niño/La Niña period. Science 303: 73–76.
14. Gill AM and Allan G (2008) Large fires, fire effects and the fire-regime concept. International J Wildland Fire 17: 688–695.
15. de Groot WJ, Pritchard JM, Lynham TJ (2009) Forest floor fuel consumption and carbon emissions in Canadian boreal forest fires. Can J For Res 39: 367–382.
16. Hollis JJ, Matthews S, Ottmar RD, Prichard SJ, Slijepcevic A, et al. (2010) Testing woody fuel consumption models for application in Australian southern eucalypt forest fires. For Ecol Manag 260: 948–964.
17. Scholes RJ, Kendall J, Justice CO (1996a) The quantity of biomass burned in southern Africa. J Geophys Res 101, D19: 23667–23676.
18. Scholes RJ, Ward DE, Justice CO (1996b) Emissions of trace gases and aerosol particles due to vegetation burning in southern hemisphere Africa. J Geophys Res 101: D19: 23677–23682.
19. Horowitz LW, Walters S, Mauzerall DL, Emmons LK, Rasch PJ, et al. (2003) A global simulation of tropospheric ozone and related tracers: Description and evaluation of MOZART, version 2. J Geophys Res 108: D24,4784.
20. Shirai T, Blake DR, Meinardi S, Rowland FS, Russell-Smith J, et al. (2003) Emission estimates of selected volatile organic compounds from tropical savannah burning in northern Australia. J Geophys Res 108: D3, 8406.
21. Hoelzemann J, Schulz MG, Brasseur GP, Granier C (2004) Global wildland fire emission model (GWEM): Evaluating the use of global area burnt satellite data. J Geophys Res 109: D14S04.
22. Ito A, Penner JE (2004) Global estimates of biomass burning emissions based on satellite imagery for the year 2000. J Geophys Res Atm 109: D14S05.
23. UNFCCC (2012) Ad Hoc Working Group on the Durban Platform for Enhanced Action. Available: http://unfccc.int/bodies/body/6645.php Accessed 8 August 2013.
24. Pan Y, Birdsey RA, Fang J, Houghton R, Kauppi PE, et al. (2011) A large and persistent carbon sink in the world's forests, 1990–2007. Science 333: 988–993.
25. Keith H, Mackey BG, Lindenmayer DB (2009) Re-evaluation of forest biomass carbon stocks and lessons from the world's most carbon dense forests. Proc Natl Acad Sci USA 106(28): 11635–11640.
26. Mackey BG, Lindenmayer DB, Gill AM, McCarthy AM, Lindesay JA (2002) Wildfire, Fire and Future Climate: A forest ecosystem analysis. CSIRO Publ., Melbourne.
27. Lindenmayer DB (2009) Forest Pattern and Ecological Processes: A Synthesis of 25 Years of Research. CSIRO Publishing, Collingwood, Australia.
28. Department of Sustainability and Environment (1996) Study of old growth forest in Victoria's Central Highlands. Available: http://www.dse.vic.gov.au/forests/publications/research-reports/study-of-old-growth-forests. Accessed 2013 Aug 8.
29. McCarthy MA, Lindenmayer DB (1998) Multi-aged mountain ash forest, wildlife conservation and timber harvesting. For Ecol Manag 104: 43–56.
30. McCarthy MA, Gill AM, Lindenmayer DB (1999) Fire regimes in mountain ash forest: evidence from forest age structure, extinction models and wildlife habitat. For Ecol Manag 124: 193–203.
31. McCarthy MA, Gill AM, Bradstock RA (2001) Theoretical fire interval distributions. Int J Wildland Fire 10: 73–77.
32. Ashton DH (1975a) The root and shoot development of *Eucalyptus regnans* F. Muell. Aust J Bot 23: 867–887.
33. Flinn D, Squire R, Waring K (2007) Victoria. In: Raison RJ, Squire RO, editors. Forest Management in Australia: Implications for Carbon Budgets. National Carbon Accounting System Technical Report No. 32, Australian Greenhouse Office, Canberra. Ch 3, 103–146.
34. Tng DYP, Williamson GJ, Jordan GJ, Bowman DMJS (2012) Giant eucalypts – globally unique fire – adapted rainforest trees? New Phytol doi: 10.1111/j.1469-8137.2012.04359.x
35. Simkin R, Baker PJ (2008) Disturbance history and stand dynamics in tall open forest and riparian rainforest in the Central Highlands of Victoria. Austral Ecol 33: 747–760.
36. Turner PAM, Balmer J, Kirkpatrick JP (2009) Stand-replacing wildfires? The incidence of multi-cohort and single-cohort *Eucalyptus regnans* and *E. obliqua* forests in southern Tasmania. For Ecol Manag 258: 366–373.
37. Banks JCG (1993) Tree ring analysis of two Mountain Ash trees from the Watts and O'Shannassy catchments, Central Highlands, Victoria. Report to the Central Highlands Old Growth Forest Project, Department of Conservation and Natural Resources, Melbourne.
38. Vivian LM, Cary GJ, Bradstock RA, Gill AM (2008) Influence of fire severity on the regeneration, recruitment and distribution of eucalypts in the Cotter River Catchment ACT. Austral Ecol 33: 55–67.
39. Lindenmayer DB, Cunningham RB, Donnelly CF, Franklin JF (2000) Structural features of old growth Australian montane ash forests. For Ecol Manag 134: 189–204.
40. Victorian Bushfires Royal Commission 2009, Australia. Available: http://www.royalcommission.vic.gov.au/Documents/HR-Chapters/. Accessed 2013 Aug 8.
41. Cruz MG, Sullivan AL, Gould JS, Sims NC, Bannister AJ, et al. (2012) Anatomy of a catastrophic wildfire: The Black Saturday Kilmore East fire in Victoria, Australia. For Ecol Manag 284: 269–285.
42. Noble WS (1977) Ordeal by Fire: The Week a State Burned Up. Hawthorn Press, Melbourne, Australia.
43. Ough K (2001) Regeneration of wet forest flora a decade after clear-felling or wildfire – is there a difference? Aust J Bot 49: 645–664.
44. Lindenmayer DB, Ough K (2006) Salvage harvesting in the montane ash forests of the Central Highlands and its potential impacts on biodiversity. Conserv Biol 20: 1005–1015.
45. Lindenmayer DB, Burton PJ, Franklin JF (2008) Salvage Logging and its Ecological Consequences. CSIRO Publishing, Collingwood, Australia.
46. Keith H, Lindenmayer DB, Mackey B, Blair D, Carter L, et al. (2014) Managing temperate forests for carbon storage: impacts of logging versus forest protection on carbon stocks. Ecosphere 5(6): 75 doi 10.1890
47. Jackson WD (1968) Fire, air, water and earth – an elemental ecology of Tasmania. Proc Ecol Soc Aust 3: 9–16.
48. Lindenmayer DB, Hobbs RJ, Likens GE, Krebs CJ, Banks SC (2011) Newly discovered landscape traps produce regime shifts in wet forests. Proc Natl Acad Sci USA doi 10.1073
49. Taylor C, McCarthy MA, Lindenmayer DB (2014) Non-linear effects of stand age on fire severity. Conserv Letters doi: 10.1111/conl.12122
50. Ashton DH (1975b) Studies in litter in *Eucalyptus regnans* forests. Aust J Bot 23: 413–433.

51. Sillett SC, van Pelt R, Koch GW, Ambrose AR, Carroll AL, et al. (2010) Increasing wood production through old age in tall trees. For Ecol Manag 259: 976–994.
52. Illic J, Boland D, McDonald M, Downes G, Blakemore P (2000) Woody density phase 1– state of knowledge. National Carbon Accounting System Technical Report No. 18, Australian Greenhouse Office, Canberra.
53. Bootle KR (2005) Wood in Australia: types, properties and uses. McGraw-Hill, Sydney.
54. Chafe SC (1985) The distribution and interrelationship of collapse, volumetric shrinkage, moisture content and density in trees of *Eucalyptus regnans* F. Muell. Wood Sci Tech 19: 329–345.
55. Mackensen J, Bauhus J (1999) The decay of coarse woody debris. National Carbon Accounting System, Technical Report No. 6, Australian Greenhouse Office, Canberra.
56. Gifford RM (2000) Carbon contents of above-ground tissues of forest and woodland trees. National Carbon Accounting System Technical Report No. 22, Australian Greenhouse Office, Canberra.
57. Keith H, Leuning R, Jacobsen KL, Cleugh HA, van Gorsel E, et al. (2009a) Multiple measurements constrain estimates of net carbon exchange by a *Eucalyptus* forest. Agric For Meteorol 149: 535–558.
58. Mackowski CW (1987) Wildlife hollows and timber management in Blackbutt forest. Masters' Thesis, University of New England, Armidale, NSW.
59. Gibbons P, Lindenmayer DB (2002) Tree hollows and wildlife conservation in Australia. CSIRO Publications, Collingwood.
60. Department of Sustainability and Environment (2007) Victoria's Statewide Forest Resource Inventory – Central, Central Gippsland and Dandenong Forest Management Areas, Natural Resources Report Series 07–1, Victorian Government.
61. Feller MC (1980) Biomass and nutrient distribution in two eucalypt forest ecosystems. Aust J Ecol 5: 309–333.
62. Keith H, Barrett D, Keenan R (2000) Review of allometric relationships for estimating woody biomass for New South Wales, the Australian Capital Territory, Victoria, Tasmania and South Australia. National Carbon Accounting System Technical Report No. 5b, Australian Greenhouse Office, Canberra.
63. Snowdon P, Eamus D, Gibbons P, Khanna P, Keith H, et al. (2000) Synthesis of allometrics, review of root biomass and design of future woody biomass sampling strategies. National Carbon Accounting System Technical report No. 17, Australian Greenhouse Office, Canberra.
64. Van Wagner CE (1968) The line intersect method in forest fuel sampling. For Sci 14: 20–26.
65. McKenzie N, Ryan P, Fogarty P, Wood J (2001) Sampling, measurement and analytical protocols for carbon estimation in soil, litter ad coarse woody debris. National Carbon Accounting System Technical Report No. 14, Australian Greenhouse Office, Canberra.
66. Lindenmayer DB, Blair D, McBurney L, Banks S (2010) Forest Phoenix: How a Great Forest Recovers after Wildfire. CSIRO Publishing, Melbourne. 128 pp.
67. Lindenmayer DB, Blanchard W, McBurney L, Blair D, Banks S, et al. (2012) Interacting factors driving a major loss of large trees with cavities in a forest ecosystem. PLoS 1: 7(10), e41864.
68. Andrew MH, Noble IR, Lange RT (1979) A non-destructive method for estimating the weight of forage on shrubs. Aust Range J 1(3): 225–231.
69. Ashton D H (1975a) Studies of litter in *Eucalyptus regnans* forests. Aust J Bot 23: 413–433.
70. Polglase PJ, Attiwill PM (1992) Nitrogen and phosphorus cycling in relation to stand age of *Eucalyptus regnans* F. Muell. I. Return from plant to soil in litterfall. Pl Soil 142: 157–166.
71. Jacobs M (1955) Growth Habits of the Eucalypts. Forestry and Timber Bureau, Canberra.
72. Keith H (1991) Effects of fire and fertilization on nitrogen cycling and tree growth in a subalpine eucalypt forest. PhD Thesis, Australian National University, Canberra. 369 pp.
73. Keith H, Mackey BG, Berry S, Lindenmayer D, Gibbons P (2010) Estimating carbon carrying capacity in natural forest ecosystems across heterogeneous landscapes: addressing sources of error. Glob Change Biol 16: 2971–2989.
74. Mackey B, Berry S, Brown T (2008) Reconciling approaches to biogeographic regionalization: a systematic and generic framework examined with a case study of the Australian continent. J Biogeog 35: 213–229.
75. Berry S, Mackey B, Brown T (2007) Potential applications of remotely sensed vegetation greenness to habitat analysis and the conservation of dispersive fauna. Pac Conserv Biol 13(2): 120–127.
76. Ashton DH (1976) The development of even-aged stand of *Eucalyptus regnans* F. Muell. in Central Victoria. Aust J Bot 24: 397–414.
77. Watson FGR, Vertessy RA (1996) Estimating leaf area index from stem diameter measurements in Mountain Ash forest. Cooperative Research Centre for Catchment Hydrology, Report 96/7.
78. Attiwill PM (1994) Ecological disturbance and the conservative management of eucalypt forests in Australia. For Ecol Manag 63: 301–346.
79. Wood SW, Hua Q, Allen KJ, Bowman DMJS (2010) Age and growth of a fire prone Tasmanian temperate old-growth forest stand dominated by *Eucalyptus regnans*, the world's tallest angiosperm. For Ecol Manag 260, 438–447.
80. Grove SJ, Stamm L, Barry C (2009) Log decomposition rates in Tasmanian *Eucalyptus obliqua* determined using an indirect chronosequence approach. For Ecol Manag 258: 389–397.
81. IPCC (2006) Guidelines for National Greenhouse Gas Inventories. Eggleston S, Buendia L, Miwa K, Ngara T, Tanabe K, editors. Institute for Global Environmental Strategies, Hayama, Japan. Available: http://www.ipcc-nggip.iges.or.jp/public/2006gl/index.html. Accessed 2013 Oct 30.
82. DCCEE (2012) Department of Climate Change and Energy Efficiency, Australian National Greenhouse Gas Accounts, National Inventory Report 2010, Australian Government Submission to the United National Framework Convention on Climate Change, April 2012.
83. Harmon ME, Ferrell WK, Franklin JF (1990) Effects on carbon storage of conversion of old growth forests to young forests. Science 247, 699–702.
84. Gough CM, Vogel CS, Harrold KH, George K, Curtis PS (2007) The legacy of harvest and fire on ecosystem carbon storage in a north temperate forest. Glob Change Biol 13, 1935–1949.
85. King KJ, deLigt RM, Cary GJ (2011) Fire and carbon dynamics under climate change in south-eastern Australia: insights from FullCAM and FIRESCAPE modelling. Int J Wildland Fire 20: 563–577.
86. Kurz WA, Apps MJ (1999) A 70-year retrospective analysis of carbon fluxes in the Canadian forest sector. Ecol Appl 9(2): 526–547.
87. French NHF, de Groot WJ, Jenkins LK, Rogers BM, Alvarado E, et al. (2011) Model comparisons for estimating carbon emissions from North American wildland fire. J Geophys Res 116: G00K05.
88. IPCC (2003) Good practice guidance for land use, land use change and forestry. Penman J et al., editors. National Greenhouse Gas Inventories Programme, Institute for Global Environmental Strategies, Kanagawa, Japan. Available: http://www.ipcc-nggip.iges.or.jp/public/gpglulucf/gpglulucf-files. Accessed 2013 Oct 30.
89. Gould JS, Cheney NP, Hutchings PT, Cheney S (1996) Final Report on Prediction of Bushfire Spread for Australian Co-ordination Committee International Decade of Natural disaster Reduction. Project 4/95. CSIRO Forestry and Forest Products.
90. Seiler W, Crutzen PJ (1980) Estimates of gross and net fluxes of carbon between the biosphere and the atmosphere from biomass burning. Clim Change 2: 207–247.
91. Gould JS, Cheney NP (2007) Fire management in Australian forests. In: Forest management in Australia: implications for carbon budgets. Raison RJ and Squire RO, editors. 341–371. National Carbon Accounting System Technical Report No.32.
92. Amiro BD, Barr AG, Barr JG, Black TA, Bracho R, et al. (2010) Ecosystem carbon dioxide fluxes after disturbance in forests of North America. J Geophys Res 115: G00K02.
93. Lindenmayer DB, Wood JT (2010) Long-term patterns in the decay, collapse and abundance of trees with hollows in the mountain ash (*Eucalyptus regnans*) forests of Victoria, southeastern Australia. Can J For Res 40: 48–54.
94. van Gorsel E, Berni JAJ, Briggs P, Cabello-Leblic A, Chasmer L, et al. (2013) Primary and secondary effects of climate variability on net ecosystem carbon exchange in an evergreen *Eucalyptus* forest. Agric For Meteorol 182–183: 248–256.
95. Manies KL, Harden JW, Bond-Lamberty BP, O'Neill KP, (2005) Woody debris along an upland chronosequence in boreal Manitoba and its impact on long-term carbon storage. Can J For Res 35: 472–482.
96. Lindenmayer DB, Incoll RD, Cunningham RB, Donnelly CF (1999) Attributes of logs on the floor of Australian Mountain Ash (*Eucalyptus regnans*) forests of different ages. For Ecol Manag 123: 195–203.
97. UNFCCC (2011) 17th Conference of the Parties to the UNFCCC in Durban, CMP.7 Land Use, Land Use Change and Forestry. Available: http://unfccc.int/files/meetings/durban_nov_2011/decisions/application/pdf/awgkp_lulucf.pdf. Accessed 2013 Aug 8.
98. UNFCCC (2013b) National Inventory Submissions. Available: http://unfccc.int/national_reports/annex_i_ghg_inventories/national_inventories_submissions/. Accessed 2013 Aug 8.
99. UNFCCC (2013a) Revised Supplementary Methods and Good Practice guidance arising from the Kyoto Protocol. Good Practice Guidance. Available: http://www.ipcc-nggip.iges.or.jp/home/docs/kpsg/KP_Supplement_precopyedit.pdf. Accessed 2013 Aug 8.
100. Australian Treasury (2011) Strong growth, low pollution: modelling a carbon price. Australian Government, Canberra.
101. Mühle J, Lueker TJ, Su Y, Miller BR, Prather KA, et al. (2007) Trace gas and particulate emissions from the 2003 southern Californian wildfires. J Geophys Res 112,D03307.
102. Williams A, Karoly D, Tapper N (2001) The sensitivity of Australian fire danger to climate change. Clim Change 49: 171–191.
103. Pitman A, Narisma G, McAneney J (2007) The impact of climate change on the risk of forest and grassland fires in Australia. Clim Change 84: 383–401.
104. Clarke H, Smith P, Pitman A (2011) Regional signatures of future fire weather over eastern Australia from global climate models. Int J Wildland Fire 20: 550–562.
105. Macintosh A (2012a) Are forest management reference levels incompatible with robust climate outcomes? A case study on Australia. Carbon Manag 2(6): 691–707.
106. Esplin B, Gill AM, Enright N (2003) Report of the Inquiry into the 2002–2003 Victorian Bushfires. Office of the Emergency Services Commissioner, State

Government of Victoria. Available: http://www.oesc.vic.gov.au/home/reviews+and+inquiries/report+of+the+inquiry+into+the+2002-2003+Victorian+bushfires. Accessed 2013 Oct 15.

107. Murphy B, Timbal B (2008) A review of recent climate variability and climate change in southeastern Australia. Int J Clim 28: 859–879.

108. Canadell J, Kirschbaum M, Kurz W, Sanz M, Schlamadinger B, et al. (2007) Factoring out natural and indirect human effects on terrestrial carbon sources and sinks. Environ Sci Policy 10, 370–384.

109. Schlamadinger B, Johns T, Ciccarese L, Braun M, Sato A, et al. (2007) Options for including land use in a climate agreement post-2012: improving the Kyoto Protocol approach. Environ Sci Policy 10, 295–305.

110. Macintosh A (2011) Potential carbon credits from reducing native forest harvesting in Australia. ANU Centre for Climate Law and Policy, Working Paper Series 2011/1.

111. Marsden-Smedley JB, Slijepcevic A (2001) Fuel characteristics and low intensity burning in *Eucalyptus obliqua* wet forest at the Warra LTER site. Tasforests 13(2): 261–277.

112. Slijepcevic A (2001) Loss of carbon during controlled regeneration burns in *Eucalyptus obliqua* forest. Tasforests 13(2): 281–290.

113. Tolhurst KG, Anderson WR, Gould JS (2006) Woody fuel consumption experiments in an undisturbed forest. In: Conference Proceedings of the V International Conference on Forest Fire Research. Viegas DX, editor.

114. Hollis JJ, Anderson WR, McCaw WL, Cruz MG, Burrows ND, et al. (2011) The effect of fireline intensity on woody fuel consumption in southern Australian eucalypt forest fires. Aust For 74: 81–96.

115. Turetsky MR, Kane ES, Harden JW, Ottmar RD, Manies KL, et al. (2011) Recent acceleration of biomass burning and carbon losses in Alaskan forests and peatlands. Nature Geosciences 4: 27–31.

116. Campbell J, Donato D, Azuma D, Law B (2007) Pyrogenis carbon emissions from a large wildfire in Oregon, United States. J Geophys Res 122: G04014.

117. Barrett K, Kasischke ES, McGuire AD, Turetsky MR, Kane ES (2010) Modelling fire severity in black spruce stands in the Alaskan boreal forest using spectral and non-spectral geospatial data. Remote Sens Environ 114: 1494–1503.

118. Barrett K, Kasischke ES (2013) Controls on variation in MODIS fire radiative power in Alaskan boreal forests: Implications for fire severity conditions. Remote Sens Environ 130: 171–181.

6

Ecological Shifts in Mediterranean Coralligenous Assemblages Related to Gorgonian Forest Loss

Massimo Ponti[1]*, Rossella Angela Perlini[1], Vincenzo Ventra[1], Daniele Grech[2], Marco Abbiati[1,3], Carlo Cerrano[2]

1 Dipartimento di Scienze Biologiche, Geologiche e Ambientali, University of Bologna, UO CoNISMa, Ravenna, Italy, **2** Dipartimento di Scienze della Vita e dell'Ambiente, Polytechnic University of Marche, UO CoNISMa, Ancona, Italy, **3** ISMAR, Consiglio Nazionale delle Ricerche - Istituto di Scienze Marine, Bologna, Italy

Abstract

Mediterranean gorgonian forests are threatened by several human activities and are affected by climatic anomalies that have led to mass mortality events in recent decades. The ecological role of these habitats and the possible consequence of their loss are poorly understood. Effects of gorgonians on the recruitment of epibenthic organisms were investigated by manipulating presence of gorgonians on experimental panels at 24 m depth, for *Eunicella cavolinii*, and at 40 m depth, for *Paramuricea clavata*, at two sites: Tavolara Island (Tyrrhenian Sea) and Portofino Promontory (Ligurian Sea). After 4 months, the most abundant taxa on the panels were encrusting green algae, erect red algae and crustose coralline algae at 24 m depth and encrusting brown algae and erect red algae at 40 m depth. Assemblages on the panels were significantly affected by the presence of the gorgonians, although effects varied across sites and between gorgonian species. Species diversity and evenness were lower on panels with gorgonian branches. Growth of erect algae and recruitment of serpulid polychaetes were also affected by the presence of the gorgonians, primarily at Tavolara. Crustose coralline algae and erect sponges were more abundant on *E. cavolinii* panels at 24 m depth, while encrusting bryozoans were more abundant on *P. clavata* panels at 40 m depth. Effects of gorgonians on recruited assemblages could be due to microscale modification of hydrodynamics and sediment deposition rate, or by a shading effect reducing light intensity. Gorgonians may also intercept settling propagules, compete for food with the filter-feeders and/or for space by producing allelochemicals. Presence of gorgonians mainly limits the growth of erect algae and enhances the abundance of encrusting algae and sessile invertebrates. Therefore, the gorgonian disappearances may cause a shift from assemblages characterised by crustose coralline algae to filamentous algae assemblages, decreasing complexity and resilience of coralligenous bioconstructions.

Editor: Sebastian C. A. Ferse, Leibniz Center for Tropical Marine Ecology, Germany

Funding: This work has been partially funded by University funds to CC and has been co-funded by the Italian Ministry of Education, Universities and Research within the Flagship Project RITMARE "The Italian Research for the Sea" (belonging to the National Research Program 2012-2016 and coordinated by the Italian National Research Council), and within the scientific research program of national interest "Coastal bioconstructions: structures, functions, and management" (2010-11 PRIN prot. 2010Z8HJ5M_003). The funders had no role in study design, data collection and analysis, decision to publish, or preparation of the manuscript.

Competing Interests: The authors have declared that no competing interests exist.

* Email: massimo.ponti@unibo.it

Introduction

The coralligenous habitats are among the most typical ones in the Mediterranean Sea. They are bioconstructions, which host the highest level of benthic species diversity in the Mediterranean Sea [1], and Porifera is the richest taxon [2]. Coralligenous habitat structure has been clearly described by Riedl [3], who outlined a microscale zonation: an infaunal layer (boring and cryptic species) and an epibenthic layer with two main levels of complexity, one made by small encrusting organisms and a second made by species with vertical growth. Among the latter, several gorgonian species and sponges belonging to the genus *Axinella* mainly contributed to the building of the 3D complexity. When the density of these tree-like shaped organisms is high, environmental conditions (light, currents, sediment deposition rates, etc.) may be strongly altered, creating micro habitats that act as a shelter for many necto-benthic organisms [4,5].

Integrity of large gorgonian forests can be affected by both natural disturbances (e.g. climatic anomalies [6], ocean acidification [7], exceptional storms [8]) and human activities (e.g. anchoring, SCUBA diving, recreational and commercial fishing [9,10]). In the gorgonians, tissue injuries and physiological stress may increase the susceptibility to pathogens and facilitate settlements of epibionts [11,12]. Over recent decades, these phenomena have led to gorgonian mass mortality events in the north-western Mediterranean Sea [13,14,15,16]. Although these phenomena have been widely investigated, little is known about the consequences of gorgonian loss on the composition and abundance of other members of the coralligenous assemblages [17,18,19].

Predicting the consequences of loss of habitat forming species is critically important. Specifically, given the present threats to biological diversity such as habitat fragmentation, overharvesting and climate change.

With this aim, a field experiment was designed to investigate the effect of gorgonian forests on the settlement and early recruitment of epibenthic organisms on coralligenous habitats. Density of the two most common Mediterranean gorgonian octocorals, *Eunicella cavolinii* (Koch, 1887) and *Paramuricea clavata* (Risso, 1826), was

Table 1. Mean values (±s.e., observations number in brackets) of the main water physical and chemical parameters at Tavolara (August 2008–July 2009, from the database of the Italian Ministry of Environment and Protection of Land and Sea, http://www.sidimar.tutelamare.it/) and Portofino (August 2008–August 2009, from the database of the Liguria Region' Environmental Agency, http://www.ambienteinliguria.it).

		Tavolara		Portofino	
Parameter	**Unit**	**Surface**	**−24 m**	**Surface**	**−24 m**
Temperature	°C	18.73±1.38 (12)	17.90±1.10 (12)	19.76±2.07 (7)	17.84±1.27 (7)
Salinity	psu	37.71±0.23 (12)	37.88±0.22 (12)	37.34±0.33 (7)	37.52±0.13 (7)
pH		8.20±0.06 (10)	8.24±0.05 (10)	8.39±0.02 (7)	8.39±0.02 (7)
O_2	% sat	114.50±3.56 (11)	101.92±3.42 (11)	81.90±0.87 (7)	97.86±1.46 (7)
Chl	$\mu g\ L^{-1}$	0.26±0.01 (10)	0.25±0.01 (10)	0.30±0.05 (7)	0.46±0.08 (7)
Transparency	m	19.00±3.43 (12)		11.93±2.01 (7)	
Nitrates	$\mu M\ L^{-1}$	0.06±0.02 (5)		0.92±0.49 (7)	
Nitrites	$\mu M\ L^{-1}$	0.05±0.04 (5)		0.11±0.11 (7)	
Ammonia	$\mu M\ L^{-1}$	0.61±0.21 (5)		0.35±0.22 (7)	
N_{tot}	$\mu M\ L^{-1}$	12.70±4.08 (5)		7.53±1.18 (7)	
Phosphates	$\mu M\ L^{-1}$	0.07±0.04 (5)		0.01±0.01 (7)	
P_{tot}	$\mu M\ L^{-1}$	0.15±0.08 (5)		0.01±0.01 (7)	
Silicates	$\mu M\ L^{-1}$	2.10±0.93 (5)		2.12±0.64 (7)	

experimentally manipulated in two locations, to quantify specific changes associated with their loss.

Materials and Methods

Experimental Set-up and Study Sites

Possible effects of the presence of *Eunicella cavolinii* (Koch, 1887) and *Paramuricea clavata* (Risso, 1826) (Anthozoa, Plexauridae) on the early recruitment of epibenthic organisms were investigated by a manipulative field experiment carried out from June to October, 2010. To limit the impact of the experiments 10.5×15.0 cm foamed PVC (Forex) panels were used to quantify the recruitment. Although artificial panels cannot exactly mimic natural substrates, they are commonly used in experimental ecology to investigate early-stage recruitment and succession processes. This method is simple, inexpensive to implement, and provides results very similar to natural substrates [20,21,22,23]. Gorgonian 'forests' were simulated by transplanting three 20 cm long ramified apical branches on each artificial panel. Branches were collected from 40–50 cm high colonies to avoid detaching whole specimens from the nearby populations. The collected apical branches were maintained in continuously renewed seawater without exposing them to air during the transplant operations, which lasted a few hours on the boat. The simulated density of gorgonians was comparable to the maximum values observed in natural patches in terms of abundance (190 colonies m^{-2}) for *Eunicella cavolinii* [24] and biomass for *Paramuricea clavata* [6]. Sets of four panels were arranged in plots (Fig. 1b). For *E. cavolinii*, four forested and four non-forested plots were deployed, interspersed, at 24 m depth. Whilst for *P. clavata* the same numbers of plots were installed at 40 m depth. Depths were chosen according to the maximum densities generally occurring in natural populations. Field experiments were simultaneously conducted in two sites with contrasting environmental conditions where dense populations of these gorgonian species naturally occur (Fig. 1a): Tavolara Island (north-western Tyrrhenian Sea, Lat. 40° 54′ 22″ N–Long. 9° 44′ 00″ E) and Portofino Promontory (Ligurian Sea, Lat. 44° 17′ 54″ N–Long. 9° 13′ 09″ E). Tavolara Island is a limestone-dolomite massif with steep cliffs, while Portofino Promontory bedrock is made of a conglomerate of heterogeneous pebbles, mainly of marly limestone, in a sandy-limestone matrix. Compared to Portofino, Tavolara waters are generally more transparent and richer in oxygen, ammonia, total nitrogen and total phosphorus. Portofino waters are richer in nitrates, nitrites and phytoplankton, in terms of Chlorophyll concentration, especially at greater depth (Table 1). To reduce possible anthropogenic disturbances, study sites were selected within Marine Protected Areas.

Panels were collected after 4 months, before the storms season. Panels were brought to the surface in individual plastic zip-bags and preserved in a buffered solution of formaldehyde (4%) until observation.

Ethics Statement

Sampling at Tavolara and Portofino was done in accordance with Italian laws, and authorisations were granted by the "Marine Protected Area Tavolara - Punta Coda Cavallo" (www.amptavolara.com) and the "Marine Protected Area of Portofino" (www.portofinoamp.it). This study did not involve endangered or protected species.

Laboratory and Data Analyses

Sessile species that recruited on the panels were identified to the lowest possible taxonomic level and their percent covers on the upper side was estimated by superimposing a reference grid of 400 equal sized squares [25]. Species richness (as number of species, S), species diversity (as Hill's diversity number $N1 = \text{Exp}\ H'$, where H' is the Shannon's index based on natural logarithm) and the corresponding evenness component (as $N10 = N1/S$) were calculated for each panel [26]. Differences in assemblage structures were represented using Principal Coordinate Analysis (PCoA unconstrained ordination plot; [27]) based on Bray-Curtis similarities of square root-transformed percent cover data.

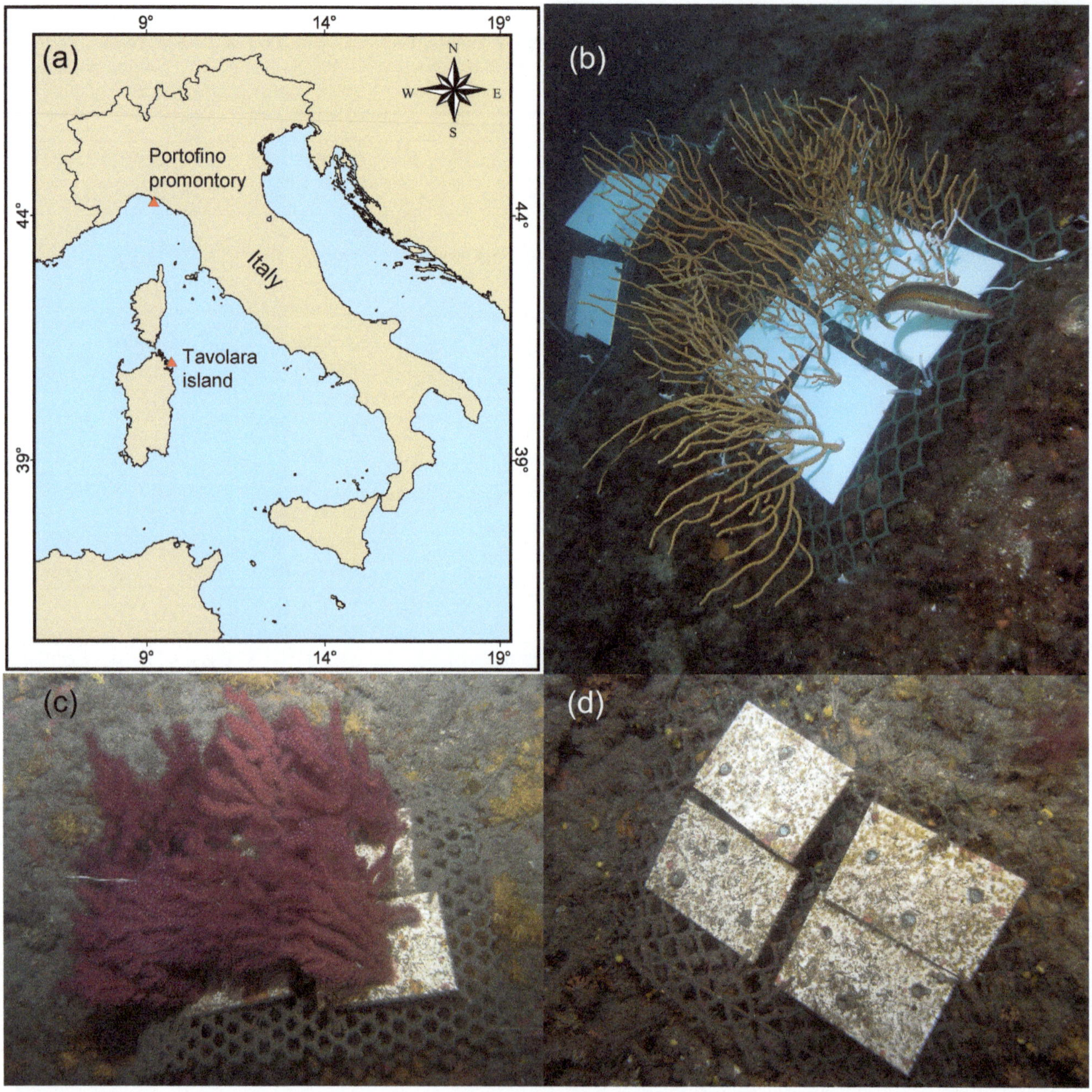

Figure 1. Study sites and experimental set-up. a) Location of the study sites; b) plots with and without (background) *Eunicella cavolinii* at 24 m depth in Portofino at the beginning of the experiment; plots with (c) and without (d) *Paramuricea clavata* at 40 m depth in Portofino at the end of the experiment.

Subsequently, taxa were pooled into main ecological and taxonomic groups and their contribution to similarity patterns of the assemblages were investigated by multivariate multiple regression using the DistLM procedure [28] and represented by correlation vectors on the PCoA ordination plots.

Differences in community structures, species abundances, ecological and taxonomic group abundances and species diversity indices between forestation treatment (Fo: fixed factor with 2 levels: presence/absence of gorgonian forest), sites (Si: fixed factor with 2 levels: Tavolara/Portofino), and plots (Pl: random factor nested in Fo×Si with 4 levels) were assessed by permutational multivariate analysis of variance (PERMANOVA, $\alpha = 0.05$; [29]). When less than 999 unique values in the permutation distribution were available, asymptotical Monte Carlo *P*-values was used instead of permutational *P*-values. Multivariate tests were performed on Bray-Curtis similarities of square root-transformed percent cover data, while univariate tests were run on untransformed data using the Euclidean distance. Significant interactions among main factors were investigated by post-hoc pair-wise tests. Statistical analyses were performed using PRIMER 6 with PERMANOVA+ add-on package [30].

Results

The experiment was monitored periodically and at the end, all transplanted gorgonian branches were alive and in good condition

Table 2. Summary of PERMANOVA test on the effects of (a) *Eunicella cavolinii* and (b) *Paramuricea clavata* forests (Fo) on the recruitment of sessile assemblages at different sites (Si) and among plots (Pl) within Fo×Si (Bray-Curtis similarities on square root transformed data).

Source	*df*	SS	MS	Pseudo-*F*	*P* (perm)	Unique perms
(a) *E. cavolinii*						
Fo	1	2'058	2'058	1.817	0.0347	9'926
Si	1	21'566	21'566	19.045	0.0001	9'914
Fo×Si	1	1601	1'601	1.413	0.1586	9'928
Pl (Fo×Si)	11	12'456	1'132	3.691	0.0001	9'719
Res	45	13'807	307			
Total	59	51'597				
(b) *P. clavata*						
Fo	1	5'345	5'345	3.690	0.0001	9'930
Si	1	33'061	33'061	22.823	0.0001	9'929
Fo×Si	1	4'738	4'738	3.271	0.0003	9'920
Pl (Fo×Si)	12	17'425	1'452	2.754	0.0001	9'703
Res	47	24'779	527			
Total	62	84'386				

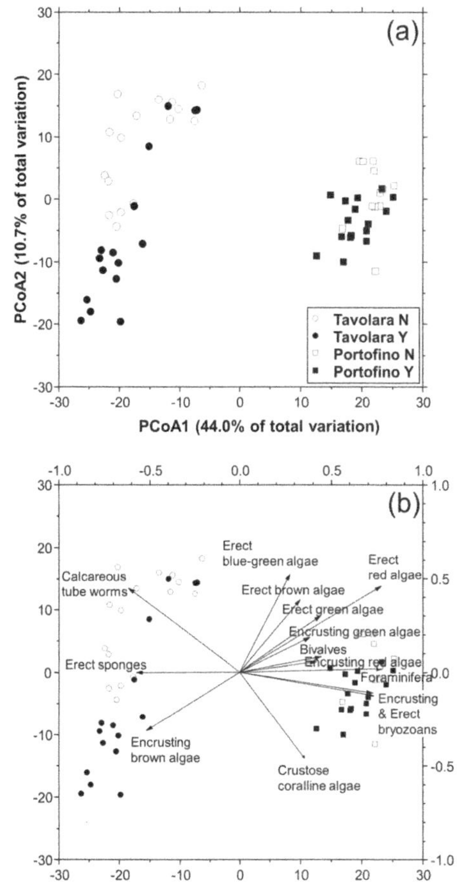

Figure 2. Effects of *Eunicella cavolinii* on benthic assemblages structures. a) PCoA unconstrained ordination plot (i.e. metric MDS) based on Bray-Curtis similarities of square rout-transformed sessile assemblages percent cover on recruitment panels with (Y) and without (N) *Eunicella cavolinii* forest after 4 months in Tavolara and Portofino; b) significant correlations of main ecological groups with PCoA Axes, represented by superimposed vectors (DistLM, marginal test *P*<*0.01*).

(Fig. 1c, d). Unfortunately, one panel with transplanted branches of *P. clavata* and an entire plot without gorgonian branches associated to the *E. cavolinii* experiment were lost at one site, Portofino. Measurements carried out at the end of the experiment confirmed that the simulated gorgonian density and biomass on the panels were consistent between forested plots and sites. In particular the mean dry mass of *P. clavata* was 827 ± 69 g m^{-2} ($\pm$s.e.), which correspond to an ash free dry mass of ~91 g m^{-2} (according to the conversion factor provided by [31]).

Overall, assemblages recruited on the experimental panels included 233 taxa, of which 146 occurred at 24 m depth, 160 at 40 m depth and 73 were common to both depths. The effect of the two gorgonian species on the recruitment of sessile assemblages was analysed separately.

Effects of *Eunicella cavolinii*

The most abundant taxa on the panels at 24 m depth were encrusting green algae (88.6%$\pm$2.0 s.e.), erect red algae (57.6%$\pm$4.1 s.e.) and crustose coralline algae (55.4%$\pm$1.4 s.e.), followed by foraminifera (29.2%$\pm$1.9 s.e.), hydroids (15.3%$\pm$1.8 s.e.), erect brown algae (11.0%$\pm$1.1 s.e.) and calcareous tube worms (10.2%$\pm$0.7 s.e.). Overall, the benthic assemblages differed between plots and sites and were significantly affected by the presence of the gorgonian *E. cavolinii* (Table 2a). The PCoA ordination plot represented 54.7% of the total variation. The PCoA1 axis showed the clear difference between the assemblages at Tavolara and Portofino, while the PCoA2 axis showed the differentiation due to the presence and absence of the gorgonian (Fig. 2a). Although the patterns of differentiation on the PCoA plot are in the same direction, the benthic assemblages at Tavolara appeared more heterogeneous and more affected by the presence of *E. cavolinii* than at Portofino. The correlation vectors, superimposed on the PCoA plot (Fig. 2b), showed the relevance of the ecological and taxonomic groups in the differentiation of the recruited assemblages. As a general trend, erect algae were more abundant on the panels without gorgonians, especially at Portofino, while calcareous tube worms were more abundant on the panels without gorgonians at Tavolara.

Table 3. Summary of PERMANOVA test on the effects of *Eunicella cavolinii* forests (Fo) on the recruitment of ecological and functional groups and diversity indices at different sites (Si) and among plots (Pl) within Fo×Si.

	Forest (Fo)			Site (Si)			Fo×Si			Plot (Fo×Si)			Res
	(1 *df*)			(1 *df*)			(1 *df*)			(11 *df*)			(45 *df*)
	MS	*F*	*P*	MS	*F*	*P*	MS	*F*	*P*	MS	*F*	*P*	MS
Erect blue-green algae	377.74	3.25	0.0949 ns	65.39	0.56	0.4905 ns	57.55	0.50	0.5193 ns	116.14	3.65	0.0006***	31.81
Erect red algae	6'238.30	4.89	0.0337*	28'074.00	22.03	0.0001***	90.40	0.07	0.8214 ns	1'274.50	4.19	0.0005***	304.05
Encrusting red algae	0.12	0.06	0.8071 ns	9.14	4.93	0.0510 ns	0.02	0.01	0.9251 ns	1.85	2.55	0.0096**	0.73
Crustose coralline algae	1'611.40	8.76	0.0133*	1'210.90	6.58	0.0266*	113.60	0.62	0.4505 ns	183.99	4.83	0.0002***	38.07
Erect brown algae	686.49	5.00	0.0475*	333.96	2.43	0.1419 ns	224.16	1.63	0.2367 ns	137.22	4.11	0.0002***	33.35
Encrusting brown algae	7.46	0.78	0.4479 ns	50.63	5.29	0.0105*	7.80	0.81	0.4364 ns	9.58	9.95	0.0001***	0.96
Encrusting green algae	1'069.00	2.42	0.1427 ns	2'025.50	4.58	0.0567 ns	16.43	0.04	0.8566 ns	442.65	2.84	0.0062**	155.64
Erect green algae	7.36	11.22	0.0057**	5.82	8.88	0.0117*	0.00	0.01	0.9317 ns	0.66	1.64	0.1153 ns	0.40
Foraminifera	703.38	3.12	0.1119 ns	7'915.90	35.06	0.0001***	406.27	1.80	0.2165 ns	225.79	4.85	0.0001***	46.60
Erect sponges	11'245.00	3.85	0.0056**	42'098.00	14.42	0.0001***	1'358.40	0.47	0.8051 ns	2'919.30	3.82	0.0001***	763.53
Hydroids	462.56	1.07	0.3229 ns	1'002.10	2.33	0.1493 ns	135.47	0.31	0.5824 ns	430.45	3.85	0.0006***	111.72
Calcareous tube worms	44.43	1.87	0.2001 ns	769.63	32.48	0.0003***	135.71	5.73	0.0358*	23.70	2.19	0.0315*	10.80
Bivalves	0.14	3.07	0.1112 ns	0.63	14.01	0.0002***	0.14	3.07	0.1008 ns	0.04	1.06	0.4171 ns	0.04
Encrusting bryozoans	10.13	1.89	0.2187 ns	234.47	43.70	0.0001***	5.48	1.02	0.3367 ns	5.37	1.78	0.0649 ns	3.01
Stoloniferous bryozoans	0.67	0.01	0.9110 ns	0.89	0.02	0.9010 ns	128.35	2.68	0.1343 ns	47.86	2.26	0.0183*	21.16
Erect bryozoans	4.09	0.36	0.5626 ns	316.64	27.66	0.0002***	15.08	1.32	0.2861 ns	11.45	4.28	0.0002***	2.67
S	513.39	3.34	0.0923 ns	1'773.60	11.54	0.0029**	27.08	0.18	0.6911 ns	153.66	6.43	0.0001***	23.90
N1	135.93	5.40	0.0370*	512.72	20.38	0.0003***	4.10	0.16	0.7033 ns	25.16	4.98	0.0001***	5.05
N10	0.01	9.91	0.0087**	0.04	39.23	0.0001***	0.00	0.81	0.3903 ns	0.00	0.79	0.6509 ns	0.00

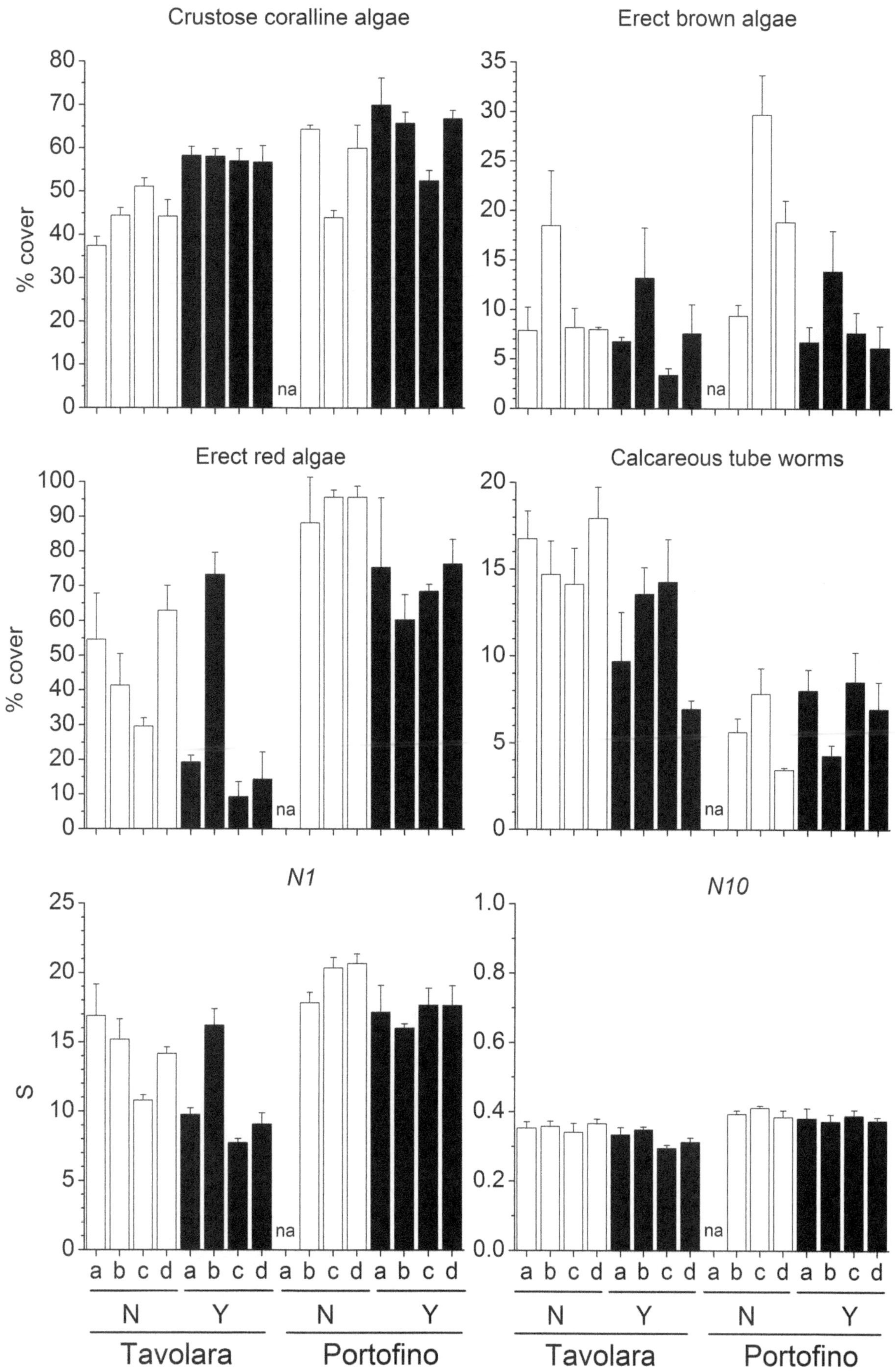
Crustose coralline algae
Erect brown algae
Erect red algae
Calcareous tube worms
N1
N10
% cover
S
na
a b c d a b c d a b c d a b c d
N Y N Y
Tavolara Portofino

Figure 3. Effects of *Eunicella cavolinii* on selected groups and species diversity indices. Mean (±s.e.) percent cover of selected groups recruited on panels and species diversity indices (Hill's diversity number *N1* and corresponding evenness component *N10*) for each experimental plot at Tavolara and Portofino, with (Y) and without (N) *Eunicella cavolinii* at 24 m depth (na = plot not available).

Despite the large local variability, the mean percent cover of several groups was significantly different between the two sites (Table 3). Erect red algae (including 44 taxa, Fig. 3), green algae (12 taxa) and crustose coralline algae (Fig. 3) were significantly more abundant at Portofino, while encrusting brown algae (2 taxa) were more abundant at Tavolara. Among invertebrates, foraminifera (4 taxa), bivalves (i.e. *Anomia ephippium* Linnaeus, 1758), encrusting bryozoans and erect bryozoans (10 and 9 taxa respectively) showed greater percent cover at Portofino. In contrast, erect sponges were more abundant at Tavolara. On panels with *E. cavolinii*, a significant reduction in percent cover of erect red algae (Fig. 3), brown algae (12 taxa, Fig. 3) and green algae was observed, whilst recruitment of crustose coralline algae and erect sponges was enhanced. The mean percent cover of calcareous tube worms (14 taxa, Fig. 3) was significantly reduced by the presence of the gorgonian, but only at one site, Tavolara. Similarly, at the same site, the serpulid polychaete *Hydroides stoichadon* Zibrowius, 1971 was also significantly less abundant in the presence of the gorgonian (pair-wise test: $P<0.05$). However, in contrast at Portofino *H. stoichadon* was occasionally observed but only on the panels which had transplanted gorgonian branches (Fig. 4).

Overall, species richness (*S*) was significantly higher at Portofino, but was not affected by the presence of the *E. cavolinii*

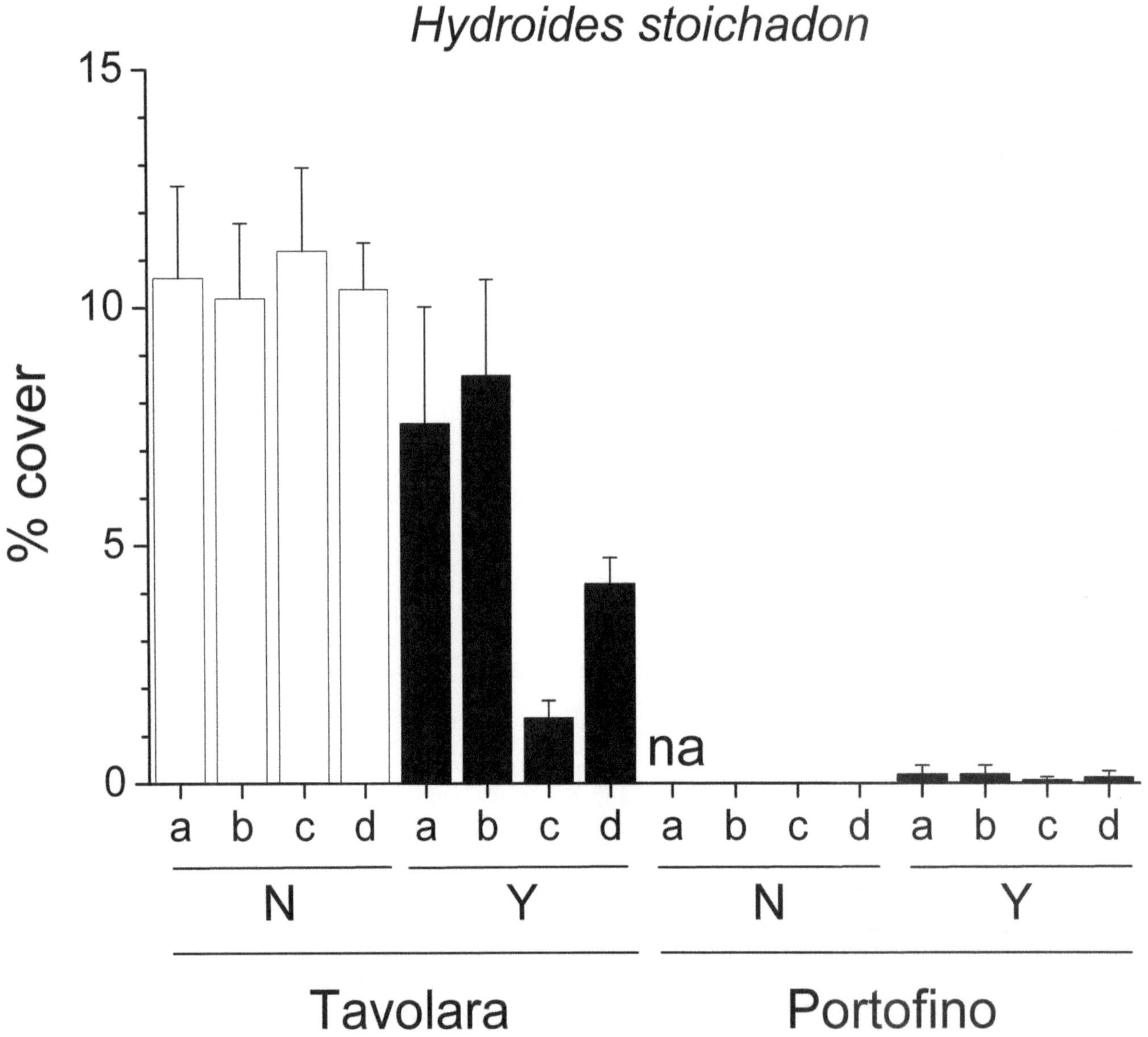

Figure 4. Effects of *Eunicella cavolinii* on the serpulid polychaete *Hydroides stoichadon*. Mean (±s.e.) percent cover of the serpulid polychaete *Hydroides stoichadon* on panels for each experimental plot at Tavolara and Portofino, with (Y) and without (N) *Eunicella cavolinii* at 24 m depth (na = plot not available).

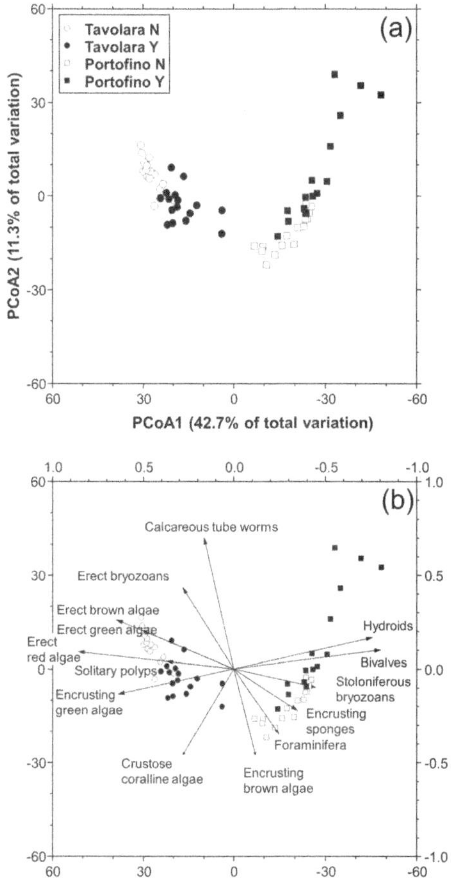

Figure 5. Effects of *Paramuricea clavata* on benthic assemblages structures. a) PCoA unconstrained ordination plot (i.e. metric MDS) based on Bray-Curtis similarities of square rout-transformed sessile assemblages percent cover on recruitment panels with (Y) and without (N) *Paramuricea clavata* forest after 4 months in Tavolara and Portofino; b) significant correlations of main ecological groups with PCoA Axes, represented by superimposed vectors (DistLM, marginal test $P<0.01$).

(Table 3). Furthermore, species heterogeneity ($N1$) and evenness ($N10$) were on average higher at Portofino, and were significantly higher in panels that did not have gorgonians present (Fig. 3).

Effects of *Paramuricea clavata*

The most abundant taxa on the panels at 40 m depth were encrusting brown algae (59.8%±3.8 s.e.) and small erect red algae (56.7%±5.2 s.e.), followed by encrusting green algae (28.6%±2.3 s.e.), foraminifera (26.5%±1.4 s.e.), crustose coralline algae (23.5%±1.8 s.e.) and hydroids (20.2%±3.3 s.e.). The recruited assemblages differed notably between plots and between sites and the assemblages were significantly affected by the combination of presence/absence of the gorgonian *P. clavata* and site (Table 2b). The PCoA plot (which displays 54.0% of the total variation) shows both the difference between Tavolara and Portofino, mainly along the PCoA1 axis, and a clear shift in the assemblage structures with relation to the presence of the gorgonian forest (Fig. 5a).

The correlation vectors, superimposed on the PCoA plot (Fig. 5b), show the relevance of the ecological and taxonomic groups in the differentiation of the recruited assemblages. At Tavolara, the groups of erect algae were more abundant on the panels without gorgonians, and crustose coralline algae tended to increase on forested panels. At Portofino, foraminifera and encrusting brown algae were more abundant on non- forested panels, while hydroids and bivalves were found in greater numbers on the forested panels.

Total percent cover of most groups showed large variability between plots, but also showed some site specific effects related to presence of gorgonians (Table 4). Erect red algae (including 38 taxa) were more abundant in Tavolara than in Portofino and were significantly lowered by the presence of gorgonian (Fig. 6). In contrast, foraminifera (3 taxa) were more abundant in Portofino but were also reduced on the forested panels. Encrusting green algae (6 taxa) were more abundant at Tavolara, while crustose coralline algae did not show any effects of site or gorgonians presence (Fig. 6). Overall, solitary polyps were found only at Tavolara, while hydroids (12 taxa), bivalves (i.e. *Anomia ephippium* Linnaeus, 1758) and stoloniferous bryozoans (2 taxa) were much more abundant at Portofino. Encrusting bryozoans (7 taxa) were facilitated by gorgonians at both sites. Even with different mean percent covers, erect brown (15 taxa, Fig. 6) and green algae (10 taxa), along with calcareous tube worms (18 taxa), were significantly more abundant on panels without gorgonians at Tavolara (pair-wise test: $P<0.01$), whilst these effects were not observed at Portofino. Conversely encrusting brown algae (5 taxa) were facilitated by the gorgonian at Tavolara (pair-wise test: $P<0.05$), and reduced at Portofino (pair-wise test: $P<0.05$). Encrusting sponges (3 taxa) were absent at Tavolara, yet were significantly more abundant on the panels without gorgonian at Portofino (pair-wise test: $P<0.05$). As observed in the experiment with *E. cavolinii*, at 24 m depth, even if less abundant at 40 m depth, the serpulid polychaete *Hydroides stoichadon* was significantly reduced by the gorgonian at Tavolara (pair-wise test: $P<0.05$) (Fig. 7).

Overall, species richness (S) was significantly higher at Tavolara, but it was not affected by the presence of the gorgonian (Table 4). Furthermore, species heterogeneity ($N1$) and evenness ($N10$) were significantly higher in absence of the gorgonian at Tavolara (pair-wise test: $P<0.01$) whilst no effects were found at Portofino (Fig. 6).

Discussion

Owing to the use of experimental panels, in 4 months of deployment it was possible to detect the effects of the two most common Mediterranean gorgonians on the early-stage recruitment of the epibenthic assemblages. The newly recruited assemblages differed amongst plots and between sites, both in *Eunicella cavolinii* (24 m depth) and in *Paramuricea clavata* (40 m depth) treatments. Analyses of the recruitment highlighted site-specific and species-specific effects of the gorgonian forests. Overall, the presence of gorgonians limited the growth of erect algae and, at Tavolara, the recruitment of serpulid polychaetes. Moreover, the presence of gorgonians, lowered species diversity and evenness of the assemblages on the experimental panels. These effects may be related to both biological and physical effects. Being passive filter feeders (they cannot create currents on their own [32]) and active predators (polyps capture preys by nematocysts) the gorgonians can intercept and feed on drifting larvae before settlement. Coma et al. [33] documented that *P. clavata* has a broad and heterogeneous diet with prey species ranging from 3.8 μm (nanoeukaryotes) to 700 μm (copepods). Protozoans and microalgae are the main items caught by the polyps. Capture of detrital and live particulate organic carbon is minimal in the summer, due to the scarcity of suspended food during this season [34]. Gorgonians may also compete with other filter-feeders for food and/or for space, producing allelochemicals [35]. On the other hand, gorgonians could affect settlement and

Table 4. Summary of PERMANOVA test on the effects of *Paramuricea clavata* forests (Fo) on the recruitment of ecological and functional groups and diversity indices at different sites (Si) and among plots (Pl) within Fo×Si.

	Forest (Fo)			Site (Si)			Fo×Si			Plot (Fo×Si)			Res
	(1 *df*)			(1 *df*)			(1 *df*)			(12 *df*)			(47 *df*)
	MS	*F*	*P*	MS	*F*	*P*	MS	*F*	*P*	MS	*F*	*P*	MS
Erect red algae	7'211.80	6.45	0.0311*	65'548.00	58.60	0.0002***	761.73	0.68	0.4255 ns	1'121.20	2.51	0.0107*	446.55
Encrusting red algae	245.90	0.78	0.3867 ns	138.87	0.44	0.5185 ns	1'093.20	3.49	0.0845 ns	314.50	3.47	0.0015**	90.68
Crustose coralline algae	33.10	0.08	0.7776 ns	521.02	1.22	0.2757 ns	903.81	2.11	0.1727 ns	428.53	3.46	0.0009***	123.75
Erect brown algae	4'513.00	14.80	0.0030**	7'803.00	25.59	0.0006***	3'803.60	12.48	0.0064**	305.48	2.01	0.0443*	152.10
Encrusting brown algae	2'006.70	1.55	0.2372 ns	6'170.20	4.75	0.0553 ns	17'609.00	13.56	0.0025**	1'302.10	4.11	0.0001***	316.47
Erect green algae	117.86	17.48	0.0020**	106.08	15.74	0.0027**	76.62	11.37	0.0078**	6.75	1.37	0.2077 ns	4.94
Encrusting green algae	2'070.00	4.48	0.0591 ns	6'343.10	13.74	0.0024**	33.45	0.07	0.7881 ns	462.91	3.03	0.0027**	152.79
Foraminifera	1'317.80	23.27	0.0010**	1'023.40	18.07	0.0026**	0.84	0.01	0.9076 ns	56.53	0.65	0.7833 ns	86.61
Encrusting sponges	0.11	12.33	0.0055**	0.44	49.32	0.0003***	0.11	12.33	0.0057**	0.01	0.18	0.9993 ns	0.05
Solitary polyps	0.01	0.62	0.4943 ns	0.12	8.28	0.0114*	0.01	0.62	0.4918 ns	0.01	1.28	0.2409 ns	0.01
Hydroids	1'305.10	1.27	0.2843 ns	20'172.00	19.61	0.0011**	757.75	0.74	0.4075 ns	1'032.10	5.39	0.0002***	191.32
Calcareous tube worms	21.72	0.30	0.6199 ns	107.14	1.47	0.2578 ns	851.29	11.71	0.0018**	72.90	3.31	0.0015**	22.04
Bivalves	49.73	1.52	0.2550 ns	1'256.40	38.30	0.0001***	72.87	2.22	0.1639 ns	32.90	4.48	0.0001***	7.35
Encrusting bryozoans	184.08	8.60	0.0121*	44.36	2.07	0.1715 ns	1.91	0.09	0.7676 ns	21.45	2.60	0.0094**	8.24
Stoloniferous bryozoans	88.99	1.74	0.2201 ns	459.42	9.00	0.0079**	116.52	2.28	0.1627 ns	51.12	1.95	0.0473*	26.26
Erect bryozoans	55.41	0.32	0.5747 ns	608.17	3.47	0.0898 ns	63.02	0.36	0.5509 ns	175.83	3.86	0.0009***	45.50
S	253.39	2.78	0.1171 ns	4'474.30	49.00	0.0003***	40.33	0.44	0.5259 ns	91.50	2.27	0.0198*	40.23
N1	524.67	27.65	0.0004***	1'487.80	78.41	0.0003***	233.77	12.32	0.0054**	19.01	1.92	0.0562 ns	9.88
N10	0.11	26.09	0.0005***	0.04	10.50	0.0112*	0.04	9.72	0.0062**	0.00	1.31	0.2454 ns	0.00

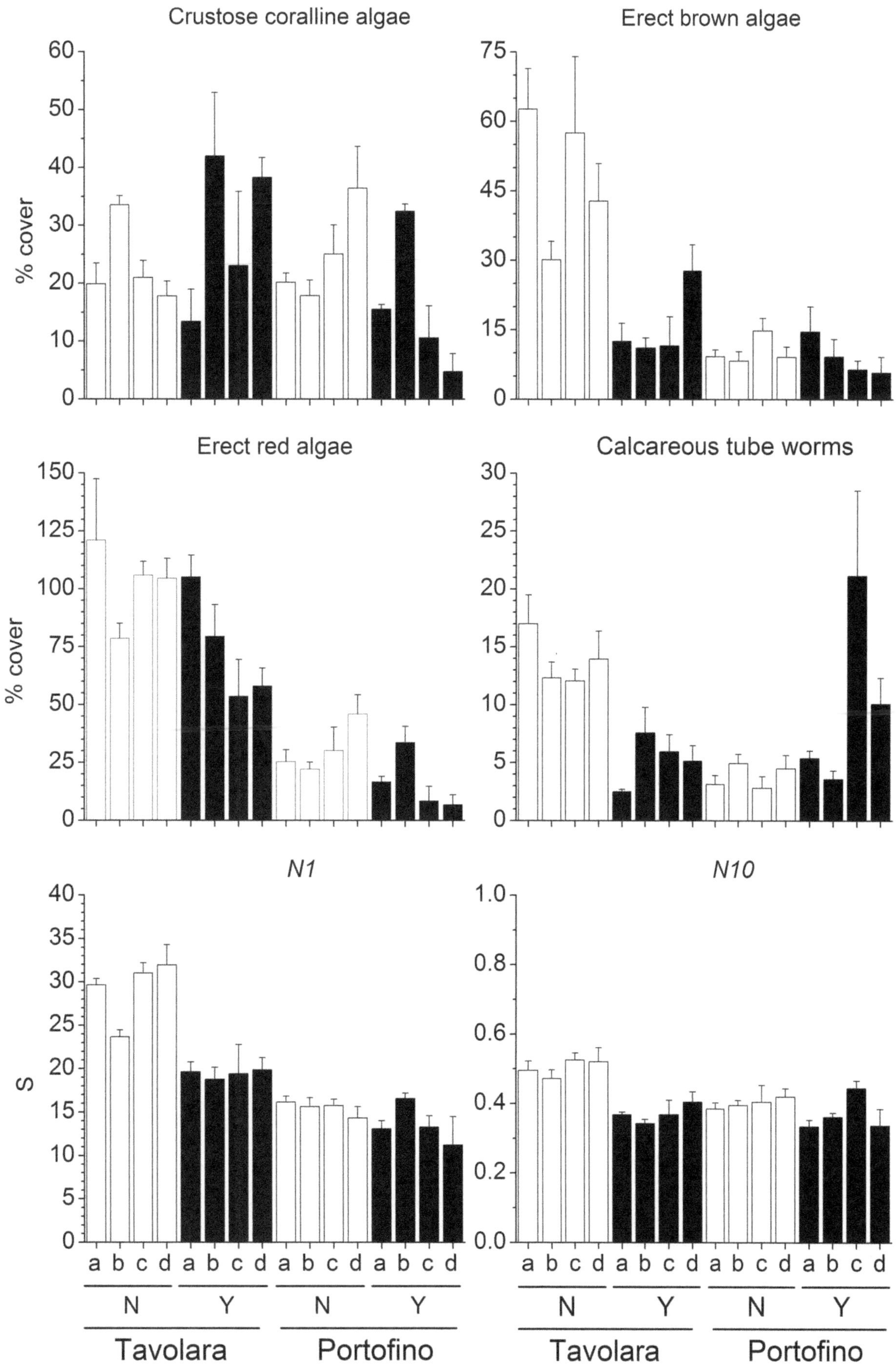
Crustose coralline algae
Erect brown algae
Erect red algae
Calcareous tube worms
N1
N10
% cover
S
a b c d a b c d a b c d a b c d
N
Y
Tavolara
Portofino

Figure 6. Effects of *Paramuricea clavata* on selected groups and species diversity indices. Mean (±s.e.) percent cover of selected groups recruited on panels and species diversity indices (Hill's diversity number *N1* and corresponding evenness component *N10*) for each experimental plot at Tavolara and Portofino, with (Y) and without (N) *Paramuricea clavata* at 40 m depth.

recruitment processes by modifying microscale hydrodynamic conditions, sediment deposition rates by limiting resuspension processes, and algal development by creating shading effects that may reduce the availability of light. In the mesophotic zone, gorgonian forests are considered to be of paramount importance to support surrounding species diversity. It was observed that they significantly increased the deposition of bioavailable substrates and enhanced benthic diversity, compared to areas without colonies [5]. As a result, organisms living in habitats characterised by the presence of these 'ecosystem engineers' (*sensu* [36]) may experience a sort of buffer zone where environmental modifications occur more slowly and within narrower ranges with respect to the surrounding 'unforested' environment.

In the present study, significant spatial variation of recruitment patterns was observed between Tavolara and Portofino. These differences could be related to variability in larval supply, chemical-physical parameters, sediment deposition and hydrodynamic conditions between sites. For example, in Portofino the value of nitrites and nitrates is higher compared to Tavolara. These differences are likely related to the fact that Tavolara is further from rivers and densely populated areas. Furthermore, at the specific study site (i.e. Punta del Faro) the Portofino Promontory receives a secondary branch of the main Ligurian

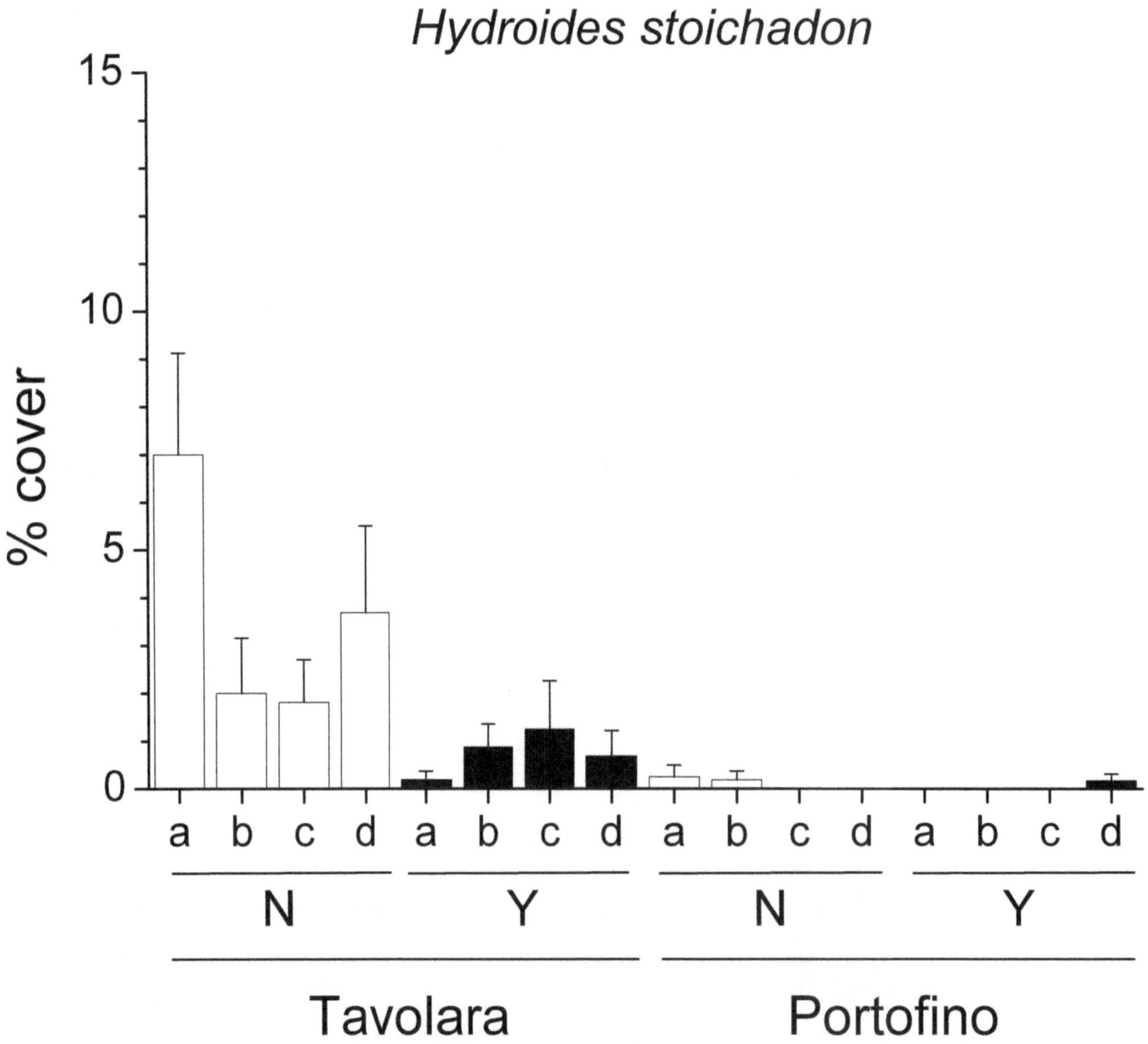

Figure 7. Effects of *Paramuricea clavata* on the serpulid polychaete *Hydroides stoichadon*. Mean (±s.e.) percent cover of the serpulid polychaete *Hydroides stoichadon* on panels for each experimental plot at Tavolara and Portofino, with (Y) and without (N) *Paramuricea clavata* at 40 m depth.

current [17], transporting urban waste-waters from the Tigullio Gulf, especially during the tourist summer season.

Longer experiments are needed to better understand the temporal dynamics of coralligenous assemblages; nevertheless, this short-time experiment has provided clear indications on the effect of gorgonians on the recruitment trajectories. In the study sites, the presence of gorgonians mainly reduces the initial growth of erect algae in favour of encrusting ones, which are the most important builders of the coralligenous frameworks [1], and some sessile invertebrates. Therefore, depending on the trophic conditions and turbidity of the water, rarefaction of gorgonians may favour a simplification of the coralligenous assemblages and a shift toward more autotrophic communities, dominated by fast growing filamentous algae instead of the slow accretion of crustose coralline algae.

Coralligenous habitats are threatened by several global stressors. The most evident, are climate change (massive mortalities related to temperature anomalies) and fishing activities (mechanical injuries and sediment re-suspension). Recently potential synergism between these stressors have been hypothesized [12], resulting in a fragmentation of the habitat that can open new space available for invasive species. Unfortunately, the dearth of information available on the interactions between octocorals and associated fauna prevents predicting any effect of gorgonian loss on the trophic structure. Owing to natural or anthropogenic causes, the loss of the 3D structure of the epibenthic layer in coralligenous assemblages [3], may affect both the established assemblages and the settlement and recruitment processes. The main question addressed by this study is whether the rapid loss of complexity due to extensive Mediterranean gorgonian mortalities [37], can affect the early-stage recruitment of coralligenous assemblages. Although the biological interaction between gorgonians and the other species deserve further studies, modifications of the edaphic conditions caused by the gorgonians forests influences larval settlement and recruitment processes of the benthic assemblages.

Coralligenous habitats are mainly built by deposition of calcareous thalli of crustose coralline algae, a process that has lasted for thousands of years [1]. These assemblages are generally characterised by high spatial heterogeneity and limited temporal variability, possibly related to the longevity and slow growth rates of the most abundant and structuring species [38], [39,40,41]. If the local disappearance of gorgonians causes a shift of the epibenthic assemblages from crustose coralline algae to filamentous algae dominated, complexity decreases. In absence of gorgonians, the resilience of coralligenous bioconstructions could be compromised due to the reduced recruitment of the main builder organisms and other key species. As another possible consequence, the establishment of non-indigenous species, such as the erect green algae *Caulerpa cylindracea* Sonder and the turf forming red algae *Womersleyella setacea* (Hollenberg) R.E.Norris, may be facilitated [42,43], irremediably altering these unique Mediterranean underwater landscapes. A regional scale assessment of coralligenous habitats taking in account not only the biological component but also the three-dimensional arrangement [44] is needed to develop adequate management plans.

Acknowledgments

We wish to thank the Tavolara-Punta Coda Cavallo and Portofino MPA Authorities for authorizations, Augusto Navone, Pier Augusto Panzalis and Egidio Trainito for logistic support at the Tavolara-Punta Coda Cavallo and Giorgio Fanciulli, Valentina Cappanera and Maurizio Pansini for the valuable field assistance at Portofino, Monica Previati, Marco Palma and Carla Huete Stauffer for diving assistance, Fabio Rindi, Federica Fava and Cristina Gioia Di Camillo for supporting in taxonomic analyses, David Cuming and Michael Sweet for English revision.

Author Contributions

Conceived and designed the experiments: MP CC. Performed the experiments: MP RAP VV DG CC. Analyzed the data: MP RAP VV DG CC. Wrote the paper: MP CC. Provided input into the manuscript: MA.

References

1. Ballesteros E (2006) Mediterranean coralligenous assemblages: a synthesis of present knowledge. Oceanogr Mar Biol, Annu Rev 44: 123–195.
2. Bertolino M, Cerrano C, Bavestrello G, Carella M, Pansini M, et al. (2013) Diversity of Porifera in the Mediterranean coralligenous accretions, with description of a new species. ZooKeys 336: 1–37.
3. Riedl R (1966) Biologie der Meereshöhlen. Hamburg: Paul Parey. 636 p.
4. Scinto A, Bertolino M, Calcinai B, Huete-Stauffer C, Previati M, et al. (2009) Role of a *Paramuricea clavata* forest in modifying the coralligenous assemblages. In: Pergent-Martini C, Brichet M, editors; 15–16 January 2009; Tabarka, Tunisa. UNEP-MAP-RAC/SPA. 136–140.
5. Cerrano C, Danovaro R, Gambi C, Pusceddu A, Riva A, et al. (2010) Gold coral (*Savalia savaglia*) and gorgonian forests enhance benthic biodiversity and ecosystem functioning in the mesophotic zone. Biodivers Conserv 19: 153–167.
6. Linares C, Coma R, Garrabou J, Díaz D, Zabala M (2008) Size distribution, density and disturbance in two Mediterranean gorgonians: *Paramuricea clavata* and *Eunicella singularis*. J Appl Ecol 45: 688–699.
7. Cerrano C, Cardini U, Bianchelli S, Corinaldesi C, Pusceddu A, et al. (2013) Red coral extinction risk enhanced by ocean acidification. Sci Rep 3.
8. Teixido N, Casas E, Cebrian E, Linares C, Garrabou J (2013) Impacts on coralligenous outcrop biodiversity of a dramatic coastal storm. PLoS ONE 8.
9. Bavestrello G, Cerrano C, Zanzi D, Cattaneovietti R (1997) Damage by fishing activities in the gorgonian coral *Paramuricea clavata* in the Ligurian Sea. Aquat Conserv 7: 253–262.
10. Coma R, Pola E, Ribes M, Zabala M (2004) Long-term assessment of temperate octocoral mortality patterns, protected vs. unprotected areas. Ecol Appl 14: 1466–1478.
11. Vezzulli L, Previati M, Pruzzo C, Marchese A, Bourne DG, et al. (2010) *Vibrio* infections triggering mass mortality events in a warming Mediterranean Sea. Environ Microbiol 12: 2007–2019.
12. Vezzulli L, Pezzati E, Huete-Stauffer C, Pruzzo C, Cerrano C (2013) 16SrDNA pyrosequencing of the Mediterranean gorgonian *Paramuricea clavata* reveals a link among alterations in bacterial holobiont members, anthropogenic influence and disease outbreaks. PLoS ONE 8: e67745.
13. Cerrano C, Bavestrello G, Bianchi CN, Cattaneo-Vietti R, Bava S, et al. (2000) A catastrophic mass-mortality episode of gorgonians and other organisms in the Ligurian Sea (Northwestern Mediterranean), summer 1999. Ecol Lett 3: 284–293.
14. Martin Y, Bonnefort JL, Chancerelle L (2002) Gorgonians mass mortality during the 1999 late summer in french Mediterranean coastal waters: the bacterial hypothesis. Water Res 36: 779–782.
15. Garrabou J, Coma R, Bensoussan N, Bally M, Chevaldonne P, et al. (2009) Mass mortality in Northwestern Mediterranean rocky benthic communities: effects of the 2003 heat wave. Glob Change Biol 15: 1090–1103.
16. Huete-Stauffer C, Vielmini I, Palma M, Navone A, Panzalis P, et al. (2011) *Paramuricea clavata* (Anthozoa, Octocorallia) loss in the Marine Protected Area of Tavolara (Sardinia, Italy) due to a mass mortality event. Mar Ecol Evol Persp 32: 107–116.
17. Cerrano C, Arillo A, Azzini F, Calcinai B, Castellano L, et al. (2005) Gorgonian population recovery after a mass mortality event. Aquat Conserv 15: 147–157.
18. Linares C, Coma R, Diaz D, Zabala M, Hereu B, et al. (2005) Immediate and delayed effects of a mass mortality event on gorgonian population dynamics and benthic community structure in the NW Mediterranean Sea. Mar Ecol Prog Ser 305: 127–137.
19. Coma R, Linares C, Ribes M, Diaz D, Garrabou J, et al. (2006) Consequences of a mass mortality in populations of *Eunicella singularis* (Cnidaria:Octocorallia) in Menorca (NW Mediterranean). Mar Ecol Prog Ser 327: 51–60.
20. Bourget E, Deguise J, Daigle G (1994) Scales of substratum heterogeity, structural complexity, and the early establishment of a marine epibenthic community. J Exp Mar Biol Ecol 181: 31–51.
21. Ben-Yosef DZ, Benayahu Y (1999) The gorgonian coral *Acabaria biserialis*: life history of a successful colonizer of artificial substrata. Mar Biol 135: 473–481.
22. Lapointe L, Bourget E (1999) Influence of substratum heterogeneity scales and complexity on a temperate epibenthic marine community. Mar Ecol Prog Ser 189: 159–170.

23. Pech D, Ardisson PL, Bourget E (2002) Settlement of a tropical marine epibenthic assemblage on artificial panels: influence of substratum heterogeneity and complexity scales. Estuar Coast Shelf S 55: 743–750.
24. Weinbauer MG, Velimirov B (1996) Relative habitat suitability and stability of the Mediterranean gorgonian coral *Eunicella cavolini* (Coelenterata: Octocorallia). Bull Mar Sci 58: 786–791.
25. Bianchi CN, Pronzato R, Cattaneo-Vietti R, Benedetti-Cecchi L, Morri C, et al. (2004) Hard bottoms. In: Gambi MC, Dappiano M, editors. Mediterranean marine benthos: a manual of methods for its sampling and study. Genova: Società Italiana di Biologia Marina. 185–216.
26. Magurran AE (2004) Measuring biological diversity: Blakwell Science Ltd. 256 p.
27. Gower JC (1966) Some distance properties of latent root and vector methods used in multivariate analysis. Biometrika 53: 325–338.
28. McArdle BH, Anderson MJ (2001) Fitting multivariate models to community data: a comment on distance-based redundancy analysis. Ecology 82: 290–297.
29. Anderson MJ, ter Braak CJF (2003) Permutation tests for multi-factorial analysis of variance. J Stat Comput Sim 73: 85–113.
30. Anderson MJ, Gorley RN, Clarke KR (2008) PERMANOVA+ for PRIMER: guide to software and statistical methods. Plymouth, UK: PRIMER-E Ltd. 214 p.
31. Coma R, Ribes M, Zabala M, Gili JM (1998) Growth in a modular colonial marine invertebrate. Estuar Coast Shelf Sci 47: 459–470.
32. Gili JM, Coma R (1998) Benthic suspension feeders: their paramount role in littoral marine food webs. Trends Ecol Evol 13: 316–321.
33. Coma R, Gili JM, Zabala M, Riera T (1994) Feeding and prey capture cycles in the aposymbiontic gorgonian *Paramuricea clavata*. Mar Ecol Prog Ser 115: 257–270.
34. Ribes M, Coma R, Gili JM (1999) Heterogeneous feeding in benthic suspension feeders: the natural diet and grazing rate of the temperate gorgonian *Paramuricea clavata* (Cnidaria: Octocorallia) over a year cycle. Mar Ecol Prog Ser 183: 125–137.
35. Rodriguez AD (1995) The natural products chemistry of West Indian gorgonian octocorals. Tetrahedron 51: 4571–4618.
36. Jones CG, Lawton JH, Shachak M (1994) Organisms as ecosystem engineers. Oikos 69: 373–386.
37. Cerrano C, Bavestrello G (2008) Medium-term effects of die-off of rocky benthos in the Ligurian Sea. What can we learn from gorgonians? Chem Ecol 24: 73–82.
38. Piazzi L, Balata D, Pertusati M, Cinelli F (2004) Spatial and temporal variability of Mediterranean macroalgal coralligenous assemblages in relation to habitat and substratum inclination. Bot Mar 47: 105–115.
39. Virgilio M, Airoldi L, Abbiati M (2006) Spatial and temporal variations of assemblages in a Mediterranean coralligenous reef and relationships with surface orientation. Coral Reefs 25: 265–272.
40. Ponti M, Fava F, Abbiati M (2011) Spatial-temporal variability of epibenthic assemblages on subtidal biogenic reefs in the northern Adriatic Sea. Mar Biol 158: 1447–1459.
41. Teixido N, Garrabou J, Harmelin J-G (2011) Low dynamics, high longevity and persistence of sessile structural species dwelling on Mediterranean coralligenous outcrops. PLoS ONE 6: e23744.
42. Piazzi L, Balata D, Cinelli F (2007) Invasions of alien macroalgae in Mediterranean coralligenous assemblages. Cryptogam Algol 28: 289–301.
43. Piazzi L, Balata D (2009) Invasion of alien macroalgae in different Mediterranean habitats. Biol Inva 11: 193–204.
44. Zapata-Ramírez PA, Scaradozzi D, Sorbi L, Palma M, Pantaleo U, et al. (2013) Innovative study methods for the Mediterranean coralligenous habitats. Adv Oceanogr Limnol 4: 102–119.

7

In Situ CO_2 Efflux from Leaf Litter Layer Showed Large Temporal Variation Induced by Rapid Wetting and Drying Cycle

Mioko Ataka[1]*, Yuji Kominami[2], Kenichi Yoshimura[2], Takafumi Miyama[2], Mayuko Jomura[3], Makoto Tani[1]

1 Laboratory of Forest Hydrology, Division of Environmental Science and Technology, Graduate School of Agriculture, Kyoto University, Kyoto, Japan, **2** Kansai Research Center, Forestry and Forest Products Research Institute (FFPRI), Kyoto, Japan, **3** College of Bioresource Sciences, Nihon University, Fujisawa, Kanagawa, Japan

Abstract

We performed continuous and manual in situ measurements of CO_2 efflux from the leaf litter layer (R_{LL}) and water content of the leaf litter layer (LWC) in conjunction with measurements of soil respiration (R_S) and soil water content (SWC) in a temperate forest; our objectives were to evaluate the response of R_{LL} to rainfall events and to assess temporal variation in its contribution to R_S. We measured R_{LL} in a treatment area from which all potential sources of CO_2 except for the leaf litter layer were removed. Capacitance sensors were used to measure LWC. R_{LL} increased immediately after wetting of the leaf litter layer; peak R_{LL} values were observed during or one day after rainfall events and were up to 8.6-fold larger than R_{LL} prior to rainfall. R_{LL} declined to pre-wetting levels within 2–4 day after rainfall events and corresponded to decreasing LWC, indicating that annual R_{LL} is strongly influenced by precipitation. Temporal variation in the observed contribution of R_{LL} to R_S varied from nearly zero to 51%. Continuous in situ measurements of LWC and CO_2 efflux from leaf litter only, combined with measurements of R_S, can provide robust data to clarify the response of R_{LL} to rainfall events and its contribution to total R_S.

Editor: Ben Bond-Lamberty, DOE Pacific Northwest National Laboratory, United States of America

Funding: Funding was provided by the Japan Society for the Promotion of Science (JSPS; grant number 25-2482 (http://www.jsps.go.jp/english/index.html)) Grant-in-Aid for Scientific Research (B) (20380182 (http://www.jsps.go.jp/english/index.html)). The funders had no role in study design, data collection and analysis, decision to publish, or preparation of the manuscript.

Competing Interests: The authors have declared that no competing interests exist.

* Email: teshimamioko@yahoo.co.jp

Introduction

Efflux of CO_2 from the soil surface (soil respiration; R_S), which is the sum of respiration by autotrophs and heterotrophs, is an important component of total CO_2 efflux from forest ecosystems [1–3]. The R_S: total ecosystem respirations varied from 58% to 76% in a mixed coniferous-deciduous forest [4], depending on interannual and seasonal changes in autotrophic and heterotrophic respiration; variability in R_S can affect the forest carbon balance on daily and seasonal time scales. To explain the cause of variability in R_S, many studies have attempted to separate differing sources of Rs and to examine factors controlling CO_2 efflux rate from each source [5–7]. Especially in forest ecosystems, heterotrophic respiration consists of CO_2 efflux from various sources (e.g., leaf and root litter, woody debris, soil organic matter) and their rates are controlled by their specific environmental condition such as water content (WC) and temperature [8], physical properties of the substrate (e.g., density and structure) [9,10], and chemical properties (e.g., labile and recalcitrant carbon) [11,12]. Moreover, CO_2 efflux from the various heterotrophic sources responds differently to these controlling factors, which illustrates the complexity of R_S. In recent decades, a variety of methods for separating components of heterotrophic respiration and for determining their contribution to total R_S have been developed [9,13].

Among heterotrophic sources of CO_2, the leaf litter layer (L-layer) is a significant reservoir of degradable carbon and a large potential source of CO_2 efflux from forest soils [14]. In temperate forests, the contribution of CO_2 efflux from the L-layer (leaf litter respiration; R_{LL}) to R_S is reported to range from 23% to 48% [13,15,16]. The L-layer is in direct contact with rainfall, solar radiation, and wind, and environmental conditions (e.g., WC and temperature) can change more dynamically in the L-layer than in lower soil layers. Rapid and transient temporal variation in WC of the L-layer has been observed, especially in warm climates [16,17]. Heterotrophic respiration responds rapidly to changes in moisture status [17,18]; therefore, rapid and transient wetting and drying cycles would produce large temporal variations in R_{LL}. This would significantly affect variation in R_S [17,19], suggesting that R_{LL} is an important controller of temporal (daily and seasonal) patterns in the carbon balance in warm regions [19,20].

Several methods for measuring R_{LL} and for calculating its contribution to R_S have been explored. Cisneros-Dozal et al. [21] used an isotope mass balance method and reported that the contribution of R_{LL} to R_S increased from 5% to 37% in response to water addition after transient drought. Deforest et al. [15]

determined that the annual contribution of R_{LL} to R_S was 48% ±12% by measuring R_S with and without the L-layer, and the ratio was consistent over a range of environmental conditions. However, there is little information about temporal variation in R_{LL} in relation to rainfall events because of the difficulty of continuous and direct measurement of R_{LL} in situ.

To continuously measure CO_2 efflux from the L-layer only, in parallel with measurement of R_S, we developed an approach for measuring R_{LL} using an automated chamber method in a treatment area from which all CO_2 sources except for the L-layer were removed. In parallel with R_{LL} and R_S measurements, we continuously measured water content of the L-layer (LWC) and soil water content (SWC). LWC was measured using a method developed by Ataka et al. [22], in which intact leaf litter was attached to surrounding capacitance sensors. Sensors were also placed on top of the L-layer and at the boundary between the L- and mineral layers. From these continuous in situ measurements, we investigated the response of R_{LL} to rainfall events by comparing R_{LL} with R_S, and examined temporal variation in the contribution of R_{LL} to R_S in a warm temperate forest in Japan.

Materials and Methods

Ethics statement

The study site (Yamashiro Experimental Forest) is maintained by the Forestry and Forest Products Research Institute. All necessary permits were obtained for the field study, and the study did not involve endangered or protected species.

Study site

Our observations of R_{LL} and R_S were conducted at the Yamashiro Experimental Forest in southern Kyoto Prefecture, Japan (34°47′N, 135°50′E). The study site is a 1.7-ha watershed characterized by an annual mean air temperature of 15.5°C (maximum, 34.8°C; minimum, −3.9°C) and annual precipitation of 1449 mm [2]. The rainy season generally occurs from early June to mid-July. Daily rates of evaporation from the forest floor are 0.4–0.8 mm day^{-1} for 1–2 days after precipitation, declining thereafter to 0.2–0.3 mm day^{-1} [23]. The soils are Regosols with sandy loam or loamy sand texture and contain fine gravel (53% by mass) composed of residual quartz crystals from granite parent material [24]. These are immature soils in which the thickness of the A horizon is 2–3 cm. Deciduous broad-leaved, evergreen broad-leaved, and coniferous tree species account for 66%, 28%, and 6% of the living tree biomass, respectively [25]. The forest is dominated by *Quercus serrata* Thunb., which accounts for approximately 33% of the biomass. The L-layer (approximately 3–4 cm thick) consists mainly of fresh *Q. serrata* litter. There is no substantial organic horizon below the L-layer.

Automated chamber method for measuring leaf litter respiration and soil respiration

We measured R_{LL} and R_S using an automated dynamic chamber system with an infrared gas analyzer (IRGA, GMP343; Vaisala Group, Vantaa, Finland) (Fig. 1A). The system consisted of two automated circular chambers for R_{LL} and R_S measurement, four solenoid valves, a pump, mass flow meter, and IRGA. The chambers (surface area 320 cm^2) were made from PVC collars with clear acrylic lids that can be opened and closed automatically using an air cylinder. Air was supplied to the cylinder from a compressor. To ensure a seal between the chamber and the closed lid, a soft rubber gasket was attached to the top edge of the chamber. Opening and closing of the chamber lid and solenoid valves of each chamber were regulated synchronously by a control unit (ZEN, OMRON, Kyoto, Japan).

The duration of measurement of CO_2 concentration inside each chamber was 6 min and was performed twice per hour. The CO_2 concentration in each chamber was recorded at 1-s intervals using a data logger (GL220, Graphtec, Kanagawa, Japan). We calculated R_{LL} and R_S from the increase in CO_2 concentration (ΔC_{CO2}) using linear regression. Data from the first 2 min were discarded to avoid effects of closing the chamber. R_{LL} and R_S were calculated using the following equation:

$$R = \frac{\Delta C_{CO_2}}{10^6} \times \frac{V}{V_{air}} \frac{273.2}{273.2+T} \times M_{CO_2} \times \frac{1}{A}, \qquad (1)$$

where R is respiration (mg CO_2 m^{-2} s^{-1}), ΔC_{CO2} is the change in CO_2 concentration per unit time (CO_2 ppm s^{-1}), V is the volume of the system (L), V_{air} is the standard gas volume (22.41 L mol^{-1}), T is temperature inside the chamber (°C), M_{CO2} is the molecular weight of CO_2 (44.01 g mol^{-1}), and A is the soil surface area covered by the chamber (m^2).

To continuously measure CO_2 efflux from the L-layer only, we developed an approach for measuring R_{LL} by using an automated chamber method in a treatment area in which all potential CO_2 sources (e.g., organic soil and fine roots) except for the L-layer were replaced with combusted granite soil (Fig. 1B). To prepare the treatment area (1 m^2), we removed surface soil (approximately 5 cm). An acrylic board was placed on the bottom and sides of the treatment area to prevent penetration of roots; a drain tube was located at the bottom of the board to prevent the treatment area from flooding with rainwater. The treatment area was then filled with granite soil combusted in a muffle furnace (500°C for 1 day). For R_{LL} measurement, we placed a PVC collar (320-cm^2 surface area) and acrylic board below the collar. The board was set at a slight incline to drain rainwater from the collar. We added 15 g of newly fallen leaf litter, which represents the average litterfall mass per unit ground surface area at this site, to the collar. We added the leaf litter to each chamber on January 2012. To acquire data on the temporal variation in R_{LL} of fresh leaf litter, we replaced the litter with newly fallen leaf litter in January 2013. The collar for measurement of R_S was placed near the treatment area for R_{LL} measurement and the L-layer inside the collar was removed and leaf litter was supplied similarly as for measurement of R_{LL}. To prevent incorporation of newly fallen litter, we placed a mesh sheet (1×1 mm mesh) on the L-layer inside the chamber, and fallen litter was removed weekly. CO_2 efflux from combusted granite soil was measured 6 months from the start of the R_{LL} measurements. The mean CO_2 flux rate (± standard deviation) was 0.00063±0.00068 mg CO_2 m^{-2} s^{-1} (n = 16) when SWC ranged from 0.05 to 0.3 m^3 m^{-3} at temperatures of 24°C. Thus, we assumed that CO_2 efflux from the combusted granite soil was negligible throughout the measurement period.

For continuous in situ measurement of LWC, we used capacitance sensors as described by Ataka et al. [22]. The measurements were performed on the top surface of the L-layer and at the boundary between the L-layer and mineral soil (Fig. 1B), to capture the large vertical distribution of WC within the L-layer. We estimated average LWC from the output voltage (V) of the two sensors using the conversion equation LWC = 12.73 V–3.42 presented by Ataka et al. [22]. LWC at the forest floor shows spatial variability associated with tree canopy conditions. Thus, to reflect the LWC of the L-layer by direct measurement, two capacitance sensors were placed on the L-layer inside the chamber. To check the validity of continuous LWC monitoring, we compared the sensor values with LWC measured

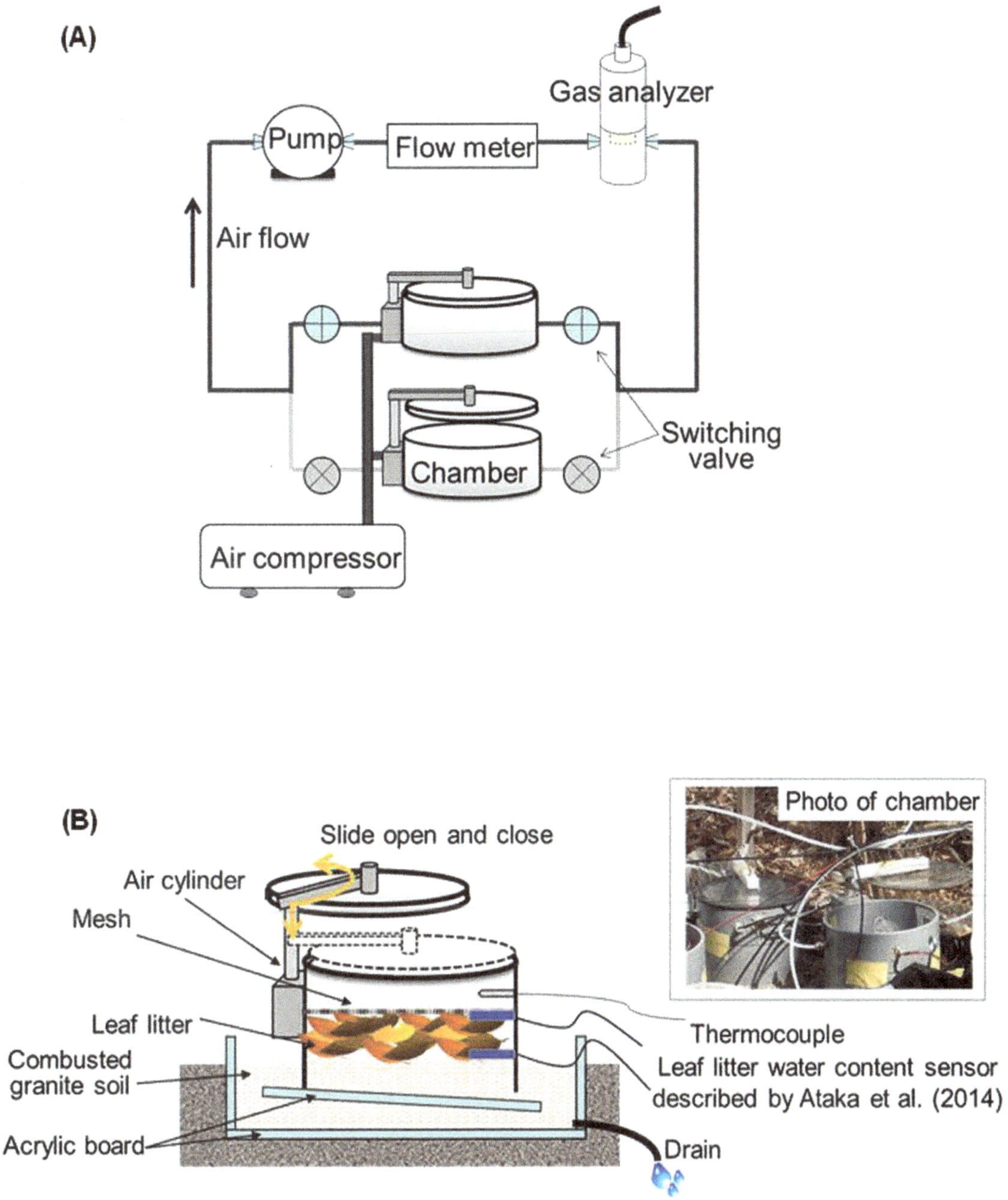

Figure 1. Schematic of the automated chamber system and the experimental design for measurement of CO_2 efflux from the leaf litter layer. A. Schematic of the automated dynamic-closed chamber system for measuring leaf litter respiration and soil respiration. **B.** The experimental design for continuous measurement of CO_2 efflux from the leaf litter layer only using automated chamber system.

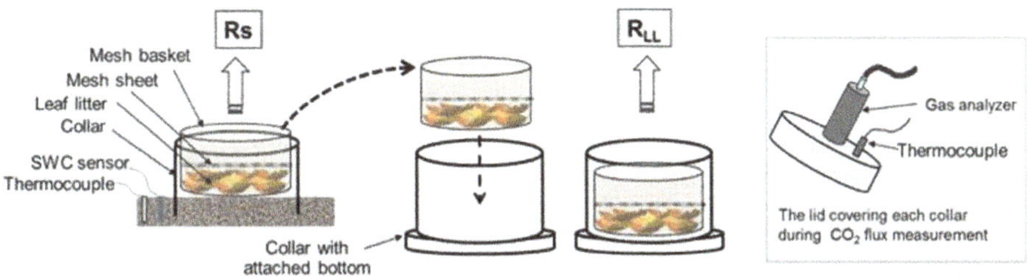

Figure 2. Schematic of the manual chamber system and the experimental design for measurement of CO_2 efflux from the leaf litter layer (R_{LL}) and soil (R_S).

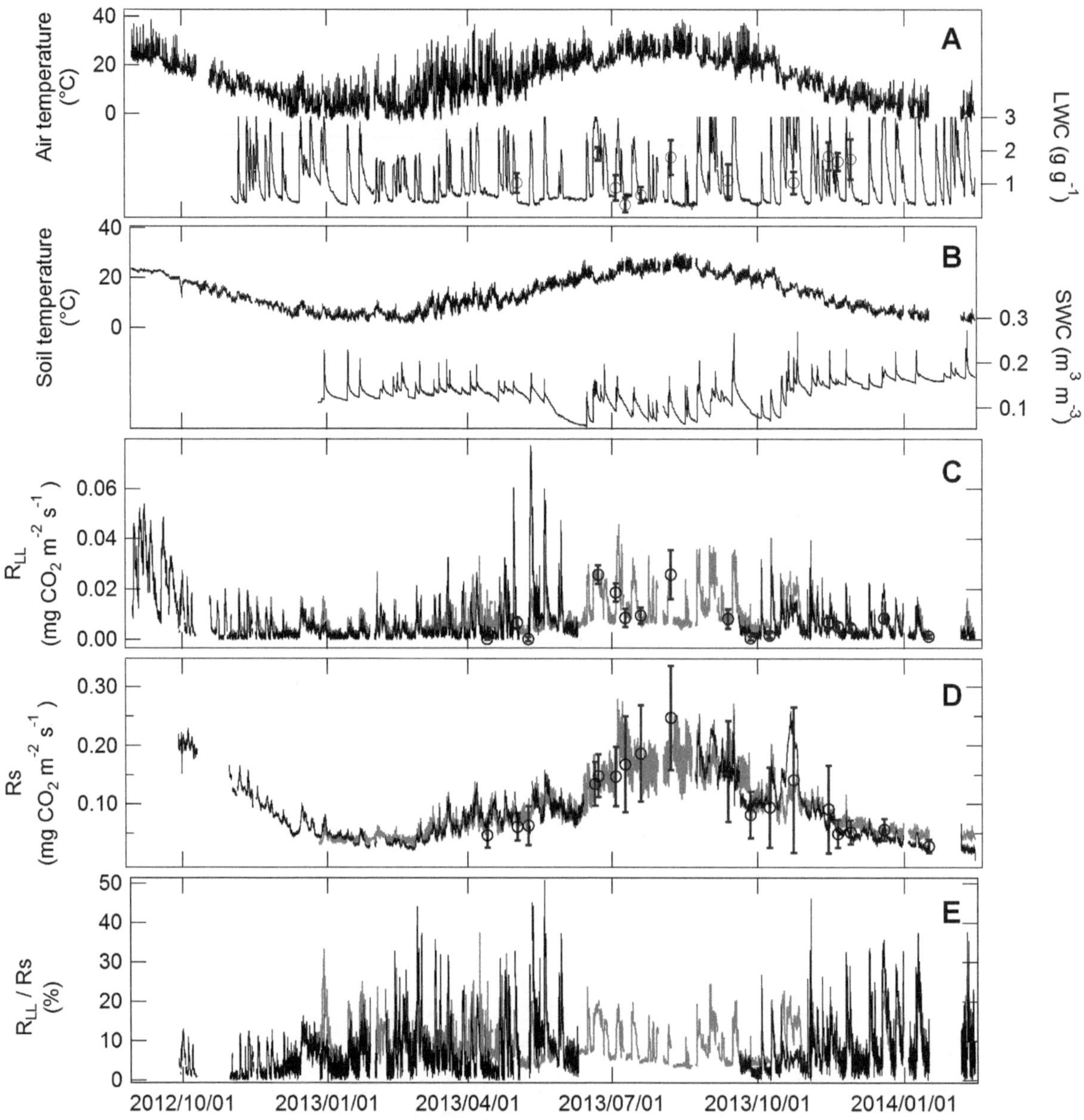

Figure 3. Seasonal variation in environmental factors, CO_2 efflux from the leaf litter layer (R_{LL}), and soil respiration (R_S). Data were measured every 30 min between September 2012 and January 2014. **A.** Bold and fine lines show air temperature and water content of the leaf litter layer (LWC), respectively. **B.** Bold and fine lines show soil temperature and soil water content (SWC), respectively. **C.** Black and grey lines show observed and estimated R_{LL}, respectively. **D.** Black and grey lines show observed and estimated R_S, respectively. **E.** Black and grey lines show the ratio of observed and estimated R_{LL} to R_S, respectively. Circles and bars show mean values and standard deviation of manual measurements. Estimated R_{LL} and R_S were calculated from regression equations using temperature (T) and water content (WC): $R_{LL} = 0.29e^{0.059T}[WC/(95.04+WC)]$ and $R_S = 0.031e^{0.10T}[WC/(0.032+WC)]$.

Table 1. Q_{10} of leaf litter respiration (R_{LL}) and soil respiration (R_s) for different water contents of the leaf litter layer (LWC) and soil (SWC).

	R_{LL}			R_s		
	LWC≤1	1<LWC≤2	2<LWC	SWC≤0.1	0.1<SWC≤0.15	0.15<SWC
Q_{10}	1.54	1.88	2.07	1.97	2.12	2.73
a	0.0019	0.0044	0.0064	0.027	0.032	0.025
b	0.043	0.063	0.073	0.068	0.075	0.10

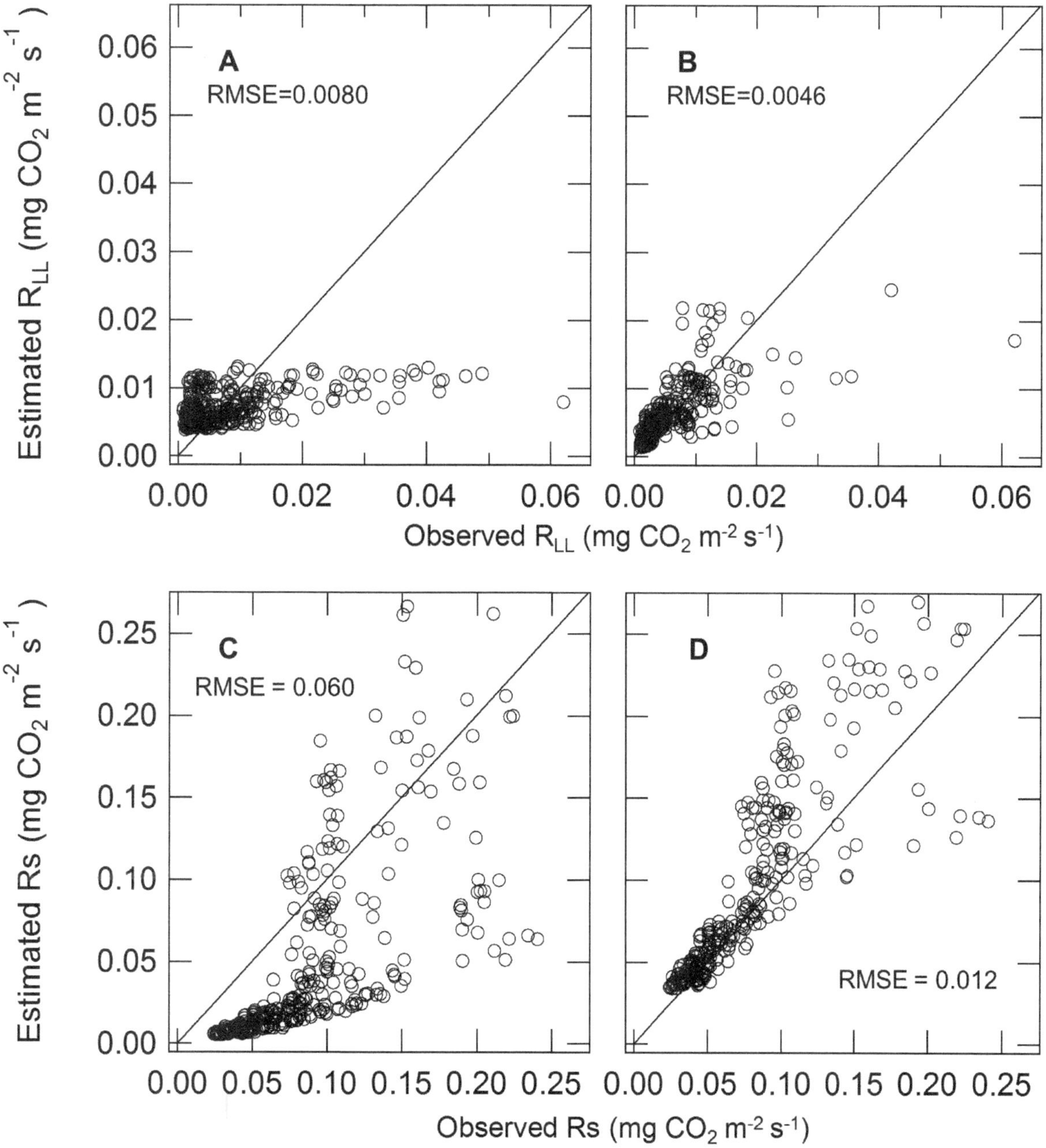

Figure 4. Relationship between observed and estimated CO_2 efflux rate from leaf litter respiration (R_{LL}) and soil respiration (R_S). R_{LL} (A, B) and *Rs* (C, D) show daily mean values. Estimated respiration rates were calculated using a function of temperature (A, C) from Eq. (5,6) and a function of temperature and water content (B, D) from Eq. (7,8) in the Results. Lines represent the 1:1 ratio. RMSE: root mean square error.

manually as described in the following section. In parallel with LWC measurement, soil temperature (copper-constantan thermocouple) and soil volumetric water content (ECH_2O EC-5 sensors; Decagon Devices, Pullman, WA, USA) were measured at 5-cm depth near each chamber. The output voltage of all environmental data was recorded every 1 min with a data logger (Datamark LS-3000 PtV; Hakusan, Japan) and average values were computed every 30 min. The environmental data, R_{LL}, and R_S were measured continuously between September 2012 and January 2014. Malfunction of IRGA resulted in a lack of data for R_{LL} and R_S for 31% of the measurements.

Manual chamber method for measuring leaf litter respiration and soil respiration

To determine the validity of R_{LL} and R_S measured using the automated chamber method, respiration was measured using the manual chamber method. We assumed that manual chamber method allow to measure under conditions that were closer to natural than the automated chamber method. We measured R_{LL} and R_S manually using a static chamber system at midday on 18 days between April 2013 and January 2014. Twelve PVC collars (320 cm^2 surface area) were placed in a 2×4 m area in January 2013. The edges of the collars were inserted approximately 1.5 cm into the soil. To measure R_{LL}, mesh baskets (1×1 mm mesh, the

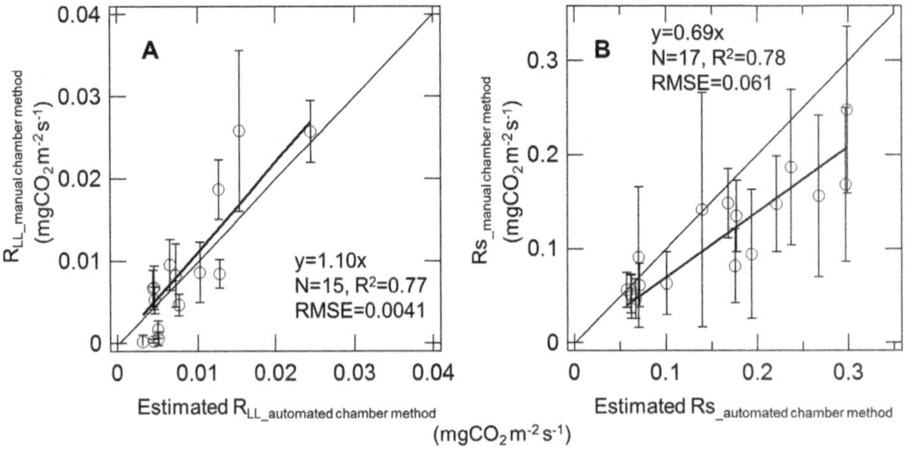

Figure 5. Relationship between respirations measured using a manual chamber method and estimated from automated chamber data. Respiration rate measured with the manual chamber method ($R_{\_manual\ chamber\ method}$) show mean value obtained from measurement of 12 collars. Bars show standard deviation. Respiration estimated from automated chamber data (estimated $R_{\_automated\ chamber\ method}$) shows daily mean respiration. The estimated R was calculated using a function based on temperature and water content (Eq. 8, 9).

same diameter as the PVC collars; 20 cm) were set into each collar and 15 g (dry weight) of newly fallen leaf litter was placed on the L-layer inside each basket (Fig. 2). To prevent supply of newly fallen litter, we placed a mesh sheet (1 × 1 mm mesh) on the L-layer inside the chamber, and fallen litter was removed weekly.

For measurement of R_S, the collars were completely covered with lids to which an IRGA and copper-constantan thermocouple were attached. Soil temperature and SWC (5 cm depth) were measured close to the collars when R_S was measured. After completing the measurements of R_S, the mesh baskets were carefully removed from the collars and placed in PVC chambers (20 cm diameter, 7 cm high; Fig. 2). We measured R_{LL} using the same methods as used for R_S measurement. The temperature and CO_2 concentrations in the chamber were recorded at 1-s intervals using a data logger (GL220). Linearity of the CO_2 flux was checked on the data logger monitor at each measurement. The measurement period for each chamber was 10 min and CO_2 data for the middle 5-min intervals were used to determine R_{LL} according to Eq. (1), excluding data from the first 3 min.

For measurement of LWC in the mesh baskets, four or five leaves were removed from each basket and immediately placed in sealed plastic bags. Fresh weight of the leaf litter was measured in the laboratory within 24 h of sampling. Leaf litter samples were oven dried at 65°C for 48 h, and water content (WC; g g^{-1}) was calculated using Eq. 2 as follows:

$$WC = \frac{(FW - DW)}{DW}, \tag{2}$$

where FW is the fresh mass of the sample (g), and DW is the dry mass of the sample (g). Samples were returned to each mesh basket within 1 week after sampling.

Leaf litter respiration and soil respiration rates as a function of environmental factors

Respiration models are fundamentally described by nonlinear functions. We used the following function to investigate the response of respiration to temperature:

$$R = a\ \exp(bT), \tag{3}$$

where T is temperature (leaf litter temperature for R_{LL} measurement or soil temperature for R_S measurement) and a and b are constants. Leaf litter temperature was assumed to be same as air temprature. b is related to the Q_{10} parameter ($Q_{10} = e^{10b}$). To determine the effects of temperature and water content on R_{LL} and R_S, we used a function that was previously applied to estimate soil respiration by Subke and Schlesinger [26]:

$$R = a\ \exp(bT)\left(\frac{WC}{c + WC}\right), \tag{4}$$

where a, b, and c are constants. LWC or SWC was used as WC in this equation. These nonlinear regressions were performed using a modified Levenberg–Marquardt method with Igor Pro 6.0 software (WaveMetrics, Lake Oswego, OR, USA). The estimated respiration values presented in this manuscript were calculated using Eq. 4.

Short-term changes in R_{LL} and LWC on wetting and drying cycle

To evaluate short-term changes in R_{LL} and LWC after rainfall events, we chose eight typical periods that included one wetting and drying cycle and had consecutive no rainfall days for at least 3 days. We used daily mean R_{LL} and LWC before the day on which precipitation occurred as the pre-wetting condition, and these values after precipitation as the post-wetting condition. Daily mean R_{LL} was calculated from R_{LL} values observed using the automated chamber method.

Effect of wetting and drying cycle of the L-layer on R_{LL} and Rs on the annual time scale

To investigate the effects of wetting and drying of the L-layer on R_{LL} on the annual time scale, we separated the estimated daily mean R_{LL} in 2013 into 'Dry' and 'Wet' periods based on daily mean LWC as a threshold value. The threshold LWC value that separated 'Dry' and 'Wet' periods for R_{LL} was estimated by the abovementioned short-term analyses. Daily mean R_{LL} was calculated from the estimated R_{LL} values because there were gaps in the continuous R_{LL} data observed using the automated

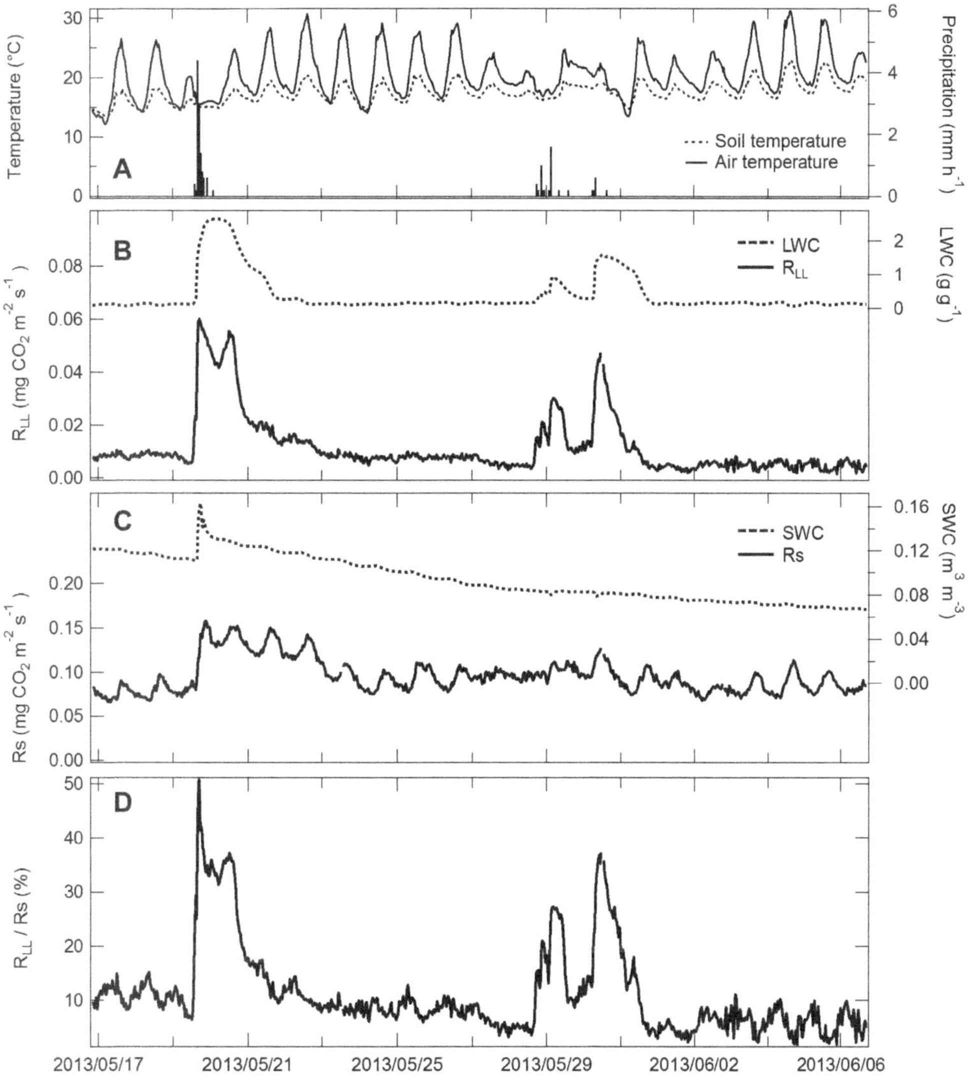

Figure 6. Temporal variation in environmental factors, CO_2 efflux from the leaf litter layer (R_{LL}), soil respiration (R_S), and the ratio of R_{LL} to R_S. Data was measured at one collar every 30 min between May 17 and June 6, 2013. **A.** Soil and air temperature. Spikes on the x-axis indicate precipitation events (mm h^{-1}). **B.** R_{LL} and water content of the leaf litter layer (LWC). **C.** R_S and soil water content (SWC). **D.** The ratio of R_{LL} to R_S (%).

chambers. We estimated the contribution of R_{LL} accumulated during the wet and dry period to total R_S.

Results

Seasonal variation in R_{LL} and R_S

The magnitude of the peak in the observed R_{LL} pulse was higher in summer than in winter (Fig. 3C). R_{LL} values were low when LWC was low (Fig. 3A, C). R_S changed substantially according to temperature (Fig. 3B, D), with higher values in summer than in winter. The relationships between respiration and temperature were described by the following functions:

$$R_{LL}(mg\ CO_2\ m^{-2}\ s^{-1}) = 0.0038\exp(0.065\times T_{LL}), \quad (5)$$

$$Rs\,(mg\ CO_2\ m^{-2}\ s^{-1}) = 0.0031\exp(0.19\times Ts), \quad (6)$$

where T_{LL} is leaf litter temperature and Ts is soil temperature (°C).

To evaluate effect of WC on the temperature sensitivity of respiration, the measured respiration data was separated into three groups based on WC (Table 1). More than 14% of total respiration data was included in each WC group. R_{LL} showed low values when WC values were low in spite of high temperature. Consequently, calculated Q_{10} values for not only R_{LL} but also R_S decreased with decreasing WC. The relationships between respiration and temperature and WC were described by the following functions:

$$R_{LL}\,(mg\ CO_2\ m^{-2}\ s^{-1}) = 0.29\exp(0.059\times T_{LL})\left(\frac{LWC}{95.04+LWC}\right), \quad (7)$$

$$Rs\,(mg\ CO_2\ m^{-2}\ s^{-1}) = 0.031\exp(0.10\times Ts)\left(\frac{SWC}{0.032+SWC}\right), \quad (8)$$

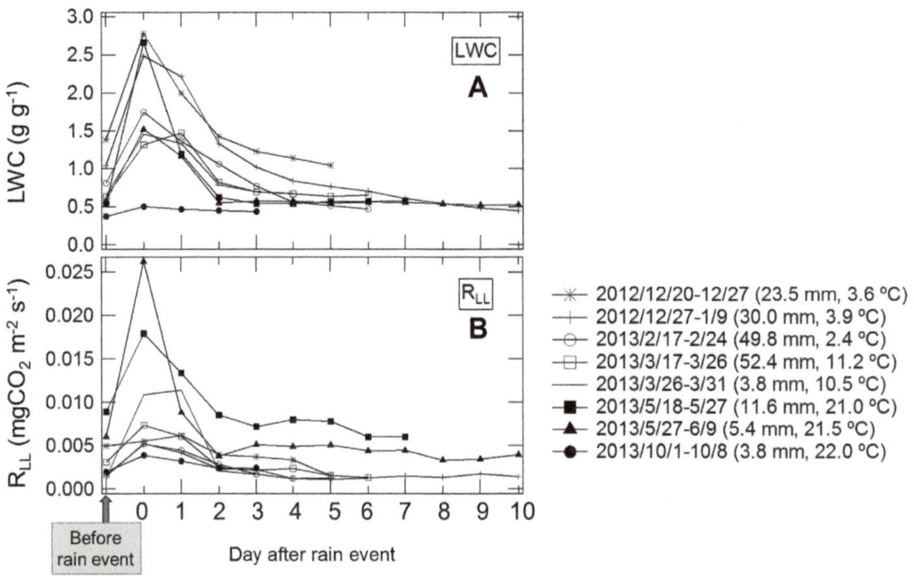

Figure 7. Temporal variation in water content of the leaf litter layer (LWC) and CO_2 efflux from the leaf litter layer (R_{LL}) after rainfall events. LWC (A) and R_{LL} (B) show the daily mean values. The rainfall intensity of each precipitation event was 23.5 mm in 2 days (2012/12/20–12/27, mean air temperature; 3.6°C); 30.0 mm in 3 days (2012/12/27–1/9, 3.9°C); 49.8 mm in 2 days (2013/2/17–2/24, 2.4°C); 52.4 mm in 3 days (2013/3/17–3/26, 11.2°C); 3.8 mm in 2 days (2013/3/26–3/31, 10.5°C); 11.6 mm in 2 days (2013/5/18–5/27, 21.0°C); 5.4 mm in 3 days (2013/5/27–6/9, 21.5°C); and 3.8 mm in 4 days (2013/10/1–10/8, 22.0°C).

where LWC ($g\ g^{-1}$) and SWC ($m^3\ m^{-3}$) are water content of leaf litter and soil, respectively. The RMSE between observed and estimated daily mean respiration based on temperature (R_{LL}, 0.0080 mg $CO_2\ m^{-2}\ s^{-1}$; R_S, 0.060 mg $CO_2\ m^{-2}\ s^{-1}$) was larger than that based on temperature and WC (R_{LL}, 0.0046 mg $CO_2\ m^{-2}\ s^{-1}$; R_S, 0.012 mg $CO_2\ m^{-2}\ s^{-1}$) (Fig. 4). Estimated respiration was calculated using the equation based on temperature and WC because of the lower RMSE. Throughout the measurement period, the contribution of observed R_{LL} to variation in R_S changed from nearly zero to 51% following a rainfall event (Fig. 3E).

To consider the validity of R_{LL} and R_S estimated from continuous measurement, we compared these values with respiration rates measured using the manual chamber method (Fig. 5). Estimated respiration was very similar to that observed using manual measurements. The RMSE between estimated and observed respiration were 0.0041 and 0.061 mg $CO_2\ m^{-2}\ s^{-1}$ for R_{LL} and R_S, respectively.

Temporal changes in R_{LL} and R_S on the short-term scale

To show clear temporal variation in R_{LL} and R_S, the period between May 17 and June 6, 2013 (Fig. 6) was chosen because this

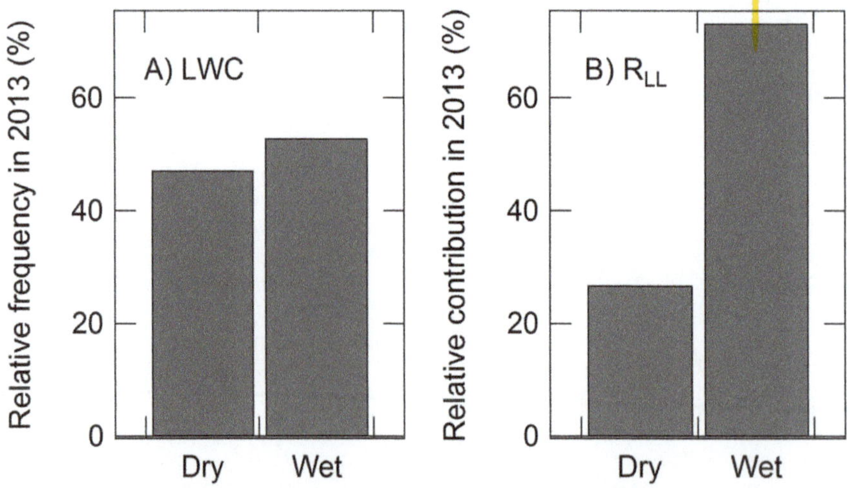

Figure 8. Histograms of the relative frequency of "Dry" and "Wet" periods in relation to water content of the leaf litter layer (LWC), and the relative contribution of estimated leaf litter respiration (R_{LL}) in 2013. The daily mean LWC (A) and R_{LL} (B) were used to present histograms. Estimated respiration rates were calculated using a function based on temperature (T) and water content (WC). $R_{LL} = 0.29e^{0.059T}[WC/(95.04+WC)]$. The daily mean LWC and R_{LL} were defined as Dry or Wet based on LWC. Days in which daily mean LWC <0.75 $g\ g^{-1}$ were defined as Dry periods, while days in which daily mean LWC ≥0.75 $g\ g^{-1}$ were defined as Wet periods.

period included two characteristic rainfall events. The rainfall intensity was 11.6 mm over 13 h during the first event and 5.4 mm over 46 h during the second event. LWC and SWC increased from 0.11 to 2.64 g g^{-1} and from 0.11 to 0.16 m^3 m^{-3}, respectively, following the first rainfall event (Fig. 6B, C). LWC increased from 0.16 to 1.58 g g^{-1} but SWC did not increase after the second rainfall event.

Temporal variation in R_{LL} measured using the automated chamber system changed according to wetting and drying of the L-layer (Fig. 6B), reaching a maximum of 0.060 and 0.047 mg CO_2 m^{-2} s^{-1} during first and second rainfall events, respectively. R_S increased following the increase in SWC and subsequently decreased gradually with diurnal variation according to temperature (Fig. 6C). Between May 17 and June 6, 2013, the contribution of R_{LL} to R_S increased from 6.5% to 51%, with a peak value of 51% during the first rainfall event and 37% during the second rainfall event (Fig. 6D).

Both R_{LL} and LWC reached a peak during or one day after rainfall events (Fig. 7). The peak of R_{LL} and LWC varied from 0.0020 to 0.026 mg CO_2 m^{-2} s^{-1} and from 0.50 to 2.66 g g^{-1}, respectively. Peak value of each rainfall event highly depended on air temperature. High peaks of R_{LL} were observed in the warm season (0.017 mg CO_2 m^{-2} s^{-1}; 2013/5/18–5/27, 0.026 mg CO_2 m^{-2} s^{-1}; 2013/5/27–6/9 in Fig. 7). Also, the peak value was related to LWC: low peak of R_{LL} was observed when LWC was low (0.004 mg CO_2 m^{-2} s^{-1}; 2013/10/1–10/8 in Fig. 7). The relationship between LWC and amout of precipitation was not clear. In the cold season, peak values of R_{LL} were relatively low (e.g., 0.005 mg CO_2 m^{-2} s^{-1}; 2013/2/17–2/24, 0.006 mg CO_2 m^{-2} s^{-1}; 2012/12/20–12/27 in Fig. 7) even when the L-layer was wet enough (LWC more than 1.5 g g^{-1}). The peak values of R_{LL} were 1.2- to 8.6-fold higher than the R_{LL} values before rainfall events, and R_{LL} fell to pre-wetting levels within 2–4 days after rainfall events and peak LWC values were 1.3- to five-fold higher than LWC before rainfall, and LWC also dropped to pre-wetting levels within 2–4 days after rainfall events.We defined R_{LL} from the period just after rainfall events through 2–4 days later as the "R_{LL} pulse".

Effects of wetting and drying of the L-layer on R_{LL} and R_S on the annual time scale

Estimated daily mean R_{LL} in 2013 was separated into 'Dry' and 'Wet' periods based on daily mean LWC. Days for which mean LWC was <0.75 g g^{-1} were categorized as Dry, while days for which mean LWC ≥0.75 g g^{-1} were categorized as Wet. The threshold value (0.75 g g^{-1}) was obtained from mean LWC 3 days after a rainfall event (Fig. 7A). The relative frequency of Dry and Wet periods in 2013 were 47.2% and 52.8%, respectively, while the relative contributions of daily mean R_{LL} during the Dry and Wet periods in 2013 were 26.9% and 73.2%, respectively (Fig. 8). Annual R_{LL} and R_S in 2013 were estimated to be 0.69 and 7.94 t C ha^{-1} y^{-1}, respectively. The RMSE between continuous respiration measured and estimated based on temperature and WC was 0.011 and 0.029 t C ha^{-1} y^{-1}, respectively.

The contribution of annual R_{LL} to R_S was 8.6%. The relative frequency of LWC was similar during Dry and Wet periods, while the contribution of R_{LL} during the Wet period was approximately three-fold higher than that during the Dry period (Fig. 8).

Discussion

As seen in Fig. 6, R_{LL} immediately increased with wetting of the L-layer and decreased to pre-wetting levels within 2–4 days after rainfall events, which was consistent with observations made in previous studies [17,19]. R_{LL} showed no diurnal variation despite a diurnal temperate range >10°C. Consequently, the Q_{10} of R_{LL} increased with increasing LWC (Table. 1). The variation in Q_{10} would be directly related to water stress experienced by microorganism. This indicated that LWC can reach to adequate low value, suspected as water stress for microorganism, within several days after rainfall. On the one hand, R_S increased during rainfall and subsequently decreased, showing diurnal variation. The Q_{10} of R_S also increased with increasing SWC. Dannoura et al. [27] reported that root respiration showed little change with variation in SWC compared with changes in R_S. Therefore, the increased Q_{10} of R_S with increasing SWC might be highly affected by not only R_{LL} but also by respiration from other heterotrophic sources.

Although the relative frequency of LWC was similar during Dry and Wet periods, the contribution of annual R_{LL} during the Wet period was approximately three-fold higher than that during the Dry period (Fig. 8), indicating strong effect of rainfall on R_{LL}. Although the R_{LL} pulse can last for only 3–4 days after a rainfall event, this pulse would determine a large part of annual R_{LL}. This suggests that the magnitude of total R_{LL} may be influenced by the frequency of rainfall events, especially in summertime, rather than the intensity of rainfall. Still, the cumulative R_{LL} in the Dry period contributed 26.9% of annual R_{LL} in 2013, even though instantaneous R_{LL} was very low. There may be large vertical variability in WC and R_{LL} within the L-layer, indicating that higher WC and R_{LL} occur in lower parts of the L-layer during the drying process because the upper L-layer dries more rapidly [28]. In that case, although the mean WC of the L-layer was very low, local wetting in lower sections would produce small CO_2 fluxes. Despite low instantaneous R_{LL}, the accumulation of R_{LL} over a long time period (approximately 6 mo) resulted in a substantial contribution (27%) of Dry-period respiration to annual R_S.

Raindrops first reach the L-layer and then percolate to the soil layers below. Small amounts of precipitation caused no change in SWC or R_S, but R_{LL} increased rapidly with increasing LWC (Fig. 6). In semi-arid and arid ecosystems, wetting of the L-layer and surface soil by small fog-drop pulses during the dry season can contribute up to 35% of R_S [29]. Although such small water inputs (e.g., brief rain showers and fog), which mainly affect the surface of the forest floor, can be significant drivers of temporal variation in R_S, the soil water content sensors (generally inserted at depths > 5 cm) could not capture these inputs. Continuous measurement of LWC allowed for realistic modeling of the effects of rapid changes in LWC on R_{LL}.

Although the annual contribution of R_{LL} to R_S was relatively small (8.6%), this contribution showed large temporal variation according to rainfall, ranging from nearly zero to 51%. Several other studies have described similar results [17,21]. For example, Borken et al. [17] reported that peaks in R_{LL} during addition of water ranged from 0.031 to 0.071 mg CO_2 m^{-2} s^{-1} in vitro, which represented 11–26% of maximum in situ R_S in the Harvard forest, although R_{LL} before addition of water was nearly zero. These findings indicate that R_{LL} is a significant component of rapid and transient temporal variation in R_S in relation to rainfall events. Although numerous studies have examined CO_2 efflux from mineral soils in relation to the intensity, duration, and frequency of rainfall [30,31], few studies have focused on R_{LL} because of the difficulty in measuring this dynamic. Here, R_{LL} pulses were observed only during and several days after rainfall events. Thus, periodic sampling (e.g., twice per week) might be insufficient to capture the contribution of the R_{LL} pulse to R_S. Moreover, manual flux measurements are usually not performed during precipitation events because of difficulties that can occur with electronic instruments and sampling methods. In our view, conducting in situ measurements of CO_2 efflux from the L-layer only over short time intervals (e.g., up to 1 h) produces robust data for understanding the response of R_{LL} to rainfall events and its contribution to R_S.

The contribution of R_{LL} to annual R_S was 8.6% in our site. In an oak forest, the contribution of R_{LL} to R_S was 23%, according to model simulation based on temperature and LWC by Hanson et al. [13]. Ngao et al. [32] reported a lower contribution (8%) in a beech forest, estimated using an isotope mass balance approach, which was close to the value observed at our site (8.6%). However, simple quantitative comparisons between studies are difficult because of the use of different methods. In addition, some technical problems remain at our site. First, we performed R_{LL} measurements in the treatment area in which the mineral soil below the L-layer was replaced with combusted granite soil. This treatment may have affected the microbial community and environmental conditions in the L-layer. Secondly, each continuous measurement of R_{LL} and R_S was performed with single chambers, so spatial heterogeneity in R_{LL} and R_S were not considered. Automated chamber methods allowed high-interval measurements of temporal variation in respiration but had poorer spatial distribution compared with the manual chamber method. The balance of trade-offs between automated and manual chamber method is subject to the relative importance of characterizing temporal and spatial variability of individual CO_2 sources. The number of chambers used can enhance the accuracy of measured mean values. Loescher et al. [33] reported that the number of chambers needs to be >100 to adequately represent spatial variability. However, this is not a feasible experimental design because of practical limitations to sampling efforts. To improve estimation of R_{LL} and R_S at the forest stand level, and to better understand the soil carbon budget, a comprehensive comparison of the diverse C pools and fluxes in forest soils is required.

Conclusions

In our study, the rapid and transient variation in R_{LL} induced by rainfall; the peak R_{LL} was observed during or one day after rainfall, and R_{LL} subsequently decreased to pre-wetting levels within 2–4 days after rainfall events, following the decrease in LWC. On the one hand, CO_2 efflux from coarse woody debris found in our site decreased during rainfall events, and subsequently, a gradual increase in CO_2 efflux continued for at least 14 days until next rainfall [34]. Therefore, coarse woody debris was a CO_2 efflux source over longer time scales, while R_{LL} approached nearly zero within a few days after rainfall events, even at high temperatures. Such specific temporal CO_2 efflux patterns for each heterotrophic source when subjected to wetting and drying cycles would be a result of substrate properties (e.g., specific surface area). In our view, continuous and direct measurements of CO_2 efflux and environmental conditions characterized by substrate properties of individual CO_2 sources could improve understanding of the processes that regulate variation in heterotrophic respiration and R_S and enable progress beyond empirical models that are primarily based on simple temperature and SWC relationships.

Moreover, the magnitude of heterotrophic respiration under wetting and drying cycles is strongly related to microbial physiology and community composition. For example, Schnurer et al. [35] showed that longer-duration wetting could promote microbial biomass, causing an increase in basal respiration. Fierer et al. [36] showed the influence of drying and rewetting frequency on microbial (fungi and bacteria) community composition. To improve understanding of heterotrophic respiration associated with response and adaptation of microorganisms under climatic changes, collected continuous in situ data for CO_2 efflux and environmental conditions (e.g., temperature and WC) of individual CO_2 sources should be combined with analyses of microbial physiology and community composition.

Acknowledgments

We greatly thank Dr. Yoshiko Kosugi and the staff of the Forest Hydrology Laboratory of Kyoto University for assistance in the field and for helpful advice.

Author Contributions

Conceived and designed the experiments: MA YK TM. Performed the experiments: MA KY MT. Analyzed the data: MA YK MT. Contributed reagents/materials/analysis tools: MA MJ. Contributed to the writing of the manuscript: MA.

References

1. Curtis PS, Hanson PJ, Bolstad P, Barford C, Randolph JC, et al. (2002) Biometric and eddy-covariance based estimates of annual carbon storage in five eastern North American deciduous forests. Agricultural and Forest Meteorology 113: 3–19.
2. Kominami Y, Jomura M, Dannoura M, Goto Y, Tamai K, et al. (2008) Biometric and eddy-covariance-based estimates of carbon balance for a warm-temperate mixed forest in Japan. Agricultural and Forest Meteorology 148: 723–737.
3. Keith H, Leuning R, Jacobsen KL, Cleugh HA, van Gorsel E, et al. (2009) Multiple measurements constrain estimates of net carbon exchange by a Eucalyptus forest. Agricultural and Forest Meteorology 149: 535–558.
4. Yuste JC, Nagy M, Janssens IA, Carrara A, Ceulemans R (2005) Soil respiration in a mixed temperate forest and its contribution to total ecosystem respiration. Tree Physiology 25: 609–619.
5. Hanson PJ, Edwards NT, Garten CT, Anderson JA (2000) Separating root and soil microbial contributions to soil respiration: A review of methods and observations. Biogeochemistry 48: 115–146.
6. Kuzyakov Y (2006) Sources of CO_2 efflux from soil and review of partitioning methods. Soil Biology & Biochemistry 38: 425–448.
7. Moyano FE, Kutsch WL, Rebmann C (2006) Soil respiration fluxes in relation to photosynthetic activity in broad-leaf and needle-leaf forest stands. Agricultural and Forest Meteorology 148: 135–143.
8. Suseela V, Conant RT, Wallenstein MD, Dukes JS (2012) Effects of soil moisture on the temperature sensitivity of heterotrophic respiration vary seasonally in an old-field climate change experiment. Global Change Biology 18: 336–348.
9. Jomura M, Kominami Y, Tamai K, Miyama T, Goto Y, et al. (2007) The carbon budget of coarse woody debris in a temperate broad-leaved secondary forest in Japan. Tellus B 59: 211–222.
10. Matsumoto A, Kominami Y, Ishii H (2010) Field measurement of heterotrophic respiration of root litter using a small chamber system. Journal of Forest Research 92(5): 269–272.
11. Tewary CK, Pandey U, Singh JS (1982) Soil and litter respiration rates in different microhabitats of a mixed oak-conifer forest and their control by edaphic conditions and substrate quality. Plant and Soil 65: 233–238.
12. Kirschbaum MUF (2013) Seasonal variations in the availability of labile substrate confound the temperature dependence of organic matter decomposition. Soil Biology & Biochemistry 57: 568–576.
13. Hanson PJ, O'Neill EG, Chambers MLS, Riggs JS, Joslin JD, et al. (2003) Soil respiration and litter decomposition. In: North America Temperate Deciduous Forest Responses to Changing Precipitation Regimes (eds Hanson PJ and Wullschleger SD). Springer, New York.
14. Andersson M, Kjoller A, Struwe S (2004) Microbial enzyme activities in leaf litter, humus and mineral soil layers of European forests. Soil Biology & Biochemistry 36: 1527–1537.
15. DeForest JL, Chen J, McNulty SG (2009) Leaf litter is an important mediator of soil respiration in an oak-dominated forest. International Journal of Biometeorology 53(2): 1432–1254.
16. Wilson TB, Kochendorfer J, Meyers TP, Heuer M, Sloop K, et al. (2014) Soil respiration in a mixed temperate forest and its contribution to total ecosystem respiration. Agricultural and Forest Meteorology 192–193: 42–50.

17. Borken W, Davidsona EA, Savagea K, Gaudinskib J, Trumborec SE (2003) Drying and wetting effects on carbon dioxide release from organic horizons. Soil Science Society of American Journal 67: 1888–1896.
18. Orchard VA, Cook FJ (1983) Relationship between soil respiration and soil moisture. Soil Biology & Biochemistry 15(4): 447–453.
19. Lee X, Wu HJ, Sigler J, Oishi C, Siccama T (2004) Rapid and transient response of soil respiration to rain. Global Change Biology 10: 1017–1026.
20. Goulden ML, Miller SD, da Rocha HR, Menton MC, de Freitas HC, et al. (2004) Diel and seasonal patterns of tropical forest CO_2 exchange. Ecological Application 14: 42–54.
21. Cisneros Dozal LM, Trumbore S, Hanson PJ (2007) Effect of moisture on leaf litter decomposition and its contribution to soil respiration in a temperate forest. Journal of Geophysical Research 112: 148–227.
22. Ataka M, Kominami, Miyama T, Yoshimura K, Jomura M, et al. (2014) Using capacitance sensors for the continuous measurement of the water content in the litter layer of forest soil. Applied and Environmental Soil Science 2014:
23. Tamai K, Hattori S (1994) Modeling of evaporation from forest floor in a deciduous broad-leaved forest and its application to basin. Journal of the Japanese Forestry Society 76: 233–241 (in Japanese with English summary).
24. Kaneko S, Akieda N, Naito F, Tamai K, Hirano Y (2007) Nitrogen budget of a rehabilitated forest on a degraded granitic hill. Journal of Forest Research 12: 38–44.
25. Goto Y, kominami Y, Miyama T, Tamai K, Kanazawa Y (2003) Aboveground biomass and net primary production of a broad-leaved secondary forest in the southern part of Kyoto prefecture, central Japan. Bulletin of FFPRI 387: 115–147 (in Japanese with English summary).
26. Subke JA, Reichstein M, Tenhunen JD (2003) Explaining temporal variation in soil CO_2 efflux in a mature spruce forest in Southern Germany. Soil Biology & Biochemistry 35: 1467–1483.
27. Dannoura M, Kominami Y, Tamai K, Jomura M, Miyama T, et al. (2006) Development of an automatic chamber system for long-term measurements of CO_2 flux from roots. Tellus 58B: 502–512.
28. Ataka M, Kominami Y, Jomura M, Yoshimura K, Uematsu C (2014) CO_2 efflux from leaf litter focused on spatial and temporal heterogeneity of moisture. Journal of Forest Research 19: 295–300.
29. Carbone MS, Still CJ, Ambrose AR, Dawson TE, Williams AP, et al. (2011) Seasonal and episodic moisture controls on plant and microbial contributions to soil respiration. Oecologia 167: 265–278.
30. Borken W, Matzner E (2009) Reappraisal of drying and wetting effects on C and N mineralization and fluxes in soils. Global Change Biology 15: 808–824.
31. Birch HF (1958) The effect of soil drying on humus decomposition and nitrogen availability. Plant and Soil 10: 9–31.
32. Ngao J (2005) Estimating the contribution of leaf litter decomposition to soil CO_2 efflux in a beech forest using 13C-depleted litter. Global Change Biology 11(10): 1768–1776.
33. Loescher HW, Law BE, Mahrt L, Hollinger DY, Campbell J, et al. (2006) Uncertainties in, and interpretation of, carbon flux estimates using the eddy covariance technique. Journal of Geophysical Research 111: D21S90.
34. Jomura M, Kominami Y, Kanazawa Y (2005) Long-term measurements of the CO_2 flux from coarse woody debris using an automated chamber system. Journal of the Japanese Forest Society 87(2): 138–144 (in Japanese with English summary).
35. Schnurer J, Clarholm M, Bostrom S, Rosswall T (1986) Effects of moisture on soil microorganisms and nematodes: A field experiment. Microbial Ecology 12(2): 217–230.
36. Fierer N, Schimel JP, Holden PA. (2003) Influence of drying-rewetting frequency on soil bacterial community structure. Microbial Ecology 45(1): 63–71.

8

Historical, Observed, and Modeled Wildfire Severity in Montane Forests of the Colorado Front Range

Rosemary L. Sherriff[1]*, Rutherford V. Platt[2], Thomas T. Veblen[3], Tania L. Schoennagel[3], Meredith H. Gartner[3]

1 Department of Geography, Humboldt State University, Arcata, California, United States of America, **2** Department of Environmental Studies, Gettysburg College, Gettysburg, Pennsylvania, United States of America, **3** Department of Geography, University of Colorado, Boulder, Colorado, United States of America

Abstract

Large recent fires in the western U.S. have contributed to a perception that fire exclusion has caused an unprecedented occurrence of uncharacteristically severe fires, particularly in lower elevation dry pine forests. In the absence of long-term fire severity records, it is unknown how short-term trends compare to fire severity prior to 20th century fire exclusion. This study compares historical (i.e. pre-1920) fire severity with observed modern fire severity and modeled potential fire behavior across 564,413 ha of montane forests of the Colorado Front Range. We used forest structure and tree-ring fire history to characterize fire severity at 232 sites and then modeled historical fire-severity across the entire study area using biophysical variables. Eighteen (7.8%) sites were characterized by low-severity fires and 214 (92.2%) by mixed-severity fires (i.e. including moderate- or high-severity fires). Difference in area of historical versus observed low-severity fire within nine recent (post-1999) large fire perimeters was greatest in lower montane forests. Only 16% of the study area recorded a shift from historical low severity to a higher potential for crown fire today. An historical fire regime of more frequent and low-severity fires at low elevations (<2260 m) supports a convergence of management goals of ecological restoration and fire hazard mitigation in those habitats. In contrast, at higher elevations mixed-severity fires were predominant historically and continue to be so today. Thinning treatments at higher elevations of the montane zone will not return the fire regime to an historic low-severity regime, and are of questionable effectiveness in preventing severe wildfires. Based on present-day fuels, predicted fire behavior under extreme fire weather continues to indicate a mixed-severity fire regime throughout most of the montane forest zone. Recent large wildfires in the Front Range are not fundamentally different from similar events that occurred historically under extreme weather conditions.

Editor: Ben Bond-Lamberty, DOE Pacific Northwest National Laboratory, United States of America

Funding: Data collection and analysis were funded by the National Science Foundation (awards BCS-0540928, 0541480, 0802667, 0541594 to TTV, RLS and RVP) and Humboldt State University Sponsored Programs Foundation and College of Arts, Humanities, and Social Sciences (awards to RLS). The funders had no role in the study design, decision to publish, or preparation of the manuscript.

Competing Interests: The authors have declared that no competing interests exist.

* Email: sherriff@humboldt.edu

Introduction

The social, environmental and fiscal costs of wildfire have escalated dramatically over the last few decades [1–2]. The costs associated with recent wildfires are particularly high in the arid mountain West, where residential structures abut or intermingle with wildland vegetation (Wildland-Urban Interface - WUI) and the exurban population has grown rapidly in recent decades [3]. Large fire events in the 1990s and early 2000s in the western U.S., particularly in lower elevation, relatively dry-pine forests, have contributed to widespread concern that fire exclusion has caused an unprecedented threat of uncharacteristically severe fires in these ecosystems [4–5]. Broad-scale monitoring of fire severity from satellite imagery since ca. 1984 shows a significant trend towards increased severity only in parts of the Southwest [6], yet findings are varied in other studies depending on the spatial scale, selected data types, and the location of study (e.g. Pacific West) [7–12]. In the absence of longer broad-scale records of fire severity, it is unknown how such short-term trends compare to fire severity prior to fire exclusion. In the context of debate about the potential effects of fire exclusion on modern fire regimes and their departures from historical fire regimes (i.e. prior to fire exclusion) [4,13–15], there is a critical need for research on whether the severity and other characteristics of modern fires depart from historical fire regimes for specific ecosystem types at broad landscape scales. The current study compares reconstructed historical fire severity, observed fire severity in recent fires and modeled fire severity from current fuel structures, and discusses the consequences of fire regime changes for management options under expected future climate in the Colorado Front Range.

Development of management goals and adaptation options in fire-prone ecosystems interfacing with the WUI is inextricably related to quantifying the range of variation of a set of ecological patterns and processes exhibited naturally or under human influences during a specified historical period [16–17]. This widely used approach, called the historical range of variability (HRV) framework uses historical ecological data to test hypotheses about the drivers and mechanisms of contemporary and future ecological change [18]. Management using the HRV framework assumes that ecosystem resilience (the ability to recover quickly) is reflected in observed ranges of past vegetation and fire dynamics. Retrospective ecological studies are considered essential for understanding likely consequences of climate change on future fire and landscape dynamics [18–19]. While there has been a shift away from using reference conditions as default for fire and forest management goals in relation to future climate change [19–20], HRV still provides the most viable framework for understanding the sensitivity of ecosystem resilience and ecological integrity to changes in fire regimes [21]. Ecological integrity has been defined as a measure of the composition, structure, and function of an ecosystem in relation to the system's natural or historical range of variation, as well as changes caused by humans [22–23]. A key research objective in the context of fire-prone landscapes in the U.S. West is the assessment of the historical role of fires of varying severities on the resilience and integrity of current ecosystems and their potential consequences for ecosystem services valued by humans [21].

Fuel reduction is currently the dominant management tool for reducing the likelihood of high-severity fire in the contexts of ecological restoration and mitigation of climate change impacts [24–26]. This dual approach (forest restoration and fire mitigation) assumes that the probability of severe fire occurrence has increased to uncharacteristic levels during decades of fire suppression in western forests [4,27–34]. Guided by this assumption, over 190 million acres of public land have been identified as "unnaturally dense" with an increased likelihood of catastrophic wildfires [24], [26]. In response to such concerns, both forest managers and communities have begun to develop strategies to alleviate the potential impacts of wildfire [35], and millions of hectares of forest lands have been treated in recent decades [36]. The goals of fuels reduction to decrease the likelihood of severe wildfires and restore historical forest structure and species composition are complementary in ecosystems where fuels and fire severity have increased, yet are incompatible elsewhere and threaten ecosystem integrity and ecosystem services [37–43].

In the Colorado Front Range, tree-ring evidence, historical landscape photographs, and General Land Office surveys demonstrate that the historical (i.e. pre-1920) fire regime of ponderosa pine and mixed-conifer forests included low-severity fires (i.e. non-lethal to large fire-resistant trees) as well as high-severity fires (i.e. killing >70% of canopy trees) [13,44–54]. There is a broad consensus that most of the montane zone of Colorado was characterized by fire regimes of mixed severity, including some component of high-severity fires [55]. However, a better understanding of which habitats across this heterogeneous landscape were explicitly affected predominantly by low-severity or higher-severity fires, or a combination of both, is necessary for determining where woody fuels have become uncharacteristically abundant as a consequence of fire exclusion [37,56].

Thus, in this research we examine changes in fire regimes across 564,413 ha of the montane forest zone of the Colorado Front Range (Figure 1). We compile and analyze new and existing datasets to refine the spatial resolution and expand the geographic scope of retrodicted (reconstructed) historical fire regimes in comparison with present wildfire potential. We address two main questions about the montane forests of the Colorado Front Range: 1) What areas and landscape characteristics were characterized by an historical fire regime of predominantly low-severity or mixed-severity (which includes moderate- or high-severity fires)? 2) How does historical fire severity compare with observed severity of large wildfires (since 2000) and modeled potential wildfire behavior across the landscape? Finally, we discuss how these comparisons can inform fire mitigation and ecological restoration under expected climate change.

Methods

Study area

The study area is located in the north-central Front Range of Colorado, bounded by six counties and the extent of the montane zone between 1800–3000 m to the east and west (Figure 1). Within the lower to upper montane zones, the mean annual precipitation ranges are ~35.6 to 51 cm and the mean annual temperature ranges are 11°C to 2.4°C, respectively [57]. The lower montane zone (~1800 to 2200 m) comprises primarily pure ponderosa pine (*Pinus ponderosa*) on south-facing slopes and a mixture of ponderosa pine and Douglas-fir (*Pseudotsuga menziesii*) on north-facing slopes. The upper montane zone (~2200 to 3000 m) is composed of ponderosa pine stands on south-facing slopes and more dense stands of ponderosa pine and Douglas-fir on north-facing slopes along with lodgepole pine (*Pinus contorta*), aspen (*Populus tremuloides*) and dispersed limber pine (*Pinus flexilis*) trees at higher elevations. The LANDFIRE existing vegetation type (EVT) layer was used to delineate the montane study area (564,413 ha) within 1800–3000 m because it represents the most up-to-date and detailed cover type classification available across land ownerships in the region, which is proportionally represented by the following EVT types: ponderosa pine (24.2%); lodgepole pine (23.1%); mixed montane conifer forest and woodland types (49.2%); and intermixed with pixels classified as lower montane-foothill grassland (<0.1%) or shrubland (3.5%) (Figure 1; EVT types included in the study are listed in Table S1). However, the accuracy of the LANDFIRE EVT classification at field sites was relatively low (see RESULTS).

During the period from 1980 to 2011, 48 fires burned over 100,000 ha along the Front Range causing severe damage to property and infrastructure in the WUI [2,58–59]. These large fires can occur any time of year and typically burn when wind speeds are high and weather conditions are dry, which is common along the Front Range. During these types of conditions, fire suppression is typically ineffective and fires often escape initial suppression efforts [2].

Field sampling

We used a combination of existing datasets (141 sites) [52–53,60–62] and newly sampled datasets (91 sites) for a total of 232 sites with information on forest structure (tree age structure) and fire history (tree-ring fire-scar records) in montane forest types.

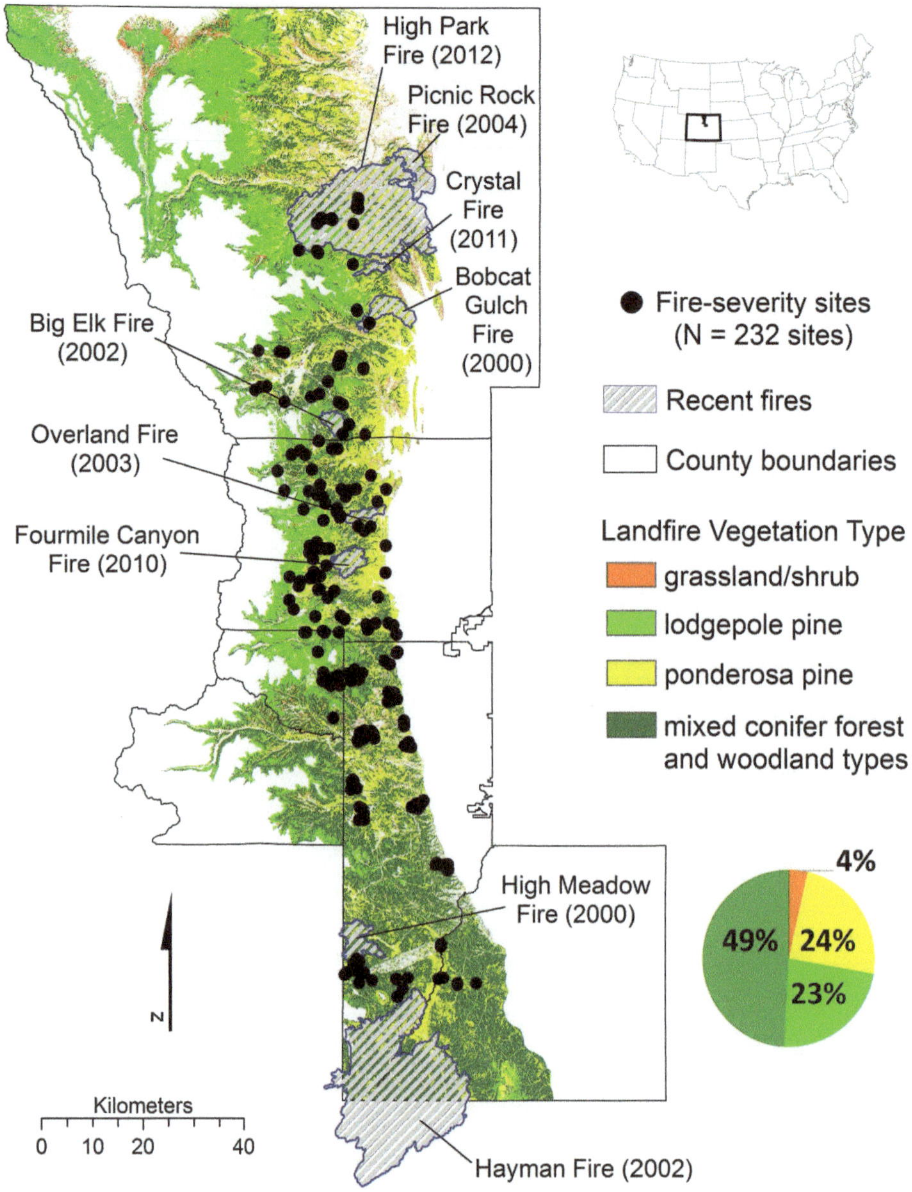

Figure 1. Study area in north-central Colorado, USA. The map includes the montane zone by cover type (LANDFIRE Existing Vegetation Type), 232 sites sampled for historical fire severity, and recent wildfires used for comparison and verification of fire behavior modeling. County boundaries (solid black lines) from north to south are of Gilpin and Clear Creek counties and the mountainous western regions of Larimer, Boulder, and Douglas counties, and Jefferson County.

The majority of sites were sampled in a stratified-random design and all sites were selected in relative proportion to their cover type within the study area, allowing inferences from our site-level datasets to reflect trends across the 564,413 ha montane study area. Because sites were dispersed over the broad regional area of montane forests we do not extrapolate the spatial spread of fire severity between sites, or determine if fires in the same year at multiple sites were from a single ignition or multiple ignition sources. We rejected sites with evidence of past logging or signs of other major anthropogenic disturbance (i.e. mining). Our sampling goal, similar to previous site-level studies of historic fire regimes in the region [51,53], was to identify the predominant structural influence of historical fires at each site (fire-severity regime as explained below). Understanding the historical fire regime and the effects on forest structure without major Euro-American influences provides the natural range of conditions. The interactive effects of logging and/or other anthropogenic disturbances are important, but beyond the scope of our study. Laboratory (tree-ring) analysis of new sites followed the same procedures used in prior studies for analyzing fire-scar samples, stand age and size structure [51–53,60–62]. The criteria used to characterize the historical fire severity at each site were derived from the results of these prior studies and are described below (*Site-level classification of historical fire severity*).

Stand-level sites. At 120 of the 232 sites [51–53,60–62] (and unpublished new data), we sampled in areas of relatively consistent (unvarying) forest structure and physical environment with the extent of sampling area varying in size from 10 to 232 ha according to tree density and the extent of the homogeneous stand structure. At 44 of the stand-level sites, fire-scarred trees were systematically sampled [60,61], fire-scarred trees and fire scars on each fire-scarred tree were tallied, and forest structure was evaluated based on the same protocol used by Sherriff and Veblen [52] (see Tables 1–2). At the remaining 76 stand-level sites, the sampling goal was more intensive. In addition to systematically sampling fire-scarred trees throughout the site, at least 50 of the closest trees (≥4 cm dbh) at a constant distance along randomly located transects were cored, and at least five of the largest and oldest characteristic live or dead trees were selectively cored to ensure large (and the oldest characteristic) trees were represented in our sample. In addition, forest structure (size, decay class, standing/down, distance from transect point), seedling (<30 cm height) and sapling (>30 cm height and <4 cm diameter) data were tallied by species along belt transects. To estimate the ages of trees too small to core (<4 cm in diameter) and to estimate the number of rings missed due to coring height (approximately 20 cm above the root-shoot boundary), 341 juvenile trees, reflecting the range of tree species within the sampling sites (in order of abundance - ponderosa pine, Douglas-fir, lodgepole pine, aspen and limber pine), were cut at sites of different elevation and aspect. Relatively open sites were selected to mimic post-fire growth conditions. Permits for sampling were issued by Rocky Mountain National Park, USDA Forest Service, and county and city open space land management agencies. Sampling did not involve protected or endangered species. For sites where seedlings were not collected, the median age-to-coring height was used from a site of similar elevation and aspect (across all sampled sites: median of 4 years – ponderosa pine; 9 years – Douglas-fir; 8 years – lodgepole pine; 1 year – aspen; 9 years – limber pine). To correct for years missed due to missing the pith, we followed Duncan's [63] procedure. Cores with more than 20 years estimated to pith were excluded.

Plot-level sites. The remaining 112 sites with information on forest structure and fire were smaller in size (3-ha plots) and randomly located from a 1-km grid throughout the CFR study area [52] (and new unpublished data). Sites with abundant logging or other human disturbances were rejected and an adjacent location was sampled. At each random plot-level site the number of fire-scarred trees and fire scars on each fire-scarred tree were tallied, fire-scarred trees were cored to estimate fire dates [64], forest structure was documented using an existing protocol [52] (see Tables 1–2), and approximately 10 cores were taken for reconstructing tree ages.

Analytical procedures

Site-level classification of historical fire severity. For this analysis, we use common and broad definitions of fire-severity regimes for montane forests of the Colorado Front Range that are relevant to forest structure and management (Table 1). Low-severity fire regimes are dominated by frequent (Mean Fire Interval – MFI <30 years), non-stand replacing fires within a stand (~100 ha) that leave multiple fire scars on individual trees throughout the stand and kill young seedlings and subcanopy trees while maintaining open, low-density stands of fire-resistant canopy trees. Mixed-severity regimes have varied fire effects that include low-severity, non-stand replacing fire to high-severity, stand- (or canopy) replacing fire both within stands and across landscapes, often in relation to topography [40,47–48,51–53,65,66]. High-severity fires occur less frequently (MFI >35–100+ years) than non-stand replacing fires, leave only small or no patches of pre-fire remnant trees, result in few to no fire scars on individual trees throughout the stand, and often initiate recruitment of a new cohort of canopy trees [51–53,55].

First, we classified the severity of fires at each (stand- and plot-level) site based on their influence on forest structure. We focused on fire dates recorded by fire scars on at least two trees per site (spreading fires) in the same year prior to 1915 (ca. effective fire exclusion). Next we used two primary metrics, along with the dates and frequency of spreading fires, for characterizing the severity of fires at each site – the percentage of remnant trees (proportion of the trees older than a spreading fire date) and the percentage of tree establishment (proportion of the trees that established within 40 years after a spreading fire). These criteria follow the approach described in prior studies in the region [51,53], correspond to classification of fire severity in other studies [65–66], and are summarized in Table 1. Specifically, individual fires were classified as low severity when ≥80% of the trees that were sampled had survived the fire (based on remnant tree ages) and ≤20% of the trees established following the fire. Fires were classified as moderate severity when 21–79% of the trees survived/established following the fire. Fires were classified as high severity when < 20% of the trees that were sampled had survived the fire and > 80% of the trees established following the fire. We excluded fires where no tree establishment followed within 40 years, assuming there was no structural impact or burning within the site [53]. Individual fires that occurred prior to subsequent mixed-severity fire, and were not associated with a pulse of establishment (>20% trees/site), were not included in the analysis of fire severity. We recognize that the estimate of fire effects for fires that occurred prior to a mixed-severity fire is less reliable because of subsequent

Table 1. Definitions of historical fire severity terms.

Term	Definition of fire effects
Historical high-severity fire	A fire that had high mortality of live, standing vegetation (<20% of the sampled trees survived the fire) and high tree establishment (>80% of the sampled trees) following the fire.
Historical low-severity fire	A fire that had low to no mortality of live, standing vegetation (>80% of the sampled trees survived) and low to no establishment (<20% of the sampled trees established following the fire).
Historical low-severity fire regime	Dominated by frequent (Mean Fire Interval <30 years), non-stand replacing fires within a stand (~100 ha) that leave multiple fire scars on individual trees throughout the stand and kill young seedlings and subcanopy trees while maintaining open, low-density stands of fire-resistant canopy trees.
Historical moderate-severity fire	A fire that had effects that were intermediate between low and high severity.
Historical mixed-severity fire regime	Varied fire effects that included low-severity, non-stand replacing fire to high-severity, stand- (or canopy) replacing fire both within stands and across landscapes, often in relation to topography.

These terms may have different meanings in the literature depending on the context in which they are used.

Table 2. Large fires in the montane study area from 2000–2012.

Fire	Year	Size	Fire Severity (%)*			
		(ha)*	Unburned/Low	Low	Moderate	High
Bobcat Fire	2000	3688	30.5	16.8	22.6	30.1
High Meadows Fire	2000	3884	21.7	37.8	32.0	8.5
Big Elk Fire	2002	1741	42.1	20.0	17.1	20.8
Hayman Fire	2002	52,016	21.1	12.7	22.5	43.7
Overland Fire	2003	1244	27.9	12.9	22.6	36.6
Picnic Rock Fire	2004	3626	27.6	37.7	22.6	12.1
Four Mile Fire	2010	2285	11.0	25.6	34.3	29.1
Crystal Fire	2011	914	22.9	38.5	26.9	11.7
High Park Fire	2012	34,905	20.8	29.7	26.9	22.5

* Total area burned and fire severity percentages within each fire perimeter (MTBS values 1–4; 1 = unburned/low, 2 = low, 3 = moderate, and 4 = high). See Figure 1 for large fire perimeters used to compare historical and observed fire severities within the study area.

mortality of older trees with time and/or subsequent disturbance. Post-fire tree establishment is not a direct measure of fire severity, and can vary significantly among species and in relation to seed availability and climate conditions, but it provides corroborative evidence of fire severity in combination with remnant trees, which provide a more direct measure of fire severity. Both metrics, pre-fire remnant trees and post-fire tree establishment, have uncertainties for evaluating fire severity and are not a perfect complement, but used in conjunction with fire dates and integrated into measureable metrics represent a robust way to reconstruct historical fire severity in montane forests (see [53] for more on methodological limitations and recommendations).

Based on the fire severity classifications, we assigned a predominant fire regime at each (stand- and plot-level) site based on the following criteria of cumulative fire effects over time. Sites characterized by frequent, non-stand replacing fires (MFI <30 years) that left multiple fire scars on individual trees throughout the site and without evidence of higher-severity fire were classified as having only low-severity fire regimes. Low-severity sites identified without historical evidence of moderate- or high-severity fire are assumed to have had only low-intensity (surface) fire and low to no canopy mortality. Sites characterized by varied fire effects that included canopy-replacing fire within sites were classified as having mixed-severity regimes (see [51–53] for within site sampling protocols). Mixed-severity fire regimes include low-, moderate- and high-severity fires where sites may illustrate different severities over time or space depending on weather, climate and stand conditions. Individual sites were differentiated based on evidence of low- and moderate-severity fire effects (no evidence of high-severity fire with <20% the stand age preceding a single fire) or low- to high-severity fire effects. Sites identified with either (low- and moderate-severity fire effects or low- to high-severity fire effects) were classified as a mixed-severity fire regime. Mixed-severity sites identified with moderate- or high-severity fire are assumed to have had components of intense fire behavior and canopy mortality, but varied in the extent of canopy mortality.

Landscape-level modeling of historical fire severity. Based on the fire severity classifications at the 232 field sites, we modeled historical fire-severity regime across the study area using potential predictor variables derived from two sources: 1) terrain variables from a 30-m digital elevation model (DEM); and 2) LANDFIRE biophysical variables. Environmental conditions at each of the 232 sites were classified in terms of the median elevation, slope steepness, aspect, cover type, distance to grassland, distance to ravine and the associated fire regime type. Elevation, slope steepness, aspect (arcsine and sine transformations) and slope curvature were all derived from the 30-m DEM. Slope curvature (concavity to convexity) and hillslope position are related to soil depth, texture and potential soil moisture [67]. Ravine drainages were delineated from the 30-m DEM using Arc GRID (ESRI 2002) hydrologic terrain modeling [68] and distance to ravine drainage was calculated. Distance to grassland was calculated as the distance to the edge of grassland areas of at least 0.1 ha.

Classification and regression trees (CART; $n = 5000$ trees) were used to identify explanatory variables with the most power to predict fire severity type (RandomForests Software, Salford Systems). CART has been increasingly used in ecological studies that require nonparametric techniques to explore complex, hierarchical interactions among variables [69–71]. We evaluated predictor variables for three response variables (historical low-, moderate- and high-severity fire) and a binary response variable (historical low- or mixed-severity fire) using all 232 sites to identify complex and non-linear interactions between variables that could be mapped across the study area (e.g. [71]). A 10-fold cross-validation process was used to avoid over fitting. The regression tree size with the smallest error was recorded for each run. The average tree size was used to prune the original tree (from the complete dataset), which tends to minimize complexity and results in a more generalizable tree [72]. There was strong overlap in the range of environmental conditions at sites with evidence of historical moderate and high fire severities. For this reason, we used the results evaluating the binary response variable (historical low- or mixed-severity fire) across all 232 sites for this study to produce a predictive map of the historical fire severity across the CFR study area in a GIS (ESRI ArcGIS 10.1). We identified the optimal model as the one with the fewest variables that best predicted the occurrence of sites with historical low and mixed severities. Model performance was evaluated using the overall percentage correctly classified (PCC) and area under the receiver operating characteristic curve (AUC) statistic. An AUC value of 0.5 indicates the prediction accuracy of sites picked at random, whereas 1.0 indicates perfect classification accuracy. AUC values of 0.7, 0.8, and 0.9 indicate fair, good, or excellent accuracy, respectively.

Comparison of historical fire severity to observed fire severity. Observed fire severities from nine large fires that occurred since 2000 (Table 2; Figure 1), including a range of fire severity, were compared with the spatially-explicit map of historical fire severity (Table 3). The 'observed' severity data came from the Monitoring Trends in Burn Severity (MTBS) program, which provides maps of the perimeters and severity of all fires larger than 1000 ha in the western United States from 1984-present [73]. MTBS thematic burn severity classification maps for each fire were used to evaluate the observed fire severity (unburned/low, low, moderate, and severe) and compared with the spatially-explicit map of historical fire severity within the study area (ESRI ArcGIS 10.1). We recognize these two datasets (observed fire severity and historical fire severity) are derived from different sources of information, each with their own limitations. Thus, our analysis focused on a summary comparison of the proportion of area burned by observed fire severity type compared to historical fire severity of the same area as a broad index of departure from historical fire regime reflecting landscape-level changes rather than a fine-scale (pixel-by-pixel) comparison. For each of the nine fires, we tested for differences (in the number of pixels) between the observed amount of low-severity fire (MTBS unburned/low and low severity classes) compared to the expected (historical low-severity) using χ^2 tests (IBM SPSS 20).

Comparison of historical fire severity to modeled fire severity. The spatially-explicit model of historical fire severity was overlaid with a model of fire potential under extreme (99th percentile) weather conditions to identify possible landscape-scale changes of historical low- and mixed-fire severity in the entire montane study area (564,413 ha; Table 3). We modeled potential fire behavior using FlamMap 3.0, a fire modeling system that predicts instantaneous fire behavior under fixed weather and fuel conditions (i.e. it is a static rather than dynamic model). FlamMap requires the following inputs: elevation, slope, aspect (derived from a 30 m USGS DEM); and percent canopy cover, stand height, canopy base height, canopy bulk density, and Scott and Burgan [74] fuel models (all from the LANDFIRE project [75]). Canopy cover and canopy base height values were adjusted downward on the recommendation of the LANDFIRE project [75] and a comparison of field-derived fuel layers in the study area [76]. The modeling procedure takes place in two steps (described in detail in [77–78] and summarized here). First, fuels are conditioned based on specified weather conditions. We used wind/weather conditions during the primary fire season (June–September) derived from four Remote Automated Weather Stations (RAWS) that span the study area and have the longest record: Redfeather (station 050505; 1964–2007), Estes Park (station 050507; 1964–2007), Corral Creek (station 051804; 1968–2007) and Bailey (station 052001; 1970–2007). In the second step, FlamMap calculates fireline intensity and crown fire activity using a suite of surface and crown fire models [79–82]. Fireline intensity is a measure of energy released per unit length along the flaming front of a fire (kW/m). Crown activity (a binary variable representing surface fire or active/passive crown fire) was calculated using the Scott and Reinhardt [83] method. To address how historical fire severity compares with modeled potential wildfire behavior, we overlaid the areas modeled as historical low-severity with potential wildfire behavior (fireline intensity and crown fire) assuming that areas of higher fireline intensity (e.g.>33,000 kW/m) and crown fire today indicate a shift from the historical fire regime to one of higher potential for crown fire and canopy mortality (>33,000 kW/m and >40,000 kW/m are the approximate median fireline intensity values for moderate and high severity fire, respectively, for all verification fires with variable severity described in the next section; Table S1). A crown fire may or may not be lethal to all dominant vegetation, and a crown fire may be continuous or may occur in patches within a mixed-severity burn. Mixed-severity fires, by definition, have components of intense fire behavior and canopy mortality, and high fireline intensity and active crown fire would be within the HRV. Our focus is on the areas of greatest change that may have shifted from low severity to potential high fire intensity or crown fire today. These fire behavior outputs were chosen because they are commonly used, interpretable, and can be generally compared to historical fire severity.

The wildfire modeling utilized the 99th percentile monthly weather conditions from ca. 1964–2007 during the June-September fire season derived from four RAWS stations (listed above). Starting with a fixed southwest wind of 74 km/h, which is the 99.9th percentile daily wind speed during the primary fire season (June–September, ca. 1964–2007), we used the WindNinja program to refine wind speeds based on topography. FlamMap's input variables and modeling procedure follow Platt et al. [78] and were verified for a level of realism by comparing observed fire severity from seven major fires that occurred since 2002 (see the next section; Table 2).

To place historical 99th percentile conditions into context, we first compared them to the conditions of past fires and then to predicted future climate. During the Hayman Fire (06/08/2002), temperature and relative humidity reached up to the 98th and 99th percentiles, respectively (Cheesman RAWS station). Similarly, during the High Park Fire (06/09/2012), both temperature and humidity were up to the 99th percentile (Redstone RAWS station). The RAWS stations recorded wind gusts over 56 kph during the High Park Fire and over 80 kph during the Hayman Fire. We make the assumption that the 99th percentiles are of comparable weather conditions between historical and contemporary time periods.

We then compared average monthly temperature in the study area during the June-September fire season to two climate model runs used in the IPCC Fourth Assessment Report [84–85]: (1) climate conditions in 1964–1999 from the Climate of the 20th Century (20C3M) model run and (2) climate conditions in 2064–2099 from the SRESA2 scenario. During the June–September fire season in 1964–1999 (20C3M model run), the 99th percentile

Table 3. Comparisons of historical fire severity, observed fire severity, and modeled potential fire behavior.

Historical	Observed (MTBS thematic burn severity classifications)	Modeled (FlamMap output)
Low-severity fire only	Unburned to low severity, Low severity	Surface fire, e.g. median fireline intensity <33,000 kW/m
Mixed-severity fire: evidence of varied severities from low-, moderate- to high-severity fire	Moderate severity, High severity	Crown, torch, e.g., median fireline intensity > 33,000 kW/m

monthly temperatures were 24°C, and the 50th percentile monthly temperatures were 19°C. During the June–September fire season from 2064–2099 (SRESA2 scenario), the 99th percentile temperatures are predicted to increase to 30°C, and the 50th percentile temperatures are predicted to increase to 26°C. The 99th percentile temperatures from 1964–1999 are predicted to become 39th percentile temperatures under the SRESA2 scenario. It is thus reasonable to say that the 99th percentile conditions described in this study may become average conditions in 50 or more years from now.

Verification of modeled fire behavior. To assess the realism of the FlamMap models described above, we compared modeled fireline intensity and crown fire class to observed fire severity of the seven regional fires that occurred since 2002, the year for which the LANDFIRE (fuels) data is valid (Table 2). Within the perimeter of each fire, we generated randomly placed points, and at each point calculated median fireline intensity and percentage active and passive crown fire within four MTBS classes (unburned to low, low, moderate, and high severity; Table S1; ESRI ArcGIS 10.1). There are limitations to such a comparison. In an actual fire, intensity varies minute by minute depending on factors such as wind, relative humidity, fuel, and fire suppression. In contrast, FlamMap estimates intensity for an instant in time across an entire landscape for the generic "extreme conditions" described in the paper. Thus, while we cannot comprehensively validate a fire we can check whether, overall, areas identified as severely burned by MTBS correspond to higher modeled fireline intensity and crown fire. Kruskal-Wallis tests (IBM SPSS 20) were used to determine if the distribution of fireline intensity and percentage crown fire were significantly different between MTBS classes at the $p = 0.01$ level.

Results

Across all 232 field sites, cover type was dominated by a single species (≥80% of the canopy trees >4 cm dbh) of ponderosa pine (42.2%–98 sites) and lodgepole pine (8.6%–20 sites), and co-dominated (<80% of the canopy trees a single species) by mixed-conifer types (49.2%–114 sites). Considering pure ponderosa pine together with mixed-conifer sites, 61.2% (142) of the 232 sites were dominated by ponderosa pine and the remaining sites were dominated by Douglas-fir (20.6% and 48 sites), lodgepole pine (15.5% and 36 sites), and aspen (2.6% and 6 sites). The overall accuracy of the LANDFIRE EVT classification was relatively low with 39.2% of the 232 sites misclassified: 31 pure ponderosa pine, 5 lodgepole pine and 55 co-dominated mixed-conifer sites. Thus, the proportional values of EVT cover types across the study area, and our results defined by EVT cover types (the best available cover type at the landscape scale), should be interpreted cautiously (i.e. Figure 1 pie chart). Nevertheless, field observations support the overall LANDFIRE cover type trends illustrating ponderosa pine throughout the entire study area and an increase of lodgepole pine and mixed-conifer stands in the higher elevation portions of the study area.

Site-level fire-severity

A total of 7680 tree cores and 1262 fire-scar samples were used to delineate fire dates and fire severity across all sites. Across 120 stand-level sites, there was evidence of 322 spreading fires (fires with ≥2 trees scarred per site) between 1597 and 1995 with only 71 (22%) of the spreading fire dates unique to one site; all other fire dates were recorded at two or more sites. Fire years of 1654, 1786, and 1859–60 were particularly extensive with 36%, 43% and 48% of the available recorder sites (with ≥2 fire-scarred trees) recording each fire year, respectively (Figure 2). These fires were recorded at multiple sites that extended over 9 km (1654), 7.5 km (1786) and up to 30 km (1859–60) distance away from one another (Figure 3). Other widespread fire dates recorded at four or more sites (recording on average 22% of the available recorder sites per year; ≥2 trees scarred per site) were 1624, 1712, 1793, 1809, 1813, 1842, 1850–52, 1857, 1863, 1866, 1871, 1880–81, 1886, and 1910 (Figure 2; multiple years indicate difficulty differentiating the exact fire year due to dormant season fires and/or missing rings before or after scarring).

Historical fire severity regime was delineated at individual sites using 297 fires that occurred prior to 1920 (pre-fire exclusion) and tree age data (remnant and establishment ages). Eighteen (7.8%) sites were characterized by only low-severity fires and 214 (92.2%) were characterized by mixed-severity fires. Across the montane region, the majority of trees established between 1820 and 1920 (66% of tree ages; Figure 4) primarily because of past mixed-severity fires during this period (this study; and also see [44–45,48,50–51,53,55]). Low-severity sites tended to have younger trees than mixed-severity sites (average date of tree establishment of 1907 compared to 1873, respectively; Figure 4). The low-severity fire regime was most common in pure ponderosa pine sites (12 sites; ≥80% of the canopy trees as ponderosa pine), but also occurred in mixed-conifer sites (6 sites with co-dominance of ponderosa pine and Douglas-fir). Sites with the historical low-severity regime tended to occur at lower elevations than sites with mixed-severity fires with median elevations of 2101 m and 2445 m, respectively.

Of the 214 sites with evidence of mixed-severity fire (Figure 5), 108 sites had evidence of low and moderate-severity fires (no evidence of high-severity fire) and 106 sites had evidence of high-severity fires (evidence of up to ~50 ha patches of high-severity in some stand-level sites). Interpretation of the spatial extent of historical fire severity is limited by our sampling unit size (sites ranged between ~3–232 ha in size). Nevertheless, there is strong evidence that in most years before 1915 fires were much more extensive than the size of a single site (only 15.5% of the fire years were unique to one site; 46 of 297 fire dates; Figure 2). Many sites burned at moderate or high severity during the same year with an average of 1.1 km distance between sites (and up to 7 km distance between nearest sites burning in the same year with evidence of higher-severity fire). Evidence of mixed-severity fires occurred in all dominant cover types sampled, including sites of pure (≥80% of the canopy trees) lodgepole pine (20 sites) and ponderosa pine (89 sites), as well as mixed-conifer types with dominance of aspen (6 sites), lodgepole pine (16 sites), Douglas-fir (40 sites), and ponderosa pine (43 sites). Across sites classified with mixed-severity fire regimes, moderate- and high-severity fire effects were evident in all dominant cover types including pure lodgepole pine (8 and 12 sites) and ponderosa pine (48 and 41 sites), as well as mixed-conifer types with dominance of aspen (2 and 4 sites), lodgepole pine (8 and 8 sites), Douglas-fir (21 and 19 sites), and ponderosa pine (18 and 25 sites), respectively.

Landscape-level historical fire severity

Elevation and slope steepness were the most important predictive variables delineating locations of only low-severity versus mixed-severity sites. The CART model with the best predicted occurrence of sites with historical low- and mixed-severity fire had a PCC of 80.6% (187 of 232 sites predicted correctly) and an AUC value of 0.77 for the cross-validation test. The model indicated low-severity sites tended to occur at or below 2263 m, or on slopes equal or less than 4 degrees above 2263 m (17 of 18 low-severity sites fit this model, 94.4%). Mixed-severity

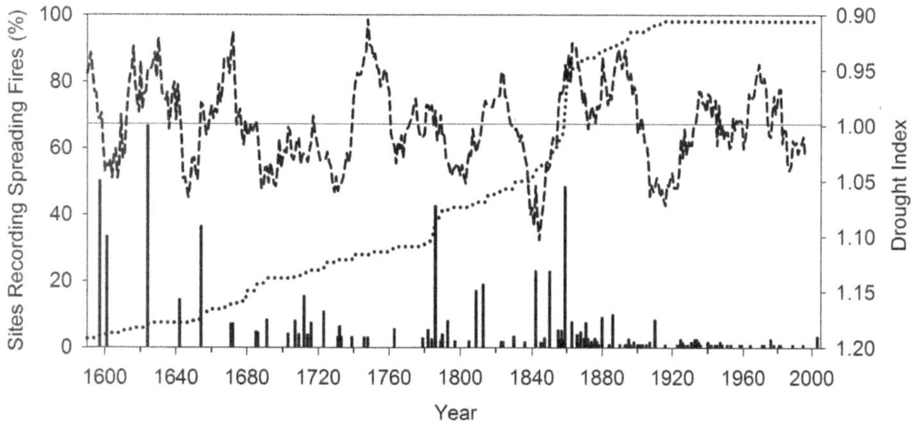

Figure 2. Percentage of sites recording spreading fires. Percentage of sites with spreading fires (minimum of 2 scarred trees per year at each stand-level site represented as histogram bars, n = 120 sites; left axis), sample depth of the percentage of sites available to record fires (minimum of 2 scarred trees available to record fires at each site; dotted line; left axis), and a 20-year smoothing average of a regional tree-ring drought index [60] with a mean value of 0.99 (dashed line with an inverted axis where smaller values are towards the top and indicate dry conditions; right axis).

sites occurred across a broad topographic gradient of the montane zone, but tended to occur on slopes greater than 4 degrees above 2263 m (170 of 214 mixed-severity sites fit this model, 79.4%). Mixed-severity sites with evidence of high-severity fire tended to occur at higher elevation, on greater slope steepness, and at greater distances from streams compared to sites with evidence of only low-severity fires. There were no strong generalizable differences between locations of mixed-severity sites with evidence of low- to high-severity fire and those without evidence of high-severity fire (only low- and moderate-severity fires). Cover type was not a significant predictor of fire severity.

The best-fit model from the CART analysis (described above) was used to map historical fire severity across the montane study area (564,413 ha). Across all cover types, 27.8% (156,198 ha) of the area was mapped with an historical regime of low-severity fire and 72.2% (406,173 ha) was mapped as mixed-severity fire (Figure 6a). Within the low-severity fire regime (27.8% of the study area) almost equal areas were classified by LANDFIRE EVT as ponderosa pine (45.4%) and mixed conifer (48.8%), intermixed with grassland/shrubland (5.7%). In comparison, areas mapped as the historical mixed-severity fire regime (72.2% of the study area) were classified with a lower percentage of ponderosa pine (16%) and grassland/shrubland (2.7%), and a higher percentage of mixed conifer (81.3%) cover types. Of areas mapped as pure ponderosa pine, 52.4% was classified as low-severity and 47.6% as mixed severity fire (Figure 6b). Of areas mapped as mixed conifer types, 18.8% was classified as low-severity and 81.2% as mixed severity fire (Figure 6c). The decrease in pure ponderosa pine dominance above 2263 m is consistent with the elevation shift in co-dominance with other species, primarily Douglas-fir, in the upper montane zone.

Comparison of historical fire severity to observed fire severity

For the nine recent fires in the study area since 2000 (Table 2; Figure 1), there was a 4% difference between the average observed (47%) and historical (51%) areas of low-severity fire severity, and a 7% difference between the total observed (39%) and historical (46%) areas of mixed-severity fire (Figure 7). The greatest differences between observed (MTBS) and expected (historical) low-severity fire was at the lowest elevations (median elevations <2155 m; Figure 7). Three fires (Picnic Rock Fire, Bobcat Fire, Crystal Fire) had lower than expected observed low-severity fire compared to the historical fire severity map (X^2 value of 6.39 for Bobcat Fire and 5.49 for Crystal Fire, $p<0.05$; the X^2 test was invalid for the Picnic Rock Fire with no pixels classified as historical mixed-severity fire). For the remaining six fires there were either no significant differences in low-severity fire (High Park Fire, Four Mile Fire, Overland Fire, Hayman Fire, X^2 value of 1.04, 2.84, 1.56, and 2.43, respectively, $p>0.05$) or higher than expected low-severity fire observed compared to the expected historical fire severity map (High Meadows Fire and Big Elk Fire, X^2 values of 5.14, and 8.25, respectively, $p<0.05$).

Verification of modeled fire behavior

Overall, we found a positive association between modeled wildfire behavior (fireline intensity and percentage crown fire) and MTBS severity class (Table S1; Figures S1 and S2). The exception was the Picnic Rock Fire, for which the modeled percentage crown fire was consistently very low across observed severity classes. This is not surprising, as the Picnic Rock Fire occurred at the lower elevation grassland ecotone with over 50% of the fire classified as unburned/low or low severity and only 21% as high severity that occurred primarily in grassland fuel types (73% of high-severity area). Overall, severely burned areas in observed fires of variable fire severities (Table 2) corresponded to higher modeled fireline intensity and proportion of crown fire (Table S1), providing confidence in the model of potential wildfire behavior and its comparison with historical fire severity.

Comparison of historical fire severity to modeled current fire behavior

Across the entire montane study area present-day potential fire behavior was 36.1% surface and 73.9% crown (1.3% torch and 62.6% crown) fire occurrence under extreme (99th percentile) weather conditions (Figure 6a). Historical fire severity for the entire study area was mapped as 27.8% low-severity only and 72.2% mixed-severity fire regimes. For the pure ponderosa pine cover type, under the extreme fire weather scenario (99th percentile), present-day fuels predicted 40.2% of the area would burn as surface fire and 59.8% as crown (3.1% torch and 56.7% crown) fire (Figure 6b). For the mixed conifer cover type, the extreme fire weather scenario and present-day fuels predicted

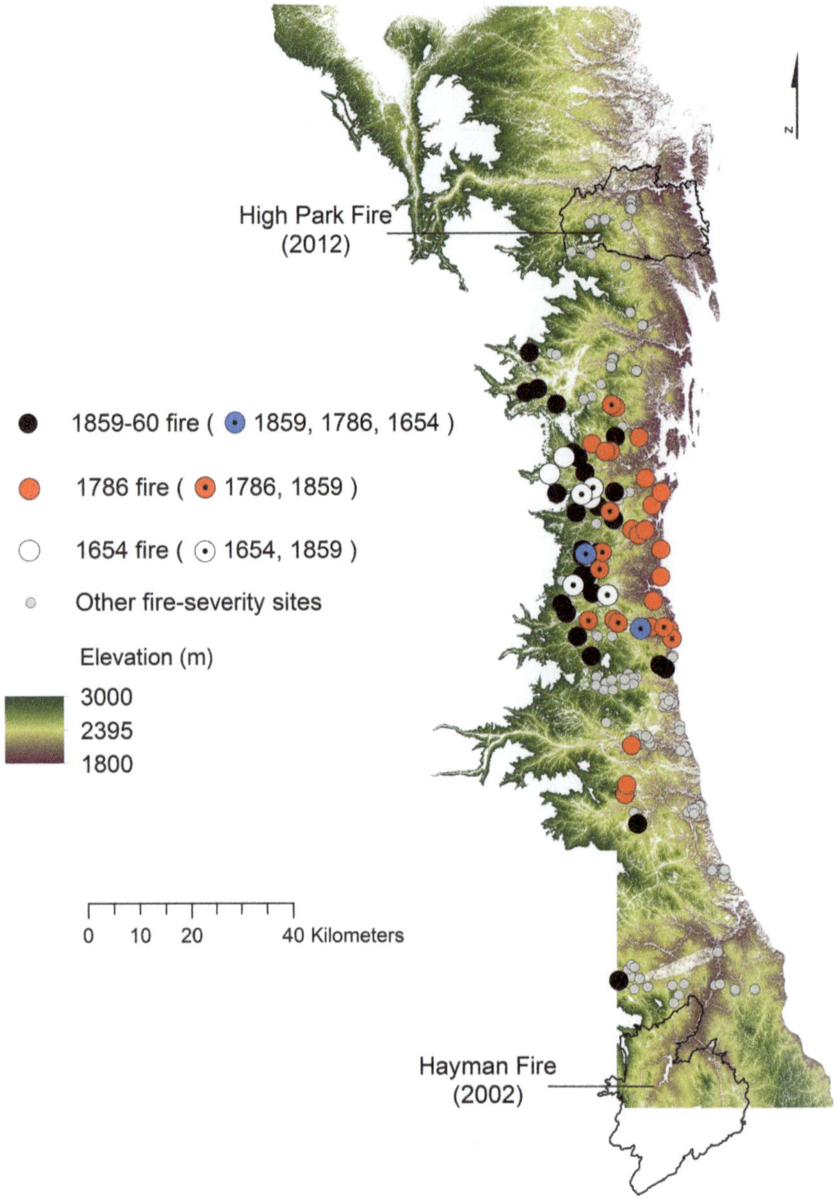

Figure 3. Historical fire-severity sites recording widespread fire dates in 1654, 1786, and 1859–60. Sites are shown across the elevation gradient of the study area. Symbols with centered dots indicate two or three of those fire years listed were recorded at that site. The two largest fires in recorded history for the region (Hayman Fire and High Park Fire) are shown for spatial comparison.

31.8% as surface and 68.2% as crown (0.7% torch and 67.5% crown) fire (Figure 6c).

Almost 12% (66,394 ha) of the entire study area showed little change in the low-severity fire regime (Figure 8 – green), based on a high spatial coincidence of historical low-severity and present-day potential for surface fire under extreme (99th percentile) weather conditions. These areas were predominantly in the lower montane zone (mean 2136 m elevation and 12 degree slope steepness), and had an average modeled fireline intensity of 14,020 kW/m. About 16% of the entire montane study area (Figure 8 – red) is now susceptible to crown fire but historically had lower-severity fire. These areas were also concentrated in the lower montane zone (mean 2177 m elevation and 16 degree slope steepness), had an average modeled fireline intensity of 43,620 kW/m, and were classified by LANDFIRE EVT as grassland or shrubland (0.5%), lodgepole pine (5.2%), ponderosa pine (43.7%), and mixed conifer (50.7%). Of areas mapped only as the historical low-severity fire regime (27.8% and 156,198 ha of the study area; Figure 8 – green and red combined), 42.5% of the area showed little change in the fire regime (Figure 8 – green), whereas 57.5% had the potential for higher-severity fire (Figure 8 – red).

Above the lower montane zone (>2263 m), the historical and modeled fire behavior under extreme conditions showed high variability from surface to crown fire, as expected in a mixed-severity fire regime. Areas above 2263 m that were mapped as historical mixed severity fire, but today have potential for surface fire and crown fire were 24.8% and 47.4% of the entire montane study area, respectively (Figure 8 – yellow and orange). Of areas mapped only with the historical mixed-severity fire regime (72.2%

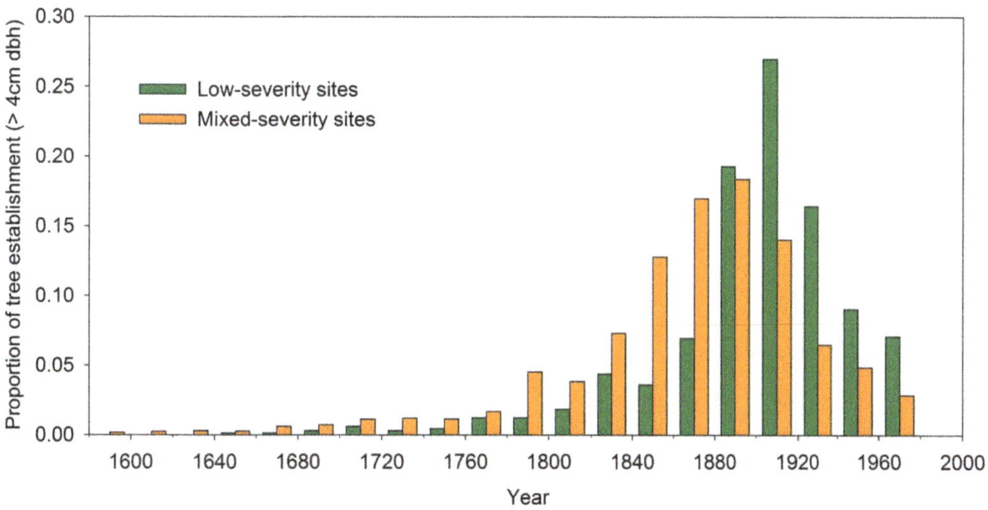

Figure 4. Proportion of tree establishment dates. Tree establishment (≥4 cm diameter) in 20-year bins for low-severity (green) and mixed-severity (orange) sites sampled at the stand-scale (120 sites). Total number of trees included are 663 and 5703, respectively.

and 406,173 ha of the study area; Figure 8 – yellow and orange), 34.4% has a low probability of crown fire today (mapped as potential for surface fire in Figure 8 (yellow); average fireline intensity of 15,861 kW/m), whereas 65.6% of the area has a high probability of crown fire activity today (mapped as potential for torch or crown fire in Figure 8 (orange); average fireline intensity of 42,290 kW/m).

Discussion

A key finding of our study of fire regime changes in the montane forest zone of the Colorado Front Range is that only 16% of the total study area recorded a shift from historical low-severity to a higher potential for crown fire today. This area of increased fire severity occurs in over half (57.5%) of the area mapped with the historical low-severity fire regime and is concentrated in the lower montane zone. A substantial portion (42.5%) of the area mapped as the historical low-severity fire regime (11.8% of the study area) shows little change in the fire regime (Figure 8), and is expected to support only surface fire even under extreme (99th percentile) weather conditions based on our modeling of present day fuels. Both areas occur primarily in the lower montane zone below 2263 m, but the areas with little change in the low-severity fire regime are on average slightly lower in elevation and slope steepness and closer to grasslands and ravine drainages than areas with higher potential for crown fire today. Observations from recent large wildfires and modeling of potential fire behavior under extreme weather conditions are consistent with the historical evidence of a varied fire regime of primarily low-severity fires at the lowest elevations to a mixed-severity fire regime at higher elevations in montane forests. Three regional fires predominantly in the lower montane forests showed the greatest differences between the observed (MTBS of recent large fires) and expected (historical fire severity) areas burned by low-severity fire. However, the differences between the observed and expected proportions of low-severity fire for the six other fires were non-significant, or showed higher than expected low-severity fire.

A decline in fire frequency over the past 100 years leading to substantial increases in stand density is supported only for the lowest elevations of forest below approximately 2200 m in the Colorado Front Range [13–14,45,51,60,86]. These areas were characterized mainly by frequent (average return intervals <30 years) low-severity fires that maintained open forests by killing mostly juvenile trees, resulting in low densities of mature trees. Greater proximity to grassland in lower elevation areas probably promoted more frequent fire due to more abundant fine, herbaceous fuels [62]. The cessation of formerly frequent fires coincides with increased stand densities broadly throughout the lower montane zone. This pattern is especially evident below 2200 m, but also occurred at some sites at higher elevations on less steep slopes most likely where montane grasslands occurred. However, overall this represents a relatively small proportion of the montane forest of the northern Colorado Front Range (27.8% of the study area is mapped with the historical low-severity fire regime).

The dominant (72.2% of the 564,413 ha study area) historical fire regime of the northern Colorado Front Range consisted of a mixed-severity regime in which stand structures were shaped primarily by moderate-severity (46.5% of sites) and high-severity (45.7% of sites) fires; only 7.8% of the sites recorded predominantly low-severity fires. At higher elevation (>2263 m), spreading fires were typically less frequent (>30 year fire intervals), and had varied fire effects (mixed-severity) that included non-stand replacing fire to canopy-replacement fire both within sites and across broad landscapes, often in relation to topographic variability. Many sites experienced intervals between successive widespread fires that would have been sufficient for conifer seedlings to reach sizes that would survive low-severity surface fires. Specifically, evidence of high-severity fire tended to occur further from grasslands and on steeper sites than those with evidence of only low-severity fire, although there is overlap between sites that show moderate- or high-severity effects particularly above the lower montane zone. Evidence of mixed-severity fires occurred in all dominant cover types sampled including sites of pure (≥80% of the canopy trees) lodgepole pine and ponderosa pine, as well as mixed-conifer types with dominance of aspen, lodgepole pine, Douglas-fir, and ponderosa pine. Consistent site- to landscape-scale evidence indicates an historical mixed-severity fire regime in which moderate- and high-severity fire effects shaped current forest age structures in the mid- and upper montane zone [13,46,51,53] (and this study).

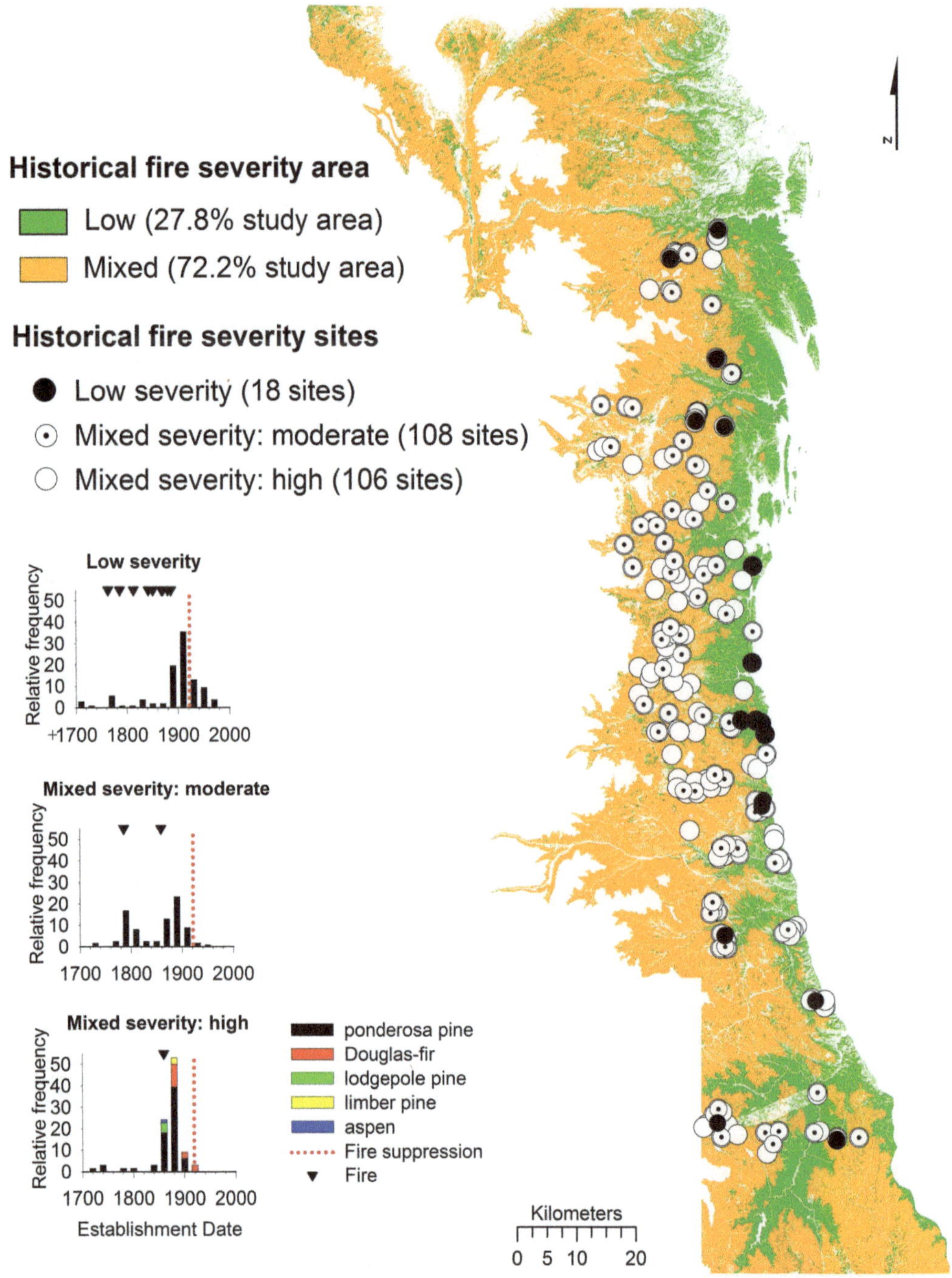

Figure 5. Distribution of 232 sites with historical (pre-1920) evidence of low-severity and mixed-severity fires. Sites with historical evidence of mixed-severity fires are differentiated with evidence of low- and moderate-severity fire (no evidence of high-severity fire; 108 sites) and low- to high-severity fire (106 sites). Areas mapped as historical low-severity fire only (27.8%) and mixed-severity fire (72.2%) are shown in green and orange, respectively. Three example graphs are also shown to illustrate the evidence (remnant trees, tree establishment and spreading fires) used to classify sites with evidence of low and mixed-severity fire effects at individual sites.

Most spreading fires recorded at our sites extended beyond a single site (88% of 322 fires with ≥2 trees scarred per site from 1597–1995). Moderate- and high-severity fires prior to the 20th century were documented at many sites, as illustrated by fires in 1654, 1786 and 1859–60 that extended at least 7–30 km between nearest sampling sites, are within the HRV for montane forests at a regional scale (Figure 3). For example, the spatial extent of sites recording fires in the same year is within the spatial scale of the largest modern wildfires on record in the study area (e.g. High Park Fire and Hayman Fire). Our ability to interpret the spatial extent of historical fire severity patches from individual fires is limited by our sampling unit size (3 ha–232 ha), subsequent fire events, and sample depth (although >20% of our sites record fires since 1700), but the evidence indicates mixed-severity fires occurred in patches up to at least 200 ha, with evidence of approximately 50 ha of high-severity in some sites. Direct

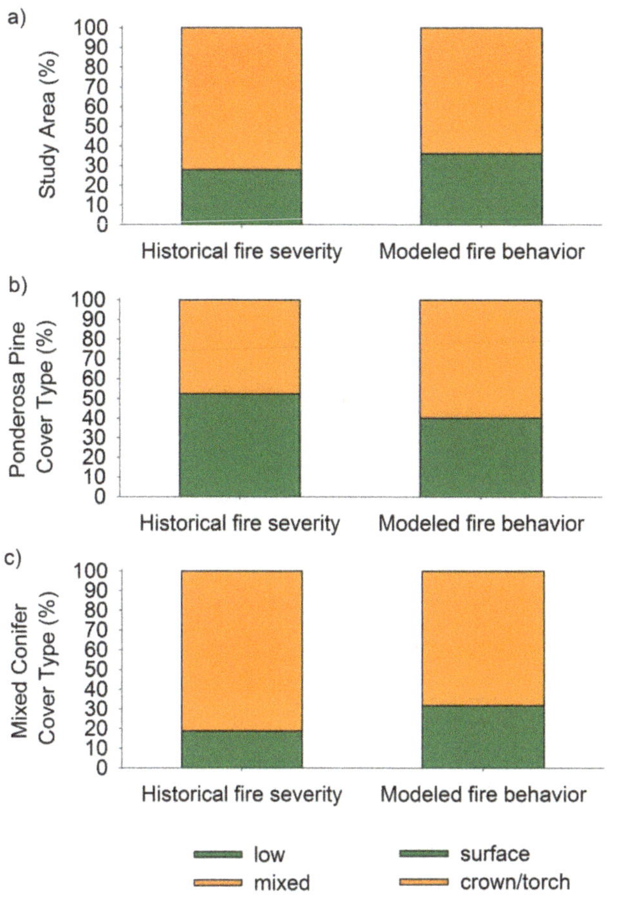

Figure 6. Percent study area and cover type classified with historical fire type and modeled fire behavior. A comparison of the percentage of the a) study area, b) ponderosa pine cover type, and c) mixed conifer cover type classified as historical fire (low-severity only or mixed-severity) and modeled current fire behavior (surface or crown/torch) under extreme weather conditions (99th percentile for 1964–2007).

comparison of patch sizes of fires is difficult given the presence of major highways and other land uses that have fragmented fuel continuity at a landscape scale. It is highly likely that variance in patch size of historical fires was great, similar to modern fire observations, rendering quantitative comparisons of rather limited value. Nevertheless, high-severity fire and high-stand densities are within the HRV for the mid- and upper montane forests (e.g. [44,51,53–54], this study), which needs to be taken into account when forest management goals consider restoring forests to pre-fire-exclusion conditions.

In the Colorado Front Range, at higher elevations (>2263 m) mixed-severity fire regimes appear to have been the predominant fire regime historically and today. The modeled potential wildfire behavior under current fuel conditions and extreme fire weather shows a mixed-severity fire regime throughout the montane forests of the Colorado Front Range. High fireline intensity and active crown fire are within the HRV of a mixed-severity fire regime particularly under extreme conditions, but would vary in extent of intensity and canopy mortality depending on existing conditions. Site-level evidence indicates high overlap in biophysical conditions that support moderate and high-severity fires (this study and [52]). These results provide a snapshot of the expectations into the near future for fire behavior in the Colorado Front Range. The advantage of comparing historical and observed fire severity with fire behavior modeling across the study area under realistic, but also extreme, weather conditions is that it provides an important present-day comparison to the past. Recent large wildfire years in the Colorado Front Range are not without precedent and similar events occurred historically (i.e. 1654, 1786, 1859–60), under similar exceptional (interannual to multi-decadal scale) climate conditions [61]. The largest fires since the year 2000 – the Hayman and High Park Fires – were both characterized by up to 98–99th percentile conditions for both temperature and relative humidity. The situation will likely be exacerbated under climate change; under the 2007 IPCC A2 scenario for the study area, the 99th percentile temperatures of the past are expected to become the 39th percentile temperatures by 2064–2099 (also see [87]). Thus, the range of weather conditions during local to more widespread fire years in the last 300+ years likely represents a similar range of potential fire-climate conditions presently, and informs expectations of potential fire behavior that may occur under severe, yet increasingly more common weather conditions.

We recognize that each of our spatial datasets (mapping of historical, observed, and potential fire) have their own set of limitations. For example, distance to grassland was not a predictor in the best-fit model used to map the historical fire severity landscape. However, our prior research indicates areas adjacent to grasslands experience more frequent fires than sites farther from grasslands, owing to the proximity to prevalent fine fuels that increase the sensitivity of fire activity to interannual climate variability [62]. Thus, the historical model likely under-estimates the amount of low-severity fire in some areas at higher elevation within or adjacent to grasslands where we have limited or no sample sites (e.g. high elevation grassland areas in the northern portion of the study area - Figures 1 and 8). This could explain why some of the study area (Figure 8) shows a shift from historical mixed-severity fire to potential low-severity fire in the contemporary forest even under the extreme weather scenario. The observed fire behavior dataset (MTBR Thematic Burn Severity) has its own limitations. To classify fire severity into 'high' 'moderate', and 'low' severity classes, image analysts determine thresholds in dNBR values based on visual interpretation of imagery, field plot data, and expert knowledge. Though dNBR is sensitive to initial vegetation conditions [88], alternatives such as RdNBR are less parsimonious and in practice may not capture burn severity with greater accuracy than dNBR [89]. Also, while the fire behavior modeling from FlamMap operates on a fixed set of fuels and weather conditions, the historical severity model is based on the fire regime over a long period of time during which fires of different severity have the opportunity to burn. Because of these limitations, we report proportional area of landscape-level changes rather than finer-scale interpretations, and emphasize broad trends evident from multiple lines of inquiry. Additionally, while our model was based on the most recently available fuels data from LANDFIRE, characterizing vegetation changes across the landscape due to wildfire and management through 2010, mortality from recent insect outbreaks, for example, are not reflected. Inferring the effect of fuel and fire behavior changes in our model due to tree mortality from insects is challenging as such fuel changes are: 1) highly dependent on the timing and severity of the outbreak, 2) not characterized well in standard fuel models, and 3) not validated for fire behavior as few empirical studies have examined how the timing and extent of insect outbreaks in montane forests affects observed fire behavior.

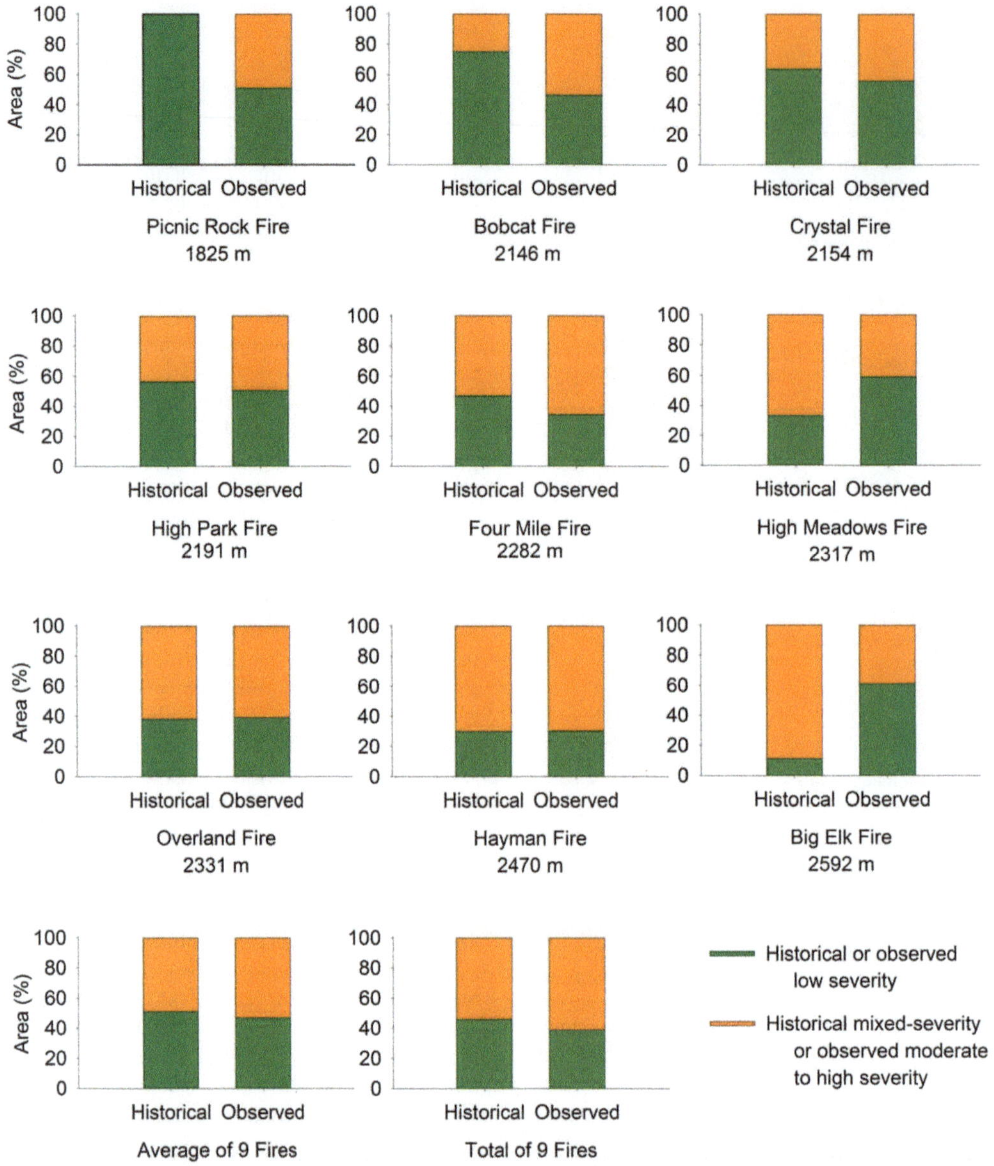

Figure 7. Comparisons of the historical and observed fire severities for nine recent fires (2000–2012). The comparison shows the proportion of low- (unburned/low and low) and mixed- (moderate and high) severity fire within the perimeter of nine recent fires (2000–2012), and the average and total (all pixels) proportions of fire severities across all nine fires in the study area. The median elevation for each fire is given under the fire name.

Implications for fuels management and ecological restoration

A clear delineation of the spatial extent of past fire regime types is a major concern for ecosystem managers in the context of wildfire risk and ecological restoration [90–92]. Evidence from previous studies in ponderosa pine and other montane forests in the Colorado Front Range have shown that the historical fire regime was variable over time and space, represented by a mixed-severity regime [44,46–48,50–54]. Goals of ecological restoration and wildland fire hazard mitigation are both compatible with management practices, like prescribed fire and thinning to reduce fuels, below approximately 2200 m in our study area, which experienced the greatest increase in fire severity, and likely fuels, since fire exclusion [93]. Disturbance from grazing and logging as well as periods of favorable climate probably also contributed to increased tree establishment in the late 19th and early 20th century, but seedling survival clearly depended on the long fire-free periods. Even at low elevations, however, some sites had historical fire regimes dominated by infrequent rather than frequent fires (e.g. steep and north-facing slopes [86]). These infrequent fires, inferred to be high-severity fires from age structure data, killed high percentages of trees within a fire perimeter, and promoted the establishment of naturally dense forest patches [44,51,86]. This suggests that some of the areas (e.g. north-facing and steep slopes) with potential for crown fire today may support characteristic fire behavior and are not necessarily out of the HRV even in the lower montane zone.

Mixed-severity fire regimes where spreading fires occurred at lower frequencies (>30 year fire intervals) are less clearly candidates for thinning than are low-severity fire regimes and a

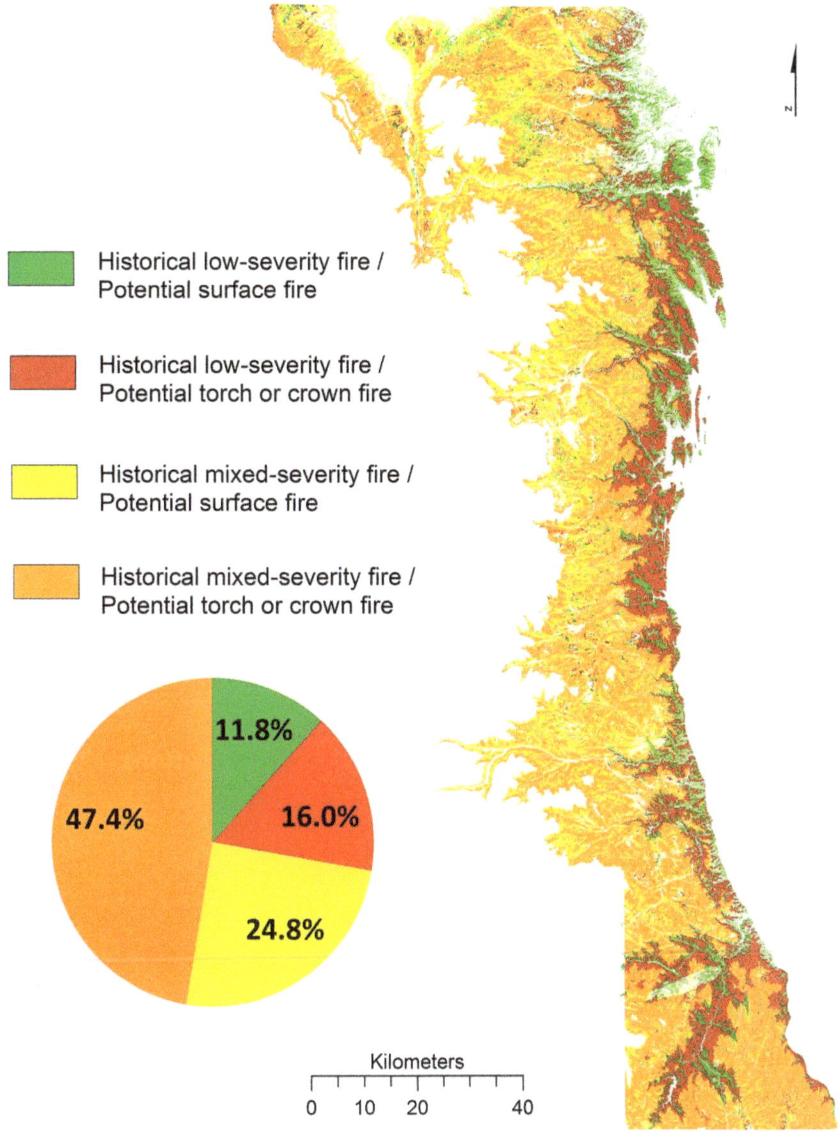

Figure 8. Historical fire severity overlaid with a model of fire potential under extreme (99th percentile weather). The areas and proportion of the study area mapped as historical low-severity fire with current potential for surface fire (green), historical low-severity fire with current potential for torch or crown fire (red), historical mixed-severity with current potential for surface fire (yellow), and historical mixed-severity with current potential for torch or crown fire (orange). Current potential fire behavior is modeled under extreme (99th percentile) weather (1964–2007).

cautious approach to restoration efforts has been recommended [15,40]. In areas naturally characterized by lower frequencies of moderate- to high-severity fires, fuel reduction through prescribed fire and thinning will likely not achieve both ecological restoration and fire hazard mitigation goals. Restoration thinning treatments will not return the fire regime to one of low severity across the Front Range montane zone, which was historically predisposed to periodic fires of varying severities, and are of questionable effectiveness in preventing severe wildfires [2]. Where extreme fire behavior appears within HRV, high-severity fire may largely be explained by extreme weather conditions (for example, high winds and low humidity during severe drought) rather than quantity of woody fuels [38]. This is illustrated by the 2002 Hayman Fire in Colorado in which more than 24,000 ha burned at high severity throughout ponderosa pine and mixed-conifer forests of variable stand structures in a single day [94]. The fire-weather conditions presented here (99th percentile for the late 20th century and early 21st century) represent only the projected average expected conditions by the mid-21st century, which is a pressing issue for existing and future WUI development [95] and fire management [2]. Learning from recent experiences where wildfire damage has been high in the WUI [2], along with considering the costs of suppressing future wildland fires (i.e. reducing forest resiliency, unsustainable federal costs to society [1]) and the sociopolitical expectations of wildland fire management are critical for managing future fire risks and forest integrity.

For the ponderosa pine cover type, an increasing number of studies in the Pacific Northwest [15,96–100] have also documented fire regimes of variable severity. These studies and ours illustrate the importance of collecting evidence on the severity of past fires (i.e. remnant tree and cohort ages, growth releases, tree mortality) because fire severity cannot be reliably interpreted from fire interval data alone, which has often the approach applied to the study of fire history in ponderosa pine forests. Although current implementation of national programs (Collaborative Forest Landscape Restoration Program) and databases (LANDFIRE) recognize variation in fire regimes within the same cover type across different geographical regions, fire regimes of a cover type within a region can still be considered relatively uniform. This is problematic in the Front Range and potentially elsewhere where the historical fire regime varied from low-severity at the lowest elevations to mixed-severity over most of the montane zone regardless of forest type. Additionally, the accuracy of the current cover type (EVT) layer for the region is low. Thus, the spatial and proportional reliability of the LANDFIRE classifications of fire regimes and vegetation do not accurately reflect the historical or current landscape, and therefore will not be highly effective in prioritizing locations for fire management and forest restoration objectives. Further corroboration of these issues are also documented at a national scale (see p. 19 in [92]) "The accuracy of various aspects of the LANDFIRE data is questionable, even when used at intended scale...Without accurate data, many assumptions and actions based on this data will be compromised. There is a need for more realistic and accurate depiction of where wildland fire hazard/risk actually occurs across the country, which can be used to base decisions upon."

Our study, along with prior studies, shows that an assumption of fire regime uniformity is not valid for the ponderosa pine and other montane cover types in the Colorado Front Range and elsewhere (i.e. [15]). These findings challenge ecologists and managers to re-consider the degree of variation in fire regimes within broadly distributed forest types. Thus, management efforts to create large areas of open woodlands in the higher elevation areas of the montane zone of the Colorado Front Range would not be consistent with historical fire regimes and stand structures.

Supporting Information

Figure S1 MTBS fire severity classification (observed) and median fire line intensity (modeled). The comparison was used as a measure of verification of the modeled fire behavior.

Figure S2 MTBS fire severity class (observed) and percentage of crown fire (modeled). The comparison was used as a measure of verification of the modeled fire behavior.

Table S1 Summary table of biophysical factors for each observed fire and potential fire model in analysis.

Table S2 List of plot-level sites. The fire severity classification for each site was documented from forest structure and fire scars in the field using an existing protocol [52].

Acknowledgments

We thank student assistants from Humboldt State University (Alicia Iverson, Kelly Muth, Madelinn Schriver) and Gettysburg College (Erin Palmer) for help with compiling datasets. We also thank Cameron Naficy, Ben Bond-Lamberty, and three anonymous reviewers for insightful comments that improved the manuscript.

Author Contributions

Conceived and designed the experiments: RLS RVP TTV TLS. Performed the experiments: RLS RVP TTV TLS MHG. Analyzed the data: RLS RVP. Wrote the paper: RLS RVP TTV TLS MHG.

References

1. Gorte R (2013) The rising cost of wildfire protection. Headwaters Economic Research Report. http://headwaterseconomics.org/wildfire/fire-cost-background. Accessed 2014 August 23.
2. Calkin DE, Cohen JD, Finney MA, Thompson MP (2013) How risk management can prevent future wildfire disasters in the wildland-urban interface. Proceedings of the National Academy of Science 756–751, doi: 10.1073/pnas.1315088111.
3. Mackun P, Wilson S (2011) Population distribution and change: 2000 to 2010. U.S. Department of Commerce Economics and Statistics Administration. U.S. Census Bureau. http://www.census.gov/prod/cen2010/briefs/c2010br-01.pdf. Accessed 2014 Aug 23.
4. Williams J (2013) Exploring the onset of high-impact mega-fires through a forest land management prism. Forest Ecology and Management 294: 4–10.
5. Stephens SL, Agee JK, Fulé PZ, North MP, Romme WH, et al. (2013) Managing forests and fire in changing climates. Science 342 (6154): 41–42, doi 10.1126/science.12402994.
6. Dillon GK, Holden ZA, Morgan P, Crimmins MA, Heyerdahl EK, et al. (2011) Both topography and climate affected forest and woodland burn severity in two regions of the western US, 1984–2006. Ecosphere 2: Article 130. http://www.esajournals.org/doi/pdf/10.1890/ES11-00271.1.
7. Odion DC, Hanson CT (2008) Fire severity in the Sierra Nevada revisited: conclusions robust to further analysis. Ecosystems 11: 12–15.
8. Miller JD, Safford HD, Crimmins M, Thode AE (2009) Quantitative evidence for increasing forest fire severity in the Sierra Nevada and southern Cascade Mountains, California and Nevada, USA. Ecosystems 12: 16–32.
9. Miller JD, Skinner CN, Safford HD, Knapp EE, Ramirez CM (2012) Trends and causes of severity, size, and number of fires in northwestern California, USA. Ecological Applications 22: 184–203.
10. Miller JD, Collins BM, Lutz JA, Stephens SL, van Wagtendonk JW, et al. (2012) Differences in wildfires among ecoregions and land management agencies in the Sierra Nevada region, California, USA. Ecosphere 3: Article 80.
11. Miller JD, Safford HD (2012) Trends in wildfire severity: 1984 to 2010 in the Sierra Nevada, Modoc Plateau, and southern Cascades, California, USA. Fire Ecology 8: 41–57.
12. Hanson CT, Odion DC (2014) Is fire severity increasing in the Sierra Nevada mountains, California, USA? International Journal of Wildland Fire. 23: 1–8.
13. Williams MA, Baker WL (2012) Comparison of the Higher-Severity Fire Regime in Historical (A.D. 1800s) and Modern (A.D. 1984–2009) Montane Forests Across 624,156 ha of the Colorado Front Range. Ecosystems 15: 832–847.
14. Fulé PZ, Swetnam TW, Brown PM, Falk DA, Peterson DL, et al. (2013) Unsupported inferences of high-severity fire in historical dry forests of the western United States: response to Williams and Baker. Global Ecology and Biogeography. doi: 10.1111/geb.12136.
15. Odion DC, Hanson CT, Arsenault A, Baker WL, DellaSalla DA, et al. (2014) Examining historical and current mixed-severity fire regimes in ponderosa pine and mixed-conifer forests of western North America. PLoS ONE. doi: 10.1371/journal.pone.0087852.
16. Morgan P, Aplet GH, Haufler JB, Humphries HC, Moore MM, et al. (1993) Historical Range Of Variability: A Useful Tool for Evaluating Ecosystem Change. In: Sampson RN, Adams DL, editors. Assessing Forest Ecosystem Health in the Inland West. Sun Valley: Haworth Press. pp. 87–109.
17. Keane RE, Hessburg PF, Landres PB, Swanson FJ (2009) The use of historical range of variability (HRV) in landscape management. Forest Ecology and Management 258: 1025–1037.
18. Hayward GD, Veblen TT, Suring LH, Davis B (2012) Challenges in the application of historical range of variation to conservation and land management. *In* Historical Environmental Variation in Conservation and Natural Resource Management. *Eds.* Wiens JA, Hayward GD, Safford HD, Giffen CM. John Wiley & Sons, Ltd.
19. Millar CI, Stephenson NL, Stephens SL (2007) Climate change and forests of the future: managing in the face of uncertainty. Ecological Applications 17: 2145–2151.
20. Wiens JA, Hayward GD, Safford HD, Giffen C (2012) Historical environmental variation in conservation and natural resource management. John Wiley & Sons.
21. Moritz MA, Hurteau MD, Suding KN, D'Antonio CM (2013) Bounded ranges of variation as a framework for future conservation and fire management. Annals of the New York Academy of Sciences 1286: 92–107.
22. DeLeo GA, Levin S (1997) The Multifaceted Aspects of Ecosystem Integrity. Conservation Ecology 1: 3.
23. Parrish JD, Braun DP, Unnasch RS. (2003) Are we conserving what we say we are? Measuring ecological integrity within protected areas. Bioscience 53: 851–860.

24. Healthy Forests Restoration Act. HR 1904 (2003) http://www.forestsandrangelands.gov/resources/overview/hfra-implementation12-2004.shtml. Accessed 2014 August 23.
25. Peterson DL, Millar CI, Joyce LA, Furniss MJ, Halofsky JE, et al. (2011) Responding to climate change in National Forests: A guidebook for developing adaptation options. USDA Forest Service General Technical Report PNW-GTR-855.
26. Wildland Fire Executive Council (2014) The National Strategy: The Final Phase in the Development of the National Cohesive Wildland Fire Management Strategy. http://www.forestsandrangelands.gov/strategy/documents/strategy/CSPhaseIIINationalStrategyApr2014.pdf. Accessed 2014 August 23.
27. Brown JK, Oberheu RD, Johnston CM (1982) Handbook for inventorying surface fuels and biomass in the Interior West. USDA Forest Service General Technical Report. INT-129.
28. Covington WW, Moore M (1994) Southwestern Ponderosa Forest Structure: changes since Euro-American settlement. Journal of Forestry 92: 39–47.
29. Covington WW (2000) Helping western forests heal. Nature 408: 135–136.
30. Agee JK (2002) The Fallacy of Passive Management Managing for Firesafe Forest Reserves. Conservation in Practice 3: 18–26.
31. Allen CD, Savage M, Falk DA, Suckling KF, Swetnam TW, et al. (2002) Ecological restoration of southwestern ponderosa pine ecosystems: a broad perspective. Ecological Applications 12: 1418–1433.
32. Agee JK, Skinner CN (2005) Basic principles of forest fuel reduction treatments. Forest Ecology and Management 211: 83–96.
33. Stephens SL, Ruth LW (2005) Federal forest-fire policy in the United States. Ecological Applications 15: 532–542.
34. Roos CI, Swetnam TW (2012) A 1416-year reconstruction of annual, multidecadal, and centennial variability in area burned for ponderosa pine forests of the southern Colorado Plateau region, Southwest USA. The Holocene 22: 281–290.
35. United States Forest Service (2002) National Database of State and Local Wildfire Hazard Mitigation Programs, http://www.wildfireprograms.usda.gov. Accessed 2014 August 23.
36. Mell WE, Manzello SL, Maranghides A, Butry DT, Rehm RG (2010) The wildland–urban interface fire problem – current approaches and research needs. International Journal of Wildland Fire 19: 238–251.
37. Veblen TT (2003) An introduction to key issues in fire regime research for fuels management and ecological restoration. In: Omi P, Joyce L, Eds. Fire, fuel treatments and ecological restoration: conference proceedings 2002, 16–18 April, Fort Collins, CO. USDA, Forest Service, RMRS-P-29.
38. Schoennagel T, Veblen TT, Romme WH (2004) The Interaction of Fire, Fuels, and Climate across Rocky Mountain Forests. BioScience 54: 661–676.
39. Nagel TA, Taylor AH (2005) Fire and persistence of montane chaparral in mixed conifer forest landscapes in the northern Sierra Nevada, Lake Tahoe Basin, California, USA. Journal of the Torrey Botanical Society 132: 442–457.
40. Baker WL, Veblen TT, Sherriff RS (2007) Fire, fuels and restoration of ponderosa pine–Douglas fir forests in the Rocky Mountains, USA. Journal of Biogeography 34: 251–269.
41. Klenner W, Walton R, Arsenault A, Kremsater L (2008) Dry forests in the Southern Interior of British Columbia: Historic disturbances and implications for restoration and management. Forest Ecology and Management 256: 1711–1722.
42. Hutto RL (2008) The ecological importance of severe wildfires: some like it hot. Ecological Applications 18: 1827–1834.
43. Colombaroli D, Gavin DG (2010) Highly episodic fire and erosion regime over the past 2,000 y in the Siskiyou Mountains, Oregon. Proceedings of the National Academy of Sciences 107: 18909–18915.
44. Veblen TT, Lorenz DC (1986) Anthropogenic disturbance and recovery patterns in montane forests, Colorado Front Range. Physical Geography 7: 1–24.
45. Veblen TT, Lorenz DC (1991) The Colorado Front Range: A Century of Ecological Change. Salt Lake: University of Utah Press. 186 p.
46. Mast JN, Veblen TT, Linhart YB (1998) Disturbance and climatic influences on age structure of ponderosa pine at the pine/grassland ecotone, Colorado Front Range. Journal of Biogeography 25: 743–755.
47. Brown P, Kaufmann M, Shepperd W (1999) Long-term, landscape patterns of past fire events in a montane ponderosa pine forest of central Colorado. Landscape Ecology 14: 513–532.
48. Kaufmann MR, Regan CM, Brown PM (2000) Heterogeneity in ponderosa pine/Douglas-fir forests: age and size structure in unlogged and logged landscapes of central Colorado. Canadian Journal of Forest Research 30: 698–711.
49. Huckaby LS, Kaufmann MR, Stoker JM, Fornwalt PJ (2001) Landscape Patterns of Montane Forest Age Structure Relative to Fire History at Cheesman Lake in the Colorado Front Range. U.S.D.A. Forest Service Proceedings RMRS-P-22.
50. Ehle DS, Baker WL (2003) Disturbance and stand dynamics in ponderosa pine forests in Rocky Mountain National Park, USA. Ecological Monographs 73: 543–566.
51. Sherriff RL, Veblen TT (2006) Ecological effects of changes in fire regimes in Pinus ponderosa ecosystems in the Colorado Front Range. Journal of Vegetation Science 17: 705–718.
52. Sherriff RL, Veblen TT (2007) A spatially-explicit reconstruction of historical fire occurrence in the ponderosa pine zone of the Colorado Front Range. Ecosystems 10: 311–23.
53. Schoennagel T, Sherriff RL, Veblen TT (2011) Fire history and tree recruitment in the Colorado Front Range upper montane zone: implications for forest restoration. Ecological Applications 21: 2210–2222.
54. Williams MA, Baker WL (2012) Spatially extensive reconstructions show variable-severity fire and heterogeneous structure in historical western United States dry forests. Global Ecology and Biogeography 21: 1042–1052.
55. Romme WH, Veblen TT, Kaufmann MR, Sherriff RL, Regan CM (2003) Ecological effects of the Hayman fire, Part I: Historical (pre-1860) and current (1860–2002) fire regimes. In: Graham RT, editor. Hayman Fire Case Study. Fort Collins CO, Rocky Mountain Research Station: USDA Forest Service, General Technical Report RMRS-GTR-114. pp. 181–195.
56. Kaufmann MR, Veblen TT, Romme WH (2006) Historical fire regimes in ponderosa pine forests of the Colorado Front Range, and recommendations for ecological restoration and fuels management. Front Range Fuels Treatment Partnership Roundtable, findings of the Ecology Workgroup. www.frftp.org/roundtable/pipo.pdf.
57. Colorado Climate Center (2013) http://climate.atmos.colostate.edu/. Accessed 2014 August 23.
58. Graham RT (2003) Hayman Fire Case Study. Gen. Tech. Rep. RMRS-GTR-114. Ogden, UT: U.S. Department of Agriculture, Forest Service, Rocky Mountain Research Station. 396 p.
59. National Fire Protection Association Codes and Standards (1989) www.nfpa.org. Accessed 2014 August 23.
60. Veblen TT, Kitzberger T, Donnegan J (2000) Climatic and Human Influence on Fire Regimes in Ponderosa Pine Forests in the Colorado Front Range. Ecological Applications 10: 1178–1195.
61. Sherriff RL, Veblen TT (2008) Variability in fire–climate relationships in ponderosa pine forests in the Colorado Front Range. International Journal of Wildland Fire 17: 50–59.
62. Gartner MH, Veblen TT, Sherriff RL, Schoennagel TL (2012) Proximity to grasslands influences fire frequency and sensitivity to climate variability in ponderosa pine forests of the Colorado Front Range. International Journal of Wildland Fire 21: 562–571.
63. Duncan (1989) An evaluation of errors in tree age estimates based on increment cores in Kahikatea (Dacrycarpus dacrydioides). New Zealand Natural Sciences Journal 16: 31–37.
64. Barrett SW, Arno SE (1998) Increment-borer methods for determining history in coniferous forests. USDA Forest Service, General Technical Report INT-244.
65. Agee JK (1993) Fire Ecology of the Pacific Northwest Forest. Island Press, Washington DC. pp. 493.
66. Hessburg P, Salter RB, James K (2007) Re-examining fire severity relations in pre-management era mixed conifer forests: inferences from landscape patterns of forest structure. Landscape Ecology 22: 5–24.
67. Wilson JP, Gallant JC (2000) Terrain analysis: principles and applications. New York: Wiley.
68. Jenson SK, Domingue JO (1988) Extracting topographic structure from digital elevation data for geographic information system analysis. Photogramm Eng Remote Sens 54: 1593–1600.
69. Breiman L, Friedman JH, Olshen RA, Stone CJ (1984) Classification and Regression Trees. Monterrey: Wadsworth and Brooks/Cole. 358 p.
70. Breiman L (2001) Random Forests. Machine Learning 45: 5–32.
71. Moisen GG, Frescino TS (2002) Comparing five modelling techniques for predicting forest characteristics. Ecological Modelling 157: 209–225.
72. Qian SS (2010) Environmental and Ecological Statistics. Boca Raton: RC Press.
73. Eidenshink J, Schwind B, Brewer K, Zhu Z, Quayle B, et al. (2007) A project for monitoring trends in burn severity. Fire Ecology 3: 3–21.
74. Scott JH, Burgan RE (2005) Standard fire behavior fuel models: A comprehensive set for use with Rothermel's surface fire spread model. In: Fort Collins, CO: USDA Forest Service Rocky Mountain Research Station, General Technical Report RMRS-GTR-153. 72 p.
75. LANDFIRE (2010) U.S. Department of Agriculture, Forest Service, U.S. Department of Interior. http://www.landfire.gov/index.php. Accessed 2014 August 23.
76. Krasnow K, Schoennagel T, Veblen TT (2009) Forest fuel mapping and evaluation of LANDFIRE fuel maps in Boulder County, Colorado, USA. Forest Ecology and Management 257: 1603–1612.
77. Finney MA (2006) An overview of FlamMap fire modeling capabilities. In: Fuels management-how to measure success: conference proceedings. 2006 March 28–30; Portland, Oregon. Proceedings RMRS-P-41. Fort Collins, CO: U.S. Department of Agriculture, Forest Service, Rocky Mountain Research Station: 213–220.
78. Platt RV, Schoennagel T, Veblen TT, Sherriff RL (2011) Modeling wildfire potential in residential parcels: A case study of the north-central Colorado Front Range. Landscape and Urban Planning 102: 117–126.
79. Rothermel RC (1972) A mathematical model for predicting fire spread in wildland fuels. Ogden, UT: USDA Forest Service Intermountain Forest and Range Experiment Station, General technical report INT-115.
80. Van Wagner CE (1977) Conditions for the start and spread of crown fire. Canadian Journal of Forest Research 7: 23–34.

81. Rothermel RC (1991) Predicting behavior and size of crown fires in the Northern Rocky Mountains. Ogden, UT: USDA Forest Service Intermountain Forest and Range Experiment Station, Research paper INT-438. 46 p.
82. Nelson RM (2000) Prediction of diurnal change in 10-hour fuel moisture content. Canadian Journal of Forest Research 31: 1071–1087.
83. Scott JH, Reinhardt ED (2001) Assessing crown fire potential by linking models of surface and crown fire behavior. In: Fort Collins CO: U.S. Department of Agriculture, Forest Service, Rocky Mountain Research Station, Research Paper RMRS-RP-29. 59 p.
84. Intergovernmental Panel on Climate Change (2007) https://www.ipcc.ch/publications_and_data/publications_and_data_reports.shtml. Accessed 2014 August 23.
85. NCAR Science and Research Data for IPCC AR4 Data Guide (2012) http://eos-earthdata.sr.unh.edu/data/dataGuides/ci_science_dg.pdf. Accessed 2014 August 23.
86. Sherriff RL (2004) The historic range of variability of ponderosa pine in the northern Colorado Front Range: past fire types and fire effects. Ph.D. Dissertation. University of Colorado, Boulder, Colorado.
87. Rocca ME, Brown PM, MacDonald LH, Carrico CM (2014) Climate change impacts on fire regimes and key ecosystem services in Rocky Mountain forests. Forest Ecology and Management 327: 290–305.
88. Miller JD, Thode AE (2007) Quantifying burn severity in a heterogeneous landscape with a relative version of the delta Normalized Burn Ratio (dNBR). Remote Sensing of Environment 109: 66–80.
89. Soverel NO, Perrakis DDB, Coops NC (2010) Estimating burn severity from Landsat dNBR and RdNBR indices across western Canada. Remote Sensing of Environment 114: 1896–1909.
90. Front Range Fuels Treatment Partnership (2006) Living with Fire: Protecting Communities and Restoring Forests – Findings and Recommendations of the Front Range Fuels Treatment Partnership Roundtable. http://frftp.org/roundtable/report.pdf. Accessed 2014 August 23.
91. Schoennagel T, Nelson CR (2011) Restoration relevance of recent National Fire Plan treatments in forests of the western United States. Frontiers in Ecology and the Environment 9: 271–277.
92. The National Cohesive Wildland Fire Management Strategy and Risk Analysis – Phase III Report (2013) http://www.forestsandrangelands.gov/. Accessed 2014 August 23.
93. Platt RV, Veblen TT, Sherriff RL (2006) Are Wildfire Mitigation and Restoration of Historic Forest Structure Compatible? A Spatial Modeling Assessment. Annals of the Association of American Geographers 96: 455–470.
94. Finney MA, Bartlette R, Bradshaw L, Close K, Collins BM, et al. (2003) Fire behavior, fuel treatments, and fire suppression on the Hayman Fire. *In* Hayman Fire case study. Ed. Graham RT. USDA Forest Service, Rocky Mountain Research Station, General Technical Report RMRS-GTR-114, pp. 33–179.
95. Theobald DM, Romme WH (2007) Expansion of the US wildland-urban interface. Landscape and Urban Planning 83: 340–354.
96. Arno SF, Scott JH, Hartwell MG (1995) Age-class structure of old growth ponderosa pine/Douglas-fir stands and its relationship to fire history. USDA Forest Service, Research Paper INT-481.
97. Wright CS, Agee JK (2004) Fire and vegetation history in the eastern Cascade mountains. Ecological Applications 14: 443–459.
98. Everett RL, Schellhaas R, Keenum D, Spurbeck D, Ohlson P (2000) Fire history in the ponderosa pine/Douglas-fir forests on the east slope of Washington Cascades. Forest Ecology and Management 129: 207–225.
99. Hessburg PF, Salter RB, James KM (2007) Re-examining fire severity relations in pre-management era mixed conifer forests: inferences from landscape patterns of forest structure. Landscape Ecology 22: 5–24.
100. Perry DA, Hessburg PF, Skinner CN, Spies TA, Stephens SL, et al. (2011) The ecology of mixed severity fire regimes in Washington, Oregon, and northern California. Forest Ecology and Management 262: 703–717.

9

Unmanned Aerial Survey of Fallen Trees in a Deciduous Broadleaved Forest in Eastern Japan

Tomoharu Inoue[1]*, Shin Nagai[1], Satoshi Yamashita[2], Hadi Fadaei[1], Reiichiro Ishii[1], Kimiko Okabe[2], Hisatomo Taki[2], Yoshiaki Honda[3], Koji Kajiwara[3], Rikie Suzuki[1]

1 Department of Environmental Geochemical Cycle Research, Japan Agency for Marine-Earth Science and Technology (JAMSTEC), Yokohama, Japan, **2** Department of Forest Entomology, Forestry and Forest Products Research Institute (FFPRI), Tsukuba, Ibaraki, Japan, **3** Center of Environmental Remote Sensing, Chiba University, Chiba, Japan

Abstract

Since fallen trees are a key factor in biodiversity and biogeochemical cycling, information about their spatial distribution is of use in determining species distribution and nutrient and carbon cycling in forest ecosystems. Ground-based surveys are both time consuming and labour intensive. Remote-sensing technology can reduce these costs. Here, we used high-spatial-resolution aerial photographs (0.5–1.0 cm per pixel) taken from an unmanned aerial vehicle (UAV) to survey fallen trees in a deciduous broadleaved forest in eastern Japan. In nine sub-plots we found a total of 44 fallen trees by ground survey. From the aerial photographs, we identified 80% to 90% of fallen trees that were >30 cm in diameter or >10 m in length, but missed many that were narrower or shorter. This failure may be due to the similarity of fallen trees to trunks and branches of standing trees or masking by standing trees. Views of the same point from different angles may improve the detection rate because they would provide more opportunity to detect fallen trees hidden by standing trees. Our results suggest that UAV surveys will make it possible to monitor the spatial and temporal variations in forest structure and function at lower cost.

Editor: Dale A. Quattrochi, NASA Marshall Space Flight Center, United States of America

Funding: This study was funded by the Environment Research and Technology Development Fund (S-9) of the Ministry of the Environment of Japan. The funders had no role in study design, data collection and analysis, decision to publish, or preparation of the manuscript.

Competing Interests: The authors have declared that no competing interests exist.

* Email: tomoharu@jamstec.go.jp

Introduction

Fallen trees are an ecologically relevant indicator of forest biodiversity [1], since they provide habitat for many species, such as small animals (e.g., birds, mammals, and insects) and fungi [2–4]. In addition, the decomposition of fallen trees is an important mechanism driving biogeochemical cycles such as nutrient and carbon cycling [5–6]. Thus, information about their spatial distribution is of use in indicating species distribution, biodiversity, and biogeochemical cycles in forest ecosystems [7–8]. Fallen trees are typically assessed by field surveys [9]. However, because field surveys are time consuming and labour intensive [10], it is expensive to assess the spatial distribution of fallen trees over a wide area. One way to reduce costs is to use remote-sensing technologies such as airborne and satellite imagery (e.g., [10–14]). For example, some studies have utilized the remote-sensing technologies for assessment of post-hurricane forest damage (e.g., [15–16]).

Recent advances in the technology of unmanned aerial vehicles (UAVs) have made UAVs ideal for remote sensing [17–20]. Equipped with sensors such as digital cameras, UAVs can gather aerial photographs with fine spatial and temporal resolution [20]. They also have greater flexibility in flying height and schedule and lower operating costs than manned aircraft [18,19,21]. For these reasons, aerial surveys using UAVs have been suggested for low-cost ecological research (e.g., vegetation monitoring and wildlife surveys) [19,20,22–26].

In this study, we photographed a deciduous broadleaved forest from a UAV, compared the positions of fallen trees identified in the images by eye with those recorded on the ground, and compared the number of the fallen trees detected between the original images and an orthorectified mosaic. The purpose of this study was to show the applicability and limits of this technique.

Materials and Methods

Study site

The study was conducted in a 300-m × 200-m plot in the Ogawa Forest Reserve (OFR) in Kitaibaraki, Ibaraki, Japan (36°56′10″N, 140°35′18″E, 650–700 m above sea level; Figs. 1, 2). The forest was dominated by deciduous broadleaved trees such as oak (*Quercus serrata*) and beech (*Fagus japonica* and *F. crenata*) [27], with a patchy distribution of dwarf bamboos on the forest floor [28]. The OFR is described in detail by Nakashizuka and

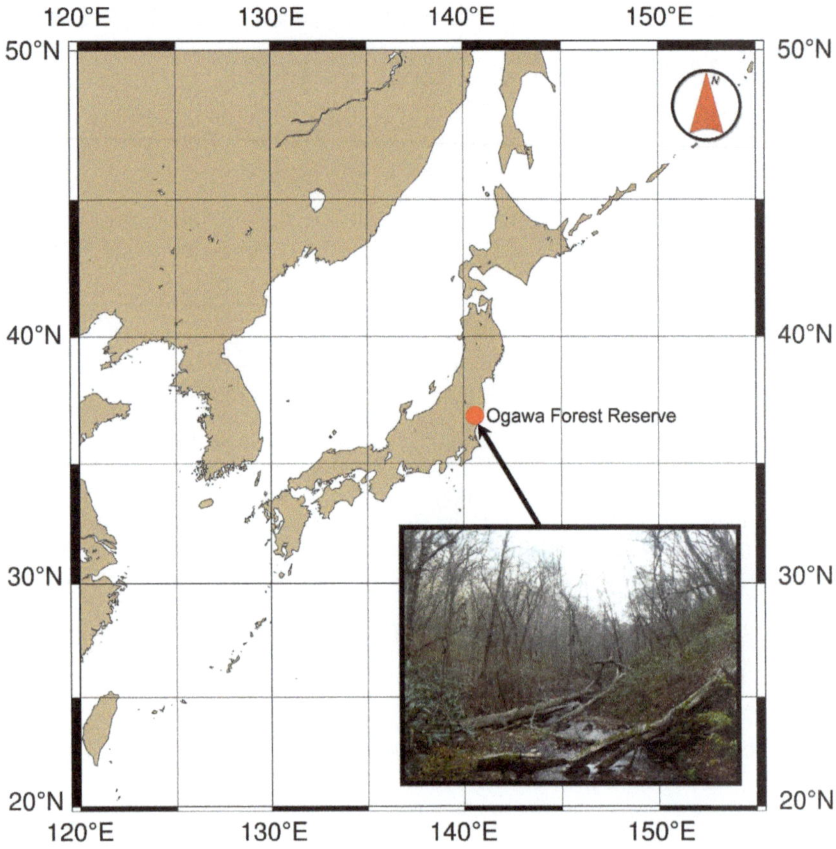

Figure 1. Location of the Ogawa Forest Reserve (OFR).

Matsumoto [29]. This study was conducted including a national forest under the permission of the Ibaraki district forest office of the Forestry Agency in Japan.

Collection of low-altitude aerial photographs from a UAV

On 29 November 2011, when the trees were bare, we flew a UAV (RMAX-G1 helicopter, Yamaha-Motor Co. Ltd., Shizuoka, Japan; Fig. 3; a model commonly used in Japan (e.g., [30–32])) over the OFR plot at 30 to 70 m above the ground in a north–south orientation at 3 m s^{-1}. A consumer-grade digital camera with a 35-mm lens (EOS Kiss X5, Canon, Tokyo, Japan; image sensor 14.9 mm × 22.3 mm) mounted beneath the UAV pointing straight down took images (5184 × 3456 pixels) every 5 s. The UAV was also equipped with a global positioning system (GPS) detector that recorded altitude, latitude, and longitude every 0.1 s. The GPS position with the timestamp closest to that of each photograph was used as the position of the UAV.

Generation of a digital elevation model (DEM)

The UAV was also equipped with a laser range finder (LRF; SkEyesBOX MP-1, SkEyes Unlimited Corp., Washington, PA, USA), which assembled a 3D point cloud. We divided the plot (300 m × 200 m) into 1-m × 1-m cells, and the lowest elevation in each cell was used in the creation of a digital elevation model (DEM) of the site.

Table 1. Relationship of maximum diameter between ground-surveyed and visually identified fallen trees.

Maximum diameter* of fallen tree (m)	Number of ground-surveyed fallen trees	Number of visually identified fallen trees	Identification rate (%)
$0.30 \leq x$	8	7	88
$0.20 \leq x < 0.30$	7	2	29
$0.10 \leq x < 0.20$	28	2	7
$0.05 \leq x < 0.10$	1	0	0

*Maximum of diameters at each end and middle (see Table S1).

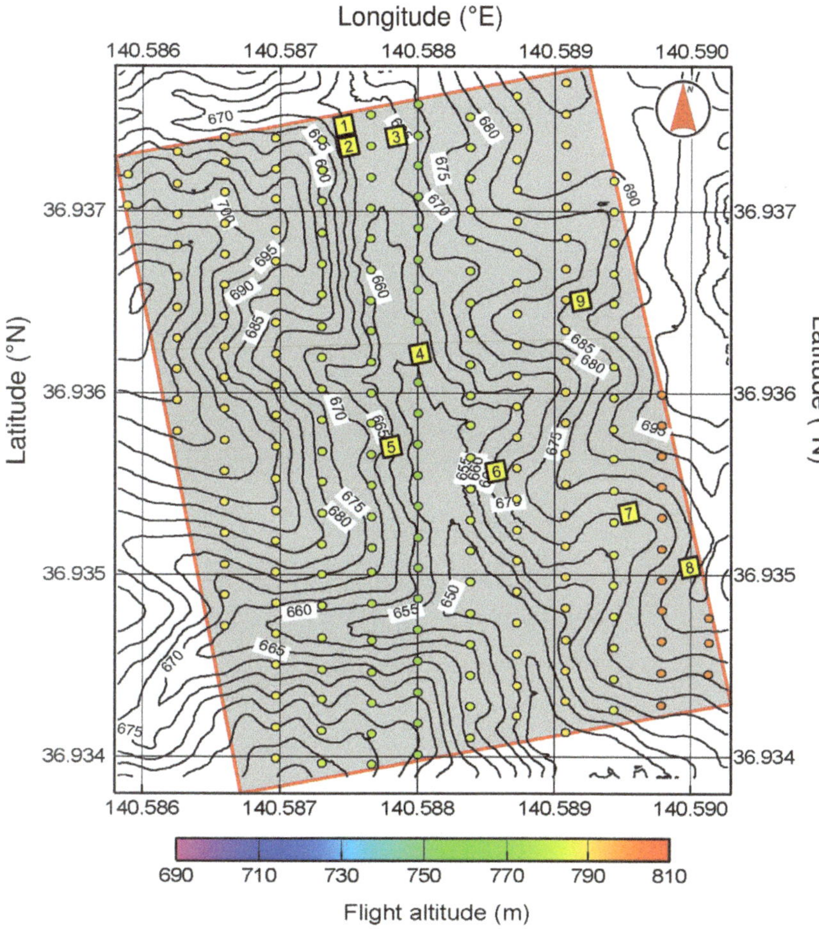

Figure 2. OFR plot (grey area). Yellow squares mark nine sub-plots for ground survey. Circles indicate UAV photograph points. Circle colour indicates flight altitude.

Orthorectification and mosaic of aerial photographs

The aerial photographs were orthorectified according to the DEM and the position (longitude and latitude) and attitude (pitch, roll, and heading) of the UAV, and assembled into one mosaic image with the help of "tie points" in overlapping photographs.

Ground survey

On 31 May 2012 in nine 10-m × 10-m sub-plots (Fig. 2), ground observers recorded the position, stem diameters (each end and midpoint) and length of all fallen trees with a diameter of > 5 cm. The ground survey was conducted in leafy season because it was easier to check the tree species in the study site when there were leaves on the trees, as opposed to a leaf-off season.

Detection of fallen trees in aerial photographs

Fallen trees were identified by eye in the original aerial photographs on a computer monitor. Positions were compared with those of fallen trees mapped in the ground survey and those identified by eye in the orthorectified mosaic.

Table 2. Relationship between lengths of ground-surveyed and visually identified fallen trees.

Length of fallen tree (m)	Number of ground-surveyed fallen trees	Number of visually identified fallen trees	Identification rate (%)
10≤ *x*	9	7	78
5≤ *x* <10	15	3	20
0≤ *x* <5	20	1	5

Figure 3. UAV (RMAX-G1) equipped with a digital camera (EOS KISS X5).

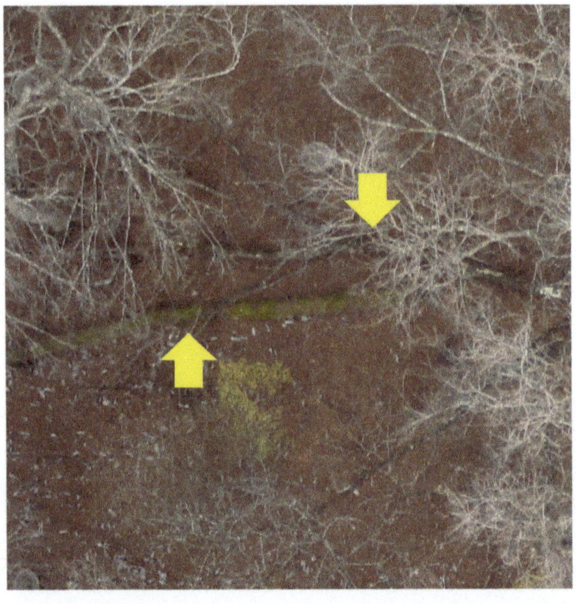

Figure 4. Close-up image of part of the forest floor. Two fallen trees are detectable.

Results

The OFR plot was covered by 211 aerial photographs (Fig. 2) with a spatial resolution of 0.5 to 1.0 cm per pixel. Since the trees were bare in late November, fallen trees were visible (Fig. 4). A DEM of the plot was generated from a cloud of 5445612 points (Fig. 5). The DEM showed a valley with an elevation range of 640 m (in the south) to 720 m (in the north). From the DEM and the position of the UAV, an orthorectified mosaic image was generated (Fig. 6).

In the ground survey, we found a total of 44 fallen trees in the sub-plots (Table S1). By eye, however, we identified only 11 fallen trees in the sub-plots on the original images (Table S1). We identified 80% to 90% of fallen trees which were >30 cm in diameter or >10 m in length, but few that were thinner or shorter (Tables 1, 2). Over the whole plot, we detected 244 fallen trees on the original photographs and 209 on the orthorectified mosaic (Fig. 6).

Discussion

Because fallen trees are generally defined as being >2.5 cm in diameter [2], we assumed that the high spatial resolution of our aerial photographs (0.5–1.0 cm per pixel) would allow us to detect

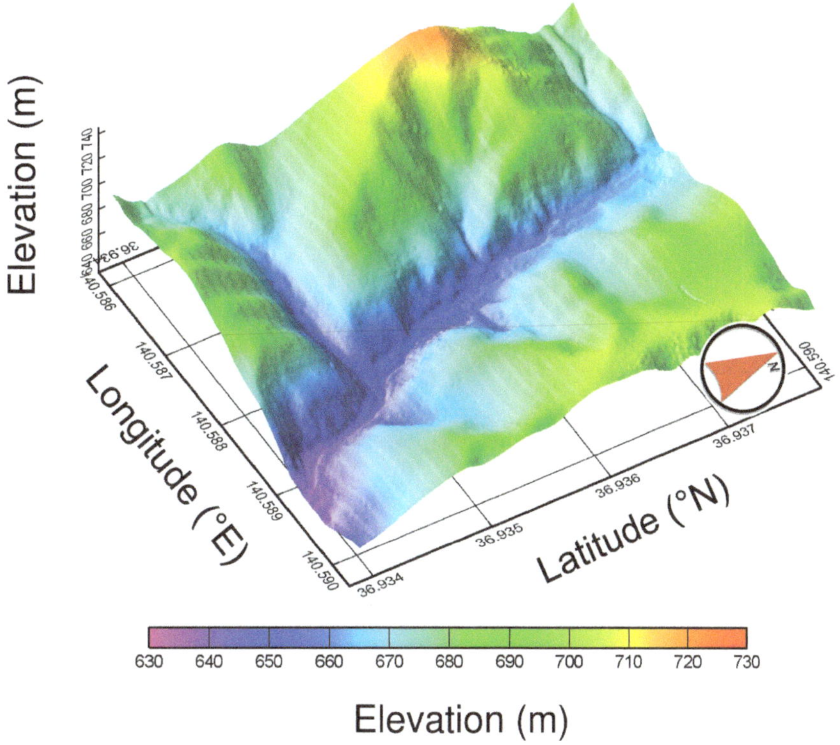

Figure 5. Digital elevation model (DEM) of the OFR plot. This DEM was used for orthorectification of the aerial photographs.

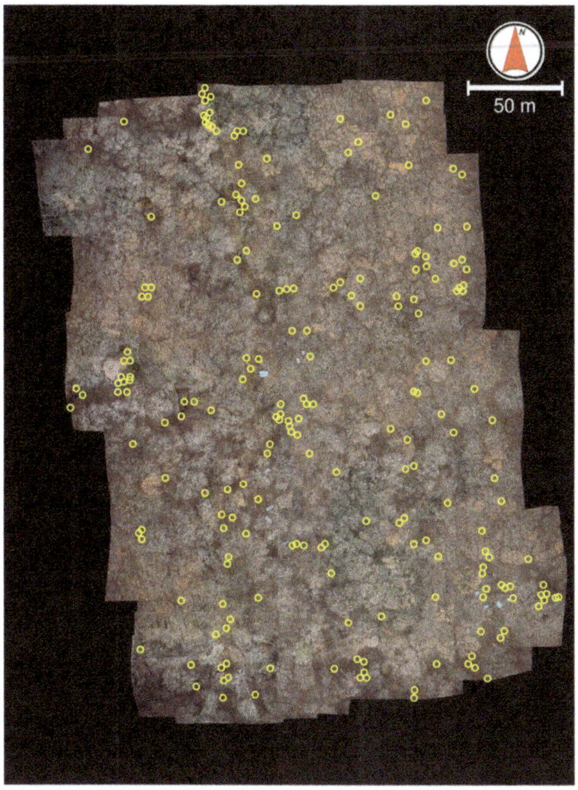

Figure 6. Fallen trees (yellow circles) detected by eye in the orthorectified mosaic.

them. However, our results suggest difficulty in identifying narrow or short fallen trees (Tables 1, 2). This failure may be due to the similarity of fallen trees to trunks and branches of standing trees, and masking of fallen trees by the branches of standing trees and forest floor vegetation. Hodgson et al., who surveyed marine mammals by UAV in Australia, reported the usefulness of overlaps between photographs for detecting animals that are masked by sun glitter [24]. For a similar reason, views of the same point from different angles may provide more opportunity to detect fallen trees hidden by standing trees (Fig. 7). The memory capacity of the camera [24] will determine the balance between coverage and overlap. The optimal degree of overlap will also depend on flying speed, flight altitude, and camera specifications (e.g., frames per second).

Another factor contributing to the poor rate of visual identification of fallen trees might be ambiguity in colour. One of the clues we used in detecting fallen trees was the colour of mosses growing on them. Trees with mosses are easy to detect visually, but freshly fallen trees with no mosses might be confused with tree branches or fallen leaves.

The orthorectification and mosaicking of aerial photographs according to the DEM require much labour and time, even for experts, and can also require specialized and expensive software. However, we identified fewer fallen trees from the orthorectified mosaic than from the original photographs. Thus, non-orthorectified photographs of the same point from different angles would allow better identification of fallen trees, at no additional cost, than orthorectified mosaics.

Conclusions

We showed the applicability of aerial photographs captured from a UAV for the detection of large fallen trees in a deciduous

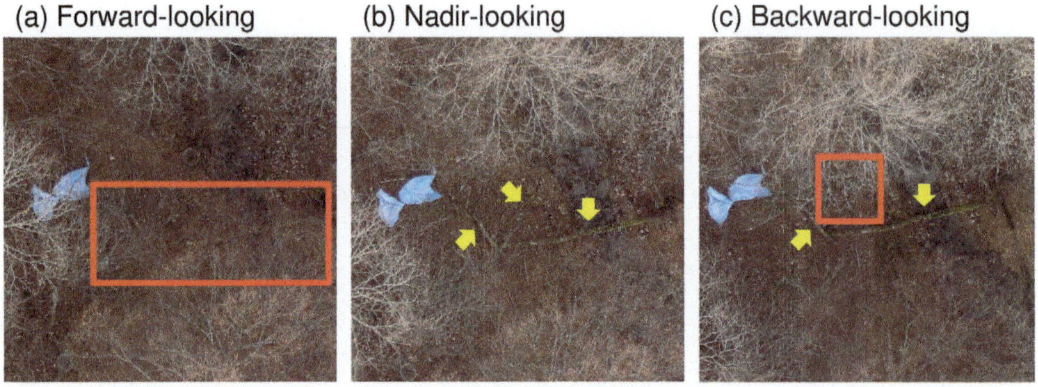

Figure 7. Example of overlapping images of the same point taken from different angles. In this example, the visual detection of three fallen trees from the forward- and backward-looking images may be difficult owing to masking by tree branches (within red boxes). The fallen trees are clearly visible in the nadir-looking image (yellow arrows).

broadleaved forest in eastern Japan. Because much tree death is episodic and irregular [33], high-frequency monitoring at multiple points is necessary for the detection of newly fallen trees and for the understanding of species distribution, biodiversity, and nutrient and carbon cycling in forest ecosystems. UAVs now permit high-frequency monitoring at low cost. This approach has great potential for forest ecology, especially for measuring temporal and spatial variations in forest structure and functioning. Furthermore, as UAVs are advancing and the payload of them is increasing, new sensors for forest monitoring on an UAV will become more common in the future [26]. Installation of a multi-angle photographing system, like a PRISM (Panchromatic Remote-sensing Instrument for Stereo Mapping) on-board the Japanese satellite ALOS (Advanced Land Observing Satellite), on an UAV would potentially help to find more opportunity for detection of fallen trees hidden by standing trees and it may increase the accuracy of fallen tree identification.

Acknowledgments

The authors are thankful to the administrators of the OFR site of the Forestry and Forest Products Research Institute (FFPRI) for their cooperation during the field data collection. We also thank the anonymous reviewers and editor for their helpful and constructive comments.

Author Contributions

Conceived and designed the experiments: RS SY KK HT SN RI HF KO YH. Performed the experiments: RS SY KK HT SN RI HF KO YH. Analyzed the data: KK RS TI SN HF. Contributed to the writing of the manuscript: TI SN RS.

References

1. Biała K, Condé S, Delbaere B, Jones-Walters L, Torre-Marín A (2012) Streamlining European biodiversity indicators 2020: Building a future on lessons learnt from the SEBI 2010 process. EEA Tech Rep No.11/2012. doi: 10.2800/55751.
2. Harmon ME, Franklin JF, Swanson FJ, Sollins P, Gregory SV, et al. (1986) Ecology of coarse woody debris in temperate ecosystems. Adv Ecol Res 15: 133–302. doi: 10.1016/S0065-2504(08)60121-X.
3. Bunnell FL, Houde I (2010) Down wood and biodiversity – implications to forest practices. Environ Rev 18: 397–421. doi: 10.1139/A10-019.
4. Stokland JN, Siitonen J, Jonsson BG (2012) Biodiversity in dead wood. Cambridge University Press, Cambridge.
5. Herrmann S, Prescott CE (2008) Mass loss and nutrient dynamics of coarse woody debris in three Rocky Mountain coniferous forests: 21 year results. Can J For Res 38: 125–132. doi: 10.1139/X07-144.
6. Ohtsuka T, Shizu Y, Hirota M, Yashiro Y, Shugang J, et al. (2014) Role of coarse woody debris in the carbon cycle of Takayama forest, central Japan. Ecol Res 29: 91–101. doi: 10.1007/s11284-013-1102-5.
7. Edman M, Jonsson BG (2001) Spatial pattern of downed logs and wood-decaying fungi in an old-growth *Picea abies* forest. J Veg Sci 12(5): 609–620. doi: 10.2307/3236900.
8. Wu JB, Guan DX, Han SJ, Zhang M, Jin C (2005) Ecological functions of coarse woody debris in forest ecosystem. J For Res 16: 247–252. doi: 10.1007/BF02856826.
9. Ståhl G, Ringvall A, Fridman J (2001) Assessment of coarse woody debris: A methodological overview. Ecol Bull 49: 57–70.
10. Bütler R, Schlaepfer R (2004) Spruce snag quantification by coupling colour infrared aerial photos and a GIS. For Ecol Manage 195: 325–339. doi: 10.1016/j.foreco.2004.02.042.
11. Meentemeyer RK, Rank NE, Shoemaker DA, Oneal CB, Wickland AC, et al. (2007) Impact of sudden oak death on tree mortality in the Big Sur ecoregion of California. Biol Invasions 10: 1243–1255. doi: 10.1007/s10530-007-9199-5.
12. Chambers JQ, Fisher JI, Zeng H, Chapman EL, Baker DB, et al. (2007) Hurricane Katrina's carbon footprint on U.S. Gulf Coast forests. Science 318: 1107. doi: 10.1126/science.1148913.
13. Kupfer JA, Myers AT, McLane SE, Melton GN (2008) Patterns of forest damage in a southern Mississippi landscape caused by Hurricane Katrina. Ecosystems 11: 45–60. doi: 10.1007/s10021-007-9106-z.
14. Pasher J, King DJ (2009) Mapping dead wood distribution in a temperate hardwood forest using high resolution airborne imagery. For Ecol Manage 258: 1536–1548. doi: 10.1016/j.foreco.2009.07.009.
15. Wang F, Xu YJ (2010) Comparison of remote sensing change detection techniques for assessing hurricane damage to forests. EnvironMonit Assess 162: 311–326. doi: 10.1007/s10661-009-0798-8.
16. Wang W, Qu JJ, Hao X, Liu Y, Stanturf JA (2010) Post-hurricane forest damage assessment using satellite remote sensing. Agric For Meteorol 150: 122–132. doi: 10.1016/j.agrformet.2009.09.009.
17. Valavanis KP (ed) (2007) Advances in unmanned aerial vehicles: State of the art and the road to autonomy. Springer, Dordrecht, Netherlands.
18. Watts AC, Perry JH, Smith SE, Burgess MA, Wilkinson BE, et al. (2010) Small unmanned aircraft systems for low-altitude aerial surveys. J Wildl Manag 7: 1614–1619. doi: 10.1111/j.1937-2817.2010.tb01292.x.
19. Watts AC, Ambrosia VG, Hinkley EA (2012) Unmanned aircraft systems in remote sensing and scientific research: Classification and considerations of use. Remote Sens 4: 1671–1692. doi: 10.3390/rs4061671.
20. Anderson K, Gaston KJ (2013) Lightweight unmanned aerial vehicles will revolutionize spatial ecology. Front Ecol Environ 11: 138–146. doi: 10.1890/120150.
21. Rango A, Laliberte A, Herrick JE, Winters C, Havstad K, et al. (2009) Unmanned aerial vehicle-based remote sensing for rangeland assessment, monitoring, and management. J Appl Remote Sens 3: 033542. doi: 10.1117/1.3216822.
22. Koh LP, Wich SA (2012) Dawn of drone ecology: low-cost autonomous aerial vehicles for conservation. Trop Conserv Sci 5: 121–132.

23. Sardà-Palomera F, Bota G, Viñolo C, Pallarés O, Sazatornil V, et al. (2012) Fine-scale bird monitoring from light unmanned aircraft systems. Ibis 154: 177–183. doi: 10.1111/j.1474-919X.2011.01177.x.
24. Hodgson A, Kelly N, Peel D (2013) Unmanned aerial vehicles (UAVs) for surveying marine fauna: a dugong case study. PLoS One 8(11): e79556. doi: 10.1371/journal.pone.0079556.
25. Vermeulen C, Lejeune P, Lisein J, Sawadogo P, Bouché P (2013) Unmanned aerial survey of elephants. PLoS One 8(2): e54700. doi: 10.1371/journal.pone.0054700.
26. Getzin S, Nuske RS, Wiegand K (2014) Using unmanned aerial vehicles (UAV) to quantify spatial gap patterns in forests. Remote Sens 6: 6988–7004. doi: 10.3390/rs6086988.
27. Masaki T, Suzuki W, Niiyama K, Iida S, Tanaka H, et al. (1992) Community structure of a species-rich temperate forest, Ogawa Forest Reserve, central Japan. Vegetatio 98: 97–111. doi: 10.1007/BF00045549.
28. Nakashizuka T, Iida S, Tanaka H, Shibata M, Abe S, et al. (1992) Community dynamics of Ogawa Forest Reserve, a species rich deciduous forest, central Japan. Vegetatio 103: 105–112. doi: 10.1007/BF00047696.
29. Nakashizuka T, Matsumoto Y (eds.) (2002) Diversity and interaction in a temperate forest community: Ogawa Forest Reserve of Japan. Ecological Studies vol. 158, Springer, Tokyo.
30. Kaneko T, Koyama T, Yasuda A, Takeo M, Yanagisawa T, et al. (2011) Low-altitude remote sensing of volcanoes using an unmanned autonomous helicopter: an example of aeromagnetic observation at Izu-Oshima volcano, Japan. Int J Remote Sens 32: 1491–1504. doi: 10.1080/01431160903559770.
31. Koyama T, Kaneko T, Ohminato T, Yanagisawa T, Watanabe A, et al. (2013) An aeromagnetic survey of Shinmoe-dake volcano, Kirishima, Japan, after the 2011 eruption using an unmanned autonomous helicopter. Earth Planets Sp 65: 657–666. doi: 10.5047/eps.2013.03.005.
32. Sanada Y, Kondo A, Sugita T, Nishizawa Y, Yuuki Y, et al. (2014) Radiation monitoring using an unmanned helicopter in the evacuation zone around the Fukushima Daiichi nuclear power plant. Explor Geophys 45: 3–7. doi: 10.1071/EG13004.
33. Franklin JF, Shugart HH, Harmon ME (1987) Tree death as an ecological process: the causes, consequences, and variability of tree mortality. BioScience 37(8): 550–556. doi: 10.2307/1310665.

10

How Does Conversion of Natural Tropical Rainforest Ecosystems Affect Soil Bacterial and Fungal Communities in the Nile River Watershed of Uganda?

Peter O. Alele[1,2,3,7]*, Douglas Sheil[4,5,6,7], Yann Surget-Groba[1], Shi Lingling[1,2], Charles H. Cannon[1,8]

1 Key Laboratory of Tropical Forest Ecology, Xishuangbanna Tropical Botanical Garden (XTBG), Chinese Academy of Sciences, Kunming, Yunnan, P. R. China, **2** University of the Chinese Academy of Sciences, Beijing, P. R. China, **3** Great Nile Conservation Centre (GNCC), Lira, Uganda, **4** Department of Ecology and Natural Resource Management, Norwegian University of Life Sciences, Ås, Norway, **5** Center for International Forestry Research (CIFOR), Bogor, Indonesia, **6** Department of Ecology and Natural Resource Management, School of Environment, Science and Engineering, Southern Cross University, Lismore, New South Wales, Australia, **7** Institute of Tropical Forest Conservation (ITFC), Mbarara University of Science and Technology (MUST), Kabale, Uganda, **8** Texas Tech University, Lubbock, Texas, United States of America

Abstract

Uganda's forests are globally important for their conservation values but are under pressure from increasing human population and consumption. In this study, we examine how conversion of natural forest affects soil bacterial and fungal communities. Comparisons in paired natural forest and human-converted sites among four locations indicated that natural forest soils consistently had higher pH, organic carbon, nitrogen, and calcium, although variation among sites was large. Despite these differences, no effect on the diversity of dominant taxa for either bacterial or fungal communities was detected, using polymerase chain reaction-denaturing gradient gel electrophoresis (PCR-DGGE). Composition of fungal communities did generally appear different in converted sites, but surprisingly, we did not observe a consistent pattern among sites. The spatial distribution of some taxa and community composition was associated with soil pH, organic carbon, phosphorus and sodium, suggesting that changes in soil communities were nuanced and require more robust metagenomic methods to understand the various components of the community. Given the close geographic proximity of the paired sampling sites, the similarity between natural and converted sites might be due to continued dispersal between treatments. Fungal communities showed greater environmental differentiation than bacterial communities, particularly according to soil pH. We detected biotic homogenization in converted ecosystems and substantial contribution of β-diversity to total diversity, indicating considerable geographic structure in soil biota in these forest communities. Overall, our results suggest that soil microbial communities are relatively resilient to forest conversion and despite a substantial and consistent change in the soil environment, the effects of conversion differed widely among sites. The substantial difference in soil chemistry, with generally lower nutrient quantity in converted sites, does bring into question, how long this resilience will last.

Editor: Morag McDonald, Bangor University, United Kingdom

Funding: This work was funded by Xishuangbanna Tropical Botanical Garden (XTBG) and the Chinese Academy of Sciences. The funders had no role in study design, data collection and analysis, decision to publish, or preparation of the manuscript.

Competing Interests: The authors have declared that no competing interests exist.

* Email: alelepeter@gmail.com

Introduction

Tropical rainforests (TRF) possess most of the world's terrestrial biodiversity and deforestation is the leading cause of biodiversity loss [1,2]. Due to their high biodiversity and endemism, the tropical rainforests in Uganda's Nile river watershed are among the world's most important for their conservation values. But these areas are under pressure. The United Nations Population Division [3] predicts that the population of the Nile Basin states will increase by 57% from 2010 to 2030, reaching 647 million people. This rapid population growth, high levels of poverty and prevalent civil insecurity continue to exert severe pressure on natural resources in the region. Uganda in particular has one of the world's highest population growth rates (3.2% per year) [4]. Most of this growing population (nearly 80%) is dependent on agriculture leading to large scale and continuing conversion of natural habitats [5].

Soil communities form the foundation of any ecosystem, in terms of nutrient cycling and availability, so understanding how land conversion affects these communities is an important first step. The effect of land use change on soil microbial communities has been studied in South American and Southeast Asian forests [6,7], but not in the biodiversity hotspots of the Nile river watershed. There is considerable global concern about the loss of biodiversity and the consequences for human well-being [8]. Microorganisms in particular play a vital role in many ecological processes and environmental services [9]: these roles are not always apparent or well characterized but if all microbes died the world would rapidly become buried in undecomposed dead material. Due to their significance in maintaining ecosystem function and productivity [9,10], our study offers a vital exploratory appraisal of microbial community dynamics in natural TRF and human-converted sites. We don't know if there are reasons to be concerned unless we look. Developing such knowledge is critical at this point, because populations in the Nile river watershed are highly dependent on forests for basic requirements such as food and fuel wood, with the environment contributing between 40–60% of the gross domestic product (GDP) of the Nile riparian states [11].

Because of widespread loss of biodiversity, focus from species conservation within particular habitats has been shifted to conservation of communities [12,13]. It is therefore important to explore and understand how composition and diversity changes across spatial scales in a given context [14–16]. Changes in ecosystems caused by conversion to intensive management can lead to biotic homogenization, the increase in community similarity over time and/or space and an implied loss of rare and vulnerable taxa when examined at larger scales [17–19]. Because microorganisms are the most diverse organisms on earth with most taxa and respective functions and behaviors as yet unknown, determining their sensitivities and biogeography remains a major challenge. But in the longer term such knowledge will help us better understand the sustainability of land-use systems and associated environmental values.

This study was therefore necessary as a first step in exploring these relationships, and to enhance understanding so as to contribute to the informed and appropriate stewardship of the region's natural resources. Our objective was to establish how forest conversion and soil factors affect soil bacterial and fungal diversity and community composition in the tropical rain-forests in the Nile river watershed of Uganda. We chose four forest sites found within protected areas, with paired treatments within each forest; (1) natural and (2) converted ecosystem sites. The natural forest ecosystem at each site had suffered minimal human disturbance, while converted areas had been transformed to cropland. These matched sites found in different locations and environmental conditions each experienced different land use histories, conservation circumstances and individual challenges for management.

In each matched set of natural and converted sites, we compared soil physical and chemical properties and microbial community diversity and composition using standard PCR-based genotyping techniques. We then calculated community similarity indices between sites. This approach would allow us to examine both environmental and biotic changes in the soil community associated with conversion. Disturbances of sufficient magnitude or duration may alter an ecosystem and force a different regime of predominant processes and structures that favor some populations over others [20].

We tested the null hypotheses that there was no difference in soil properties, band-types, and diversity between treatments. The influence of soil properties on microbial community diversity was measured by discriminant analysis and canonical correspondence analysis (CCA), [21,22]. Because additive partitioning of diversity provides a useful framework for quantifying the spatial patterns of diversity across hierarchical spatial scales [23,24], we partitioned total diversity (γ) in each ecosystem type (natural and converted) into additive components representing within-community diversity (α) and between-community diversity (β). Our objective was to identify the most important sources of total diversity so as to propose conservation measures for microbial communities in the TRF ecosystems of the Nile river watershed of Uganda.

Methods

Site description

We selected four tropical rainforest (TRF) sites because of their relative size, biodiversity, socio-economic and scientific importance (Fig 1). Mabira forest is located between the highly populated and urbanized Kampala city on the western side; the extensive and mechanized Lugazi sugar and tea plantations on the Eastern; and Lake Victoria on the southern side. Budongo forest is located next to the extensive Kinyala sugar plantations on one side and a densely populated mainly subsistence population scattered around it. Maramagambo and Kaniyo Pabidi are located within Queen Elizabeth and Murchison Falls national parks (NP) respectively. These two NP forests had perhaps the best protection due to presence of Uganda Wildlife Authority (UWA) personnel. However, Maramagambo's location starting on the steep slopes of the rift valley subjected it to frequent storms with strong runoff flow that swept away most of its top soil (Table 1).

Soil sampling design

We collected 400 core soil samples within 40 plots (1000 m^2 each) in four TRF sites (Fig 1). We sampled five plots from each site of the natural TRF and five plots from the converted TRF. We established the plots at least 100 m from the ecosystem edge and 500 m apart and collected 10 evenly placed core subsamples of top soil (0–15 cm) from each plot and homogenized them into one sample per plot. We then derived a 500 g composite sample from the mixture, sieved and packed it for physical and chemical analyses and DNA extraction.

Sample preparation

We sieved 100 g of the soil on-site through a 4 mm mesh, transported it to the laboratory on ice, and stored in a freezer at $-40°C$ prior to nucleic acid extraction and analysis. We kept the rest of the soil for drying and physical and chemical analysis. We performed DNA extractions from 1 g of soil using the Ultra Clean soil DNA kit (Mo Bio Labs, Solana Beach, CA, USA) following the manufacturer's protocol. The purified DNA was detected by agarose gel electrophoresis, and the DNA was amplified by polymerase chain reaction (PCR).

Soil property analyses

We measured the soil pH in 2.5:1 water to soil suspension using a pH meter (10 g soil+ 25 ml of distilled water, shaken for 30 min and read on a calibrated pH meter). We then used the Walkley and Black method [25] to analyze soil organic carbon (SOC) and the Kjeldahl method [26] to determine soil nitrogen. We measured the soil phosphorus by the Bray and Kurtz no. 1 method [27]. The photoelectric flame photometer was used to determine the soil potassium, sodium and calcium after extraction with neutral ammonium acetate. We used the atomic absorption spectrometer to measure the soil magnesium after extraction with neutral ammonium acetate. The Bouyoucos hydrometer method adopted from Gee and Bauder [28] was used to determine soil texture. The soil copper and iron were then determined using the atomic absorption spectrometer after extraction with EDTA.

PCR amplification and DGGE analysis

Polymerase chain reaction-denaturing gradient gel electrophoresis (PCR-DGGE) method has been used extensively in microbial ecology and is a robust and cost effective method for exploratory classification of microbial communities [29]. Following soil DNA extraction, we performed a PCR for each DNA extraction to

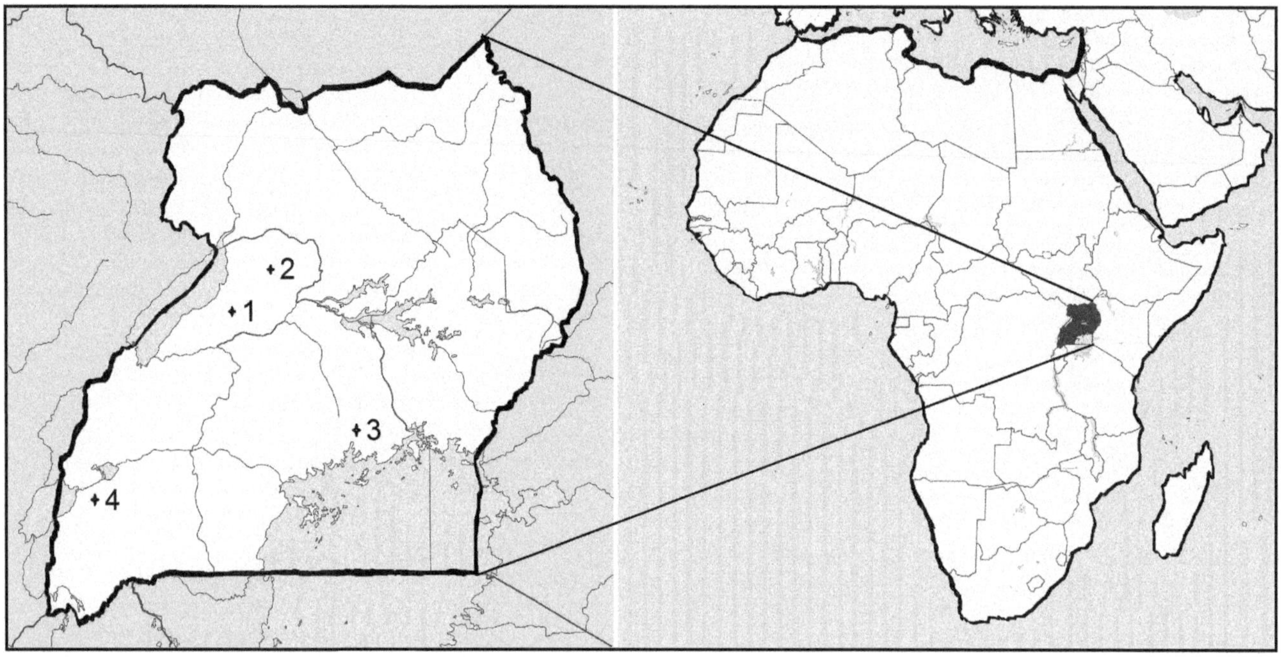

Figure 1. Map of Uganda showing the distribution of sampling sites; Budongo forest (1), Kaniyo Pabidi (2), Mabira forest (3), and Maramagambo forest (4).

amplify the 16S rRNA genes for bacteria and 18S rRNA genes for fungi using universal primers (Table 2).

PCR reactions had a final volume of 25 µl containing a final concentration of 1× TaKaRa ExTaq PCR buffer with $MgCl_2$, 300 pM of primers for bacteria. We then added 200 µM dNTPs, 2.5 U ExTaq DNA polymerase (TaKaRa Bio, Otsu, Japan) and milliQ H_2O to complete the volume, BSA was also added for the fungal community analysis. We performed PCR cycles with an initial denaturing temperature of 95 °C for 5 min, followed by 35 cycles of 95 °C for 30 sec, annealing temperature of 50 °C for 30 sec, extension of 72 °C for 1 min; and a final extension of 72 °C for 10 min. We checked the product of the PCR-rounds and quantified by agarose gel-electrophoresis.

We then performed 16S rRNA and 18S rRNA-DGGE analysis using a universal mutation detection system (Dcode Bio-Rad, Richmond, CA, USA) with a 6% and 8% acrylamide gel for bacteria and fungi respectively containing a gradient of 40–60% denaturant (100% denaturant contains 7 mol urea and 40% formamide). We applied 100 ng of PCR samples to the DGGE gel. DGGE was performed in 1 × TAE Buffer (40 mol Tris/acetate, pH 8; 1 mol ethylene diaminetetra acetic acid) at 60 °C and a constant voltage of 60 V for 16 hours. After staining with SYBR Green1, we recorded the DGGE gels as digital images and analyzed the DNA band numbers using image-processing software after subtracting background noise.

Data analysis

We used the Rolling disk method with Quantity One (Bio-Rad laboratories Inc.), which normalizes the band pattern from electrophoresis for identification of each band. We then converted the band patterns into binary data based on the presence or absence of each band for part of our analysis. The DGGE fingerprints were interpreted in terms of band richness (number of predominant DGGE bands/population). The pixel intensity of each band was detected by Quantity One software and is expressed as relative abundance (P_i) [30]. Shannon Index (H') and Simpson index (D), the most widely used diversity indices were then calculated using the richness and relative abundance data following the equations:

$$H' = -\sum_{i=1}^{R} P_i InP_i \tag{1}$$

$$D = \sum_{i=1}^{R} (P_i)^2 \tag{2}$$

Where R, the richness, is number of different bands each data set contains, $P_i = \frac{n_i}{N}$ and n_i is the abundance of the *ith* band and N the total abundance of all bands in the sample.

Band-type data of the DGGE fingerprints was then used to derive the alpha diversity (bands per sample and ecosystem type), beta diversity (total bands per site) [31]. Jaccard's similarity indices [32] between converted and natural TRF sites were determined using the equation:

Jaccard's Similarity Index $= {}^{A}\!/\!_{(A+B+C)}$

Where,

A = Total number of bands present in both converted (C) and natural (N) ecosystem samples (plots) (also β-diversity)

B = Number of bands present in C but not in N

C = Number of bands present in N but not in C

We determined the influence of site factors as revealed by soil physicochemical properties on the variation of soil microbial communities by applying discriminant analysis using Statistical Package for the Social Sciences (SPSS). This was done to assess the relative importance of each predictor variable (pH, SOC, N, P, K, Na, Ca, Mg, and soil texture). We also used the Mann-Whitney

Table 1. Summary of study site description.

Forest Site	Location	Size (km^2)	Altitude (masl)	Geology	Forest type	Habitat type	Ecosystem description
Budongo	31°N 35° E 1°S 45° N	793	700–1,270	Weathered pre-cambium with ferrallitic sandy clay loams	Ironwood forest (*Cynometra alexandri*); Mixed forest (*Maesopis*), and colonizing forest (*Entandrophragma*)	Primary forest	Consists of a medium altitude moist semi-deciduous forest with areas of savanna and woodland. Converted areas consisted of deforested agricultural land being cultivated and planted with maize, beans, sweet potatoes and cassava. This land has existed as agricultural land for at least 15years.
Mabira	33° 0.00′ E 0° 30.00′ N	300	1,070–1,340	Ferrallitic soils with mainly sandy clay loams	Mixed forest	Secondary forest, heavily influenced by humans	The forest is surrounded by a densely populated area. Converted areas were actively cultivated and used to grow maize, groundnuts, beans, yams, cassava, sweet potatoes and a few scattered plants of coffee and sugarcane. The converted land had existed as agricultural land for at least 10 years.
Maramagambo	00° 33′ 00" S and 29° 53′ 00" E	1,978 (QENP)	910–1390	Ferrallitic soils with undifferentiated dark horizons	Medium altitude, moist, semi-deciduous forest	Secondary forest influenced by wildlife and humans	Forms part of the Queen Elizabeth NP (QENP) which is 1,978 km2. Converted areas consisted of cultivated and grazed areas with gardens of sweet potatoes, beans, maize, and sorghum with areas commonly grazed by cattle and goats.
Kaniyo Pabidi	Lat 1.916667 Long 31.666667	N/A	700–1,270	Freely drained ferruginous tropical soils	Mixed forest	Primary forest	Located north of Budongo forest and part of Murchison Falls N.P. Converted areas consisted of cultivated areas with crops like maize, beans, cassava and sweet potatoes.

[5,61,70,71].

Table 2. Sequences of primers used in study.

Microorganism	Primer	Sequence (5′–3′)	Reference
Bacteria	F357	CGC CCG CCG CGC GCG GCG GGC GGG GCG GGG GCA CGG GGG GCC TAC GGG AGG CAG CAG	[72]
	907R	CCG TCA ATT CMT TTG AGT TT	
Fungi	FF390	CGA TAA CGA ACG AGA CCT	[73]
	FR1GC	AIC CAT TCA ATC GGT AIT	

test to examine differences between soil properties in natural and converted ecosystems, and microbial communities in natural and converted ecosystems.

We tested the null hypothesis that diversity is uniform at all spatial scales by additive partitioning of total diversity (γ diversity). To determine contributions of α and β diversity to overall diversity across a range of spatial scales [14,23,33], an additive relationship between diversity components (i.e., $\beta = \gamma - \alpha$) was derived (Fig 2.). The scale at which diversity is maximized was therefore identified [23,34], to facilitate planning processes and management strategies to conserve natural levels of diversity accordingly [35–38].

We used PARTITION 3.0 software [39] to calculate average diversity at each scale and diversity was measured as band richness. Individual-based randomization procedure in the software was used to test whether the observed partitions of diversity within the ecosystem could have been obtained by a random allocation of lower-level samples nested among higher-level samples [34]. Null values of β_i obtained from 1,000 randomizations were used to obtain a *p*-value for the observed β_i at each hierarchical scale. Deviations of the observed diversity from the null expectation indicated a nonrandom spatial distribution of fungi or bacteria at a given scale.

Results

Soil property variations

Soil pH comparisons using a Mann-Whitney U test of significance, between five plots of natural and five plots of converted TRF ecosystems in each of the four forest sites, found significantly higher (less acidic) pH in three of the four sites at Budongo ($p = 0.0107$), Kaniyo Pabidi ($p = 0.0112$), and Mabira ($p = 0.0269$); and non-significant difference at Maramagambo ($p = 0.1706$). Percentage soil organic carbon (SOC) was significantly higher in natural than converted ecosystems in all four sites i.e. Budongo ($p = 0.0119$), Kaniyo Pabidi ($p = 0.0212$), Mabira ($p = 0.0122$) and Maramagambo ($p = 0.0119$) with combined %SOC in natural sites more than double of that in converted sites; whereas %soil nitrogen was only significantly higher in natural forests at Budongo ($p = 0.0112$) and Kaniyo Pabidi ($p = 0.0119$), and non-significant at Mabira ($p = 0.6015$) and Maramagambo ($p = 0.0947$) (Table 3).

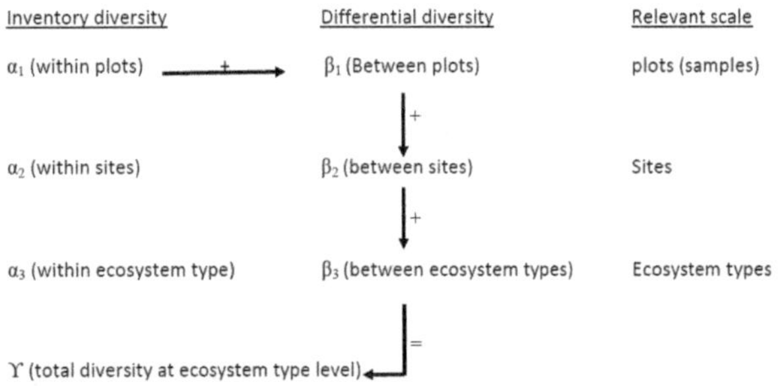

Figure 2. Illustration of hierarchical spatial scales in our additive partitioning model. The α scale is the within-level, and β scale, the between-level components. Because a diversity at a given scale is the sum of the α and β diversity at the next lower scale, the total diversity (γ) can be described by the following formula: $\alpha1+\beta1+\beta2+\beta3$ [14,22].

Ecosystem and site comparisons of microbial community diversity

Bacterial (B) communities were significantly richer ($p = 0.0304$; Mann-Whitney U) in detectable bands than fungal (F) communities in both converted (C) and natural (N) ecosystems (converted: medians; F = 36, B = 61.5; natural: medians; F = 39.5, B = 60.5). While total band richness (B+F) did not differ between natural and converted forests we observed greater fungal richness in natural than in converted forests (medians: C = 36, N = 39.5; test stat = 18.5) and more bacterial bands in converted than in natural ecosystems (medians: C = 61.5, N = 60.5; test stat = 18.5). Kaniyo Pabidi was the most diverse site overall with the highest number of bacterial and fungal bands, while Maramagambo had the least band richness (Fig 3).

Natural sites harbored more bands unique to one site than converted sites for bacteria at Kaniyo Pabidi and Maramagambo and for fungi at Maramagambo and Budongo. Mabira and Kaniyo Pabidi had higher numbers of unique bacterial bands than at Maramagambo and Budongo. There were also more unique fungal bands at Mabira and Budongo than at Maramagambo and Kaniyo Pabidi (Fig 4).

We also found that Mabira and Maramagambo had the lowest bacterial Jaccard's community similarity indices [32] between natural and converted ecosystems, whereas Budongo and Mabira had the lowest fungal community similarity between natural and converted ecosystems (Table 4). Dissimilarity between natural and converted ecosystems was nonetheless non-significant in all sites for both fungal and bacterial communities. Also, there was generally greater dissimilarity between sites of fungal communities than in bacterial communities suggesting a higher susceptibility to habitat change among fungi than bacteria (Table 4).

Ecosystem classification and importance of predictor variables

The CCA showed that, despite the relatively small amount of difference between sites, soil pH, average phosphorus, and texture (%sand) had strong influence on bacterial diversity in the TRF ecosystem (Fig 5); whereas organic carbon, sodium, pH and average phosphorus were strongly associated with fungal community variation in both natural and converted TRF ecosystems (Fig 5). The CCA also showed that bacterial communities in both Kaniyo Pabidi and Mabira were unique to bacterial communities

Table 3. Mean (Standard deviation) for diversity indices of Bacterial (B) and Fungal (F) communities and soil properties in natural (N) and converted (C) ecosystems.

	Budongo		Kaniyo Pabidi		Mabira		Maramagambo	
	Natural	Converted	Natural	Converted	Natural	Converted	Natural	Converted
Shannon (B)	2.49(0.68)	2.73(0.10)	3.25(0.21)	3.12(0.25)	3.06(0.18)	3.19(0.18)	3.04(0.16)	2.93(0.07)
Simpson (B)	0.89(0.10)	0.93(0.01)	0.95(0.13)	0.95(0.02)	0.95(0.01)	0.95(0.01)	0.95(0.01)	0.94(0.01)
Shannon (F)	2.44 (0.30)	2.49 (0.17)	2.83(0.19)	2.77(0.20)	2.57(0.62)	2.64(0.30)	2.10(0.50)	1.93(0.67)
Simpson (F)	0.90(0.29)	0.90(0.03)	0.93(0.01)	0.93(0.02)	0.90(0.06)	0.92(0.03)	0.84(0.08)	0.82(0.12)
pH	5.88(0.11)*	5.08(0.20)*	6.24(0.23)*	5.38(0.27)*	6.46(0.54)*	6.18(0.42)*	6.18(0.45)	5.80(0.24)
OC (%)	6.14(0.67)*	1.59(0.17)*	5.69(1.20)*	3.65(0.54)*	5.87(2.03)*	3.53(1.19)*	9.08(0.79)*	3.98(1.09)*
N (%)	0.43(0.05)*	0.11(0.01)*	0.43(0.08)*	0.25(0.03)*	0.21(0.10)	0.22(0.06)	0.28(0.04)	0.19(0.08)
Ca (Cmoles/kg)	8.00(2.56)*	4.00(0.58)*	13.54(3.36)*	6.74(0.98)*	8.14(1.19)*	6.20(0.76)*	12.12(2.13)*	7.34(1.16)*

*significant differences ($p<0.05$) between natural and converted ecosystems.

in the other sites and there was high contrast between bacterial communities of converted and natural ecosystems at Kaniyo Pabidi. Fungal communities at Maramagambo and Mabira were also unique to those in other sites and there was high contrast between fungal communities at Mabira's natural and converted ecosystems. Furthermore, the CCA showed that fungal communities responded more to soil pH levels than bacterial communities (Fig 5), with site-specific patterns showing that bacteria and fungi were grouping according to sites.

A discriminant analysis to predict whether bacterial or fungal communities were from natural or converted ecosystems found that only OC, Ca, N, and pH for bacterial communities; and OC, N, Ca, and pH for fungal communities (all ranked from most important to least important) were found to be significant predictors of soil physicochemical properties. All other variables were poor predictors in this context (Table 5).

Hierarchical scaling

We found 58 and 56 fungal bands in natural and converted forests respectively, from 17 plots of natural ecosystems and 20 plots of converted forests. There were also 92 and 88 bacterial bands in natural and converted ecosystems respectively found in 20 plots of converted ecosystems and 17 plots of natural ecosystem. All these were within four sites. β-diversity varied more than α-diversity between natural and converted ecosystems for both bacteria and fungi. We found higher bacterial and fungal β-diversity in converted ecosystems than in natural ecosystems at lower hierarchical scales (β_1); higher β-diversity in natural than converted at between-site scale (β_2), and higher β-diversity in converted than in natural ecosystems at the between-ecosystem type scale (β_3) (Fig 6).

We also found substantial contribution of observed β-diversity (β_1, β_2, and β_3) to total band richness (γ-diversity), while α-diversity of both bacteria and fungi in converted and natural ecosystems were similar. Spatial partitioning of total diversity also consistently showed that the beta components (β_1 and β_2) were always greater than expected by chance, whereas the alpha component (α_1) was always lower than expected. For both fungal and bacterial communities in natural and converted ecosystems, observed within plot diversity were substantially less than values expected from individual-based randomizations (Fig 7).

Discussion

Soil property variations and site differences

Studies in both tropical and temperate zones show that soils in converted or cropped areas normally have reduced soil aggregation, structural stability and organic matter, and an increase in bulk density when compared to forests [40,41]. Habitat conversion may also alter soil properties such as nutrient levels, and abiotic conditions and may affect associations between organisms. In our study there are some local details that may influence our results.

Both Maramagambo and Kaniyo Pabidi are located within Queen Elizabeth NP and Murchison Falls NP respectively and are protected by Uganda Wildlife Authority (UWA) personnel. They are well protected and there is little evidence of recent encroachment. There is significant wildlife populations including elephants, buffaloes, zebras and the areas are frequented by tourists. Protection by UWA and presence of dangerous animals (such as buffalos and lions) reduce damaging human activity at Kaniyo Pabidi and Maramagambo which should enhance the difference between natural and converted ecosystems. Maramagambo's location, in contrast, means the

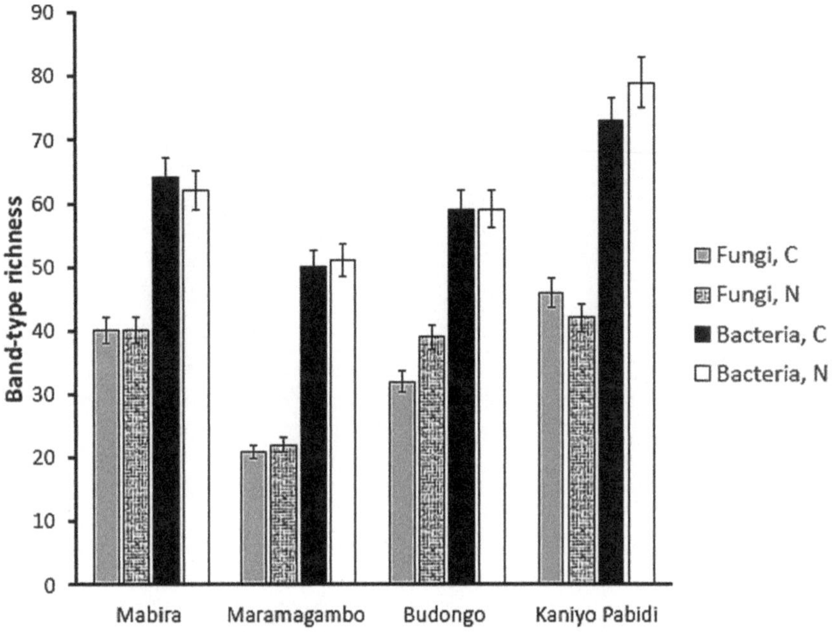

Figure 3. Band richness for fungal and bacterial communities in converted and natural ecosystems. All richness values are total bands present in five samples of each ecosystem treatment (error bars are 5% confidence interval).

forest is subjected to substantial natural disturbance from frequent storms and strong erosive runoff even within the natural forest, whereas tourist activity at Kaniyo Pabidi seemed to have little impact on soil properties. Converted areas at Kaniyo Pabidi were also sparsely populated with limited human impacts on the environment. Its sites were old and might have been cultivated for at least 20years.

In our study, Budongo is located next to a high, mainly subsistence population and resultant population activity. But even though encroachment, illegal hunting and logging in natural habitats in Budongo are not uncommon, there seems to be minimal impact of conversion on soil properties in our sample locations; whereas proximity of Mabira's natural forest to densely populated urban areas exposes it to increased human activities, likely reducing its difference with converted sites.

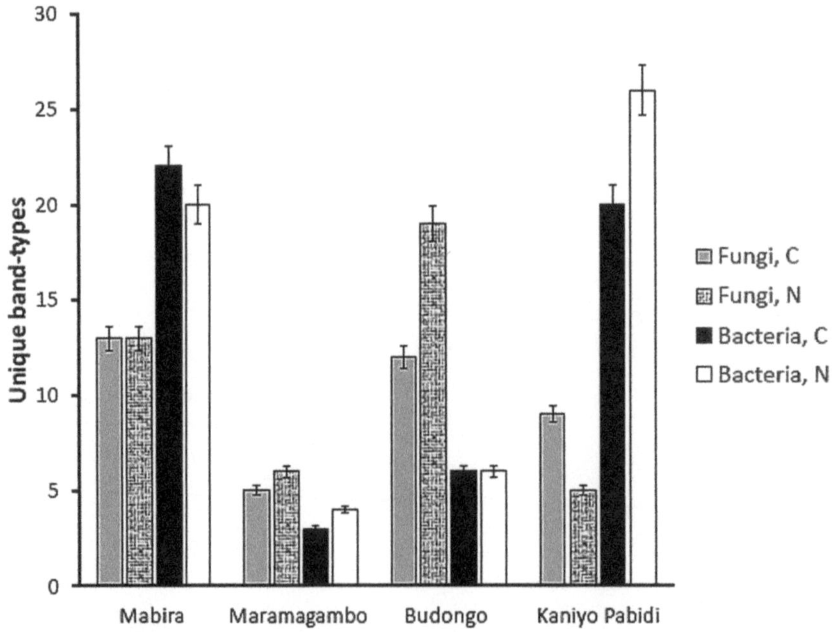

Figure 4. Bacterial and fungal bands unique to converted (C) and natural (N) ecosystems at each site (error bars are 5% confidence interval).

Table 4. Jaccard's similarity indices between bacterial and fungal communities in natural (N) and converted (C) sites of Mabira (Mb), Maramagambo (Mg), Budongo (Bd) and Kaniyo Pabidi (Kp).

			F		U		N		G		I
			Mb	*Mg*	*Bd*	*Kp*	*Mb*	*Mg*	*Bd*	*Kp*	
			N	N	N	N	C	C	C	C	
B	*Mb*	N	1	0.713	0.769	0.732	0.671	0.685	0.696	0.768	F
A	*Mg*	N	0.841	1	0.733	0.666	0.666	0.711	0.646	0.666	
C	*Bd*	N	0.803	0.840	1	0.804	0.808	0.693	0.622	0.827	U
T	*Kp*	N	0.731	0.765	0.698	1	0.774	0.670	0.794	0.785	
E	*Mb*	C	0.667	0.788	0.739	0.741	1	0.663	0.710	0.761	N
R	*Mg*	C	0.866	0.885	0.846	0.755	0.809	1	0.684	0.698	
I	*Bd*	C	0.825	0.857	0.844	0.696	0.731	0.856	1	0.786	G
A	*Kp*	C	0.805	0.869	0.716	0.683	0.759	0.776	0.733	1	I
			B	A	C	T	E	R	I	A	

Microbial community variations

Soil properties determine many aspects of soil microbial community structure [42–44]. Carbon availability [45–47], nitrogen availability [45,48,49] and soil pH [44,50,51] can all influence microbial community composition and diversity. In addition, correlation studies have shown that plant species [52–54] and soil type [45,54,55] are associated with variation in microbial communities. It has also been shown that land use indirectly affects bacterial community structure by modification of soil properties [56] but similarity between converted and natural ecosystem bacterial communities may also suggest a high number of generalists.

Nacke et al., [56] found that bacterial community composition in forests and grassland was largely determined by tree species and soil pH. Jesus et al. [6] also showed that bacterial community structure is influenced by changes linked to soil acidity and nutrient concentration. Other studies also suggest that soil pH is a major factor influencing microbial community composition [50,57–59]. This influence of soil pH has been recognized at different taxonomic levels [45,60] with most microorganisms thriving within a limited pH range. This is because acids can denature proteins and large pH changes may inhibit growth in microorganisms. Fierer and Jackson, 2006 [44] found, in contrast, that net carbon mineralization rate (an index of C availability) was the best predictor of phylum-level abundances of dominant bacterial groups, and Bisset et al., [42] found that soil microbial communities were consistent with disturbance gradients within different agricultural treatments and relatively undisturbed non-agricultural sites.

Because of widespread forest conversion in Uganda as a result of increasing population pressure, estimated at between 1.1% and 3.15% per year [61]; natural ecosystems and their inhabitant biodiversity are at risk [62]. Loss of diversity increases the likelihood of losing important functional roles and associated ecosystem processes. At landscape scale, spatial and temporal variations of microbial communities in forest soils are influenced by numerous biotic and abiotic factors. These factors may include climate, soil types, and vegetation associations [50,63,64]. Owing to this study design, many of these factors were assumed to be similar between natural and converted ecosystems. For instance, the proximity of natural and converted ecosystem sites meant that climate and geology were, we assume, similar in the two treatments. Even though there could still be a number of underlying causes of community differences, two likely influences were assumed to be soil properties [45,54,55] and vegetation types [52–54]. Clearly both of these sets of factors change when forest is converted for agriculture or range lands.

Despite substantial reductions in SOC, N, Ca and pH in converted sites in this study, differences in microbial communities were small meaning that converted sites still had sufficient SOC, N and Ca to sustain the same microbial populations. The close proximity of the matched pairs could also lead to a source-sink relationship between the natural and converted forests, with the presence of unique taxonomic groups a likely indication of habitat preference (endemism) for some taxonomic groups. It may also be an indicator of relative habitat dissimilarity. The high numbers of unique bacterial bands at Mabira and Kaniyo Pabidi and unique fungal bands at Mabira and Budongo (Fig 4) thus suggests that ecosystem alteration at these sites was sufficient to force a different regime of processes and structures enabling a new set of taxonomic groups to predominate. Mabira had high numbers of both unique bacterial and unique fungal bands that can be attributed to the extent of disturbance at its sites (Mabira

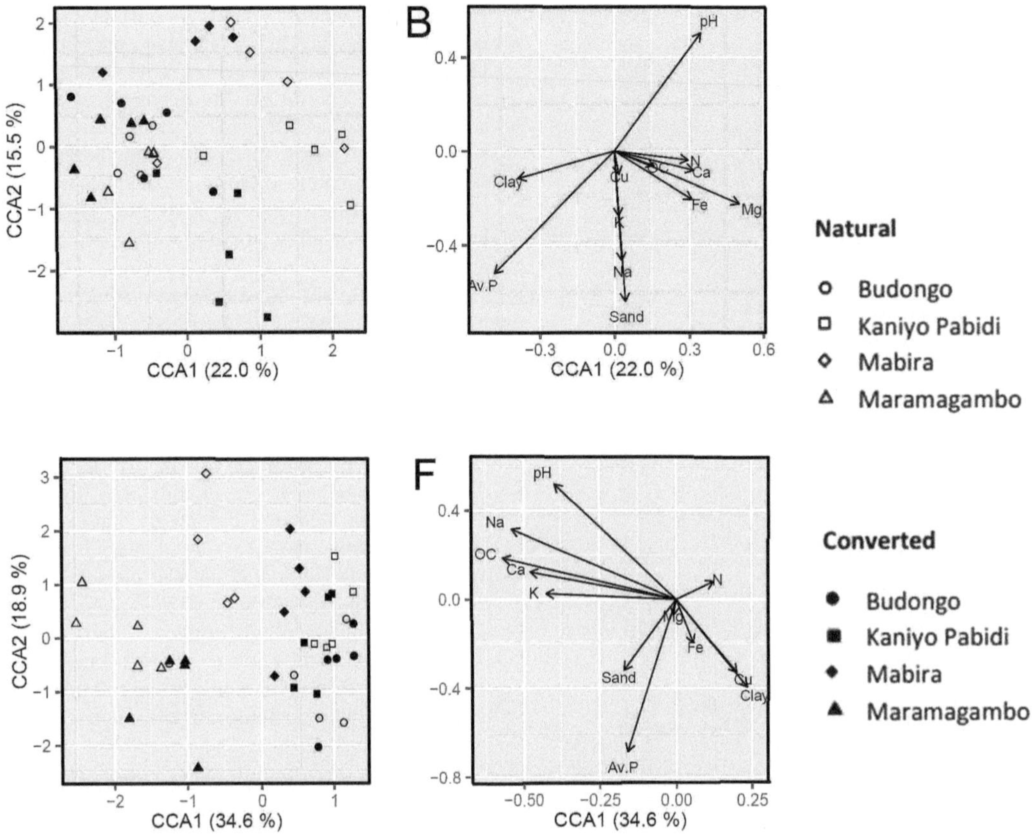

Figure 5. CCA for bacterial (B) and fungal (F) relationships using relative abundance of bands and soil physicochemical properties in natural and converted ecosystems. The symbols (left graphs) represent the similarity between each sample (plot) as defined by their diversity, and the vectors (right graphs) represent the structural matrix for soil properties and their influence on relative abundance of each band. The length of the vectors represents the relative strength of influence of the particular aspect of soil physicochemical property.

is the only peri-urban tropical rainforest site among the four selected sites).

The low numbers of unique bacterial and fungal bands at Maramagambo can be attributed to the high erosion at natural sites that reduced the contrast between the natural and converted sites. For the other sites, bacteria and fungi had different responses to ecosystem alterations. This could indicate separate influences on microbial distribution that exist when alteration is moderate. Similarity indices suggested that bacterial and fungal communities were determined by separate forces leading to distinct responses across the study locations.

Hierarchical scaling

Many studies have shown that specialist species are more negatively affected by current global changes than generalists [7,65]. The process of biotic homogenization can involve the replacement of native biota with non-natives or the introduction of generalist species [66]. In this study, the net decrease in β-diversity from natural to converted TRF ecosystem at the between-site scale (β_2) for both fungi and bacteria was an indication of biotic homogenization [18,66]. This can result from ecosystem alterations which can in-turn alter ecosystem function and reduce ecosystem resilience to disturbance [65,67].

We also showed that the β components of diversity ($\beta_1$and β_2; the average diversity between the plots and sites, respectively) were consistently higher than those expected by chance, whereas the local α_1 diversity component (α_1, the average diversity within the plots) was consistently lower than that expected (Fig 7). Such scale-dependent deviations of the observed diversity from the expected can be generally explained by aggregation at a relatively small "local" scale and, spatial differentiation of diversity at a larger "landscape" scale [33,34,68,69].

Relatively lower diversity within converted ecosystems suggests that conversion of natural TRF ecosystems results in reduced diversity for both bacteria and fungi. This is compatible with recent studies that show that conversion of TRF ecosystems threatens microbial diversity [7] and because microorganisms, like all other organisms, have habitat preferences and may be affected by land-use changes [6,64]. While we cannot be certain that such decline in diversity has led to a decline in any particular ecosystem functions or services, this is a possibility that deserves further evaluation, and we speculate that such loss of diversity will at the very least cause a reduction in functional redundancy and associated resilience.

Higher β-diversity of both bacterial and fungal communities at the between-plot scale (β_1) in converted ecosystems indicates

Table 5. Structure matrix rank showing absolute size of correlation between discriminant analysis function from most important to least important predictor variable of site factors (soil physiochemical properties) and their influence on the variation of soil microbial communities.

Bacteria		Fungi	
Predictor Variables	**Function 1**	**Predictor variables**	**Function 1**
OC	0.625*	OC	0.625*
Ca	0.471*	N	0.493*
N	0.421*	Ca	0.473*
pH	0.355*	pH	0.340*
Mg	0.298	Mg	0.281
Cu	0.197	Cu	0.221
Na	0.181	Na	0.183
K	0.169	K	0.160
Av.P	−0.077	Sand	0.064
Sand	0.070	Av.P	−0.060
Simpson	−0.065	Fe	0.024
Fe	0.029	Shannon	0.012
Shannon	−0.023	Simpson	0.011

(* = important predictor variable, with 0.30 used as the threshold).

differentiation (reduced community similarity) in converted ecosystems at this hierarchical scale. Considering the multiple land-uses and cropping systems of converted areas, this was expected. There was also substantial contribution of β-diversity to total diversity (γ). This suggests the importance of nonrandom ecological processes at the between-plot and between-site scale in determining total richness and community composition [14,34]. Differences between the observed and expected diversity components could be due to ecological processes that lead to a non-random dispersion of individuals. These processes could include intra-specific aggregation, habitat selection, and limited dispersal capacity [33].

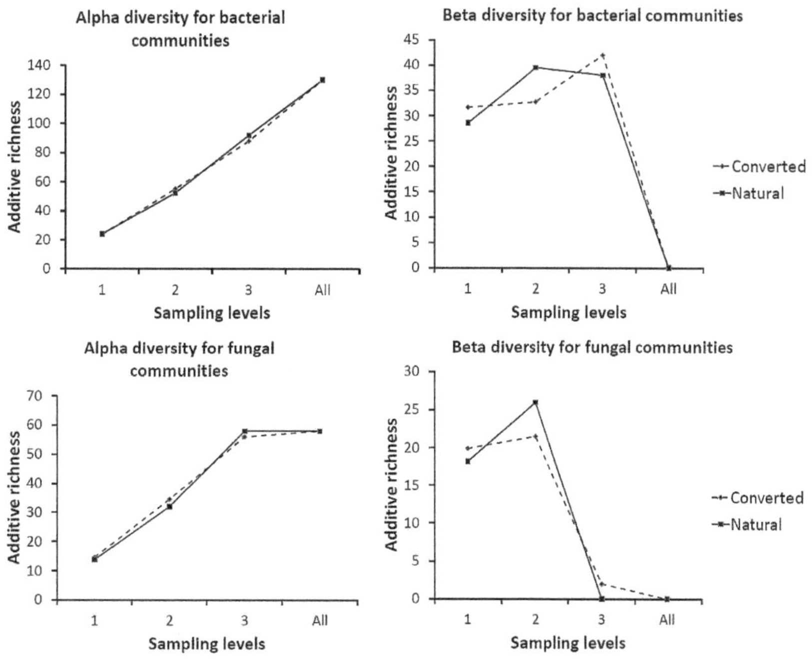

Figure 6. Additive partitioning of bacterial and fungal diversity (expressed as additive richness) across alpha and beta hierarchical spatial scales at three sampling levels (plot, site and ecosystem type) in natural and converted TRF ecosystems.

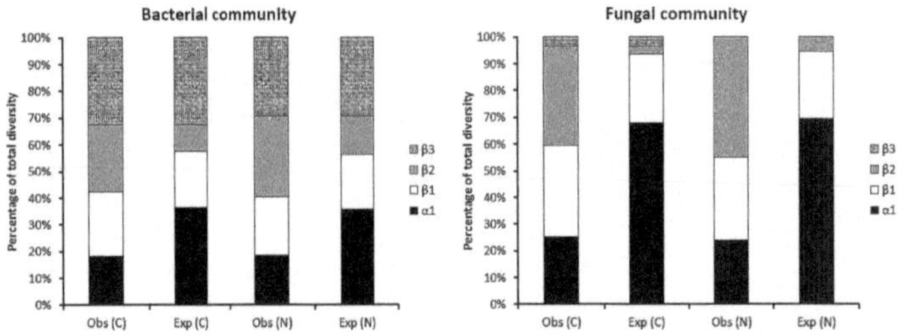

Figure 7. The additive partitioning of total bacterial and fungal community, γ -diversity into α and β components at three nested spatial scales, with each component expressing their relative contributions to γ –diversity; where γ –diversity is equal to $\alpha1+\beta1+\beta2+\beta3$. The observed (obs) partitions are compared with the expected (exp) values, as predicted by the null model based on 1000 iterations using individual-based randomization.

Conclusion

There is international concern about the threat to natural habitats in the Nile river watershed and the consequential loss of important biodiversity. Whereas aspects of microbial biogeography and influence of forest conversion in Uganda's Nile river watershed is largely unknown, this study offers an important first glimpse into indicators of spatial diversity patterns of soil fungal and bacterial communities in the Uganda's Nile river watershed. Our observations of reduced soil microbial diversity, both bacterial and fungal, in converted ecosystems though unsurprising in itself causes us some concern and would justify further work to determine the significance of the diversity lost and the wider implications.

By focusing on diversity patterns across multiple hierarchical spatial scales, we were able to identify the scale at which regional microbial diversity is maximized. We showed that there was substantial contribution of β-diversity to total ecosystem diversity (γ) which includes taxa at the between-plot, site and ecosystem scales and unique taxa, highlighting the necessity to conserve marginal habitats and ecotones. Soil microbial communities in Uganda's Nile river watershed exhibit considerable resilience to forest conversion even though SOC, N, Ca and pH were all significantly altered. This result is surprising given that these physical and chemical properties typically strongly influence microbial diversity. Additionally, the variation among sites was quite large, indicating that soil communities in this region vary considerably on a regional spatial scale. Our results do not explain this variation. Most studies suggest that biogeographic barriers play little role in the geographic structure of soil communities. Rather than a consistent general pattern of microbial community change following forest conversion we find that responses are largely site-specific and widely variable.

Acknowledgments

We confirm that our field work did not involve any endangered or protected species and that we were granted access to protected areas by Uganda Wildlife Authority (UWA) and National Forestry Authority (NFA). Our sample site coordinates included: (1) Mabira; Lat, N00°24.495'; Long, E033°02.464' (Required permission obtained from NFA); (2) Maramagambo; Lat, S00°04.421'; Long, E030°02.305' (Required permission obtained from UWA); (3) Budongo; Lat, N01°41.852'; Long, E031°29.363' (Required permission obtained from UWA/NFA); (4) Kaniyo Pabidi; Lat, N01°55.128'; Long, E031°43.116' (Required permission obtained from UWA).

We would like to thank the Institute of Tropical Forest Conservation (ITFC) of Mbarara University of Science and Technology for supporting and co-advising this research. We also thank the Genetics Lab at Faculty of Science and the soil department lab at Faculty of Agriculture both at Makerere University, Kampala, for their lab facilitation during extractions and analyses. And finally we acknowledge the ideas and insights of Masatoshi Katabuchi, Dossa Gbadamassi, and our field and lab technicians and assistants Solomon Echel, Francis Alele Ozirit and Boniface Balikuddembe.

Author Contributions

Conceived and designed the experiments: POA CHC. Performed the experiments: POA SL. Analyzed the data: POA DS CHC. Contributed reagents/materials/analysis tools: DS YSG CHC. Contributed to the writing of the manuscript: POA DS YSG CHC.

References

1. Turner IM (1996) Species loss in fragments of tropical rain forest: a review of the evidence. J Appl Ecol 33: 200–209.
2. Joseph Wright S, Muller-Landau HC (2006) The Future of Tropical Forest Species. Biotropica 38: 287–301.
3. United Nations (2013) World Population Prospects: The 2012 Revision. Highlights and Advance Tables. New York.
4. United Nations D of E and SAPD (2007) World Population Prospects: The 2006 Revision.
5. Kayanja FIB, Byarugaba D (2001) Disappearing forests of Uganda: The way forward. Curr Sci 81: 936–947.
6. Jesus da CE, Marsh TL, Tiedje, M J, de S Moreira FM (2009) Changes in land use alter the structure of bacterial communities in Western Amazon soils. ISME J 3: 1004–1011. Available: http://www.ncbi.nlm.nih.gov/pubmed/19440233. Accessed 15 November 2012.
7. Rodrigues JLM, Pellizari VH, Mueller R, Baek K, Jesus EDC, et al. (2012) Conversion of the Amazon rainforest to agriculture results in biotic homogenization of soil bacterial communities. Proc Natl Acad Sci U S A. Available: http://www.ncbi.nlm.nih.gov/pubmed/23271810. Accessed 29 December 2012.

8. Cardinale BJ, Duffy JE, Gonzalez A, Hooper DU, Perrings C, et al. (2012) Biodiversity loss and its impact on humanity. Nature 486: 59–67. Available: http://www.ncbi.nlm.nih.gov/pubmed/22678280. Accessed 25 May 2014.
9. Venail PA, Vives MJ (2013) Positive effects of bacterial diversity on ecosystem functioning driven by complementarity effects in a bioremediation context. PLoS One 8: e72561. Available: http://www.pubmedcentral.nih.gov/articlerender.fcgi?artid = 3762786&tool = pmcentrez&rendertype = abstract. Accessed 24 June 2014.
10. Torsvik V, Øvreås L (2002) Microbial diversity and function in soil: from genes to ecosystems. Curr Opin Microbiol 5: 240–245. Available: http://www.ncbi.nlm.nih.gov/pubmed/12057676.
11. Nile Basin Initiative (2012) State of the River Nile Basin. Entebbe (Uganda): Nile Basin Initiative Secretariat.
12. Olson DM, Dinerstein E, Powell GVN, Wikramanayake ED (2002) Conservation Biology for the Biodiversity Crisis. Conserv Biol 16: 1–3.
13. Ricklefs RE (2004) A comprehensive framework for global patterns in biodiversity. Ecol Lett 7: 1–15. Available: http://doi.wiley.com/10.1046/j.1461-0248.2003.00554.x. Accessed 21 January 2014.
14. Gering JC, Crist TO, Veech J a. (2003) Additive Partitioning of Species Diversity across Multiple Spatial Scales: Implications for Regional Conservation of Biodiversity. Conserv Biol 17: 488–499. Available: http://doi.wiley.com/10.1046/j.1523-1739.2003.01465.x.
15. Whittaker RJ, Willis KJ, Field R (2001) Scale and species richness: towards a general, hierarchical theory of species diversity. J Biogeogr 28: 453–470. Available: http://doi.wiley.com/10.1046/j.1365-2699.2001.00563.x.
16. Demeny P (2003) Population Policy: A Concise Summary. Internatio. Demeny P, McNicoll G, editors New York: MacMillan Reference. doi:173.
17. Olden JD (2006) Biotic homogenization: a new research agenda for conservation biogeography. J Biogeogr 33: 2027–2039. Available: http://doi.wiley.com/10.1111/j.1365-2699.2006.01572.x. Accessed 5 May 2014.
18. Olden JD, Poff NL (2004) Ecological processes driving biotic homogenization: testing a mechanistic model using fish faunas. Ecology 85: 1867–1875.
19. Olden JD, Poff NL (2003) Toward a mechanistic understanding and prediction of biotic homogenization. Am Nat 162: 442–460. Available: http://www.ncbi.nlm.nih.gov/pubmed/14582007.
20. Folke C, Carpenter S, Walker B, Scheffer M, Elmqvist T, et al. (2004) Regime Shifts, Resilience, and Biodiversity in Ecosystem Management. Annu Rev Ecol Evol Syst 35: 557–581. Available: http://www.annualreviews.org/doi/abs/10.1146/annurev.ecolsys.35.021103.105711. Accessed 26 October 2012.
21. Ter Braak CJ F. (1986) Canonical Correspondence Analysis: A New Eigenvector Technique for Multivariate Direct Gradient Analysis. Ecol Soc Am 67: 1167–1179.
22. Legendre P, Legendre L (1998) Numerical Ecology. 2nd ed. Amsterdam: Elsevier.
23. Lande R (1996) Statistics and partitioning of species diversity, and similarity among multiple communties. Oikos, Nord Soc 76: 5–13.
24. Godfray HCJ, Lawton JH (2001) Scale and species numbers. Trends Ecol Evol 16: 400–404. Available: http://www.ncbi.nlm.nih.gov/pubmed/11403873.
25. Walkley A., Black I (1934) An examination of the Degtjareff method for determining organic carbon in soils: effect of variations in digestion conditions and of inorganic soil constituents. Soil Sci: 251–263.
26. Kjeldahl J (1883) A new method for the determination of nitrogen in organic matter. Zeitschreft fur Anal Chemie 22.
27. Bray RH, Kurtz LT (1945) Determination of total, organic, and available forms of phosphorus in soils. Soil Sci 59: 39–45.
28. Gee G., Bauder J. (1986) Particle-size analysis. p. 383–411. In A Klute (ed.) Methods of Soil Analysis, Part 1. Physical and Mineralogical Methods. Agronomy Monograph No. 9 (2ed). Am Soc Agron Sci Soc Am Madison, WI.
29. Cleary DFR, Smalla K, Mendonça-Hagler LCS, Gomes NCM (2012) Assessment of variation in bacterial composition among microhabitats in a mangrove environment using DGGE fingerprints and barcoded pyrosequencing. PLoS One 7: e29380. Available: http://www.pubmedcentral.nih.gov/articlerender.fcgi?artid = 3256149&tool = pmcentrez&rendertype = abstract. Accessed 27 October 2012.
30. Reche I, Pulido-Villena E, Morales-Baquero R, Casamayor EO (2005) Does ecosystem size determine aquatic bacterial richness? Ecology 86: 1715–1722. Available: http://www.ncbi.nlm.nih.gov/pubmed/17489473.
31. Whittaker RH (1972) Evolution and Measurement of Species Diversity. Int Assoc Plant Taxon (IAPT) 21: 213–251.
32. Jaccard P (1908) Nouvelles recherches sur la distribution florale. Bull Soc Vaud Sci Nat 44: 223–270.
33. Veech JA, Summerville KS, Crist TO, Gering JC (2002) The additive partitioning of species diversity: recent revival of an old idea. Nord Soc Oikos 99: 3–9.
34. Crist TO, Veech J a, Gering JC, Summerville KS (2003) Partitioning species diversity across landscapes and regions: a hierarchical analysis of alpha, beta, and gamma diversity. Am Nat 162: 734–743. Available: http://www.ncbi.nlm.nih.gov/pubmed/14737711.
35. Chandy S, Gibson DJ, Robertson P a. (2006) Additive partitioning of diversity across hierarchical spatial scales in a forested landscape. J Appl Ecol 43: 792–801. Available: http://doi.wiley.com/10.1111/j.1365-2664.2006.01178.x. Accessed 22 January 2014.
36. Ribeiro DB, Prado PI, Brown Jr KS, Freitas AVL (2008) Additive partitioning of butterfly diversity in a fragmented landscape: importance of scale and implications for conservation. Divers Distrib 14: 961–968. Available: http://doi.wiley.com/10.1111/j.1472-4642.2008.00505.x. Accessed 3 February 2014.
37. Wu F, Yang XJ, Yang JX (2010) Additive diversity partitioning as a guide to regional montane reserve design in Asia: an example from Yunnan Province, China. Divers Distrib 16: 1022–1033. Available: http://doi.wiley.com/10.1111/j.1472-4642.2010.00710.x. Accessed 6 February 2014.
38. Sasaki T, Katabuchi M, Kamiyama C, Shimazaki M, Nakashizuka T, et al. (2012) Diversity partitioning of moorland plant communities across hierarchical spatial scales. Biodivers Conserv. doi:10.1007/s10531-012-0265-7.
39. Veech JA, Crist TO (2009) Partition: software for hierarchical partitioning of species diversity, version 3.0. Available: http://www.users.muohio.edu/cristto/partition.htm.
40. Monkiedje A, Spiteller M, Fotio D, Sukul P (2006) The effect of land use on soil health indicators in peri-urban agriculture in the humid forest zone of southern cameroon. J Environ Qual 35: 2402–2409. Available: http://www.ncbi.nlm.nih.gov/pubmed/17071911. Accessed 9 November 2012.
41. Neris J, Jiménez C, Fuentes J, Morillas G, Tejedor M (2012) Vegetation and land-use effects on soil properties and water infiltration of Andisols in Tenerife (Canary Islands, Spain). Catena 98: 55–62. Available: http://linkinghub.elsevier.com/retrieve/pii/S0341816212001270. Accessed 28 December 2012.
42. Bissett A, Richardson AE, Baker G, Thrall PH (2011) Long-term land use effects on soil microbial community structure and function. Appl Soil Ecol 51: 66–78. Available: http://linkinghub.elsevier.com/retrieve/pii/S0929139311001922. Accessed 9 November 2012.
43. Garbisu C, Alkorta I, Epelde L (2011) Assessment of soil quality using microbial properties and attributes of ecological relevance. Appl Soil Ecol 49: 1–4. Available: http://linkinghub.elsevier.com/retrieve/pii/S0929139311001089. Accessed 9 November 2012.
44. Fierer N, Jackson RB (2006) The diversity and biogeography of soil bacterial communities. Proc Natl Acad Sci U S A 103: 626–631. Available: http://www.pubmedcentral.nih.gov/articlerender.fcgi?artid = 1334650&tool = pmcentrez&rendertype = abstract.
45. Shange RS, Ankumah RO, Ibekwe AM, Zabawa R, Dowd SE (2012) Distinct soil bacterial communities revealed under a diversely managed agroecosystem. PLoS One 7: e40338. Available: http://www.pubmedcentral.nih.gov/articlerender.fcgi?artid = 3402512&tool = pmcentrez&rendertype = abstract. Accessed 1 November 2012.
46. Wang Y, Boyd E, Crane S, Lu-Irving P, Krabbenhoft D, et al. (2011) Environmental conditions constrain the distribution and diversity of archaeal merA in Yellowstone National Park, Wyoming, U.S.A. Microb Ecol62: 739–752. Available: http://www.ncbi.nlm.nih.gov/pubmed/21713435. Accessed 9 November 2012.
47. Fierer N, Bradford MA, Jackson RB (2007) Toward an ecological classification of soil bacteria. Ecology 88: 1354–1364. Available: http://www.ncbi.nlm.nih.gov/pubmed/17601128.
48. Saiya-Cork K., Sinsabaugh R., Zak D. (2002) The effects of long term nitrogen deposition on extracellular enzyme activity in an Acer saccharum forest soil. Soil Biol Biochem 34: 1309–1315. Available: http://linkinghub.elsevier.com/retrieve/pii/S0038071702000743.
49. Steenwerth K, Jackson L, Calderon F, Scow K, Rolston D (2005) Response of microbial community composition and activity in agricultural and grassland soils after a simulated rainfall. Soil Biol Biochem 37: 2249–2262. Available: http://linkinghub.elsevier.com/retrieve/pii/S0038071705001549. Accessed 9 November 2012.
50. De Vries FT, Manning P, Tallowin JRB, Mortimer SR, Pilgrim ES, et al. (2012) Abiotic drivers and plant traits explain landscape-scale patterns in soil microbial communities. Ecol Lett 15: 1230–1239. Available: http://www.ncbi.nlm.nih.gov/pubmed/22882451. Accessed 2 November 2012.
51. Hartman WH, Richardson CJ, Vilgalys R, Bruland GL (2008) Environmental and anthropogenic controls over bacterial communities in wetland soils. Proc Natl Acad Sci U S A 105: 17842–17847. Available: http://www.pubmedcentral.nih.gov/articlerender.fcgi?artid = 2584698&tool = pmcentrez&rendertype = abstract.
52. Cadotte MW, Cardinale BJ, Oakley TH (2008) Evolutionary history and the effect of biodiversity on plant productivity. Proc Natl Acad Sci U S A 105: 17012–17017. Available: http://www.pubmedcentral.nih.gov/articlerender.fcgi?artid = 2579369&tool = pmcentrez&rendertype = abstract.
53. Garbeva P, Veen JA Van, Elsas JD Van (2004) Microbial Diversity in Soil: Selection of Microbial Populations by Plant and Soil Type and Implications for Disease Suppressiveness. Annu Rev Phytopathol 42: 243–270. doi:10.1146/annurev.phyto.42.012604.135455.
54. Schulz S, Giebler J, Chatzinotas A, Wick LY, Fetzer I, et al. (2012) Plant litter and soil type drive abundance, activity and community structure of alkB harbouring microbes in different soil compartments. ISME J doi:10.103: 1763–1774.
55. Kuramae EE, Yergeau E, Wong LC, Pijl AS, van Veen J a, et al. (2012) Soil characteristics more strongly influence soil bacterial communities than land-use type. FEMS Microbiol Ecol 79: 12–24. Available: http://www.ncbi.nlm.nih.gov/pubmed/22066695. Accessed 7 November 2012.
56. Nacke H, Thürmer A, Wollherr A, Will C, Hodac L, et al. (2011) Pyrosequencing-based assessment of bacterial community structure along different management types in German forest and grassland soils. PLoS One 6: e17000. Available: http://www.pubmedcentral.nih.gov/articlerender.

fcgi?artid = 3040199&tool = pmcentrez&rendertype = abstract. Accessed 5 November 2012.
57. Waldrop MP, Balser TC, Firestone MK (2001) Linking microbial community composition to function in a tropical soil. Soil Biol Biochem 32: 1837–1846.
58. Dinsdale E a, Edwards R a, Hall D, Angly F, Breitbart M, et al. (2008) Functional metagenomic profiling of nine biomes. Nature 452: 629–632. Available: http://www.ncbi.nlm.nih.gov/pubmed/18337718. Accessed 26 October 2012.
59. Lauber CL, Hamady M, Knight R, Fierer N (2009) Pyrosequencing-based assessment of soil pH as a predictor of soil bacterial community structure at the continental scale. Appl Environ Microbiol 75: 5111–5120. Available: http://www.pubmedcentral.nih.gov/articlerender.fcgi?artid = 2725504&tool = pmcentrez&rendertype = abstract. Accessed 19 March 2014.
60. Russo SE, Legge R, Weber K a., Brodie EL, Goldfarb KC, et al. (2012) Bacterial community structure of contrasting soils underlying Bornean rain forests: Inferences from microarray and next-generation sequencing methods. Soil Biol Biochem 55: 48–59. Available: http://linkinghub.elsevier.com/retrieve/pii/S0038071712002362. Accessed 8 November 2012.
61. Winterbottom B, Eilu G (2006) Uganda Biodiversity and Tropical Forest Assessment. Washington DC.
62. Laurance WF (1999) Reflections on the tropical deforestation crisis. Biol Conserv 91: 109–117.
63. Scheckenbach F, Hausmann K, Wylezich C, Weitere M, Arndt H (2010) Large-scale patterns in biodiversity of microbial eukaryotes from the abyssal sea floor. Proc Natl Acad Sci U S A 107: 115–120. Available: http://www.pubmedcentral.nih.gov/articlerender.fcgi?artid = 2806785&tool = pmcentrez&rendertype = abstract. Accessed 1 November 2012.
64. Martiny JBH, Bohannan BJM, Brown JH, Colwell RK, Fuhrman J a, et al. (2006) Microbial biogeography: putting microorganisms on the map. Nat Rev Microbiol 4: 102–112. Available: http://www.ncbi.nlm.nih.gov/pubmed/16415926. Accessed 27 October 2012.
65. McKinney M, Lockwood J (1999) Biotic homogenization: a few winners replacing many losers in the next mass extinction. Trends Ecol Evol 14: 450–453. Available: http://www.ncbi.nlm.nih.gov/pubmed/10511724.
66. Olden JD, Rooney TP (2006) On defining and quantifying biotic homogenization. Glob Ecol Biogeogr 15: 113–120. doi:10.1111/j.1466-822x.2006.00214.x.
67. Devictor V, Julliard R, Clavel J, Jiguet F, Lee A, et al. (2008) Functional biotic homogenization of bird communities in disturbed landscapes. Glob Ecol Biogeogr 17: 252–261. Available: http://doi.wiley.com/10.1111/j.1466-8238.2007.00364.x. Accessed 21 January 2014.
68. Summerville KS, Boulware MJ, Veech J a., Crist TO (2003) Spatial Variation in Species Diversity and Composition of Forest Lepidoptera in Eastern Deciduous Forests of North America. Conserv Biol 17: 1045–1057. Available: http://doi.wiley.com/10.1046/j.1523-1739.2003.02059.x.
69. Weiher E, Howe A (2003) Scale-dependence of environmental effects on species richness in oak savannas. J Veg Sci 14: 917–920.
70. Obua J, Agea JG, Ogwal JJ (2010) Status of forests in Uganda. Afr J Ecol 48: 853–859.
71. National Environment Management Authority (NEMA) (2008) State of the Environment Report for Uganda.
72. Amann RI, Ludwig W, Schleifer K-H (1995) Phylogenetic identification and in situ detection of individual microbial cells without cultivation. Microb Rev 59: 143–169.
73. Vainio EJ, Hantula J (2000) Direct analysis of wood-inhabiting fungi using denaturing gradient gel electrophoresis of amplified ribosomal DNA. Mycol Res 104: 927–936.

11

New Insights into the Consequences of Post-Windthrow Salvage Logging Revealed by Functional Structure of Saproxylic Beetles Assemblages

Simon Thorn[1]*, Claus Bässler[1], Thomas Gottschalk[2], Torsten Hothorn[3], Heinz Bussler[4], Kenneth Raffa[5], Jörg Müller[1,6]

1 Sachgebiet Forschung und Dokumentation, Nationalparkverwaltung Bayerischer Wald, Grafenau, Germany, **2** Hochschule für Forstwirtschaft Rottenburg, Rottenburg am Neckar, Germany, **3** Abteilung Biostatistik, Universität Zürich, Zürich, Switzerland, **4** Bavarian State Institute for Forestry, Freising, Germany, **5** Department of Entomology, University of Wisconsin-Madison, Madison, United States of America, **6** Chair for Terrestrial Ecology, Department of Ecology and Ecosystem Management, Technische Universität München, Freising, Germany

Abstract

Windstorms, bark beetle outbreaks and fires are important natural disturbances in coniferous forests worldwide. Windthrown trees promote biodiversity and restoration within production forests, but also cause large economic losses due to bark beetle infestation and accelerated fungal decomposition. Such damaged trees are often removed by salvage logging, which leads to decreased biodiversity and thus increasingly evokes discussions between economists and ecologists about appropriate strategies. To reveal the reasons behind species loss after salvage logging, we used a functional approach based on four habitat-related ecological traits and focused on saproxylic beetles. We predicted that salvage logging would decrease functional diversity (measured as effect sizes of mean pairwise distances using null models) as well as mean values of beetle body size, wood diameter niche and canopy cover niche, but would increase decay stage niche. As expected, salvage logging caused a decrease in species richness, but led to an increase in functional diversity by altering the species composition from habitat-filtered assemblages toward random assemblages. Even though salvage logging removes tree trunks, the most negative effects were found for small and heliophilous species and for species specialized on wood of small diameter. Our results suggested that salvage logging disrupts the natural assembly process on windthrown trees and that negative ecological impacts are caused more by microclimate alteration of the dead-wood objects than by loss of resource amount. These insights underline the power of functional approaches to detect ecosystem responses to anthropogenic disturbance and form a basis for management decisions in conservation. To mitigate negative effects on saproxylic beetle diversity after windthrows, we recommend preserving single windthrown trees or at least their tops with exposed branches during salvage logging. Such an extension of the green-tree retention approach to windthrown trees will preserve natural succession and associated communities of disturbed spruce forests.

Editor: Marc Hanewinkel, Swiss Federal Institute for Forest, Switzerland

Funding: ST was funded by the Scholarship Programme of the German Federal Environmental Foundation. The funders had no role in study design, data collection and analysis, decision to publish, or preparation of the manuscript.

Competing Interests: The authors have declared that no competing interests exist.

* Email: simon@thornonline.de

Introduction

Forest ecosystems worldwide are periodically affected by natural disturbances, such as wind storms, fires, avalanches and insects [1,2]. Since the 1990s, disturbance events in forests of the northern hemisphere have increased, particularly in mature conifer stands, owing to both an increase in growing stock and global climate change [3–5]. After such disturbances, forest managers try to limit the economic loss by focusing on saving downed wood from fungal infestation and avoiding an increase of pest species populations [6,7]. Even if such salvage logging is broadly publicly accepted [8], both ecologists and conservationists are increasingly aware that natural disturbances conserve biodiversity in forests moulded by anthropogenic impacts [9,10]. In addition, whether salvage logging should be conducted and how it would be best conducted, particularly in coniferous forests, is highly controversial [11–13].

Between 1950 and 2000 in Europe, wind storms annually damaged an estimated average of 18.7 million m^3 of wood [3]. Such events often are followed by outbreaks of the European spruce bark beetle *Ips typographus* (Linnaeus, 1758), which damages an additional 2.9 million m^3 wood annually [3]. In contrast to our knowledge of the high value of windthrows for biodiversity [14,15], our knowledge about why particular species are affected or not by salvage logging is limited; most recent studies have focused on post-fire salvage logging [16–19] or economic consequences and bark beetles [20,21]. The relatively few studies on post-windthrow salvage logging focus mainly on the decrease in species numbers [14,22,23].

Recent studies using, for example, guild-specific analysis of bird assemblages demonstrate that species richness poorly reflects the effect of human intervention [24]. A quantification of species loss is not sufficient to guide conservation efforts and resource management directly [25,26]. Therefore, functional approaches have become increasingly important throughout broad areas of ecological research [26–28]. It has recently been proposed that the species position in a functional space can be used as a tool to reveal advanced warnings for changes in disturbed ecosystems [29].

Saproxylic beetles are highly diverse, play important roles in the decomposition of wood [30] and are sensitive to forest management, and are thereby an ideal model group to study the impact of salvage logging on biodiversity [31,32]. Here we studied the impact of salvage logging and focused on recently published ecological traits of saproxylic beetles [33]: mean body size, diameter and decay stage of wood in which larvae develop and canopy cover of forests in which the species is known to occur.

The body size of saproxylic beetles is positively correlated to the diameter of the substrate used for larval development [34]. Assemblage values should therefore decrease with the removal of tree sections of large diameter [35,36]. Similarly, a shift in assemblages toward species with a preference for wood of smaller diameters might follow the removal of major tree trunks. Since the cutting of uprooted trees in salvage-logging operations removes the trunk and the remaining branches rapidly decompose to advanced decomposition stages [37], we therefore expect that colonizers of earlier successional stages of wood decomposition also decline [35]. Finally, the removal of dead wood might decrease shady conditions provided by the cross-laminated arrangement of trees after windthrows, thereby promoting heliophilous species [38].

We calculated effect sizes of saproxylic beetle functional diversity, and mean assemblage values and diversity values of each single trait to test the predictions that salvage logging results in 1) a lower overall functional diversity, 2) lower mean body size and body size diversity, 3) a minor occurrence of species preferring dead wood of large diameter, 4) a decrease in species preferring early successional stages of decomposition and 5) an increase in species preferring open canopies.

Methods

Study area

The study was carried out in the high montane spruce forest in the Bavarian Forest National Park in south-eastern Germany. Forest stands in this area, at an elevation above 1,100 m, are naturally dominated by Norway spruce (*Picea abies*). Annual precipitation ranges from 1,300 to 1,800 mm, and annual mean air temperature ranges from 3.0 to 4.0°C [39].

On January 16, 2007, an area of approximately 1,000 ha of spruce forests was felled to various extents by the windstorm Kyrill, ranging from single trees to stands covering several hundreds of hectares. From a total affected amount of wood of about 160,000 m^3, 50,000 m^3 are concentrated on four larger windthrow areas (~170 ha). These centres were partially excluded from the overall salvage-logging operation. Such operations basically remove the main trunk to preserve it from fungal and pest species infestation. The branches are cut off the trunk and remain on the ground surface, which is covered by a grass layer of *Calamagrostis villosa* (32±11 cm height on logged plots and 32±9 cm height on non-logged plots, measured by relevés [39]). Standardized measurements of dead-wood objects per plot revealed that on salvaged-logged plots, 90% (n = 29) of dead-wood objects had direct contact with the ground surface, in contrast to only 10% (n = 35) on non-logged plots. Wind-felled trees on non-logged plots remained mostly alive after the storm in spring 2007 until they were colonized by *Ips typographus* in 2008. In contrast, salvage logging typically kill trees immediately. Salvage logging in the major windthrow areas removed about 255 m^3/ha and was completed in autumn 2007 (Fig. 1).

Beetle sampling

To reflect the emerging beetle fauna of surrounding dead wood, we used flight-interception traps [40]. Traps were established throughout windthrow centres and surrounding salvage-logged areas: 22 in logged areas and 22 in non-logged areas in spring 2008. Traps in logged areas were surrounded by at least a 50 m radius of completely salvage-logged windthrows (all trees removed); traps in non-logged areas were surrounded by at least a 50 m radius of completely non-logged windthrow (all trees of the previous stand were wind felled). Each trap consisted of a crossed pair of transparent plastic shields (40×60 cm) and contained a 3.0% copper-vitriol solution to preserve trapped specimens [41]. The shortest distance between two traps was 50 m, and the largest distance between traps was 6,500 m. Sampling was conducted during the entire growing season between May after the snow melted until September over four consecutive years until 2011. Traps were emptied monthly. All sampled beetles were identified to the species level, but only saproxylic beetles were considered in the analysis [42,43].

Trait characterization

We used four ecological traits that enabled us to link species habitat selection [33] directly to forest management: mean body size, diameter of wood in which the larvae of the species was recorded, decay stage of the wood, and canopy cover of forests in which the larvae of the species is known to occur. The single classes of niche positions were classified as follows: *wood diameter class*: 1, <15 cm; 2, 15–35 cm; 3, 35–70 cm; and 4, >70 cm; *wood decay stage*: 0, alive; 1, freshly dead (up to two years); 2, initiated decomposition with loose bark and tough sapwood; 3, advanced decomposition with soft sapwood and partly tough hardwood; and 4, extremely decomposed and mouldered; *canopy cover*: 1, open; 2, semi-open; and 3, closed (for mean niche positions of species, see Table S1).

Data analysis

Prior to the main analysis, trapped specimens were grouped according to the trap level in each year, and subsequent analyses were conducted at the trap per year level. We calculated the mean trait value of the assemblage for each of the four single traits as an arithmetic mean, weighted by the number of trapped individuals (e.g. log-transformed number of individuals) per species. The results of abundance-weighted data, abundance-weighted data based on the log-transformed number of individuals, and presence/absence data were similar; therefore, we present only abundance-weighted results, which do not overestimate singletons and represents the majority of trapped beetles [40]. To test the impact of salvage logging on saproxylic beetles, we selected these mean assemblage trait values of every single trait and their related trait diversity as target variable and year, and we selected logged/non-logged as predictor variables.

A challenge in functional diversity measures is to compare dispersion of functional traits in the functional space independent of species numbers [29]. Hence, we calculated a distance matrix based on the pairwise Euclidean distance between functional traits of all possible species pairs within an assemblage per plot for each single trait [44]. Similarly, we used the function *dist* to calculate an

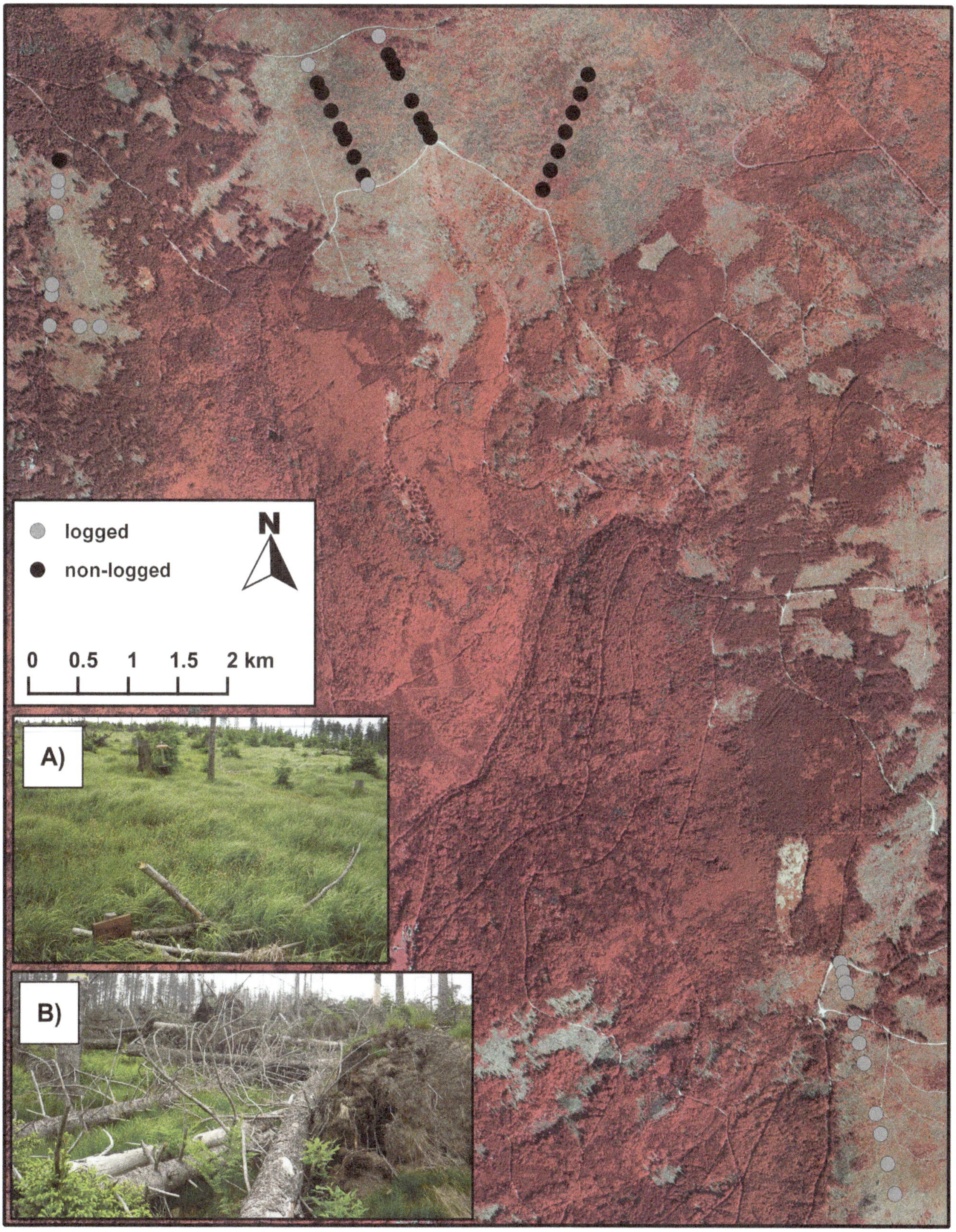

Figure 1. Locations of flight-interception traps within our study area in salvage-logged (A) and non-salvage-logged windthrows, sampled from 2008 to 2011 (B).

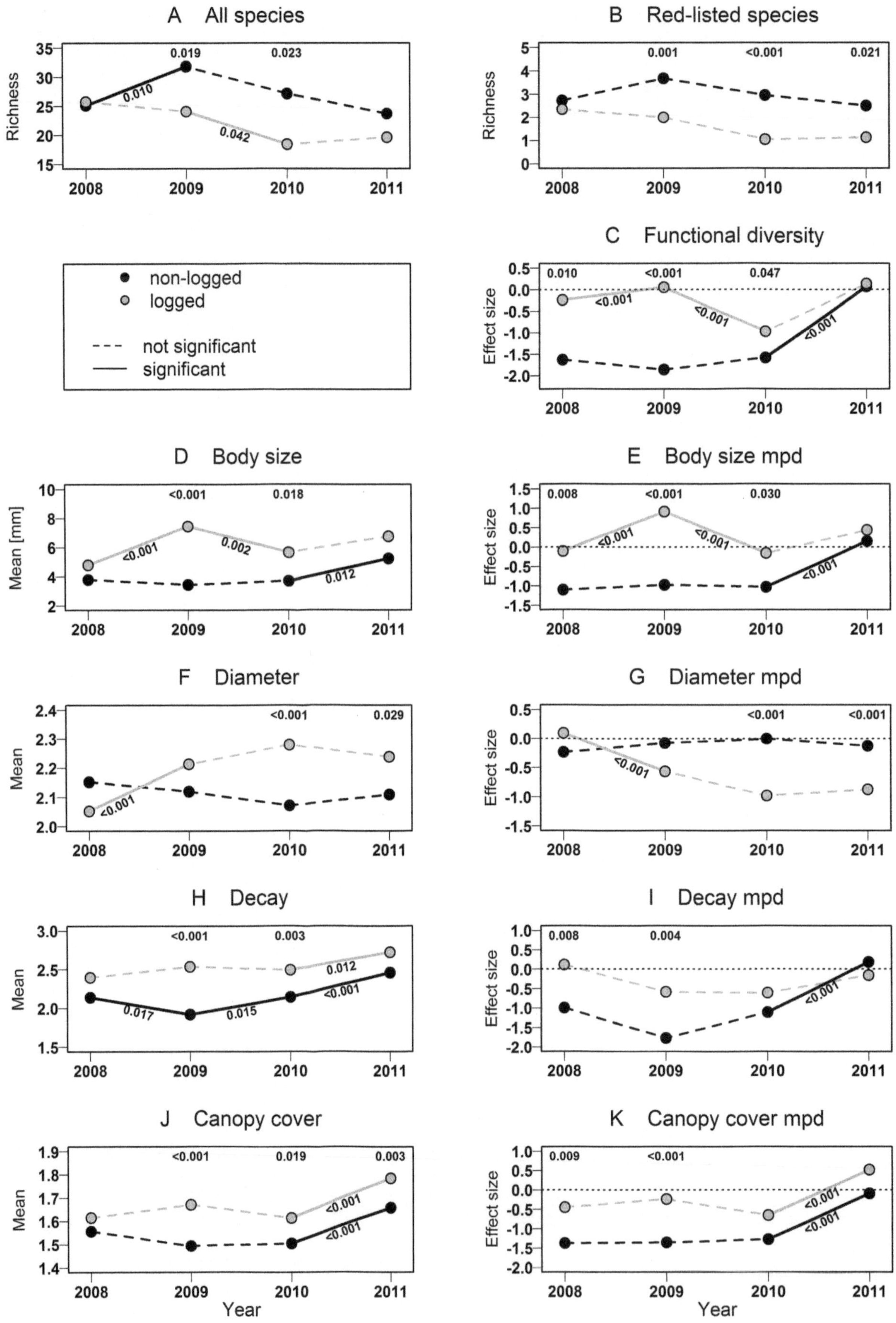
A All species
B Red-listed species
C Functional diversity
D Body size
E Body size mpd
F Diameter
G Diameter mpd
H Decay
I Decay mpd
J Canopy cover
K Canopy cover mpd
Richness
Effect size
Mean [mm]
Mean
Year
non-logged
logged
not significant
significant

Figure 2. Effects of salvage logging on saproxylic beetles. Mean species richness of saproxylic beetles (A) and red-listed saproxylic beetles (B), abundance-weighted mean niche positions and standardized effect size (based on mean pairwise distance) of functional diversity (C) and body size (D, E), niche diameter (F, G), niche decay (H, I) and niche canopy cover (J, K) in logged and non-logged windthrow areas in a spruce mountain forest based on a GLMM with treatment and year as fixed factors, and space and plot as random factors. For multiple comparisons between treatments, the adjusted p-values are drawn above the respective points; for comparisons between different years within the same treatment, p-values are drawn below respective lines. Full details on p-values and model estimators can be found in Tables S2 and S3.

overall distance matrix based on Euclidean distance between among four traits. We compared the observed value of mean niche position for each trait in each assemblage to an artificial assemblage with equal number of randomly selected species from the regional species pool (all species recorded on plots pooled). To create these random assemblages, we used null models with tip shuffling and abundance weighting in 999 randomizations using the function *ses.mpd* (abundance.weighted = TRUE) in the add-on package *picante* of R version 2.15 [45,46]. The effect size, which is provided by the null model, indicates the dispersion of a specific trait and compares it to the trait dispersion in an artificial assemblage. Values >0 indicate over-dispersion of a trait. In turn, values <0 indicate clustering and a trait dispersion in observed assemblages smaller than expected from a random assemblage. In general, over-dispersion is a result of competition or facilitation, and clustering is a result of environmental filtering [47].

The natural occurrence of windthrow patches necessitates that logged and non-logged plots are near each other in the same forest site (Fig. 1). This might reduce the independence of observations. Furthermore, we cannot treat the measurement on each plot as independent as the measurements were conducted in consecutive years. To address these issues, we analysed the data using a linear mixed random effects model with sampling plot and coordinates of sampling plot as random factors. The plot coordinates enable us to account for possible spatial autocorrelation within the arrangement of our plots, as a second-order trend surface (for the R code, see Text S1; for the method, see [48]).

To illustrate the impact of salvage logging on the primary target species, we modelled the mean number of *I. typographus* individuals per trap by using an observation-specific random factor to control the mixed model with Poisson distribution for potential over-dispersion [49]. To consider the resulting problem of testing hypothesis families, we applied a multiple post-hoc comparison with adjusted p-values using the function *glht* in the add-on package *multcomp* and constructed a matrix of coefficients [50]. This enabled us to compare the impact of salvage logging within each year and to compare changes within consecutive years on logged or non-logged plots.

Results

Effect of pest control

In total, 33,796 specimens belonging to 179 species of 37 saproxylic beetles families were sampled. Of these, 29 (89 individuals) species were found exclusively on non-logged plots, and 34 species (85 individuals) were found only on salvage-logged plots. Forty-two species (19 on non-logged plots) were caught as singletons. The most abundant species in our data on non-logged plots were *Ips typographus* (7,785 individuals), *Pityogenes chalcographus* (6,697 individuals) and *Xyloterus lineatus* (1,579 individuals), all of subfamily *Scolytinae* (Curculionidae). The most frequent species on logged plots were *P. chalcographus* (3,147 individuals), *Hylastes cunicularius* (1,101 individuals) and *I. typographus* (796 individuals).

Salvage logging significantly decreased the mean number of the primary target species *I. typographus* per trap in 2008, 2009 and 2010. The mean number of *I. typographus* individuals significantly increased from 2008 to 2009 and significantly decreased from 2009 to 2010, reflecting the colonization of the windthrow by this pest species. The species richness of saproxylic beetles was significantly lower on salvage-logged plots than on non-logged plots in 2009 and 2010 (Fig. 2A). On non-logged plots, species richness significantly increased from 2008 to 2009, whereas on logged plots, richness significantly decreased from 2009 to 2010 (Fig. 2A). A similar trend was found only for red-listed species, which displayed significantly lower species richness on salvage-logged plots in 2009, 2010 and 2011 (Fig. 2B; for complete species list, see Table S1).

Assemblage functional response

Beetle assemblages of logged and non-logged windthrows differed in their functional trait dispersion. The total functional diversity was lower on non-logged plots and revealed a clustered pattern, but was random on logged plots (Fig. 2C). From 2010 to 2011, the functional assemblage composition of non-logged plots developed significantly toward more random assemblages. The mean body sizes and the corresponding niche diversity were consistently and significantly higher on logged plots than on non-logged plots in 2009 and 2010 (Fig.2 D, E). We also found significantly higher mean diameter niche values on logged plots in 2010 and 2011, but not in 2008 and 2009 (Fig. 2F). The corresponding niche diversity of assemblages was significantly more randomly distributed on non-logged plots in 2010 and 2011 (Fig. 2G). In 2009 and 2010, the mean decay niche, which reflects the process of decomposition, was significantly higher on logged plots (Fig. 2H). The corresponding decay niche diversity (Fig. 2I) showed a significantly more clustered pattern on non-logged plots than on logged plots in 2008 and 2009. The mean canopy niche value of the local assemblages of logged and non-logged plots differed in 2009–2011. Significantly more species preferring open-canopy conditions (sunny habitats) were found on non-logged plots, as indicated by a significantly higher mean niche value of the canopy niche in 2009, 2010 and 2011 (Fig. 2J). A similar trend was found in the corresponding niche diversity, which displayed a significantly clustered pattern in the canopy cover niche in 2008 and 2009 (Fig. 2K). Full details on p-values and model estimators can be found in Tables S2 and S3.

Discussion

Our study confirmed long-held empirical and scientific findings that salvage logging after windthrows reduces populations of pest bark beetles [7,12] and has accompanying negative effects on the species richness of saproxylic beetles [51]. Surprisingly, our examination of the functional traits did not support most of our predictions: single-trait analyses revealed species on logged plots that were on average larger, preferred dead wood of larger diameter, and were adapted to shady habitats, compared to species on non-logged plots. These unexpected results underline the high sensitivity of an analysis based on species functional traits to detect complex and subtle changes in species assemblages caused by anthropogenic impact [26,52–54].

Salvage logging aims at controlling populations of one or a few pest species but reduces biodiversity per se as collateral damage

[6,55,56]. Accordingly, our results demonstrated that salvage logging dramatically decreases species richness of saproxylic beetles, including red-listed species and not only the target species *I. typographus* (Fig. 2A, B). These reductions seem to be mainly caused by the loss of species directly associated with *I. typographus* (compare [57]), e.g. by the decrease in predators of *I. typographus*, such as *Thanasimus* sp.; by the decrease in species that exploit bark beetle galleries, such as *Crypturgus cinereus*; and by the loss of species associated with a similar early decay stage of wood. Saint-Germain *et al.* [58] demonstrated that the majority of saproxylic beetle species on the conifer black spruce (*Picea mariana*) colonize the early decay stages, while saproxylic beetle species on the broadleaf tree aspen (*Populus tremula*) occur mostly on wood of later decay stages — a commonly observed pattern [41,59,60]. Our study, which focuses on the first four years of the decay process, therefore reflects the most important stage of succession within windthrown stands of coniferous trees [61]. In particular, because colonization patterns of saproxylic organisms this stage determines subsequent saproxylic communities, like.g. the early-arriving bark beetle *Hylurgops palliates* and the wood-decaying fungus *Fomitopsis pinicola*, enabled a higher colonization success of the endangered beetle *Peltis grossa* after 10 years [62].

Larger beetles on logged plots

In contrast to our prediction, our results showed a consistent separation of beetle assemblages, from on average large species on logged plots toward small species on non-logged plots. One explanation for this pattern may be found behind the functional structure of the assemblages, i.e. in the habitat specificity of the species in our study: the majority of the larger species identified are widely distributed habitat generalists (except *Ampedus auripes*; [63]). One of the largest species, *Hylobius abietis*, which was more frequent in logged stands (see Table S1), is a known pest species attracted by the odour of resin in tree stumps [64]. Such species breed well in stumps and in logging residuals. Hence, an increase in the harvesting of stumps for bioenergy might expand the negative impacts of salvage logging to include currently less-affected species [65–67].

Species preferring wood of large diameters on logged plots

After regular clear-cutting, the remaining dead wood amounts to approximately 10 $m^3\ ha^{-1}$ [68]. Post-windthrow salvage-logged sites offer much more dead wood, e.g. from 45 to >70 $m^3\ ha^{-1}$ on 90 sites in Switzerland [37]. Such high amounts of remaining dead wood on salvage-logged areas regularly surpass the critical thresholds of dead-wood amount for diversity in boreal forests [32]. Accordingly, the mean diameter niche position of beetles in our study was significantly higher in 2010 and 2011. Thereby, simply the amount of dead wood does not seem to be the limiting factor for saproxylic beetles species richness in salvage-logged plots. The limiting factor is rather the loss of small branches, as indicated by the loss of species preferring wood of small diameter (Fig. 2J). Accordingly, an alteration of the remaining dead-wood resources appears to be more crucial than the simple removal of the main wood volume by salvage logging.

This assumption is strongly supported by our findings on functional trait diversity (Fig. 2C): we found a clustering of functional assemblage structure on non-logged plots, which indicated a strong habitat-filtering (dead-wood resources) effect on dead-wood communities. Using co-occurrence null-model approaches, Azeria *et al.* [69] also found a strong habitat-filtering effect on saproxylic assemblages on burned trees. In accordance, Ding *et al.* [70] proposed that disturbance in forest ecosystems generates communities by abiotic filtering. Hence, our data suggested that anthropogenic intervention, i.e. salvage logging, of natural disturbances can disrupt the natural habitat filtering in the assembly process.

Accelerated decomposition on salvage-logged plots

The amount and diversity of dead-wood resources of salvage-logged areas does not significantly differ between salvage-logged and non-salvage-logged windthrows [37]. But, in contrast to natural windthrows, the remaining dead wood on salvage-logged windthrows in our study area is mostly scattered on the ground surface (e.g. Fig. 1). Owing to the stronger attraction of wood-inhabiting fungi, salvage-logged sites tend to harbour more advanced decomposition stages than non-salvage-logged sites [37,71,72]. This shift within decay stages of available dead-wood resources was well reflected by our finding of a mean decay niche with significantly higher mean decay niche values on salvage-logged plots in 2009 and 2010. Furthermore, the corresponding niche diversity indicated a strong habitat filtering effect towards species of early decay stages on non-salvage-logged plots.

Decrease of heliophilous species through salvage logging

Sun exposure increases the probability of the presence of red-listed species in aspen retention trees in clear-cuts in Norway [38]. Similarly, the endangered longhorn beetle *Rosalia alpina* prefers trees with a lower percentage of canopy closure and higher sun exposure than the average tree [73]. Hence, sun exposure is a major predictor determining saproxylic beetle communities in dead-wood resources [74,75]. In our study, we demonstrated a shift of assemblages comprising heliophilous species on non-logged plots toward species preferring shady habitats on logged plots. This result is contrasts our prediction that the removal of dead wood might decrease the shady conditions provided by the cross-laminated arrangement of trees after windthrow. However, the majority of logging residuals on logged plots lie on the ground surface and are covered by an extensive grass layer, both of which create a moist microclimate, which decreases the availability of sun-exposed dead wood. An accelerated growth of natural regeneration on salvage-logged plots [76] does not appear to be of great importance, since natural regeneration in our study area is still poor and not able to shade the complete surroundings of a flight-interception trap (see Fig. 1 insets). Furthermore, Priewasser [77] demonstrated for a comparable study area that local factor, such as soil pH, are the main predictors for the growth of natural regeneration and not salvage logging per se. Based on our findings, it seems necessary to experimentally test our assumptions of the importance of microclimate and to estimate the amount of retention trees in windthrows sufficient for conserving saproxylic biodiversity.

Conclusion

Spruce forests in Europe will be affected heavily by increasing storm damages in the future, which will lead to increasingly heated debates between economists and ecologists on the appropriate means to limit the negative effects of salvage logging on biodiversity [3,4]. Our analysis based on functional traits revealed an unexpected response of saproxylic beetles to salvage logging and suggested that microclimate conditions are more crucial for the use of dead-wood resources by saproxylic beetles than dead-wood diameter or diversity [32,37]. The direct relationship between species traits and logging-affected structures enables us to derive new guidelines for conservationists and managers to optimize salvage logging with a consideration of biodiversity conservation: downed tree tops unaffected by salvage logging operations and complete single windthrown trees should be preserved and allowed to naturally decay on salvage-logged areas to help sustain heliophilous species, colonizers of early decay stages and species that prefer wood of small diameter. Such a preservation of some windthrown trees in salvage-logged areas is an extension of the "green-tree retention approach" to downed trees in forests worldwide.

Acknowledgments

We are grateful to all those who contributed to this study in the field. We thank Karen A. Brune for linguistic revision of the manuscript and Ulrich Bense for beetle determination.

Author Contributions

Conceived and designed the experiments: JM CB. Performed the experiments: JM CB. Analyzed the data: ST TH JM. Contributed to the writing of the manuscript: ST TG HB CB JM TH KR.

References

1. Attiwill PM (1994) The disturbance of forest ecosystems - the ecological basis for conservative management. Forest Ecology and Management 63: 247–300.
2. Shorohova E, Kuuluvainen T, Kangur A, Jogiste K (2009) Natural stand structures, disturbance regimes and successional dynamics in the Eurasian boreal forests: a review with special reference to Russian studies. Ann For Sci 66: 20.
3. Schelhaas MJ, Nabuurs GJ, Schuck A (2003) Natural disturbances in the European forests in the 19th and 20th centuries. Glob Change Biol 9: 1620–1633.
4. Seidl R, Schelhaas M-J, Lexer MJ (2011) Unraveling the drivers of intensifying forest disturbance regimes in Europe. Glob Change Biol 17: 2842–2852.
5. Raffa KF, Aukema BH, Bentz BJ, Carroll AL, Hicke JA, et al. (2008) Cross-scale drivers of natural disturbances prone to anthropogenic amplification: the dynamics of bark beetle eruptions. Bioscience 58: 501–517.
6. Overbeck M, Schmidt M (2012) Modelling infestation risk of Norway spruce by *Ips typographus* (L.) in the Lower Saxon Harz Mountains (Germany). Forest Ecology and Management 266: 115–125.
7. Schröder LM (2011) Colonization of storm gaps by the spruce bark beetle: influence of gap and landscape characteristics. Agric For Entomol 12: 29–39.
8. Lindenmayer DB, Foster DR, Franklin JF, Hunter ML, Noss RF, et al. (2004) Ecology - Salvage harvesting policies after natural disturbance. Science 303: 1303–1303.
9. Lindenmayer D (2006) Salvage harvesting - past lessons and future issues. Forestry Chronicle 82: 48–53.
10. Noss RF, Lindenmayer DB (2006) The ecological effects of salvage logging after natural disturbance. Conservation Biology 20: 946–948.
11. Black SH (2005) Logging to control insects: the science and myths behind managing forest insect "pests". A synthesis of independently reviewed research. Portland: The Xerces Society for Invertebrate Conservation.
12. Fettig CJ, Klepzig KD (2006) The effectiveness of vegetation management practices for mitigating the impacts of insects on forest ecosystems: a science synthesis. U. S. Department of Agriculture Forest Service.
13. Stokstad E (2006) Ecology: salvage logging research continues to generate sparks. Science 311: 761–761.
14. Bouget C, Duelli P (2004) The effects of windthrow on forest insect communities: a literature review. Biological Conservation 118: 281–299.
15. Chambers JQ, Robertson AL, Carneiro VMC, Lima AJN, Smith ML, et al. (2009) Hyperspectral remote detection of niche partitioning among canopy trees driven by blowdown gap disturbances in the Central Amazon. Oecologia 160: 107–117.
16. Castro J, Moreno-Rueda G, Hodar JA (2010) Experimental Test of Postfire Management in Pine Forests: Impact of Salvage Logging versus Partial Cutting and Nonintervention on Bird-Species Assemblages. Conservation Biology 24: 810–819.
17. Hayes JP (2009) Post-fire Salvage Logging in Central Oregon: Short-term Response in Bats, Birds and Small Mammals. Fire Science Brief: 1–6.
18. Hebblewhite M, Munro RH, Merrill EH (2009) Trophic consequences of postfire logging in a wolf-ungulate system. Forest Ecol Manage 257: 1053–1062.
19. Morissette JL, Cobb TP, Brigham RM, James PC (2002) The response of boreal forest songbird communities to fire and post-fire harvesting. Can J For Res-Rev Can Rech For 32: 2169–2183.
20. Gautam S, Pulkki R, Shahi C, Leitch M (2010) Economic and energy efficiency of salvaging biomass from wildfire burnt areas for bioenergy production in northwestern Ontario: a case study. Biomass Bioenerg 34: 1562–1572.
21. Jakus R, Edwards-Jonasova M, Cudlin P, Blazenec M, Jezik M, et al. (2011) Characteristics of Norway spruce trees (*Picea abies*) surviving a spruce bark beetle (Ips typographus L.) outbreak. Trees Struct Funct 25: 965–973.
22. Lain EJ, Haney A, Burris JM, Burton J (2008) Response of vegetation and birds to severe wind disturbance and salvage logging in a southern boreal forest. Forest Ecology and Management 256: 863–871.
23. Zmihorski M (2010) The effect of windthrow and its management on breeding bird communities in a managed forest. Biodivers Conserv 19: 1871–1882.
24. Kroll AJ, Giovanini J, Jones JE, Arnett EB, Altman B (2012) Effects of salvage logging of beetle-killed forests on avian species and foraging guild abundance. J Wildl Manage 76: 1188–1196.
25. Cadotte MW, Carscadden K, Mirotchnick N (2011) Beyond species: functional diversity and the maintenance of ecological processes and services. Journal of Applied Ecology 48: 1079–1087.
26. Stuart-Smith RD, Bates AE, Lefcheck JS, Duffy JE, Baker SC, et al. (2013) Integrating abundance and functional traits reveals new global hotspots of fish diversity. Nature 501: 539–542.
27. Reich PB, Tilman D, Isbell F, Mueller K, Hobbie SE, et al. (2012) Impacts of biodiversity loss escalate through time as redundancy fades. Science 336: 589–592.
28. Yang Q, Wang X, Shen Y, Philp JNM (2013) Functional diversity of soil microbial communities in response to tillage and crop residue retention in an eroded loess soil. Soil Science and Plant Nutrition 59: 311–321.
29. Mouillot D, Graham NAJ, Villéger S, Mason NWH, Bellwood DR (2013) A functional approach reveals community responses to disturbances. Trends Ecol Evol 28: 167–177.
30. Alexander KNA (2008) Tree biology and saproxylic coleoptera: issues of definitions and conservation language. Rev Ecol-Terre Vie: 9–13.
31. Lassauce A, Paillet Y, Jactel H, Bouget C (2011) Deadwood as a surrogate for forest biodiversity: Meta-analysis of correlations between deadwood volume and species richness of saproxylic organisms. Ecol Indic 11: 1027–1039.
32. Müller J, Bütler R (2010) A review of habitat thresholds for dead wood: a baseline for management recommendations in European forests. Eur J For Res 129: 981–992.
33. Gossner MM, Lachat T, Brunet J, Isacsson G, Bouget C, et al. (2013) Current near-to-nature forest management effects on functional trait composition of saproxylic beetles in beech forests. Conservation Biology 27: 605–614.
34. Brin A, Bouget C, Brustel H, Jactel H (2011) Diameter of downed woody debris does matter for saproxylic beetle assemblages in temperate oak and pine forests. J Insect Conserv 15: 653–669.
35. Bussler H, Bouget C, Brustel H, Brandle M, Riedinger V, et al. (2011) Abundance and pest classification of scolytid species (Coleoptera: Curculionidae, Scolytinae) follow different patterns. Forest Ecology and Management 262: 1887–1894.
36. Foit J (2010) Distribution of early-arriving saproxylic beetles on standing dead Scots pine trees. Agric For Entomol 12: 133–141.
37. Priewasser K, Brang P, Bachofen H, Bugmann H, Wohlgemuth T (2013) Impacts of salvage-logging on the status of deadwood after windthrow in Swiss forests. Eur J For Res 132: 231–240.
38. Sverdrup-Thygeson A, Ims RA (2002) The effect of forest clearcutting in Norway on the community of saproxylic beetles on aspen. Biological Conservation 106: 347–357.
39. Bässler C, Muller J, Dziock F (2010) Detection of climate-sensitive zones and identification of climate change indicators: a case study from the Bavarian Forest National Park. Folia Geobotanica 45: 163–182.
40. Sverdrup-Thygeson A, Birkemoe T (2009) What window traps can tell us: effect of placement, forest openness and beetle reproduction in retention trees. J Insect Conserv 13: 183–191.
41. Hyvärinen E, Kouki J, Martikainen P (2006) Fire and green-tree retention in conservation of red-listed and rare deadwood-dependent beetles in Finnish boreal forests. Conservation Biology 20: 1711–1719.

42. Schmidl J, Buβler H (2004) Ökologische Gilden xylobionter Käfer Deutschlands. Naturschutz und Landschaftsplanung 36: 202–218.
43. Freude H, Harde K, Lohse GA (1964-1983) Die Käfer Mitteleuropas. Goecke und Evers, Krefeld.
44. Laliberte E, Legendre P (2010) A distance-based framework for measuring functional diversity from multiple traits. Ecology 91: 299–305.
45. Purves DW, Turnbull LA (2010) Different but equal: the implausible assumption at the heart of neutral theory. J Anim Ecol 79: 1215–1225.
46. Webb CO, Ackerly DD, McPeek MA, Donoghue MJ (2002) Phylogenies and community ecology. Annu Rev Ecol Syst 33: 475–505.
47. Pausas JG, Verdu M (2010) The jungle of methods for evaluating phenotypic and phylogenetic structure of communities. Bioscience 60: 614–625.
48. Hothorn T, Mueller J, Schröder B, Kneib T, Brandl R (2011) Decomposing environmental, spatial, and spatiotemporal components of species distributions. Ecological Monographs 81: 329–347.
49. Elston DA, Moss R, Boulinier T, Arrowsmith C, Lambin X (2001) Analysis of aggregation, a worked example: numbers of ticks on red grouse chicks. Parasitology 122: 563–569.
50. Hothorn T, Bretz F, Westfall P (2008) Simultaneous inference in general parametric models. Biometrical Journal 50: 346–363.
51. Cobb TP, Morissette JL, Jacobs JM, Koivula MJ, Spence JR, et al. (2011) Effects of postfire salvage logging on deadwood-associated beetles. Conservation Biology 25: 94–104.
52. Ernst R, Linsenmair KE, Rodel MO (2006) Diversity erosion beyond the species level: dramatic loss of functional diversity after selective logging in two tropical amphibian communities. Biological Conservation 133: 143–155.
53. Flynn DFB, Gogol-Prokurat M, Nogeire T, Molinari N, Richers BT, et al. (2009) Loss of functional diversity under land use intensification across multiple taxa. Ecol Lett 12: 22–33.
54. Winter M, Devictor V, Schweiger O (2013) Phylogenetic diversity and nature conservation: where are we? Trends Ecol Evol 28: 199–204.
55. Grodzki W, Jakus R, Lajzova E, Sitkova Z, Maczka T, et al. (2006) Effects of intensive versus no management strategies during an outbreak of the bark beetle *Ips typographus* (L.) (Col.: Curculionidae, Scolytinae) in the Tatra Mts. in Poland and Slovakia. Ann For Sci 63: 55–61.
56. McFarlane BL, Parkins JR, Watson DOT (2012) Risk, knowledge, and trust in managing forest insect disturbance. Can J For Res-Rev Can Rech For 42: 710–719.
57. Weslien J (1992) The arthropod complex associated with *Ips typographus* (L) (Coleoptera, Scolytidae), species composition, phenology, and impact on bark beetle productivity. Entomol Fenn 3: 205–213.
58. Saint-Germain M, Drapeau P, Buddle CM (2007) Host-use patterns of saproxylic phloeophagous and xylophagous Coleoptera adults and larvae along the decay gradient in standing dead black spruce and aspen. Ecography 30: 737–748.
59. Müller J, Noss RF, Bussler H, Brandl R (2010) Learning from a "benign neglect strategy" in a national park: Response of saproxylic beetles to dead wood accumulation. Biological Conservation 143: 2559–2569.
60. Kouki J, Hyvärinen E, Lappalainen H, Martikainen P, Simila M (2012) Landscape context affects the success of habitat restoration: large-scale colonization patterns of saproxylic and fire-associated species in boreal forests. Divers Distrib 18: 348–355.
61. Stokland JN, Siitonen J, Jonsson BG (2012) Biodiversity in Dead Wood. Octavo: Cambridge University Press.
62. Weslien J, Djupstrom LB, Schroeder M, Widenfalk O (2011) Long-term priority effects among insects and fungi colonizing decaying wood. J Anim Ecol 80: 1155–1162.
63. Jarzabek-Müller A, Müller J (2008) On the distinction between Ampedus auripes (Reitter, 1895) and Ampedus nigrinus (Herbst, 1784) (Coleoptera: Elateridae). Elateridarium 2: 199–212.
64. Pitkdnen A, Kouki J, Viiri H, Martikainen P (2008) Effects of controlled forest burning and intensity of timber harvesting on the occurrence of pine weevils, *Hylobius* spp., in regeneration areas. Forest Ecology and Management 255: 522–529.
65. Lassauce A, Lieutier F, Bouget C (2012) Woodfuel harvesting and biodiversity conservation in temperate forests: Effects of logging residue characteristics on saproxylic beetle assemblages. Biological Conservation 147: 204–212.
66. Victorsson J, Jonsell M (2012) Effects of stump extraction on saproxylic beetle diversity in Swedish clear-cuts. Insect Conservation and Diversity 6: 483–493.
67. Brin A, Bouget C, Valladares L, Brustel H (2012) Are stumps important for the conservation of saproxylic beetles in managed forests? Insights from a comparison of assemblages on logs and stumps in oak-dominated forests and pine plantations. Insect Conservation and Diversity 6: 255–264.
68. Gibb H, Ball JP, Johansson T, Atlegrim O, Hjalten J, et al. (2005) Effects of management on coarse woody debris volume and composition in boreal forests in northern Sweden. Scand J Forest Res 20: 213–222.
69. Azeria ET, Ibarzabal J, Hebert C (2012) Effects of habitat characteristics and interspecific interactions on co-occurrence patterns of saproxylic beetles breeding in tree boles after forest fire: null model analyses. Oecologia 168: 1123–1135.
70. Ding Y, Zang RG, Letcher SG, Liu SR, He FL (2012) Disturbance regime changes the trait distribution, phylogenetic structure and community assembly of tropical rain forests. Oikos 121: 1263–1270.
71. Olsson J, Jonsson BG, Hjalten J, Ericson L (2011) Addition of coarse woody debris: the early fungal succession on Picea abies logs in managed forests and reserves. Biological Conservation 144: 1100–1110.
72. Jacobs JM, Work TT (2012) Linking deadwood-associated beetles and fungi with wood decomposition rates in managed black spruce forests. Can J For Res-Rev Can Rech For 42: 1477–1490.
73. Russo D, Cistrone L, Garonna AP (2011) Habitat selection by the highly endangered long-horned beetle *Rosalia alpina* in southern Europe: a multiple spatial scale assessment. J Insect Conserv 15: 685–693.
74. Buse J, Schroder B, Assmann T (2007) Modelling habitat and spatial distribution of an endangered longhorn beetle - A case study for saproxylic insect conservation. Biological Conservation 137: 372–381.
75. Jonsell M, Nordlander G, Ehnström B (2001) Substrate Associations of Insects Breeding in Fruiting Bodies of Wood-Decaying Fungi. Ecological Bulletins 49: 173–194.
76. Fischer A, Fischer HS (2012) Individual-based analysis of tree establishment and forest stand development within 25 years after wind throw. Eur J For Res 131: 493–501.
77. Priewasser K (2013) Factors influencing tree regeneration after windthrow in Swiss forests. PhD ETH Zurich 157.

12

Impacts of Diffuse Radiation on Light Use Efficiency across Terrestrial Ecosystems Based on Eddy Covariance Observation in China

Kun Huang[1,2], Shaoqiang Wang[1]*, Lei Zhou[1], Huimin Wang[1], Junhui Zhang[3], Junhua Yan[4], Liang Zhao[5], Yanfen Wang[2], Peili Shi[1]

1 Key Laboratory of Ecosystem Network Observation and Modeling, Institute of Geographic Sciences and Natural Resources Research, Chinese Academy of Sciences, Beijing 100101, China, **2** University of Chinese Academy of Sciences, Beijing 100049, China, **3** Institute of Applied Ecology, Chinese Academy of Sciences, Shenyang 110016, China, **4** South China Botanical Garden, Chinese Academy of Sciences, Guangzhou 510650, China, **5** Northwest Plateau Institute of Biology, Chinese Academy of Sciences, Xining 810001, China

Abstract

Ecosystem light use efficiency (LUE) is a key factor of production models for gross primary production (GPP) predictions. Previous studies revealed that ecosystem LUE could be significantly enhanced by an increase on diffuse radiation. Under large spatial heterogeneity and increasing annual diffuse radiation in China, eddy covariance flux data at 6 sites across different ecosystems from 2003 to 2007 were used to investigate the impacts of diffuse radiation indicated by the cloudiness index (*CI*) on ecosystem LUE in grassland and forest ecosystems. Our results showed that the ecosystem LUE at the six sites was significantly correlated with the cloudiness variation ($0.24 \leq R^2 \leq 0.85$), especially at the Changbaishan temperate forest ecosystem ($R^2 = 0.85$). Meanwhile, the *CI* values appeared more frequently between 0.8 and 1.0 in two subtropical forest ecosystems (Qianyanzhou and Dinghushan) and were much larger than those in temperate ecosystems. Besides, cloudiness thresholds which were favorable for enhancing ecosystem carbon sequestration existed at the three forest sites, respectively. Our research confirmed that the ecosystem LUE at the six sites in China was positively responsive to the diffuse radiation, and the cloudiness index could be used as an environmental regulator for LUE modeling in regional GPP prediction.

Editor: Dafeng Hui, Tennessee State University, United States of America

Funding: This research was jointly supported by the CAS for Strategic Priority Research Program (Grant No. XDA05050602), National Basic Research Program of China (973Program) (Grant No. 2010CB833503) and the Key Project in the National Science & Technology Pillar Program of China (Grant No. 2013BAC03B00). The funders had no role in study design, data collection and analysis, decision to publish, or preparation of the manuscript.

Competing Interests: The authors have declared that no competing interests exist.

* Email: sqwang@igsnrr.ac.cn

Introduction

Terrestrial ecosystems play an increasingly important role in global carbon cycle under climate change [1]. Light use efficiency (LUE) was first presented in the context of agricultural ecosystem focusing on the linear relationship between yield and solar irradiance, and gross primary production (GPP) was defined as the overall photosynthetically fixation of carbon per unit space and time [2]. The fact that GPP represents the critical flux component driving the terrestrial ecosystem carbon cycle implies that subtle fluctuations in GPP have substantial implications for future climate warming scenarios [3,4]. With the quantification terrestrial ecosystem GPP for regions, continents, or the globe, we can gain insight into the feedbacks between the terrestrial biosphere and the atmosphere under global change and climate policy-making facilitation [5,6]. Still, GPP predictions at regional scale to global scale are a major challenge due to the spatial heterogeneity [7,8]. Moreover, with the great carbon sequestration potential of the terrestrial ecosystem of China in global carbon budget [9], large uncertainties exist in terrestrial ecosystem GPP simulation in China.

A number of modeling approaches have been developed for regional/global GPP estimations, including ecological process-based models and light use efficiency models driven by remote sensing data [10]. Among all the models, LUE models encompassing the LUE algorithm proposed by [2] may have the highest potential to identify the spatio-temporal dynamics of regional GPP due to the simplicity of concept and availability of remote sensing data [11]. With this method, GPP was defined as product of photosynthetically active radiation (PAR) absorbed by the vegetation canopy and a conversion factor, LUE [2,12]. Various LUE models have been developed for this purpose, including MODIS GPP algorithm [13], Vegetation Photosynthesis Model (VPM) [14], EC-LUE model [15], Vegetation Index (VI) model [16], C-Fix model [17], Temperature and Greenness Rectangle (TGR) model [18], Temperature and Greenness (TG) model [19] and so on. In order to acquire GPP estimations of high accuracy, the biophysical controls on the ecosystem LUE are significantly important to be fully understood [8,20]. Recent studies indicated that GPP and LUE were affected by both the quantity and composition of the incoming solar radiation [10,21–23]. With a given value of total incoming radiation, LUE of the entire canopy will increase with the increasing fraction of diffuse radiation (FDR) [23–25]. Under cloudy or aerosol-laden skies, incoming radiation was more diffuse and more uniformly distributed in the canopy with a smaller fraction of the canopy that was light saturated [10]. Consequently, canopy photosynthesis was inclined to be more light-use efficient under diffuse sunlight than under direct sunlight condition [10,21,23,24,26,27]. Evidences showed that global secondary organic aerosol in the atmosphere will increase by 36% in 2100 [28]. The aerosol influenced the cloud formation, which was the main contributor to the increment on FDR in the atmosphere [29–31]. Furthermore, an increasing trend of annual diffuse radiation in China has been proved to be 7.03 $MJ.m^{-2}.yr^{-1}$ per decade from 1981 to 2010 [32]. However, few studies on ecosystem GPP predictions took into account effects of the FDR variations of the incoming radiation on LUE based on the LUE models.

Up to now, the eddy covariance (EC) technique provides an alternative way to measure NEE continuously that can be used for GPP calculation by subtracting the modeled ecosystem respiration components [5,33–35]. Multi-sites and continuous eddy covariance (EC) flux and meteorological observation from the ChinaFLUX network provided a valuable tool for GPP and LUE calculation across ecosystems in China [34]. Therefore, in order to reveal the biophysical controls on measured ecosystem LUE for better regional GPP predictions in terrestrial ecosystems of China which is of high spatial heterogeneity, the impact of diffuse radiation resulting from cloud condition on LUE is of growing concern to be characterized by a uniform proxy. Despite the study that effect of cloudiness change on ecosystem LUE and water use efficiency was detected by the clearness index [36], it was difficult to incorporate the clearness index into LUE model for regional GPP estimates due to the specification of the highest interval of solar elevation angle in each grid. Here we employed an cloudiness index algorithm based on simple inputs [13,22], flux and metrological measurements from six sites of ChinaFLUX encompassing three forest ecosystems and three grassland ecosystems, to address the impact of diffuse radiation on light use efficiency (defined as GPP/PAR) [21,23,26]. The objectives of this study are to: (1) illustrate the seasonal dynamics of the cloudiness index and light use efficiency at different sites; (2) address the influence of fraction of diffuse radiation on ecosystem light use efficiency; (3) identify whether the cloudiness index thresholds favorable for enhancing ecosystem carbon sequestration exist or not.

Materials and Methods

Sites descriptions and measurements

In this study, flux observations were implemented at three forest ecosystems and three grassland ecosystems attached to the Chinese Terrestrial Ecosystem Flux Observational Network (ChinaFLUX). The three forest sites were comprised of the Changbaishan temperate mixed forest (CBS), Qianyanzhou subtropical evergreen needle leaf planted forest (QYZ), and Dinghushan subtropical evergreen broad-leaved forest (DHS). Subject to moosoon-influenced, temperate continental climate, CBS was located in the Jilin province of China, in which growing season ranged from May to September [37]. The QYZ site was located in the subtropical continental monsoon region, in which the mean annual air temperature was 17.9°C [38,39]. Located in the Guangdong province with a subtropical monsoon humid climate, DHS had a wet season from April to September and dry season from November to March [39]. The three grassland ecosystems were the Inner Mongolia semi-arid *L. chinensis* steppe (NMG) which is C3 grassland, Haibei alpine frigid *P. fruticosa* shrub(HB), and Damxung (DX) alpine meadow-steppe ecosystem with short sparse vegetation(about 10 cm). NMG was located in the Xilin River Basin, Inner Mongolia Autonomous Region of China with a temperate semiarid continental climate. Its growing season lasted from late April to early October [40]. HB was located in the northeast of the Qinhai-Tibet Plateau with a plateau continental climate, which was characterized by lengthy cold winters and very short warm summers. Being situated in a frigid highland, HB receives strong solar radiation, with a mean annual global radiation of up to 6000–7000 $MJ.m^{-2}$ [36,41]. The DX site was located in the Lhasa City, Tibet, categorized as plateau monsoon climate. Its growing season duration was from May to September. The PAR was usually high, similar to that in alpine meadow area located in eastern Tibetan Plateau and higher than other grassland ecosystems [42]. The locations of six sites were shown in Figure1, and the detailed information of the six sites was provided in Table 1.

Routine meteorological variables were measured simultaneously with the eddy fluxes at each site. Air humidity and air temperature were measured with shielded and aspirated probes (HMP45C, Vaisala, Helsinki, Finland) at different sites. Global radiation and net radiation were recorded with radiometers (CM11 and CNR-1, Kipp & Zonen, Delft, the Netherlands). Photosynthetically active radiation (PAR) above the canopy was measured with a quantum sensor (LI-190Sb, LiCor Inc., USA). All meteorological observations were recorded at 30-min intervals with dataloggers (Model CR10X & CR23X, Campbell Scientific Inc.) [37,39,41–44].

In the study, we only used data measured during the periods of relatively stable leaf area index (LAI) each year from 2003 to 2007 in order to eliminate the potential effect of changing LAI [36]. The LAI of temperate ecosystems (CBS, NMG, HB and DX) remain stable in the mid-growing season. DHS and QYZ were evergreen forest ecosystems, and their LAI did not vary much with season. Therefore, data of mid growing season (June-August) from all the six flux sites were used to analyze the impact of diffuse radiation on ecosystem light use efficiency and photosynthesis.

Ethics statement

Three forest ecosystems (CBS, QYZ and DHS) and three grassland ecosystems (NMG, HB and DX) attached to ChinaFLUX were maintained by different institutions of Chinese Academy of Sciences (CAS), respectively. The CBS site was maintained by the Institute of Applied Ecology, CAS; the QYZ site and DX site was maintained by the Institute of Geographic Sciences and Natural Resources Research, CAS; the DHS site was maintained by the South China Botanical Garden, CAS; the NMG site was maintained by University of CAS, and the HB site was maintained by Northwest Plateau Institute of Biology, CAS. All necessary permits were obtained for the described field study. The field study did not involve endangered or protected species. Data will be made available upon request.

Eddy flux data

Carbon flux data (GPP and NEP) observed at 6 typical sites from 2003 to 2007 across China were applied to in this study (Figure1). The raw 30-min flux data procedure included: (1) 3D coordinate rotation was applied to force the average vertical wind speed to zero and to align the horizontal wind to mean wind direction, (2) flux data was corrected according the variation of air density caused by transfer of heat and water vapor [45], (3)the storage below EC height was corrected for forest sites [46], and (4)

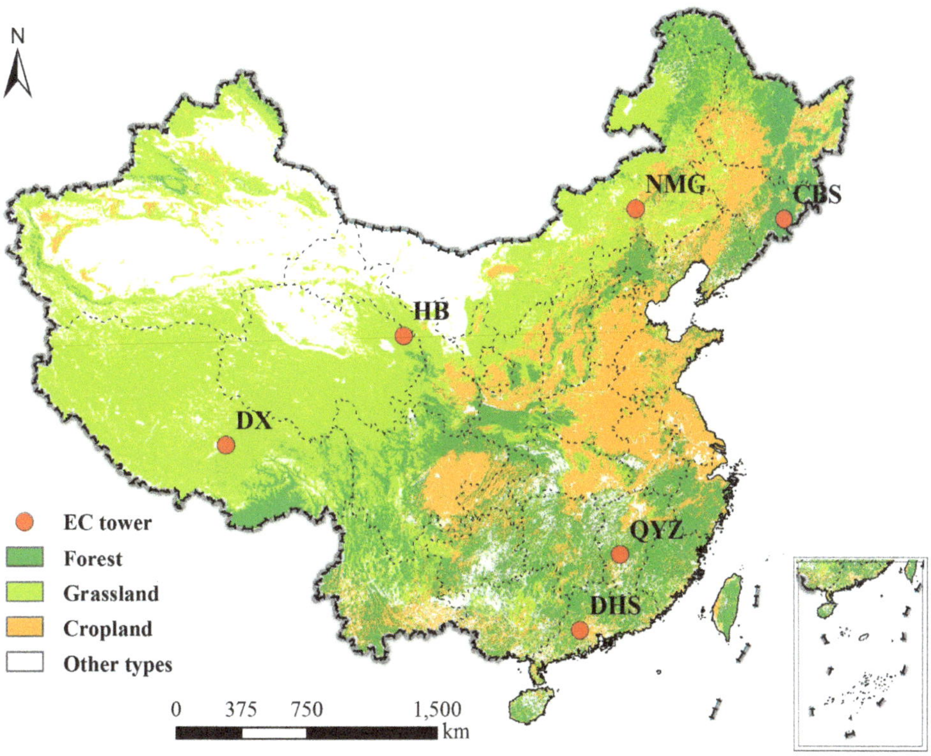

Figure 1. Distribution of the 6 eddy covariance flux sites in China in this study. The background was the MODIS land cover map.

the outlier data were filtered and data gaps were filled by using the look-up table method and mean diurnal variation(MDV) [37,39,47]. In the end, continuous 30 min flux data was performed.

The flux of net ecosystem CO_2 exchange (NEE, mg CO_2 m^{-2} s^{-1}) between the ecosystem and the atmosphere was calculated with equation (1), the net ecosystem productivity (NEP) was assigned to –NEE. Negative NEE values denote carbon uptake, while positive values denote carbon source.

$$NEP = -(\overline{w'\rho'c(zr)} + \int_0^{zr} \frac{\delta \overline{\rho c}}{\delta t} dz), \quad (1)$$

where the first term on right-hand side is the eddy flux for carbon dioxide or water vapor below the height of observation (z_r), and all advective terms in the mass conservation equation were ignored.

Daily GPP data are partitioned from NEP data measured every 30-min using the eddy covariance technique. GPP was derived from the measured NEP, which was processed using the same method as [36]. Gross primary production (GPP) was calculated employing the following equation:

$$GPP = Re + NEP \quad (2)$$

Table 1. Site descriptions.

Site (ab.)[a]	Changbaishan (CBS)	Qianyanzhou (QYZ)	Dinghushan (DHS)	Haibei (HB)	Inner Mongolia (NMG)	Damxung (DX)
Location	42°24′N	26°45′N	23°10′N	37°40′N	43°32′N	30°51′N
	128°06′E	115°04′E	112°32′E	101°20′E	116°40′E	91°05′E
Elevation(m)	738	102	300	3293	1189	4333
LAI(m^2m^{-2})	6.1	5.6	4.0	2.8	1.5	1.88
Annual mean precipitation(mm)	600–900	1489	1956	580	350–450	480
Annual mean temperature(°C)	3.6	18.6	21	−1.7	−0.4	1.3
Vegetation type	Mixed forest	Evergreen needle leaf forest	Evergreen broadleaf forest	Alpine frigid shrub	Temperate steppe	Alpine steppe-meadow

[a]Abbreviation for sites.

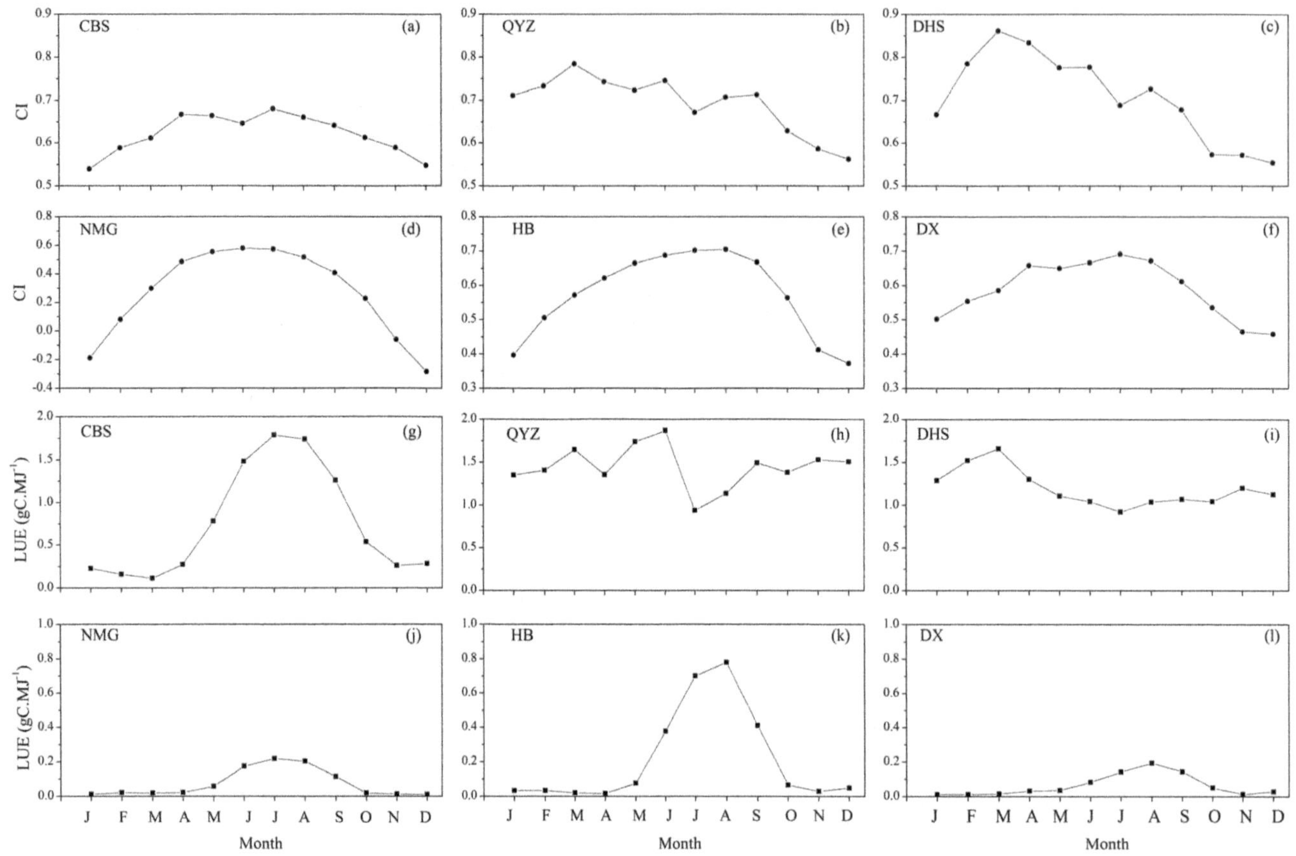

Figure 2. The seasonal variations of monthly mean cloudiness index (*CI*), monthly mean light use efficiency (LUE) from 2003 to 2007 at the five sites.

NEP was obtained directly from the eddy covariance measurement. Ecosystem respiration (Re) of the seven sites was estimated using the Lloyd-Taylor equation (1994) [41,48]. The nighttime NEP data under turbulent conditions were used to establish Re-temperature response relationship Eq.(3):

$$\mathrm{Re} = R_{ref} e^{E_0(1/(T_{ref}-T_0)-1/(T-T_0))} \tag{3}$$

where R_{ref} represents the ecosystem respiration rate at reference temperature (Tref, 10°C); E_0 is the parameter that determines the temperature sensitivity of ecosystem respiration, and T_0 is a constant and set as −46.02°C; T is the air temperature or soil temperature(°C). Eq. (3) was also used to estimate daytime Re.

Calculation of light use efficiency

In this study, LUE ($gC.MJ^{-1}$) was defined as the ratio of daily GPP ($gC.m^{-2}.d^{-1}$) to incident PAR ($MJ^{-1}.m^{-2}.d^{-1}$, using 217 kJ mol^{-1} photons),

$$\mathrm{LUE} = \frac{\mathrm{GPP}}{\mathrm{PAR}} \tag{4}$$

where PAR was directly measured by the in situ meteorological equipment simultaneous with the flux tower observation.

Cloudiness index

A cloudiness index implemented in CFLUX model was used in our model, since an increase on light use efficiency under overcast conditions at both hourly and daily time steps has been proved in previous studies [22,49]. The cloudiness index was calculated as [22]:

$$CI = 1 - \downarrow PAR / \downarrow PAR_{po} \tag{5}$$

where *CI* is the cloudiness index, ↓ PAR is incident PAR(MJd^{-1}) from daily observation input, ↓ PAR_{po} is potential incident PAR as a derivation of the algorithm of [50]. With the simple inputs of digital elevation model (DEM) data and readily available parameters, the ↓ PAR_{po} can be calculated as the global solar radiation at daily time scale in each grid. The spatial resolution of the DEM data was 500 m×500 m, provided by Institute of Geographical Sciences and Natural Resources Research, Chinese Academy of Sciences. More details of the algorithm can be found in the previous literature [50].

The clear sky LUE (LUE_{cs}) was specified for each site based on observations of LUE at eddy covariance flux towers. The clear sky LUE was based on the value when ↓ PAR/ ↓ PAR_{po} (decreasing cloud cover) approximated 1.0 by a function of LUE under low stress conditions plotted against ↓ PAR/ ↓ PAR_{po} [51].

Statistical analysis

The relationships between different variables were fitted with linear and non-linear equations. All analyses were conducted using the origin package v.8.0 (OriginLab Corporation, Northampton, MA, USA). Statistically significant differences were set with $P < 0.05$ ($\alpha = 0.05$) unless otherwise stated.

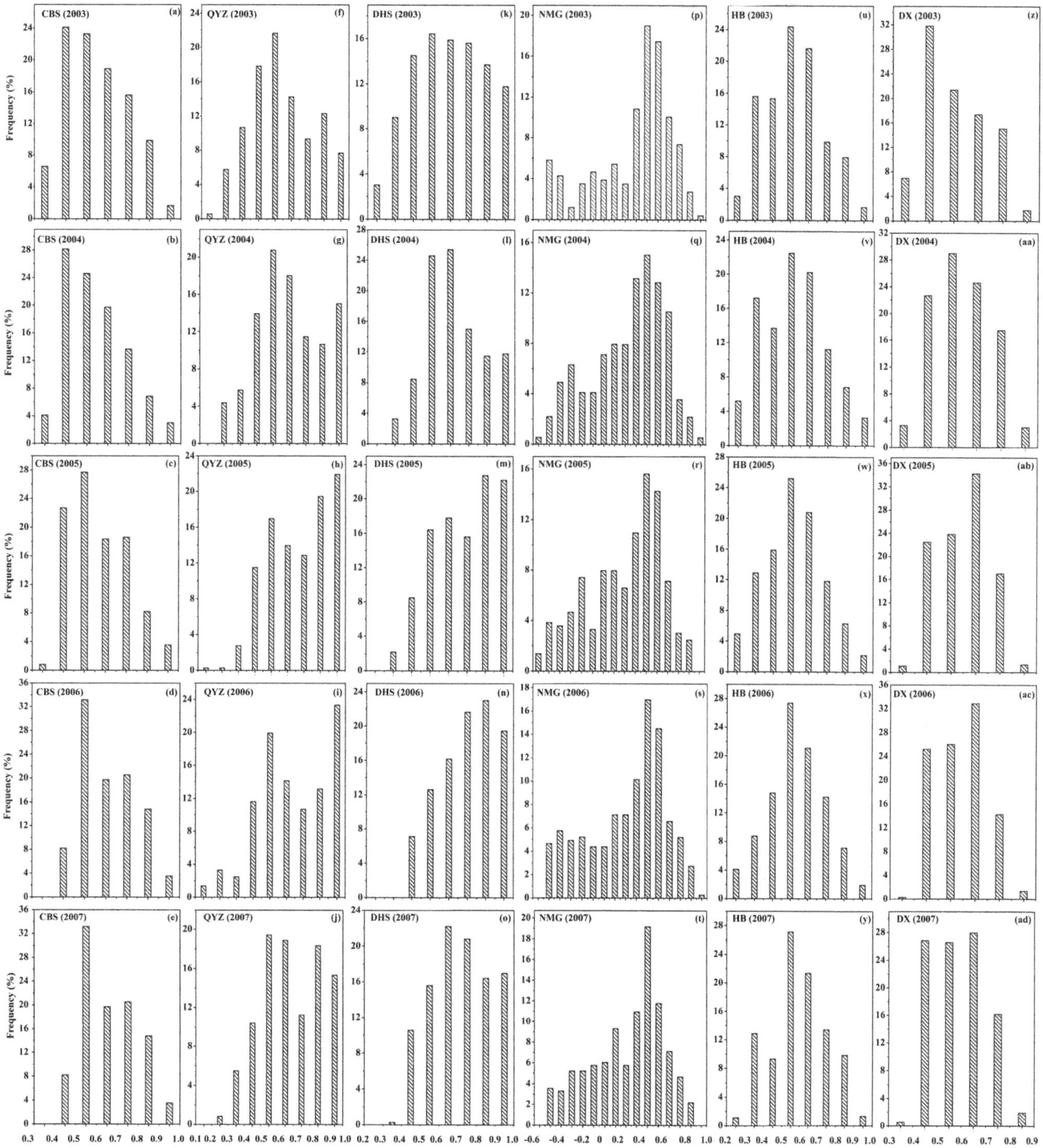

Figure 3. Histograms of the cloudiness index (*CI*) value at the six sites during the mid-growing seasons from 2003 to 2007.

Table 2. Light use efficiency under clear sky in different ecosystems.

Sites	CBS	QYZ	DHS	NMG	HB	DX
LUE_{cs}($gC.MJ^{-1}$)	0.29	0.425	0.569	0.003	0.013	0.009

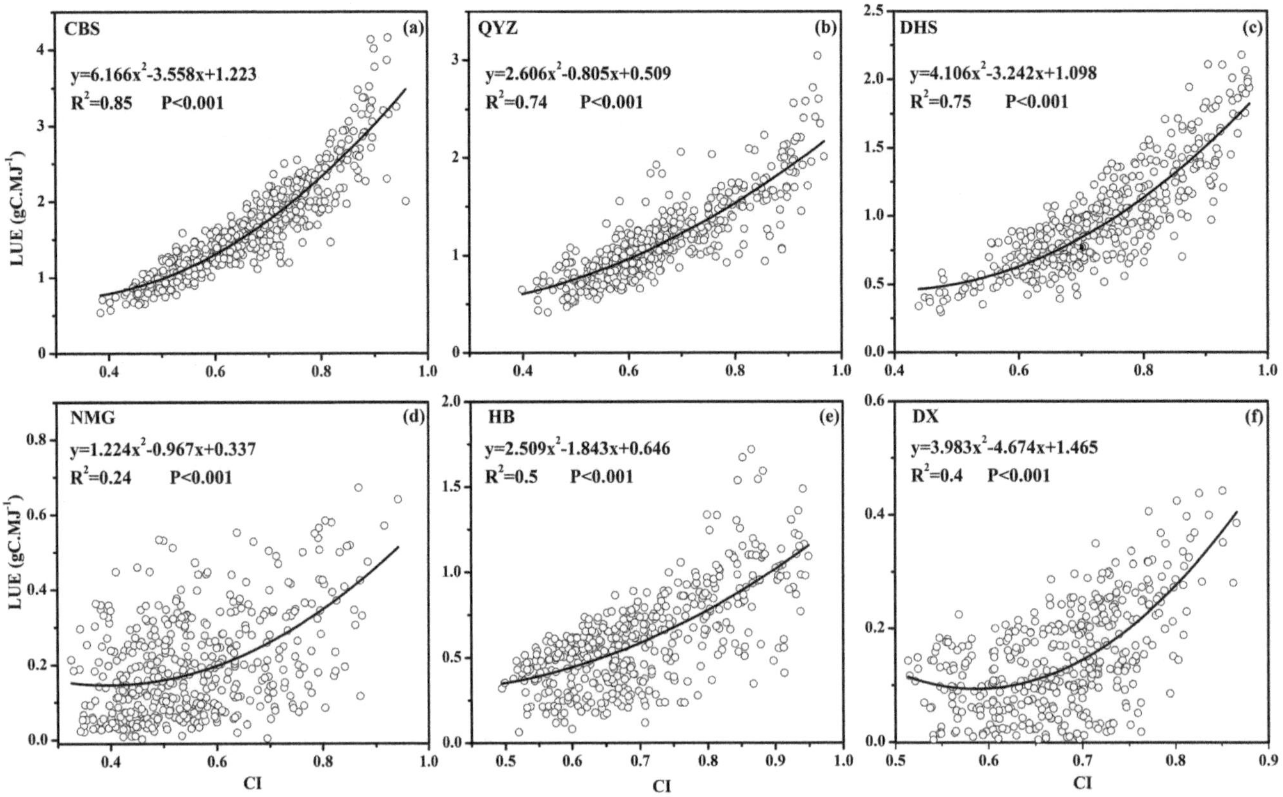

Figure 4. Relationships between LUE and the cloudiness index (*CI*, positive values) during the mid-growing season from 2003 to 2007 at the six sites.

Results and Discussion

Seasonal variation of cloudiness index and light use efficiency across ecosystems

Figure 2 showed the seasonal variations of the cloudiness index and light use efficiency of the six sites from 2003 to 2006. Mostly, cloudiness index (*CI*) was greater at QYZ and DHS than the other temperate ecosystems (CBS, NMG, HB and DX). The *CI* values of subtropical ecosystems (QYZ and DHS) reached the maximum in March, and were higher during the mid-growing season than the two ends of the year (Figure 2b, c). At the temperate ecosystems, the *CI* values peaked during the mid-growing season (Figure 2a, d, e and f), while the *CI* values of the subtropical ecosystems failed to show substantial variations with the seasonal changes. This indicated that sky conditions of two subtropical ecosystem sites were cloudier than those of four temperate ecosystem sites, and cloudy days were more during the mid-growing seasons at the temperate sites. It was also noted that negative *CI* values were found at NMG site, which was in consistency with meteorological observation that the NMG site received stronger solar radiation during the non-growing season. Meanwhile, the forest ecosystems LUE were significantly higher than grassland ecosystem LUE (Figure 2g, h, i, j, k and l). The LUE at subtropical forest sites (QYZ and DHS) failed to show significantly seasonality, while LUE of the temperate ecosystems (CBS, NMG, HB and DX) peaked during mid-growing season. Furthermore, the ecosystem LUE at QYZ site reached its turning point in July during mid-growing season, presented by a sharp fall resulting from the epidemic summer drought [38]. Among grassland sites, the LUE at HB site exhibited apparently higher values than the other two grassland sites and reached its maximal value in August, whereas the ecosystem LUE at NMG and DX site peaked in July and August, respectively (Figure 2j, k and l).

Frequency distribution of cloudiness index value across ecosystems

Apart from the seasonal dynamics of sky conditions (Figure 3a–f), the temporal patterns of cloudiness at the six sites in the mid-growing seasons were showed by the frequency distribution of *CI* values (Figure 3). Despite inter-annual variations resulting from climatic variability, common characteristics of the cloudiness pattern were found to be among the six sites. The *CI* values at CBS site occupied the largest frequency around 0.4 in 2003 and 2004(Figure 3a, b), while the *CI* value frequency took the most part around 0.5 from 2005 to 2007(Figure 3c, d and e). The peaks of *CI* value frequency at QYZ site located around 0.5 (Figure 3f, g and j) and 0.9 (Figure 3h, i). The *CI* value frequency at the DHS peaked between 0.5 and 0.7, except for 2005 and 2006, in which the largest frequency occurred around 0.8 in the mid-growing seasons (Figure 3m, n). As to the NMG site, the largest *CI* value frequency occurred between 0.4 and 0.5. Meanwhile, the *CI* frequency peaked around 0.5 at the HB site (Figure3u–z). The *CI* frequency between 0.5 and 0.7 occupied the largest proportion at DX site, except that peaked around 0.4 in 2003(Figure 3z). Overall, the *CI* frequencies occurred between 0.8 and 1.0 in the subtropical forest sites (QYZ and DHS) were much larger than what in the temperate ecosystems (CBS, NMG, HB and DX), which was verified by the report that spatial patterns of annual diffuse radiation in China showed strong regional heterogeneity, lower in the north but higher in the south [32].

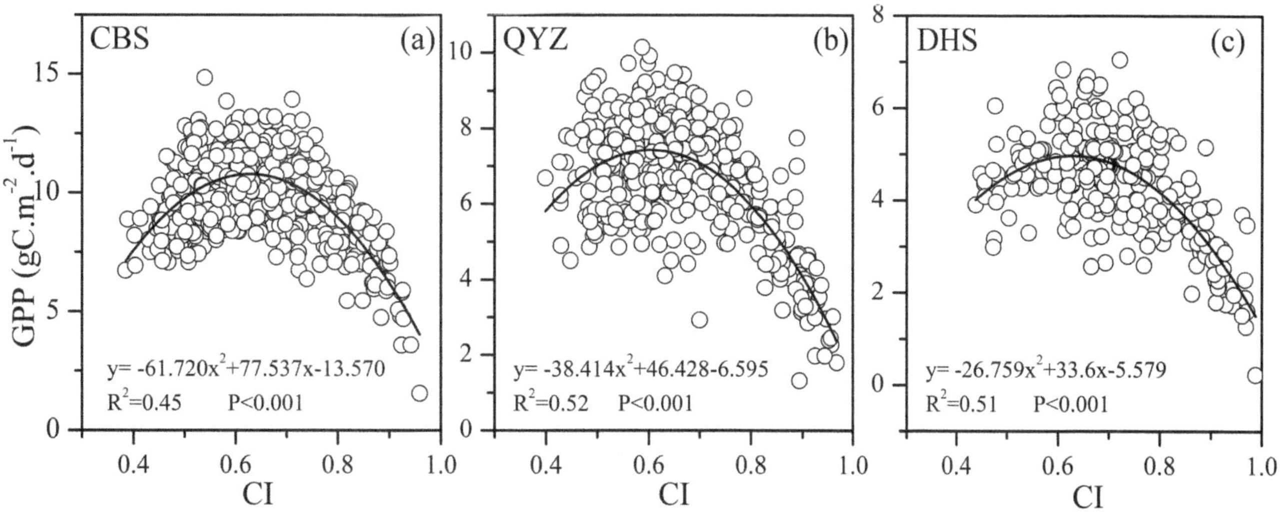

Figure 5. Relationships between ecosystem daily GPP and *CI* during the mid-growing season at three forest ecosystems.

Clear sky light use efficiency in the six ecosystems

The ecosystem LUE was plotted against the ratio of PAR to potential PAR (decreasing cloud cover), and the clear sky LUE (LUE_{cs}) was acquired when PAR/PAR_{po} was around 1.0 [51]. LUE_{cs} values were greater at forest sites than those at grassland sites (Table 2). Among the forest sites, the ecosystem LUE_{cs} was greatest at DHS site, intermediated at QYZ site and lowest at CBS site. The LUE_{cs} of forest ecosystem decreased with the degree of latitude. For grassland sites, the LUE_{cs} peaked at the HB site, followed by the DX site and NMG site, respectively. This was verified by the fact that measured ecosystem LUE was higher than the other two grassland ecosystem sites (Figure 2).

Impacts of diffuse radiation on ecosystem light use efficiency

Figure 4 exhibited the interactive responses of ecosystem LUE to the variation in the diffuse radiation fraction of incoming solar radiation (indicated by the cloudiness index) in different ecosystems. At all sites, significantly quadratic regression relationships were found between the ecosystem LUE and *CI* during the mid-growing season. Once the value of *CI* exceeded a certain one, determined by the minimal value (zero) of the first derived function of each quadratic regression function, the ecosystem LUE increased with *CI* dramatically. LUE of forest ecosystem showed more significantly positive relationship with *CI* ($R^2 \geq 0.74$), compared with three grassland ecosystem sites ($R^2 \leq 0.5$). Also, differences in enhancement on ecosystem LUE induced by the variation of diffuse PAR existed within the ecosystem type across sites. For the forest sites, the ecosystem LUE at CBS site demonstrated stronger increasing trend than the other two subtropical forest sites with the largest correlation coefficient ($R^2 = 0.85$) and quadratic term coefficient (6.166)(Figure 4a, 4b and 4c). The ecosystem LUE at NMG site exhibited least increasing potential ($R^2 = 0.24$, quadratic term coefficient = 1.224) with the variation of *CI* among the three grassland ecosystem sites. The expectation that canopy LUE could be enhanced by the diffuse components of solar radiation compared to direct radiation has been reported in previous studies [21,26,52]. Under cloudy skies, incoming radiation was more diffuse and more uniformly distributed in the canopy with a smaller fraction of the canopy that was light saturated [10]. Consequently, canopy photosynthesis was inclined to be more light-use efficient under diffuse sunlight than under direct sunlight conditions [10,21,23,24,26,27]. In addition, differences in canopy structure density across different ecosystem types was presented to contribute to the increasing rate differences of LUE to fraction of diffuse radiation [36], due to its effective penetration to the lower depths of canopy [32,53].

Cloudiness threshold for enhancing forest ecosystem carbon sequestration

Similar significances of quadratic regression relationships between daily GPP and *CI* during mid-growing season were confirmed at three forest sites (Figure 5). The quadratic regression relationships implied that the ecosystem GPP would peak at a certain value of *CI*, and then decreased with increasingly values of *CI*. Specifically, it was noted that the impact of cloudiness on ecosystem carbon exchange process was also dependent on local thermal, moisture and light conditions [36]. At the beginning stage, the forest ecosystems GPP were in positive association with the *CI*. Instead, the forest ecosystems GPP were gradually restrained by the cloudiness when the value of *CI* exceeded a

Table 3. Cloudiness thresholds for enhancing forest ecosystem LUE and GPP.

Cloudiness index	CBS	QYZ	DHS
Lower bounds	0.288	0.154	0.394
Upper bounds	0.628	0.604	0.629

certain one where the symmetric axis of the parabolic curve regression functions located. This phenomenon could partly be ascribed to the decreasing PAR absorbed by the vegetation canopy, based on the radiation conversion efficiency concept of Monteith (1972) [2]. Consequently, the cloudiness thresholds (Table 3) were calculated by a range that began from the value where the symmetric axis of the parabolic curve regression functions (response of LUE to *CI*) located (Figure 4a, 4b and 4c), and stopped at the point where symmetric axis of the parabolic curve regression functions (response of GPP to *CI*) located (Figure 5). However, the optimal cloudiness index threshold was not available for the three grassland sites because of the poor quadratic relationship between GPP and *CI* ($P>0.05$). The difference in responses of GPP to the variation of diffuse PAR received by the ecosystem between forest sites and grassland sites was likely to result from the difference in canopy structure [36]. The LAI of forest ecosystem at CBS, QYZ and DHS were higher than those of grassland ecosystem (Table 1). Previous studies reported that LUE and GPP of an ecosystem with low LAI, such as grassland and shrubs, did not increase on cloudy days [54,55]. This inconsistency was partly attributed to the differences of climate conditions of the studied ecosystems, including light, water and thermal conditions [36].

Conclusions

Eddy covariance flux observations from six sites encompassing two ecosystem types and the cloudiness index were used to detect the response of LUE and GPP to diffuse radiation during mid-growing season. Results indicated that (1) cloudiness index (*CI*) was mostly greater at two subtropical forest ecosystem sites (QYZ and DHS) than the other temperate ecosystem sites (CBS, NMG, HB and DX), and LUE in the temperate ecosystem peaked during mid-growing season;(2) LUE under clear sky were greater at forest sites than at grassland sites, and the LUE under clear sky of forest ecosystem decreased with the degree of latitude; (3) significantly quadratic regression relationships were found between the ecosystem LUE and *CI* during the mid-growing season at all sites;(4) cloudiness thresholds favorable for enhancing ecosystem carbon sequestration existed in forest ecosystem sites.

Due to the large regional heterogeneity existing in terrestrial ecosystem of China, more EC flux sites involved in the future are essential to reveal the impacts of diffuse radiation on terrestrial ecosystem LUE in China. Furthermore, the cloudiness index could be incorporated as an environmental regulator into LUE models for regional GPP simulations in China.

Author Contributions

Conceived and designed the experiments: KH SW. Performed the experiments: KH SW LZ. Analyzed the data: KH SW LZ. Contributed reagents/materials/analysis tools: KH SW LZ HW JZ JY LZ YW PS. Wrote the paper: KH SW. Discussed the manuscript: KH SW LZ HW JZ JY LZ YW PS.

References

1. Nemani RR, Keeling CD, Hashimoto H, Jolly WM, Piper SC, et al. (2003) Climate-driven increases in global terrestrial net primary production from 1982 to 1999. Science 300: 1560–1563.
2. Monteith JL (1972) Solar Radiation and Productivity in Tropical Ecosystems. J Appl Ecol 9: 747–766.
3. Cai WW, Yuan WP, Liang SL, Zhang XT, Dong WJ, et al. (2014) Improved estimations of gross primary production using satellite-derived photosynthetically active radiation. J Geophys Res-Biogeo 119: 110–123.
4. Raupach MR, Canadell JG, Le Quere C (2008) Anthropogenic and biophysical contributions to increasing atmospheric CO_2 growth rate and airborne fraction. Biogeosciences 5: 1601–1613.
5. Wu CY, Munger JW, Niu Z, Kuang D (2010) Comparison of multiple models for estimating gross primary production using MODIS and eddy covariance data in Harvard Forest. Remote Sens Environ 114: 2925–2939.
6. Xiao JF, Zhuang QL, Baldocchi DD, Law BE, Richardson AD, et al. (2008) Estimation of net ecosystem carbon exchange for the conterminous United States by combining MODIS and AmeriFlux data. Agric For Meteorol 148: 1827–1847.
7. Canadell JG, Mooney HA, Baldocchi DD, Berry JA, Ehleringer JR, et al. (2000) Carbon metabolism of the terrestrial biosphere: A multitechnique approach for improved understanding. Ecosystems 3: 115–130.
8. Wang Y, Zhou GS (2012) Light Use Efficiency over Two Temperate Steppes in Inner Mongolia, China. Plos One 7.
9. Piao SL, Fang JY, Ciais P, Peylin P, Huang Y, et al. (2009) The carbon balance of terrestrial ecosystems in China. Nature 458: 1009–U1082.
10. He MZ, Ju WM, Zhou YL, Chen JM, He HL, et al. (2013) Development of a two-leaf light use efficiency model for improving the calculation of terrestrial gross primary productivity. Agric For Meteorol 173: 28–39.
11. Ogutu BO, Dash J (2013) Assessing the capacity of three production efficiency models in simulating gross carbon uptake across multiple biomes in conterminous USA. Agric For Meteorol 174: 158–169.
12. Monteith JL, Moss CJ (1977) Climate and the Efficiency of Crop Production in Britain [and Discussion]. Philosophical Transactions of the Royal Society of London B, Biological Sciences 281: 277–294.
13. Running SW, Nemani RR, Heinsch FA, Zhao MS, Reeves M, et al. (2004) A continuous satellite-derived measure of global terrestrial primary production. BioScience 54: 547–560.
14. Xiao XM, Hollinger D, Aber J, Goltz M, Davidson EA, et al. (2004) Satellite-based modeling of gross primary production in an evergreen needleleaf forest. Remote Sens Environ 89: 519–534.
15. Yuan WP, Liu S, Zhou GS, Zhou GY, Tieszen LL, et al. (2007) Deriving a light use efficiency model from eddy covariance flux data for predicting daily gross primary production across biomes. Agric For Meteorol 143: 189–207.
16. Wu CY, Niu Z, Gao SA (2010) Gross primary production estimation from MODIS data with vegetation index and photosynthetically active radiation in maize. J Geophys Res-Atmos 115.
17. Veroustraete F, Sabbe H, Eerens H (2002) Estimation of carbon mass fluxes over Europe using the C-Fix model and Euroflux data. Remote Sens Environ 83: 376–399.
18. Yang YT, Shang SH, Guan HD, Jiang L (2013) A novel algorithm to assess gross primary production for terrestrial ecosystems from MODIS imagery. J Geophys Res-Biogeo 118: 590–605.
19. Sims DA, Rahman AF, Cordova VD, El-Masri BZ, Baldocchi DD, et al. (2008) A new model of gross primary productivity for North American ecosystems based solely on the enhanced vegetation index and land surface temperature from MODIS. Remote Sens Environ 112: 1633–1646.
20. Garbulsky MF, Penuelas J, Papale D, Ardo J, Goulden ML, et al. (2010) Patterns and controls of the variability of radiation use efficiency and primary productivity across terrestrial ecosystems. Global Ecol Biogeogr 19: 253–267.
21. Gu LH, Baldocchi DD, Wofsy SC, Munger JW, Michalsky JJ, et al. (2003) Response of a deciduous forest to the Mount Pinatubo eruption: Enhanced photosynthesis. Science 299: 2035–2038.
22. Turner DP, Ritts WD, Styles JM, Yang Z, Cohen WB, et al. (2006) A diagnostic carbon flux model to monitor the effects of disturbance and interannual variation in climate on regional NEP. Tellus B 58: 476–490.
23. Mercado LM, Bellouin N, Sitch S, Boucher O, Huntingford C, et al. (2009) Impact of changes in diffuse radiation on the global land carbon sink. Nature 458: 1014–U1087.
24. Oliphant AJ, Dragoni D, Deng B, Grimmond CSB, Schmid HP, et al. (2011) The role of sky conditions on gross primary production in a mixed deciduous forest. Agric For Meteorol 151: 781–791.
25. Roderick ML, Farquhar GD, Berry SL, Noble IR (2001) On the direct effect of clouds and atmospheric particles on the productivity and structure of vegetation. Oecologia 129: 21–30.
26. Gu LH, Baldocchi D, Verma SB, Black TA, Vesala T, et al. (2002) Advantages of diffuse radiation for terrestrial ecosystem productivity. J Geophys Res-Atmos 107.
27. Misson L, Lunden M, McKay M, Goldstein AH (2005) Atmospheric aerosol light scattering and surface wetness influence the diurnal pattern of net ecosystem exchange in a semi-arid ponderosa pine plantation. Agric For Meteorol 129: 69–83.
28. Heald CL, Henze DK, Horowitz LW, Feddema J, Lamarque JF, et al. (2008) Predicted change in global secondary organic aerosol concentrations in response to future climate, emissions, and land use change. J Geophys Res-Atmos 113.
29. Kim SW, Jefferson A, Yoon SC, Dutton EG, Ogren JA, et al. (2005) Comparisons of aerosol optical depth and surface shortwave irradiance and their

effect on the aerosol surface radiative forcing estimation. J Geophys Res-Atmos 110.
30. Feddema JJ, Oleson KW, Bonan GB, Mearns LO, Buja LE, et al. (2005) The importance of land-cover change in simulating future climates. Science 310: 1674–1678.
31. Schiermeier Q (2006) Oceans cool off in hottest years. Nature 442: 854–855.
32. Ren XL, He HL, Zhang L, Zhou L, Yu GR, et al. (2013) Spatiotemporal variability analysis of diffuse radiation in China during 1981–2010. Ann Geophys-Germany 31: 277–289.
33. Baldocchi D, Falge E, Gu LH, Olson R, Hollinger D, et al. (2001) FLUXNET: A new tool to study the temporal and spatial variability of ecosystem-scale carbon dioxide, water vapor, and energy flux densities. B Am Meteorol Soc 82: 2415–2434.
34. Yu GR, Wen XF, Sun XM, Tanner BD, Lee XH, et al. (2006) Overview of ChinaFLUX and evaluation of its eddy covariance measurement. Agric For Meteorol 137: 125–137.
35. Yu GR, Zhu XJ, Fu YL, He HL, Wang QF, et al. (2013) Spatial patterns and climate drivers of carbon fluxes in terrestrial ecosystems of China. Glob Change Biol 19: 798–810.
36. Zhang M, Yu G-R, Zhuang J, Gentry R, Fu Y-L, et al. (2011) Effects of cloudiness change on net ecosystem exchange, light use efficiency, and water use efficiency in typical ecosystems of China. Agric For Meteorol 151: 803–816.
37. Guan DX, Wu JB, Zhao XS, Han SJ, Yu GR, et al. (2006) CO2 fluxes over an old, temperate mixed forest in northeastern China. Agric For Meteorol 137: 138–149.
38. Wen XF, Wang HM, Wang JL, Yu GR, Sun XM (2010) Ecosystem carbon exchanges of a subtropical evergreen coniferous plantation subjected to seasonal drought, 2003–2007. Biogeosciences 7: 357–369.
39. Yu GR, Zhang LM, Sun XM, Fu YL, Wen XF, et al. (2008) Environmental controls over carbon exchange of three forest ecosystems in eastern China. Glob Change Biol 14: 2555–2571.
40. Fu YL, Yu GR, Wang YF, Li ZQ, Hao YB (2006) Effect of water stress on ecosystem photosynthesis and respiration of a Leymus chinensis steppe in Inner Mongolia. Sci China Ser D 49: 196–206.
41. Fu YL, Yu GR, Sun XM, Li YN, Wen XF, et al. (2006) Depression of net ecosystem CO_2 exchange in semi-arid Leymus chinensis steppe and alpine shrub. Agric For Meteorol 137: 234–244.
42. Shi PL, Sun XM, Xu LL, Zhang XZ, He YT, et al. (2006) Net ecosystem CO2 exchange and controlling factors in a steppe - Kobresia meadow on the Tibetan Plateau. Sci China Ser D 49: 207–218.
43. Fu Y, Zheng Z, Yu G, Hu Z, Sun X, et al. (2009) Environmental influences on carbon dioxide fluxes over three grassland ecosystems in China. Biogeosciences 6: 2879–2893.
44. Hu ZM, Yu GR, Fu YL, Sun XM, Li YN, et al. (2008) Effects of vegetation control on ecosystem water use efficiency within and among four grassland ecosystems in China. Glob Change Biol 14: 1609–1619.
45. Webb EK, Pearman GI, Leuning R (1980) Correction of flux measurements for density effects due to heat and water vapour transfer. Quarterly Journal of the Royal Meteorological Society 106: 85–100.
46. Carrara A, Kowalski AS, Neirynck J, Janssens IA, Yuste JC, et al. (2003) Net ecosystem CO_2 exchange of mixed forest in Belgium over 5 years. Agric For Meteorol 119: 209–227.
47. Falge E, Baldocchi D, Olson R, Anthoni P, Aubinet M, et al. (2001) Gap filling strategies for defensible annual sums of net ecosystem exchange. Agric For Meteorol 107: 43–69.
48. Yu GR, Song X, Wang QF, Liu YF, Guan DX, et al. (2008) Water-use efficiency of forest ecosystems in eastern China and its relations to climatic variables. New Phytol 177: 927–937.
49. Turner DP, Urbanski S, Bremer D, Wofsy SC, Meyers T, et al. (2003) A cross-biome comparison of daily light use efficiency for gross primary production. Glob Change Biol 9: 383–395.
50. Fu P, Rich PM (1999) Design and implementation of the Solar Analyst: an ArcView extension for modeling solar radiation at landscape scales. Proceedings of the 19th annual ESRI user conference, San Diego, USA.
51. King DA, Turner DP, Ritts WD (2011) Parameterization of a diagnostic carbon cycle model for continental scale application. Remote Sens Environ 115: 1653–1664.
52. Farquhar GD, Roderick ML (2003) Atmospheric science: Pinatubo, diffuse light, and the carbon cycle. Science 299: 1997–1998.
53. Urban O, Janous D, Acosta M, Czerny R, Markova I, et al. (2007) Ecophysiological controls over the net ecosystem exchange of mountain spruce stand. Comparison of the response in direct vs. diffuse solar radiation. Glob Change Biol 13: 157–168.
54. Letts MG, Lafleur PM, Roulet NT (2005) On the relationship between cloudiness and net ecosystem carbon dioxide exchange in a peatland ecosystem. Ecoscience 12: 53–59.
55. Niyogi D, Chang H-I, Chen F, Gu L, Kumar A, et al. (2007) Potential impacts of aerosol–land–atmosphere interactions on the Indian monsoonal rainfall characteristics. Nat Hazards 42: 345–359.

13

Is Matang Mangrove Forest in Malaysia Sustainably Rejuvenating after More than a Century of Conservation and Harvesting Management?

Arnaud Goessens[1,2◑¶], Behara Satyanarayana[1,2,3*◑¶], Tom Van der Stocken[1,3], Melissa Quispe Zuniga[1,3], Husain Mohd-Lokman[2], Ibrahim Sulong[2], Farid Dahdouh-Guebas[1,3◑]

1 Laboratory of Systems Ecology and Resource Management, Université Libre de Bruxelles - ULB, CPI 264/1, Brussels, Belgium, **2** Mangrove Research Unit (MARU), Institute of Oceanography and Environment (INOS), Universiti Malaysia Terengganu - UMT, Kuala Terengganu, Malaysia, **3** Laboratory of Plant Biology and Nature Management, Vrije Universiteit Brussel - VUB, Brussels, Belgium

Abstract

Matang Mangrove Forest Reserve (MMFR) in Peninsular Malaysia is under systematic management since 1902 and still considered as the best managed mangrove forest in the world. The present study on silvimetrics assessed the ongoing MMFR forest management, which includes a first thinning after 15 years, a second thinning after 20 years and clear-felling of 30-year old forest blocks, for its efficiency and productivity in comparison to natural mangroves. The estimated tree structural parameters (e.g. density, frequency) from three different-aged mangrove blocks of fifteen (MF15), twenty (MF20), and thirty (MF30) years old indicated that *Bruguiera* and *Excoecaria* spp. did not constitute a significant proportion of the vegetation (<5%), and hence the results focused majorly on *Rhizophora apiculata*. The density of *R. apiculata* at MF15, MF20 and MF30 was 4,331, 2,753 and 1,767 stems ha^{-1}, respectively. In relation to ongoing practices of the artificial thinnings at MMFR, the present study suggests that the first thinning could be made earlier to limit the loss of exploitable wood due to natural thinning. In fact, the initial density at MF15 was expected to drop down from 6,726 to 1,858 trees ha^{-1} before the first thinning. Therefore the trees likely to qualify for natural thinning, though having a smaller stem diameter, should be exploited for domestic/commercial purposes at an earlier stage. The clear-felling block (MF30) with a maximum stem diameter of 30 cm was estimated to yield 372 t ha^{-1} of the above-ground biomass and suggests that the mangrove management based on a 30-year rotation is appropriate for the MMFR. Since Matang is the only iconic site that practicing sustainable wood production, it could be an exemplary to other mangrove locations for their improved management.

Editor: Martin Heil, Centro de Investigación y de Estudios Avanzados, Mexico

Funding: The travel grant of AG was sponsored by the Interuniversity Council of the French Community of Belgium - University Commission for Development (CIUF-CUD). BS FDG were supported by the Belgian National Science Foundation (FNRS-MIS F.4508.11). TVdS MQZ were supported by the Flemish Interuniversity Cooperation - University Development Cooperation (VLIR-UOS). The funders had no role in study design, data collection and analysis, decision to publish, or preparation of the manuscript.

Competing Interests: The authors have declared that no competing interests exist.

* Email: satyam2149@gmail.com

◑ These authors contributed equally to this work.

¶ These authors are co-first authors on this work.

Introduction

Mangrove forests are considered as one of the most productive ecosystems in the world and have a well-established ecological, economic and cultural importance [1,2,3,4,5,6,7,8,9,10]. However, the constant pressure exerted by anthropogenic (more than natural) events is responsible for its decline at a faster rate than that of tropical rainforests [11,12,13,14,15]. Along with mangrove cover depletion, the loss of its biodiversity and economic value are the other perturbing issues [8,16]. In particular, mangroves should be treated carefully without underestimating their role for local livelihoods, and see their long-term benefits reach future generations via appropriate conservation and management practices [17,18,19].

The Matang Mangrove Forest Reserve (hereafter referred to as 'MMFR'), located on the northwest coast of Peninsular Malaysia (Fig. 1), is under concerted scientific management since the beginning of the 20th century and still considered as the best managed mangrove forest in the world [2,20]. The management here is based on a 30-year rotation cycle with two artificial thinnings in 15 and 20-year old blocks [21] (Fig. 2A). These artificial thinnings are based on a spacing technique called the 'stick method': whereas the first thinning is carried out for any tree within a 1.2 m radius of the selected central tree (measured by a stick this long), the second thinning will be done for any tree within a 1.8 m radius. The aim of these thinnings is not only to obtain poles [22], but also to promote better growth of the remaining trees [23]. After the clear felling in a 30-year old block (for charcoal production), the area is replanted with the seedlings of *Rhizophora apiculata* Blume at 1.2 m and *R. mucronata* Lamk. at 1.8 m intervals [21]. However, the local management authorities at Matang are still focused on an improved productivity because of declining yields in recent years [24]. While management decisions

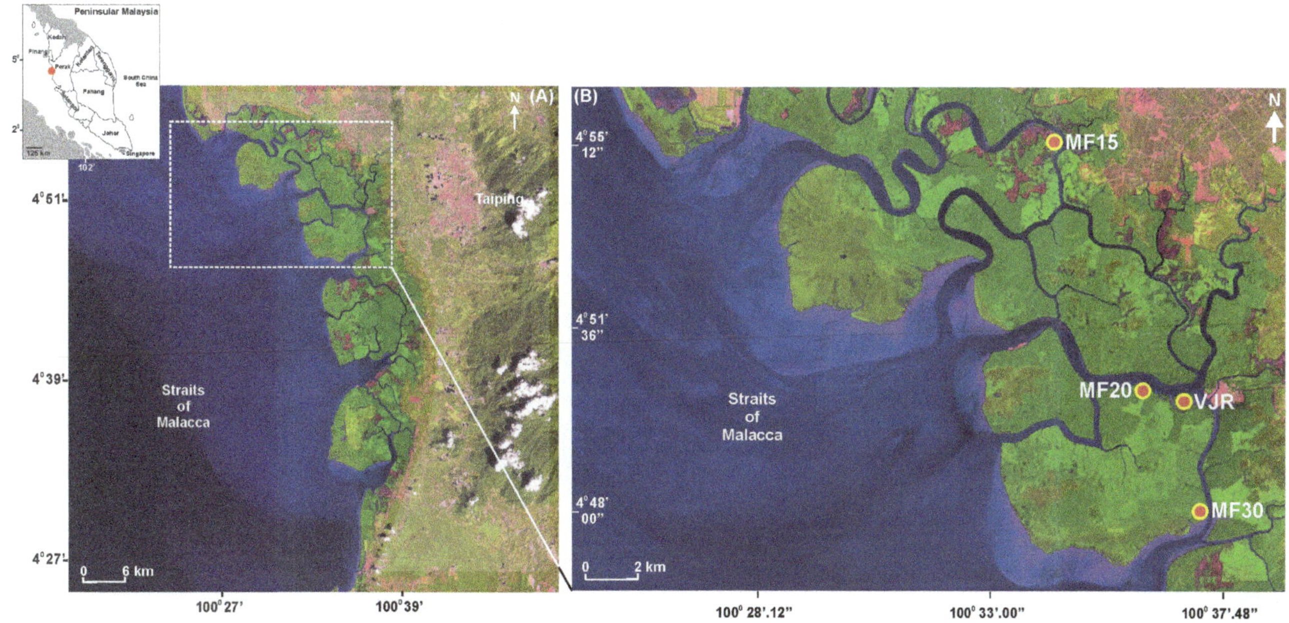

Figure 1. Matang Mangrove Forest Reserve in the state of Perak on the West coast of Peninsular Malaysia (A) (dotted square represents the study zone); (B) Location (yellow circle with red dots) of the Virgin Jungle Reserve (VJR) and the Managed Mangrove Forest (MF with 15, 20 and 30 year old vegetation) blocks considered for silvimetric measurements in the present study (image source: Landsat 7 dated 27 Dec 1999 from the NASA's Earth Observatory).

are often based on poor or incomplete information [25], the State Forestry Department of Perak is devoted to accomplish scientific studies, and update, if necessary, the existing production and conservation policy at Matang.

Although few studies have attempted to validate the 30-year forest rotation plan for Matang, the availability of (field-based) tree structural measurements was a constraint. For example, the first and only published study on silviculture vis-à-vis management system was Gong and Ong in 1995 [26]. They have studied the demography of the forest by analyzing size distribution and biomass in different aged blocks. However, in view of their data collected in November 1980 (from just four adjacent 10 m×10 m plots), they have expressed the low reliability of their results and recommended further studies with a more intensive regime. Apart from the literature on plant biomass and nutrient fluxes [27,28,29]; sediment and recycling organic matter [30,31,32]; canopy gaps and regeneration [23], etc., and despite its exemplary potential, there was no (published) scientific field-based study on the tree structural assessment in relation to conservation and management guidelines at the MMFR before the present study. Even the recent innovative paper by Fontalvo-Herazo et al. [33] on silviculture management using individual-based model relied upon the 20 year-old literature data from Gong and Ong [26]. Hence, there is a need for understanding whether or not the harvesting regime leads to a sustainable rejuvenation i.e., an efficient and value-added long-term use of the forest resources while having a minimum environmental impact.

The present paper aimed at sustainable rejuvenation validation in Matang, both from an ecological and a silvicultural point of view. In addition, the assessment of silvimetric parameters (juvenile, young and adult tree density, vegetation biomass) from different aged forest blocks offers field-based recommendations to the local authorities for a better mangrove management. Considering that mangroves occur in 123 countries on all continents with tropical climates, and that no other mangrove forest on the planet has been managed for over a century (not even for half that period) our study also serves as a reference for long-term mangrove forest management. It also opens doors to investigate mangrove charcoal trade and its ecological and socio-economic implications.

Materials and Methods

Study area

The MMFR covers 40,288 ha in the state of Perak (4°45′N, 100°35′E) (Fig. 1) and is a typical riverine mangrove forest in Malaysia [21]. The reserve is shaped like a crescent moon along the 51.5 km coastline from Kuala Gula in the north to Bagan Panchor in the south. The major portion of the reserve lies on seven deltaic islands separated by a network of creeks and canals [34]. We recall that this MMFR is under sustainable management by the State Forestry Department of Perak for timber production since 1902 [21,23]. As required by the Malaysian National Forestry Policy, the Matang management also has other objectives such as shoreline protection from coastal erosion, to assure functionality of the forest as breeding/nursery ground and wildlife habitat, and as support for forest conservation, research, education and ecotourism [21,34].

The latest version of the Matang management plan covers the period 2010–2019, and is also based on a 30-year rotation cycle [35]. It should be noticed that the MMFR has four management zones namely, 'protective', 'productive', 'restrictive productive' and 'unproductive' forests of which the productive zone got 110 compartments/blocks with different aged *Rhizophora* stands to implement both the thinning and the clear-felling operations on a regular basis. This working plan also indicated several research needs such as monitoring of the forest transitions (i.e. from *Avicennia* to *Rhizophora*, *Rhizophora* to dryland and reversion from dryland to *Rhizophora*), the scattered circular zones of dead trees (due to lightning strikes), the trees collapsing onto navigating

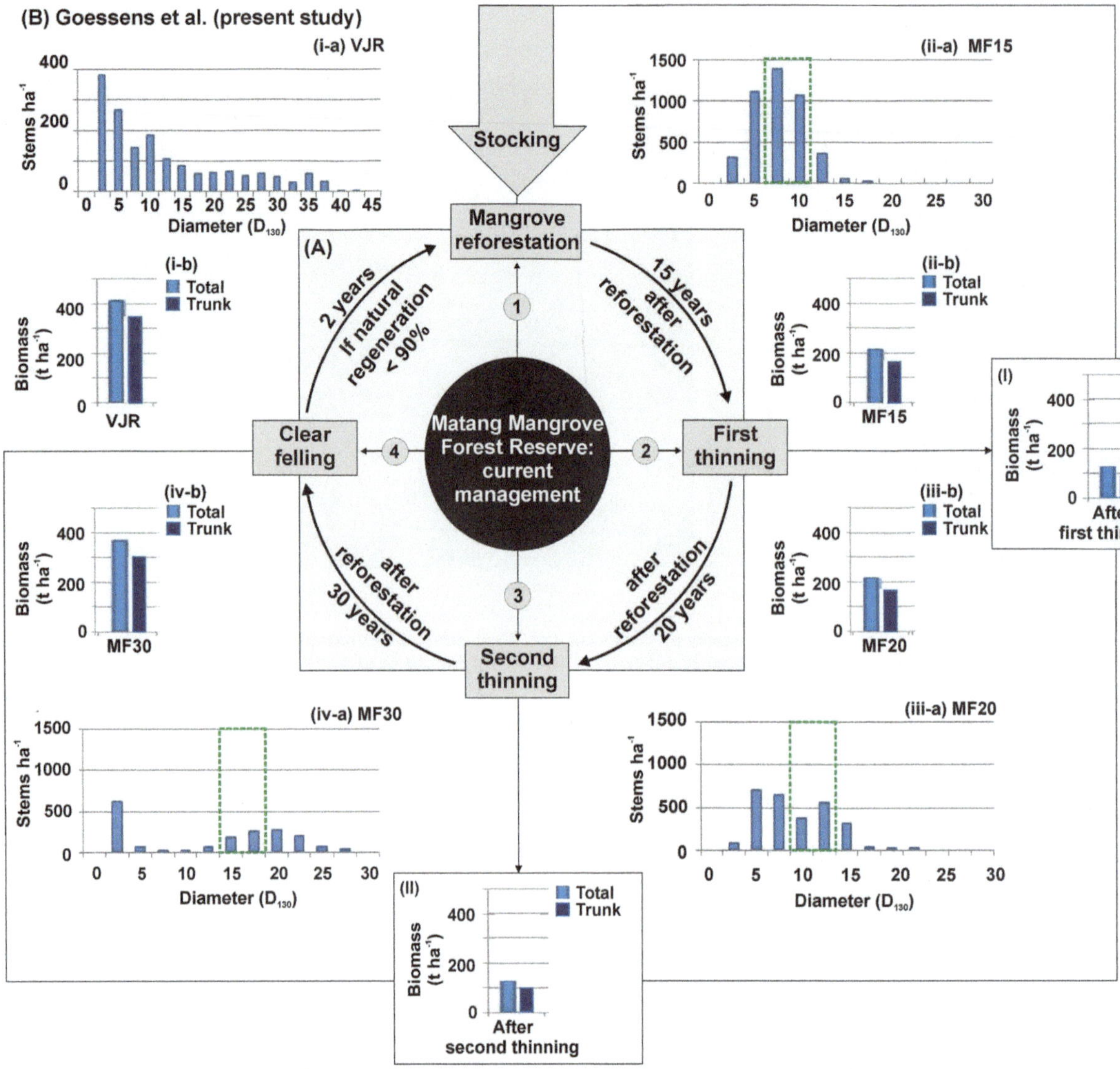

Figure 2. Schematic representation showing the Matang mangrove management system (A); (B) the present observations on *Rhizophora apiculata* – (a) stem size (D_{130}) and, (b) biomass distribution at different forest blocks: (i) VJR: Virgin Jungle Reserve, (ii) MF15: Managed Forest at 15 years old, (iii) MF20: Managed Forest at 20 years old and, MF30: Managed Forest at 30 years old; B (I–II): the computed biomass of *R. apiculata* after the first and second thinnings. The green coloured rectangle (dashed box) in B (ii-a to iv-a) shows the expected/required stem size by the State Forestry Department of Perak.

boats (buffer zones), and the pharmacological use of mangrove species. Along with *R. apiculata* and *R. mucronata*, the other mangrove species found in the area were *Avicennia officinalis* L., *Bruguiera gymnorrhiza* (L.) Lamk., *B. parviflora* Wight and Arnold ex Griffith, *B. cylindrica* (L.) Blume, *Ceriops tagal* (Perr.) C.B. Rob., *Excoecaria agallocha* L., and *Sonneratia alba* J. Smith.

MMFR is influenced by an equatorial climate (warm and humid) with a mean annual temperature of 23.7–33.4°C and humidity of 76.5–83.5%. The rainfall, varying between 2000 and 3000 mm, occurs throughout the year [24]. The tides are semi-diurnal with an amplitude in the range of 1.60–2.98 m [36].

Fieldwork and data analysis

The present study was carried out from February to April, 2011. To obtain the tree structural parameters (e.g. girth, height), both the Virgin Jungle Reserve (VJR: 4°50′3″N, 100°37′0″E), and the Managed Forest (MF) were visited with a prior permission from the State Forestry Department of Perak. While the former has not been exploited for at least 80 years, and therefore considered 'natural', the latter comprises the productive mangrove forest blocks with three different ages of fifteen (MF15: 4°55′16″N, 100°34′9″E), twenty (MF20: 4°50′22″N, 100°36′12″E), and thirty (MF30: 4°48′10″N, 100°37′25″E) years old (Fig. 1B). Tree girth (G_{130}) was measured at 130 cm height above the ground and along the stem using a measuring tape, while the tree height was estimated with the help of a MDL LaserAce 300 (accuracy: 10 cm). For tree measurements, the suggestions offered by Dahdouh-Guebas and Koedam [37] were followed. It is worth mentioning that there are also multiple-stemmed trees (MST) in the forest. Therefore we estimated both the number of trees ha^{-1} considering the MST as a unique tree and, the number of stems ha^{-1} considering each stem of the MST as a separate tree. Although single and MST are commonly found in the mangroves [37], there was not much debate on this issue and the related

Table 1. Adult tree density (stems ha^{-1}) and frequency (%) at Matang Mangrove Forest Reserve.

Species	VJF	MF15	Thinning-I	MF20	Thinning-II	MF30
Rhizophora apiculata	1,633	4,331	2,599	2,753	1,652	1,767
	(100%)	(100%)		(100%)		(100%)
R. mucronata	10*	44*	26	-	-	-
	(5%)	(12%)				
Bruguiera parviflora	271	13*	-	20	-	87
	(33%)	(6%)		(13%)		(47%)
B. cylindrica	291	6*	-	-	-	-
	(38%)	(6%)				
B. gymnorrhiza	5*	-	-	-	-	-
	(5%)					
Excoecaria agallocha	119	-	-	33*	-	-
	(14%)			(7%)		
Avicennia officinalis	19	-	-	-	-	-
	(10%)					
Sonneratia alba	5*	-	-	-	-	-
	(5%)					
Total:	2,352	4,394	-	2,807	-	1,853
MST proportion (%)	5.25	77.82		22.98		5.21
No. stems per MST	2.25	2.71		2.26		2.30

The values under thinnings I and II are the computed stem density which is likely to be present after the thinning events at MF15 and MF20. VJR is Virgin Jungle Reserve; MF15, MF20 and MF30 are the Managed Forest blocks at 15, 20 and 30 years old. MST is multiple-stemmed tree. Except *Rhizophora*, all other species encountered during inspection visits and/or thinning operations at the managed forest blocks will be clear-felled.
*found only at single plot.

research, such as the valuable work of Gong and Ong [26], did not refer to it.

At each forest block, two parallel transects (50 m apart) were laid from the waterfront to the inland mangrove and we counted as well as measured the trees within 10 m×10 m plots at 50 m intervals along each transect. Altogether 67 plots (VJR: 21, MF15: 16, MF20: 15 & MF30: 15) were sampled. In addition to adult trees (≥1.3 m height with a stem diameter D_{130}≥2.5 cm or G_{130}≥8 cm) measurements, the data (nos.ha^{-1}) on mangrove juveniles (=fallen propagules and saplings with three leaf pairs or less), and young trees (<1.3 m height with a D_{130}<2.5 cm or G_{130}<8 cm) (divisions comparable to Kairo et al. [38]), were also collected to examine the natural recruitment in the blocks. The taxonomic identification of mangroves was based on using Tomlinson [39]. A handheld global positioning system (Garmin, GPS III) was used for navigation.

Tree density (no. ha^{-1}), basal area (m^2 ha^{-1}), relative density (%), relative dominance (%), relative frequency (%), and species' individual rankings were estimated by following the standard protocols [37,40,41,42,43]. In view of the prevailing MST in each forest block, the actual density of trees was calculated by the following equation:

$$\text{Density of trees (trees ha}^{-1}\text{)} (De_{trees}) = \left(\frac{De_{stems} \times p_{MST}}{100 \times r}\right) + \left(\frac{p_{SST}}{100}\right) \quad (1)$$

where De_{stems} is the density of stems (see Eq. 2); p_{MST} – the percentage of multiple-stemmed trees; r – mean number of stems per multiple-stemmed tree; p_{SST} – the percentage of single-stemmed trees.

At the same time, the possible loss of stems (Q) before the first thinning at MF15 was also estimated using the equation below:

$$\text{Loss of stems} (Q) = \left(E \times \frac{p_{MST}}{100} \times r\right) + \left(E \times \frac{p_{SST}}{100}\right) \quad (2)$$

where E is the expected loss of tree density (see eq. 1 for other terms).

And finally, the trunk and above-ground biomass of *Rhizophora* trees were calculated through the allometric relationships published by Ong et al. [44] for Matang:

$$W_{\text{trunk (total trunk weight)}} (\text{kg}) = 0.0067 \times (G_{130})^{2.5414} (\text{cm}) \quad (3)$$

$$W_{\text{ag (total above-ground weight)}} (\text{kg}) = 0.0135 \times (G_{130})^{2.4243} (\text{cm}) \quad (4)$$

Since the managed forest blocks MF15 and MF20 are scheduled for their first and second thinnings in the year 2011 (not yet started at the time of this study), we also tried to compute the density as well as biomass of *R. apiculata* that is likely going to be present in the two blocks after thinning. For this purpose, we followed Ong [22] who has estimated that 35 to 50% of the trees will be removed after each thinning process. To simplify, it has been averaged to 40% reduction at each block which also coincides with the self-thinning percentage (40%) of Gong and Ong [26]. These projected values are supposed to complement the field data

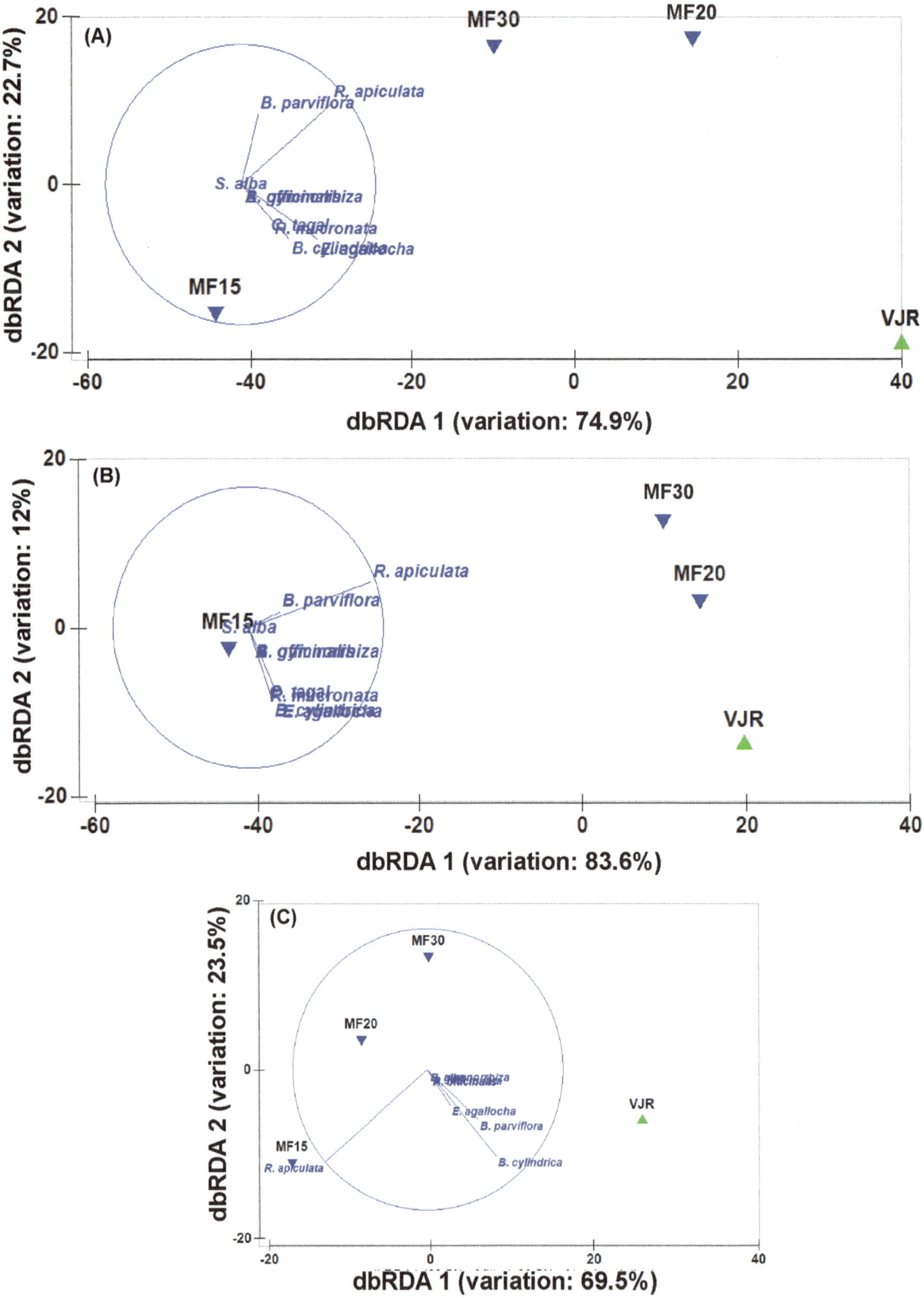

Figure 3. The distance-based redundancy analysis (dbRDA) showing variations between virgin and managed mangrove forest blocks in relation to their – (A) juvenile, (B) young and, (C) adult vegetation at the Matang Mangrove Forest Reserve. While density of the juvenile and the young vegetation was estimated for nos. ha^{-1}, the adult tree density was estimated for no. stems ha^{-1} (VJR: Virgin Jungle Reserve; MF15, MF20 and MF30: Managed Forest blocks at 15, 20 and 30 years old) (circles in all panels represent correlation circles, and the orientation of mangrove species' lines approximate their correlation to the ordination axes).

collected by the forest authorities after thinnings, and provide an indication on the forest structure intervened with the ongoing silvicultural practices. However, the authors found that there is no such practice of taking tree structural measurements officially and hence the present results were compared only with the available literature on Matang management [26,33].

Statistical analysis

Tree structural variables were estimated using the standard as well as above mentioned equations in Microsoft Excel. In addition, the non-parametric one-way analysis of variance (One-way ANOVA) was used to find differences between the stem size (D_{130}) distribution and the tree density counts at 'virgin' and 'managed' forest blocks (using GraphPad Prism *v*.5 software). The distance-based redundancy analysis (dbRDA) - a method of constrained ordination which can display the relationships among samples points from a fitted model [45,46,47], was used to know the percentage (%) variation between the virgin and the managed mangrove blocks in relation to their juvenile, young and adult vegetation (using PRIMER *v*.6.1.12 software). In this context, the analyses were conducted on a Bray-Curtis similarity matrix calculated from the vegetation counts (root-transformed data for approximating poisson to normal distribution).

Results

Stand structure

In all forest blocks, *R. apiculata* was dominant whereas *B. parviflora*, *B. cylindrica*, *R. mucronata* and *E. agallocha* did not constitute more than 5% of the total density (Table 1). Hence the results of this study were focused majorly on *R. apiculata*. In fact, the variability between plots was also high and in most cases the species like *R. mucronata*, *E. agallocha* and *B. cylindrica* occurred only in single plots (Table 1). The computed values show that the density of *R. apiculata* would decrease from 4,331 to 2,599 stems ha^{-1} at MF15 after the first thinning and, from 2,753 to 1,652 stems ha^{-1} at MF20 after the second thinning (Table 1). The (total) basal area was ranged between 27.11 and 37.62 m^2 ha^{-1} in the managed forest blocks of MF20 and MF30, respectively. In the case of VJR, the stem density was 2,352 ha^{-1} (Table 1), with a basal area of 42.17 m^2 ha^{-1}. Followed by *R. apiculata*, *B. cylindrica* and *B. parviflora* were the most frequently seen species in this block (Table 1). The proportion of MST was higher at MF15 (78%) followed by MF20 (23%) (Table 1). The distance-based redundancy analysis (dbRDA) (Fig. 3) indicated that the variance between VJR and MF blocks (i.e. MF15, MF20 and MF30) was higher for young trees (variation along axis-1: 83.6%), than to juvenile (variation: 74.9%) and/or adult (variation: 69.5%) vegetation. It also shows a decreasing trend of adult *R. apiculata* density from MF15>MF20>MF30 to VJR (Fig. 3C).

Size distribution

At VJR, although the majority of the *Rhizophora* trees come under adult vegetation category, the density of stems holding 2.5–10 cm diameter (D_{130}) was high (971 stems ha^{-1}) (max. diameter, 42.5 cm) (Fig. 2B-i-a). In the case of managed mangrove forests, the MF15 had more stems with 5–12.5 cm diameter (3,587 stems ha^{-1}) (max. diameter, 20 cm) (Fig. 2B-ii-a), while MF20 with 5–15 cm (1,733 stems ha^{-1}) (max. diameter, 25 cm) (Fig. 2B-iii-a), and MF30 with 15–22.5 cm (906 stems ha^{-1}) (max. diameter, 30 cm) (Fig. 2B-iv-a). At the above given ranges of stem size (D_{130}) distribution, the tree densities in the VJR and the MF blocks were significantly different (One-way ANOVA, $F = 8.7$, $R^2 = 0.67$, $P = 0.002$).

Natural recruitment

Data on young and juvenile vegetation at the managed forest blocks also revealed the higher abundance of *R. apiculata* (Table 2). Among the managed blocks, MF20 supported maximum number of young (4,227 ha^{-1}) and juvenile (2,867 ha^{-1}) vegetation.

Vegetation biomass

The trunk and above-ground (total) biomass at MF15 and MF20 were almost equal (216–217 t ha^{-1}), whereas MF30 had 372 t ha^{-1} with a difference of 43 t ha^{-1} in relation to VJR (i.e. 415 t ha^{-1}) (Figs. 2B-i-b to 2B-iv-b). In the case of MF15, the biomass after the first thinning is expected to come down from 216 to 130 t ha^{-1} and, at MF20 after the second thinning it could reduce from 217 to 130 t ha^{-1} (Figs. 2B-I and 2B-II).

Discussion

Initial stocking

Since the density of *R. apiculata* at MF15 (4,331 ha^{-1}) (Table 1) was a stem-based count, it does represent the number of stems exploitable as poles, but it does not reflect the actual number of trees planted/present in this block. Therefore, according to the fact that 78% of the trees at MF15 are MST (with a mean no. of 3 stems per tree) (Table 1), the actual density of *R. apiculata* trees is expected to be *ca.* 1,858 trees ha^{-1}.

If the mangrove management objective is only to have a dense forest cover (and an ecologically functional forest), all naturally recruited (mangrove) seedlings should be considered for effective stocking [38]. However, if forests are meant for commercial exploitation (also with selective species) plantations such as in MMFR seem to be the rule. Under the actual management system at Matang, artificial planting is carried out within two years after clear felling (at MF30), provided the density of the natural recruitment in the clear-cut areas is less than 90% [26,35]. However, in most cases, afforestation was said to be necessary (pers. comm. with the authorities of the State Forestry Department of Perak) due to the limited number of propagules stranding in the clear-felled locations after hydrochorous dispersal. The recommended as well as the implemented space for artificial (*R. apiculata*) regeneration was 1.2 m×1.2 m, which allows a planting density of 6,726 seedlings ha^{-1} [48]. This information enables us to compare the actual planting density with the trees available after 15 years of management (i.e. 1,858), to conclude that there was a loss of juvenile, young and/or adult trees of as much as 4,868 individuals (72%). In natural environments, the juvenile/young/adult tree ratios reflect particular stages in distribution and survival [16]. While juveniles represent a possibility for a species to reach a particular location through simple fall and/or hydrochory [49,50,51], young trees represent survival from propagule predation [52,53,54], and adult trees represent survival from unfavorable environmental conditions [53,55]. Since MF15 is an afforested block, the loss of vegetation could be due to the above mentioned factors as well as natural thinning. Meanwhile, the natural recruitment, playing an important role for the rise of seedlings at each forest block (Table 2), should not be ignored. Therefore, in agreement with Gong and Ong [26], to counter propagule loss, artificial regeneration should be conducted if the naturally recruited density is less than 50%, instead of 90%. In addition, we propose the space for artificial regeneration (1.2 m×1.2 m) to be increased not only to reduce the loss of propagules due to competition, but also to allow the trees to better grow and develop. However, in view of the species planted as well as naturally recruited (with different growth patterns) over the

Table 2. Density (no. ha^{-1}) and frequency (%) of the juvenile and young vegetation at Matang Mangrove Forest Reserve.

	VJR				MF15				MF20				MF30			
	Juvenile		Young		Juvenile		Young		Juvenile		Young		Juvenile		Young	
Species	**Den**	**Freq**	**Den**	**Freq**	**Den**	**Freq**	**Den**	**Freq**	**Den**	**Freq**	**Den**	**Freq**	**Den**	**Freq**	**Den**	**Freq**
Rhizophora apiculata	4,238	91	2,795	86	94	56	138	31	2,620	100	2,973	100	1,040	93	3,453	100
R. mucronata	190	14	91	14	-	-	-	-	-	-	-	-	-	-	-	-
Bruguiera parviflora	1,286	24	367	48	19	12	125	50	200	13	967	13	60	13	347	20
B. cylindrica	571	19	143	29	-	-	-	-	-	-	-	-	-	-	-	-
B. gymnorrhiza	-	-	5	5	-	-	-	-	-	-	-	-	-	-	-	-
Excoecaria agallocha	-	-	248	10	-	-	13	6	47	7	287	7	-	-	-	-
Avicennia officinalis	-	-	5	5	-	-	-	-	-	-	-	-	-	-	-	-
Sonneratia alba	-	-	-	-	-	-	-	-	-	-	-	-	-	-	-	-
Ceriops tagal	857	5	81	5	-	-	-	-	-	-	-	-	-	-	-	-
Total:	7,142		3,735		113		276		2,867		4,227		1,100		3,800	

VJR, MF15, MF20 and MF30 are the Virgin Jungle Reserve and the Managed Forest blocks at 15, 20 and 30 years old. Den is density, Freq is frequency.

period of time, further studies and/or field-based experiments are necessary to determine the most appropriate space for this artificial regeneration. By looking at this outcome, the other mangrove locations elsewhere may also follow the suggestion of increased space for their plantation efforts.

Stand density

With the ongoing management at the MMFR, except for *Rhizophora*, all other species encountered during inspection visits and/or thinning operations at the managed forest blocks are clear-felled [35]. Therefore the productive forest areas are mostly composed of *Rhizophora* trees [23]. At the time of first thinning in MF15, mangrove stems are expected to reach to a diameter of 7.5–10 cm [21]. However, taking into account the loss of vegetation (4,868 nos.) and the present status of MST (78%), there could be a loss of 11,352 stems ha^{-1} in the 15-year old forest block. Therefore, implementation of the first thinning before a stand age of 15 years may help reducing loss of exploitable biomass due to natural thinning or unfavourable conditions (as explained in the previous section). In addition, the smaller-sized diameter poles resulting from early thinning could be used for domestic or commercial purposes, including firewood or white charcoal production [35]. Earlier, also Gong and Ong [26] suggested more frequent thinning events at shorter intervals. It was further supported by Fontalvo-Herazo et al. [33] who used the pattern oriented modeling (POM) in KiWi to reproduce forest (i.e. *R. apiculata*) density and size class distribution, and thereby the selection of thinning strategies within the 30-year harvesting cycle. Among the four tested types of thinning activities, Fontalvo-Herazo et al. (op. cit.) found that a management plan with three artificial thinnings at the age of 7, 13 and 20 years was most productive. However, the legitimacy of these predictions is known only when they are tested in the field. The raised densities of *R. apiculata* after the first and second thinnings in the present study (Table 1) are attributable to young vegetation present in both forest blocks (i.e. MF15 and MF20) (Table 2), but beyond the area cleared by the stick method.

In addition, the tree density values in the present study were found to be different from the simulation-based predictions by Fontalvo-Herazo et al. [33] (Fig. 4A): the density was more at MF15 (2,593 instead of 1,885 trees ha^{-1} predicted), and less at both MF20 (1,308 instead of 2,177 trees ha^{-1} predicted), and MF30 (around 1,450 instead of 1,735 trees ha^{-1} predicted) forest blocks. Also, the computed densities of *R. apiculata*, after the first and second thinnings, were dissimilar. These discrepancies are nothing but due to two different approaches (field-based and literature-based), limited data available for a precise parameterization in mangrove model, different stocking density and reduction percentages, etc.

Size distribution

Despite the fact that age and growth relationship can be well established for afforested forests than for natural, the literature on mangrove stem size distribution in relation to age is still limited [38,56,57,58]. Coinciding with the (Matang) Forest Department's required stem size (D_{130}) of 7.5–10 cm before first thinning, the majority of the *Rhizophora* stems at MF15 have had 5–10 cm diameter (Fig. 2B-ii-a). A similar trend was also observed in Kenya where 12-year old *R. mucronata* stands, though younger by 3 years, showed a stem size range between 2.5 and 12.4 cm [58]. In the case of the second thinning at MF20, where the actual requisite of stem size was 10–13 cm [21], the stems had grown to a diameter of up to 22.5 cm (Fig. 2B-iii-a). In addition, the prevailing high density of stems with 5–10 cm diameter at

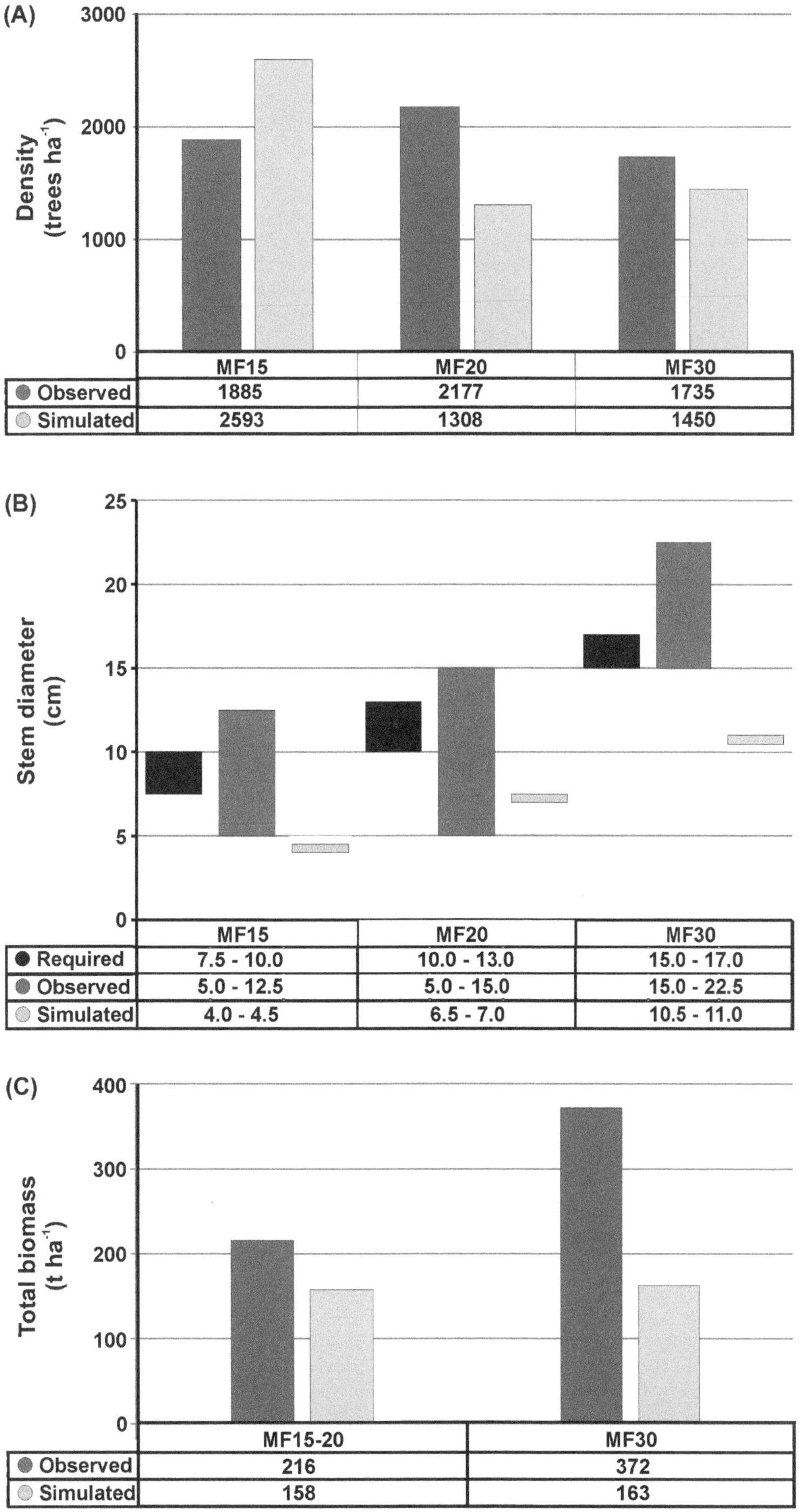
(A)
Density (trees ha^{-1})
3000
2000
1000
0
MF15 MF20 MF30
Observed 1885 2177 1735
Simulated 2593 1308 1450
(B)
Stem diameter (cm)
25
20
15
10
5
0
MF15 MF20 MF30
Required 7.5 - 10.0 10.0 - 13.0 15.0 - 17.0
Observed 5.0 - 12.5 5.0 - 15.0 15.0 - 22.5
Simulated 4.0 - 4.5 6.5 - 7.0 10.5 - 11.0
(C)
Total biomass (t ha^{-1})
400
300
200
100
0
MF15-20 MF30
Observed 216 372
Simulated 158 163

Figure 4. A comparative account on – (A) tree density, (B) Stem diameter (D_{130}) and, (C) harvestable (total) biomass range, as observed by the field- and the simulation-based investigations at Matang Mangrove Forest Reserve (MF15, MF20 and MF30: Managed Forest blocks at 15, 20 and 30 years old). While 'observed' stands for the present (field-based) study, 'simulated' indicates the results obtained from Fontalvo-Herazo et al. (2011). Similarly, 'required' is the stem size as expected by the State Forestry Department of Perak.

MF20 was possibly due to non-removal of the young trees at the time of first thinning. Finally, in the 30-year old forest block (MF30), the diameter reached by the stems (i.e. 15–25 cm) was higher than the required size (15–17 cm) (Fig. 2B-iv-a). One possible reason for this situation is that this forest block may be older than expected. According to local forest authorities, the agenda for thinning/clear-felling cannot be respected sometimes due to administrative difficulties (e.g. work tender finalization, officers' availability, etc.) and hence they conduct the thinning or felling activities earlier or later than scheduled. Another reason could be that the first and second thinnings in this forest block had been very intensive to allow the remaining trees to grow better. The role of natural recruitment, with numerous stems holding 2.5 cm diameter (must have originated from MF20, Table 2), was also evident in this block (Fig. 2B-iv-a). The decreased stem densities are balanced by an increasing diameter (D_{130}) between MF15 and MF30 (Figs. 2B-i-a to 2B-iv-a), which is reassuring in the light of a steady-state mangrove resource management [59], but expected to be less functional and less resilient ecologically.

On the other hand, the stem diameter (D_{130}) observed by both simulated- and field-based investigations was different (Fig. 4B), possibly due to the lack of a required stem size parameterization in the mangrove model [33] as well as the reasons mentioned above under the 'Stand density'.

Natural recruitment

The successfully reforested mangrove sites would be able to improve local hydrodynamics and soil physicochemical conditions that in turn can enhance the natural colonization of both plants and animals [1,55,56,57,60,61]. In fact, the witnessed secondary succession of non-planted mangrove species into the managed forest blocks at Matang (Table 2) suggests an effective reforestation such as in Gazi Bay in Kenya [56,57,58]. However, the lack of persistence of these recruits into the older vegetation layers (due to the thinning practices) makes it evidently less exemplary as a 'natural' forest. Based on density as well as frequency of juvenile and young vegetation at the managed forest blocks (Table 2), MF30 (clear-felling block) is still supporting 4,493 *R. apiculata* individuals (juvenile+young), though it was less than the recommended initial (stocking) density of 6,726 no. ha^{-1} [48]. Despite the fact that natural recruitment should be also considered as a key factor in mangrove restoration [57,61], it is perhaps difficult to apply of the ongoing MMFR management. Hence, most of the propagules, seedlings and young trees naturally recruited or regenerated at MF30 are lost during the clearing operation [26]. However, post-clearing habitat recovery via natural recruitment (propagule stranding) can still be a pre-planting option.

Next to *R. apiculata*, *B. gymnorrhiza* was abundant and had a chance to grow as co-dominant species in the vicinity (Table 2). However, its young tree density which decreased from 45 to 9% between 15 and 30-year old blocks was due to its removal during the first and second thinnings. On the other hand, *E. agallocha* never represented more than 6.5% of the total density and thus did not constitute an effective competitor for *R. apiculata*, which is not surprising as their optimal elevations within the intertidal zone are different [16,54]. The lower density of juvenile (113 nos. ha^{-1}) and young (276 stems ha^{-1}) vegetation at MF15 has made this block distinct from other blocks (Fig. 3A–B). Overall, the juvenile/young vegetation data analysis based on *R. apiculata* as well as other mangrove species (i.e. *A. officinalis*, *B. cylindrica*, *B. gymnorrhiza*, *B. parviflora*, *C. tagal*, *E. agallocha*, *R. mucronata* and *S. alba*) was found to be important for testing how ecologically close the VJR and the MF blocks in terms of their species composition/abundances. The present observations are however in contrast to Fontalvo-Herazo et al. [33] whose model did not allow natural recruitment or self-thinning to happen in both MF15 and MF20 stands.

Biomass increment

Being the richest carbon pool in the world [62], mangrove forests are important for climate change mitigation via Reduced Emissions from Deforestation and Degradation (REDD+) [63]. However, in view of the long-lasting man-mangrove linkage on one hand and the increased population/over exploitation of the resources on the other, the sustainable management of these forests should be a priority [2,8,11,17].

Gong and Ong [28] compared the biomass of a virgin mangrove stand with the second generation yields of 1967–69 and 1970–77 in the MMFR. While the difference of 163 t ha^{-1} between virgin and managed forest blocks was justified because of the 50–70 years older trees at VJR, they also reported a decline of 22 t ha^{-1} in between the two i.e., 1967–69 and 1970–77 managed blocks. In order to increase the biomass, Gong and Ong (op. cit.) have suggested implementing an earlier silvicultural thinning (for the removal of non *Rhizophora* trees) at the age of 8–9 years, followed by the first thinning at 12–13 years and the second thinning at 17–18 years. In addition, they proposed a 25-year rotation (instead of 30 years) in view of the no marked increase in stem diameter/biomass after 18 years. Although Fontalvo-Herazo et al. [33] have used the same observations from Gong and Ong (op. cit.), they found a total gain of 22.95 t ha^{-1} by following the 7-13-20 year old thinning practices in 30 years rotation. According to Fontalvo-Herazo et al. (op. cit.), the 7-13-20 thinning practice would be able to provide best profit possibility for medium sized poles (D_{130}: >10 to <13 cm) at those respective managed mangrove forest blocks.

However, the present estimates on tree density, stem diameter, and harvestable biomass are different from the above two investigations. In fact, the increased biomass from 216 t ha^{-1} at MF15 to 317 t ha^{-1} at MF30 (at the rate of 15.5 t ha^{-1} $year^{-1}$) (Figs. 2B-ii-b to 2B-iv-b) coincides well with the earlier estimates [44], and also suggests that the mangrove management based on a 30-year rotation cycle is still appropriate for the MMFR. Moreover, the change in 30 years rotation period was said to be difficult due to commitment of the Forestry Department, mind-set of the pole/charcoal contractors, etc. [35]. As mentioned under the 'initial stocking' and the 'stand density' sections, the earlier thinning before 15 years is necessary, but the optimal frequency and age of thinning(s) remains uncertain due to no consecutive (field-based) tree structural measurements. The lower rate of biomass increment (0.2 t ha^{-1} $year^{-1}$) between MF15 and MF20, which could be due to limited time interval for the trees (having 73% MST) to grow better after the first artificial thinning, is further adjudicating the necessity of earlier thinning before 15 years. In this context, the increased space of artificial regeneration (see 'initial stocking'), which ultimately leads to decreased stocking density of the seedlings, might even reduce or compensate the loss of trees at MF15. Therefore a practical validation of the recommendations is warranted.

The mangrove biomass, which varies with age, species and location, is usually higher in the tropics than in temperate areas [64]. When the present biomass estimates are compared with the published literature on natural and replanted *Rhizophora* sites (Table 3), the Matang is contributing a highest biomass for which the ongoing management with a sustainable rejuvenation should also be acknowledged. The trunk biomass, which is principally used for charcoal industry, followed the same trend as that of total above-ground biomass (Figs. 2B-ii-b to 2B-iv-b). In view of the existing differences in stem size distribution, the harvestable biomass was also different between the field- and the simulation-based studies (Fig. 4C).

Century-long management as a global reference

For efficient conservation and management of the mangrove forests, the cooperation from local people is necessary [2,65,66,67]. Besides the community-based mangrove restoration projects, the locals can be part of carbon financing schemes, eco-tourism development, sustainable sale of commercially valuable timber and non-timber products, etc. [67,68]. In this context, an appropriate sharing of 'Revenue, Rights, Responsibility and Relationship' (4Rs) among the people/stakeholders - whoever interested in that particular mangrove ecosystem, would be able to deliver long-term benefits [8]. The chief objective of the State Forestry Department of Perak is to maximize wood production (for pole and charcoal, both for local use and export) as well as to improve the (local) people's quality of life in the vicinity of Matang [21,34,35,48]. There are as many as 214 mangrove pole and charcoal contractors [35], and every year the MMFR authorities will allocate a minimum 2.2 ha of the productive forest area to each contractor to see their pole and/or charcoal trading are met regularly [35,69]. This activity is also benefiting several workers under the each contactor for their livelihood.

In light of the necessity to develop alternative and sustainable energy technologies for personal use and power generation units, the mangrove charcoal was found to be a source of renewable energy [70]. Though it is produced in Thailand, Vietnam and on the east coast of Sumatra of Indonesia, Matang is a unique place where the local people became guardians of the forest for over 100 years and it is also the only example to other mangrove countries for learning/practicing long-term sustainable forest resource utilization [66]. Despite the fact that forest product utilization (in relation to species' availability and local knowledge) varies between the countries [71,72], the areas with higher human intervention (especially in South and Southeast Asia and African regions) may want to adapt similar guidelines and manage the remaining mangrove forests (Fig. 5). However, the silvicultural management would be more efficient if the given recommendations (with reference to initial stocking, stand density and natural recruitment) are considered by all parties. From the ecological point of view, the VJR and other functional forest categories (i.e. old growth, eco-tourism, education and research) in the vicinity are serving as a mangrove seed bank for Matang as well as other locations in Peninsular Malaysia. The 'protective', 'productive' and 'restrictive productive' forest zones within the reserve are also delivering several eco-socio-economic benefits which are sustainable (Fig. 5).

We also recall that the main emphasis of the MMFR lies on a steady-state resource management of *Rhizophora apiculata* charcoal, which from the ecological perspective is probably less functional and less resilient than a mangrove forest under a

Table 3. A comparative account on the above ground biomass of *Rhizophora* spp. from representative mangrove sites.

	Forest type	Age (years)	Species	Stem diameter (cm)	Mean height (m)	Biomass (t ha^{-1})	Source
Matang, Malaysia	Planted	15	*R. apiculata*	5.0–12.5	12.8	216	Present study
Matang, Malaysia	Planted	20	*R. apiculata*	5.0–15.0	13.3	217	Present study
Matang, Malaysia	Planted	30	*R. apiculata*	15.0–22.5	14.8	372	Present study
Halmahera, Indonesia	Primary	-	*R. apiculata*	-	21.2	357	[74]
Ranong, Thailand	Primary	-	*Rhizophora* spp.	-	-	299	[75]
Yingluo Bay, China	Natural	Oldest	*R. stylosa*	7.4	3.4	271	[76]
Andaman, India	Primary	-	*Rhizophora* spp.	-	22.5	124	[77]
Sri Lanka	Fringe	-	*Rhizophora* spp.	-	7.2	240	[78]
Gazi Bay, Kenya	Planted	5	*R. mucronata*	>5	3.9	20	[79]
Gazi Bay, Kenya	Planted	12	*R. mucronata*	2.5–12.4	8.5	107	[58]
Florida, America	Dwarf	-	*R. mangle*	>2	1.1	12.5	[80]

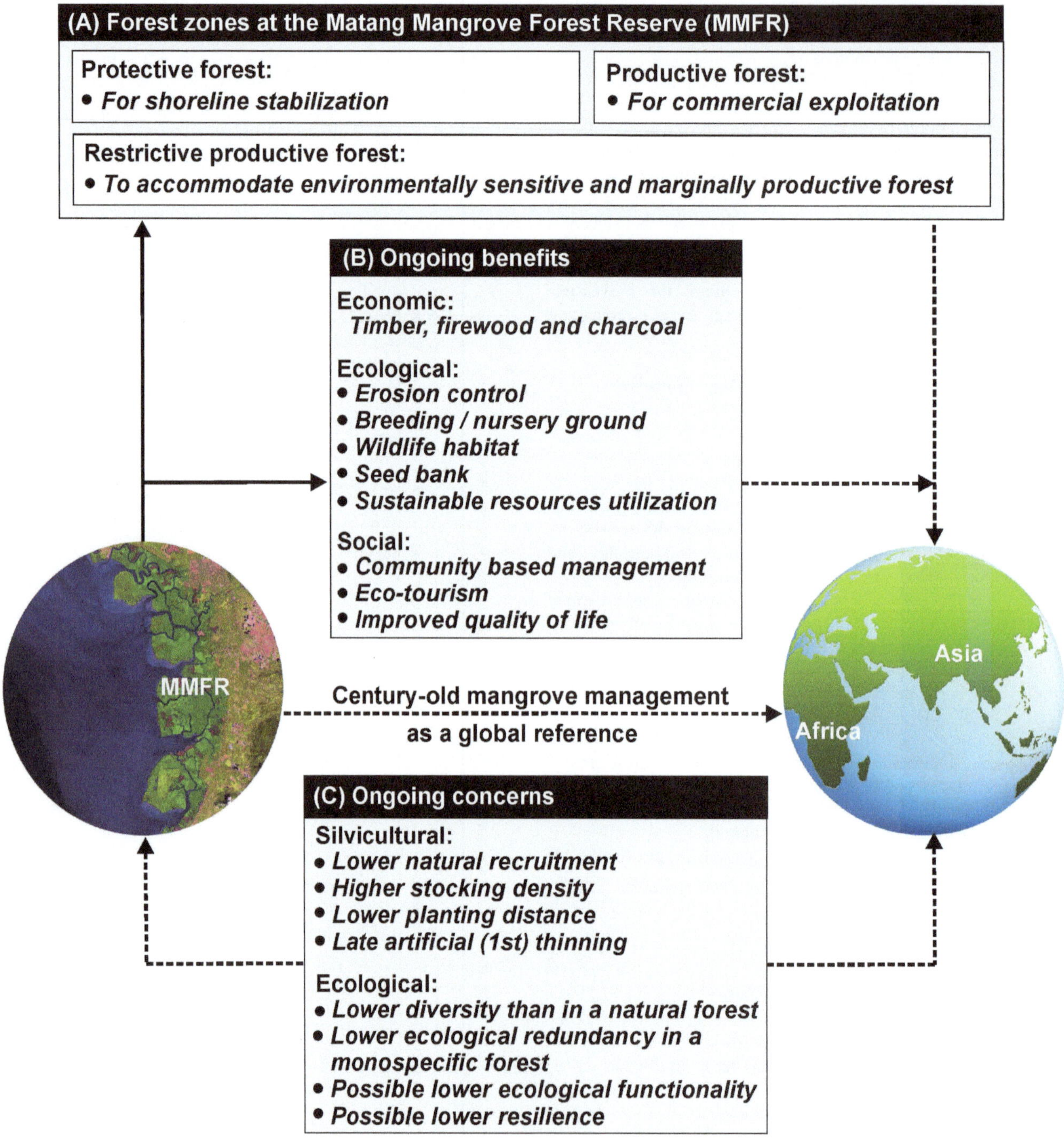

Figure 5. Schematic chart showing the century-old mangrove management at the Matang Mangrove Forest Reserve (MMFR) as a global reference for sustainable silviculture. While the bold-line arrows indicate the features available for Matang (A–B), the dotted-line arrows show the features that could be considered by other mangrove locations for their improved/sustainable mangrove management. Some of the ongoing silvicultural and ecological concerns (C) represented by dotted-arrows, are applicable to the both MMFR and other mangrove locations elsewhere.

resilience-based ecosystem stewardship [59]. For instance, when managing for a single species there is hardly any ecological redundancy in what is naturally already a species-poor ecosystem. Any severe perturbation on a mangrove ecosystem only composed of *R. apiculata* is therefore unlikely to be absorbed by a congeneric species such as *R. mucronata* or even a family representative like *Bruguiera* spp. Although the highly dynamic nature of mangroves generally observed elsewhere [16,73] may provide an input of propagules from a variety of species, the current management procedure indicates that this is insufficiently the case for MMFR (and also not aimed at). A management aiming at fostering variability within the ecosystem in the light of human-induced and climatic uncertainty and change is therefore lacking. World-wide many exploited and non-exploited mangrove forests suffer severe degradation due to over-exploitation, land conversion, pollution, etc. What MMFR may be exemplary at, is at demonstrating that exploitation does not need to result in degradation, or at least not within a century.

Conclusions and future perspectives

The present study provided invaluable information for a better conservation and management policy at the MMFR. The results strongly support earlier thinning at the MF15 to reduce the loss of mangrove trees exploitable as fuelwood, both subsistence-based and commercial. At the same time, the reduced initial (stocking) density of the seedlings might be able to compensate additional loss of trees at MF15. Is Matang mangrove forest in Malaysia sustainably rejuvenating after more than a century of conservation and harvesting management? In general, the 30-year rotation cycle with a higher yield of biomass is indeed expected to be sustainable as a silvicultural practice. However, monitoring of wood extraction volumes through socio-ecologic surveys of local cutters should be able to confirm or reject whether or not this expectation is met.

Since the mangrove management plan for 2010–2019 is officially declared, the MMFR authorities may focus on the issues highlighted in this paper (along with their identified research needs), and incorporate necessary changes in the next working plan (i.e. 2020–2029) at least. Meanwhile, the suggestions given for increasing the space of (*R. apiculata*) afforestation as well as time for early thinning(s) could be tested practically by allocating few experimental sites at the MMFR. After witnessing the differences between stem- and tree-based counts, the density calculation/expression should be made appropriately. Despite the fact that MMFR is still an excellent example for sustainable (forest) resources use and wood production in the world, its ecological status and ecosystem services are also in need of urgent and regular monitoring (under appropriate scientific direction), as the functionality in comparison to a natural mangrove may be suboptimal as suggested by the lower diversity in the managed blocks as opposed to VJR. Therefore the term 'sustainably' in our title is to be interpreted in a context of silviculture, not necessarily in a context of ecological functionality. A possible solution to maintain an eased recolonization through hydrochory, is to assure a healthy mix between managed blocks (and within management block patches in varied successional stages) and unmanaged blocks. In this way the monospecific managed blocks could bath in a matrix of unmanaged blocks that contain the full local diversity and that can function as a reservoir to renew the adjacent managed blocks after perturbation. This approach may in fact also increase the likelihood of having clear-cut blocks colonised naturally.

The present study/assessment based on silvimetric parameters is of great help not only to the local Forestry Department to manage the resources wisely, but also to the mangrove managers elsewhere who consider Matang as an exemplary for their improved management aiming at silviculture. In addition, the present results are useful to the researchers using mangrove modelling software (e.g. KiWi, Mangal, Forman, NetLogo) for a precise parameterization. The differences encountered between field- and simulation-based studies are due to two different working platforms, limited data for a precise parameterization in mangrove models, different stocking density and reduction percentages. Perhaps a follow-up study to Fontalvo-Herazo et al. [33], integrating present silvimetric observations, would be able to revalidate as well as offer better management options to the MMFR. Meanwhile, our recent findings on the mangrove charcoal production/trade at Matang [69] are also unveiling an eco-socio-economic model for its sustainable long-term management in which wood extraction is matched to the forest's wood productivity, and to the socio-economical sustainability of the mangrove wood trade system in the light of uncertainty and change [Quispe Zuniga, Satyanarayana, Grüters, Berger, Mohd-Lokman, Sulong & Dahdouh-Guebas, unpublished data].

Acknowledgments

Authors are very grateful to the administrative authorities at ULB, Interuniversitary Council of the Flemish Community of Belgium (VLIR-UOS), the Belgian National Science Foundation (FNRS and FWO), the EC-funded Marie-Curie International Research Staff Exchange Scheme entitled 'Coastal Research Network on Environmental Changes - CREC' (Grant Agreement n° 247514), Universiti Malaysia Terengganu (UMT), and the State Forestry Department of Perak, for their kind cooperation to complete this study successfully. Special thanks are due to Mrs. Martha Liliana Fontalvo-Herazo for supplying the simulation-based results of tree density, diameter and biomass at different managed forest blocks in the Matang Mangrove Forest Reserve. This paper in part was presented at the 6-yearly international conference entitled 'Meeting on Mangrove ecology, functioning and Management (MMM3)' held in Sri Lanka (2–6 July 2012). We do appreciate the objective criticism and valuable suggestions offered by the anonymous referees.

Author Contributions

Conceived and designed the experiments: FDG BS AG. Performed the experiments: AG. Analyzed the data: BS AG. Contributed reagents/materials/analysis tools: FDG BS HML IS. Wrote the paper: BS FDG AG. Supervised the research: FDG BS HML IS. Provided background data: TVdS MQZ. Provided logistical support: HML IS. Provided substantial revisions: BS FDG AG TVdS MQZ HML IS.

References

1. Ellison AM (2008) Managing mangroves with benthic biodiversity in mind: moving beyond roving banditry. J Sea Res 59: 2–15.
2. Walters BB, Rönnbäck P, Kovacs JM, Crona B, Hussain SA, et al. (2008) Ethnobiology, socio-economics and management of mangrove forests: a review. Aquat Bot 89: 220–236.
3. Nagelkerken I, Blaber SJM, Bouillon S, Green P, Haywood M, et al. (2008) The habitat function of mangroves for terrestrial and marine fauna: a review. Aquat Bot 89: 155–185.
4. Cannicci S, Bartolini F, Dahdouh-Guebas F, Fratini S, Litulo C, et al. (2009) Effects of urban wastewater on crab and mollusc assemblages in equatorial and subtropical mangroves of East Africa. Estuar Coast Shelf Sci 84: 305–317.
5. Feagin RA, Mukherjee N, Shanker K, Baird AH, Cinner J, et al. (2010) Shelter from the storm? Use and misuse of coastal vegetation bioshields for managing natural disasters. Conservation Letters 3: 1–11.
6. Bayas JCL, Marohn C, Dercon G, Dewi S, Piepho HP, et al. (2011) Influence of coastal vegetation on the 2004 tsunami wave impact in west Aceh. Proc Natl Acad Sci USA 108: 18612–18617.
7. Cunha-Lignon M, Coelho C Jr, Almeida R, Menghini RP, Schaeffer-Novelli Y, et al. (2011a) Characterisation of mangrove forest types in view of conservation and management: a review of mangals at the Cananéia region, São Paulo State, Brazil. J Coast Res 64: 349–353.
8. Satyanarayana B, Bhanderi P, Debry M, Maniatis D, Foré F, et al. (2012) A socio-ecological assessment aiming at improved forest resource management and sustainable ecotourism development in the mangroves of Tanbi Wetland National Park, the Gambia, West Africa. Ambio 41: 513–526.
9. Zhang K, Liu H, Li Y, Xu H, Shen J, et al. (2012) The role of mangroves in attenuating storm surges. Estuar Coast Shelf Sci 102–103: 11–23.
10. Das S, Crépin A-S (2013) Mangroves can provide protection against wind damage during storms. Estuar Coast Shelf Sci 134: 98–107.
11. Duke NC, Meynecke J-O, Dittmann S, Ellison AM, Anger K, et al. (2007) A world without mangroves? Science 317: 41–42.
12. Alongi DM (2008) Mangrove forests: resilience, protection from tsunamis, and responses to global climate change. Estuar Coast Shelf Sci 76: 1–13.

13. Bosire JO, Dahdouh-Guebas F, Walton M, Crona BI, Lewis III RR, et al. (2008) Functionality of restored mangroves: a review. Aquat Bot 89: 251–259.
14. Nfotabong Atheull A, Din N, Essomè Koum LG, Satyanarayana B, Koedam N, et al. (2011) Assessing forest products usage and local residents' perception of environmental changes in peri-urban and rural mangroves of Cameroon, Central Africa. J Ethnobiol Ethnomed 7: 41.
15. Nfotabong Atheull A, Din N, Dahdouh-Guebas F (2013) Qualitative and quantitatice characterization of mangrove vegetation structure and dynamics in a peri-urban settings of Douala (Cameroon): An approach using airborn imagery. Estuaries Coast 36: 1181–1192.
16. Satyanarayana B, Koedam N, De Smet K, Di Nitto D, Bauwens M, et al. (2011) Long-term mangrove forest development in Sri Lanka: early predictions evaluated against outcomes using VHR remote sensing and VHR ground-truth data. Mar Ecol Prog Ser 443: 51–63.
17. Satyanarayana B, Mulder S, Jayatissa LP, Dahdouh-Guebas F (2013) Are the mangroves in the Galle-Unawatuna area (Sri Lanka) at risk? A social-ecological approach involving local stakeholders for a better conservation policy. Ocean Coast Manag 71: 225–237.
18. Lee SY, Primavera JH, Dahdouh-Guebas F, McKee K, Bosire JO, et al. (2014) Ecological role and services of tropical mangrove ecosystems: A reassessment. Global Ecol Biogeogr 23: 726–743.
19. Mukherjee N, Sutherland WJ, Khan MNI, Berger U, Schmitz N, et al. (2014) Using expert knowledge and modelling to define mangrove composition, functioning and threats and estimate time-frame for recovery. Ecology and Evolution (in press).
20. Okamura K, Tanaka K, Siow R, Man A, Kodama M, et al. (2010) Spring tide hypoxia with relation to chemical properties of the sediments in the Matang mangrove estuary, Malaysia. Jpn Agric Res Q 44: 325–333.
21. Azahar M, Nik Mohd Shah NM (2003) A Working Plan for the Matang Mangrove Forest Reserve, Perak: the third 10- year period (2000–2009) of the second rotation (5th revision). Malaysia: State Forestry Department of Perak. 319 p.
22. Ong JE (1993) Mangroves – a carbon source and sink. Chemosphere 27: 1097–1107.
23. Amir AA (2012) Canopy gaps and the natural regeneration of Matang mangroves. For Ecol Manage 269: 60–67.
24. Ashton EC, Hogarth PJ, Ormond R (1999) Breakdown of mangrove leaf litter in a managed mangrove forest in Peninsular Malaysia. Hydrobiologia 413: 77–88.
25. Cunha-Lignon M, Kampel M, Menghini RP, Schaeffer-Novelli Y, Cintrón G, et al. (2011) Study of mangrove forests submitted to depositional processes and salinity variation, using satellite images and vegetation structure surveys. J Coast Res 64: 344–348.
26. Gong WK, Ong JE (1995) The use of demographic studies in mangrove silviculture. Hydrobiologia 295: 255–261.
27. Putz FE, Chan HT (1986) Tree growth, dynamics, and productivity in a mature mangrove forest in Malaysia. For Ecol Manage 17: 211–230.
28. Gong WK, Ong JE (1990) Plant biomass and nutrient flux in a managed mangrove forest in Malaysia. Estuar Coast Shelf Sci 31: 519–530.
29. Ong JE, Gong WK, Wong CH (2004) Allometry and partitioning of the mangrove, *Rhizophora apiculata*. For Ecol Manage 188: 395–408.
30. Alongi DM, Sasekumar A, Tirendi F, Dixon P (1998) The influence of stand age on benthic decomposition and recycling of organic matter in managed mangrove forests of Malaysia. J Exp Mar Biol Ecol 225: 197–218.
31. Alongi DM, Chong VC, Dixon P, Sasekumar A, Tirendi F (2003) The influence of fish cage aquaculture on pelagic carbon flow and water chemistry in tidally dominated mangrove estuaries of peninsular Malaysia. Mar Environ Res 55: 313–333.
32. Alongi DM, Sasekumar A, Chong VC, Pfitzner J, Trott LA, et al. (2004) Sediment accumulation and organic material flux in a managed mangrove ecosystem: estimates of land–ocean–atmosphere exchange in peninsular Malaysia. Mar Geol 208: 383–402.
33. Fontalvo-Herazo ML, Piou C, Vogt J, Saint-Paul U, Berger U (2011) Simulating harvesting scenarios towards the sustainable use of mangrove forest plantations. Wetlands Ecol Manage 19: 397–407.
34. Chong VC (2006) Sustainable utilization and management of Mangrove ecosystems of Malaysia. Aquat Ecosyst Health Manag 9: 249–260.
35. Roslan A, Nik Mohd Shah NM (2014) A Working Plan for the Matang Mangrove Forest Reserve, Perak: the first 10- year period (2010–2019) of the third rotation (6th revision). Malaysia: State Forestry Department of Perak. 229 p.
36. JUPEM (2004) A hand book on Tide Tables. Malaysia: Department of Survey and Mapping.
37. Dahdouh-Guebas F, Koedam N (2006) Empirical estimate of the reliability of the use of the Point-Centred Quarter Method (PCQM): solutions to ambiguous field situations and description of the PCQM+ protocol. For Ecol Manage 228: 1–18.
38. Kairo JG, Dahdouh-Guebas F, Gwada PO, Ochieng C, Koedam N (2002) Regeneration status of mangrove forests in Mida creek, Kenya: a compromised or secured future? Ambio 31: 7–8.
39. Tomlinson PB (1986) The Botany of Mangroves. Cambridge: Cambridge University Press. 419 p.
40. Curtis JT (1959) The Vegetation of Wisconsin: An Ordination of Plant Communities. Wisconsin: University of Wisconsin Press. 704 p.
41. Cox GW (1996) Laboratory Manual of General Ecology (7th edition). Boston: Wm. C Brown Publishers. 278 p.
42. Kathiresan K, Bingham BL (2001) Biology of mangrove and mangrove ecosystems. Adv Mar Biol 40: 81–251.
43. Odum EP, Barett GW (2005) Fundamentals of Ecology (5th edition). Singapore: Brooks/Cole. 598 p.
44. Ong JE, Gong WK, Wong CH (1985) Seven years of productivity studies in a Malaysian managed mangrove forest then what? In: Bardsley KN, Davie JDS, Woodroffe CD, editors. Coasts and Tidal wetlands of the Australian monsoon region. Australia: Australian National University. pp. 213–223.
45. Anderson MJ, Gorley RN, Clark KR (2008) Permanova+ for Primer: Guide to Software and Statistical Methods. United Kingdom: Primer–E Ltd, Plymouth. 214 p.
46. Gates GM, Mohammed C, Wardlaw T, Davidson NJ, Ratkowsky DA (2011) Diversity and phenology of the macrofungal assemblages supported by litter in a tall, wet Eucalyptus obliqua forest in southern Tasmania, Australia. Fungal Ecology 4: 68–75.
47. Sangil C, Sansón M, Clemente S, Afonso-Carrillo J, Hernández JC (2014) Contrasting the species abundance, species density and diversity of seaweed assemblages in alternative states: Urchin density as a driver of biotic homogenization. J Sea Res 85: 92–103.
48. Haron HA (1981) A working plan for the second 30 years rotation of the Matang Mangrove Forest Reserve, Perak: the first 10-year period 1980–1989. Malaysia: State Forestry Department. 109 p.
49. Di Nitto D, Dahdouh-Guebas F, Kairo JG, Decleir H, Koedam N (2008) Digital terrain modelling to investigate the effects of sea level rise on mangrove propagule establishment. Mar Ecol Prog Ser 356: 175–188.
50. De Ryck DJR, Robert EMR, Schmitz N, Van der Stocken T, Di Nitto D, et al. (2012) Size does matter, but not only size: Two alternative dispersal strategies for viviparous mangrove propagules. Aquat Bot 103: 66–73.
51. Van der Stocken T, De Ryck DJR, Balke T, Bouma TJ, Dahdouh-Guebas F, et al. (2013). The role of wind in hydrochorous mangrove propagule dispersal. Biogeosciences 10: 3635–3647.
52. Bosire JO, Kazungu J, Koedam N, Dahdouh-Guebas F (2005) Predation on propagules regulates regeneration in a high-density reforested mangrove plantation. Mar Ecol Prog Ser 299: 149–155.
53. Cannicci S, Burrows D, Fratini S, Smith TJ III, Offenberg J, et al. (2008) Faunistic impact on vegetation structure and ecosystem function in mangrove forests: a review. Aquat Bot 89: 186–200.
54. Dahdouh-Guebas F, Koedam N, Satyanarayana B, Cannicci S (2011) Human hydrographical changes interact with propagule predation behaviour in Sri Lankan mangrove forests. J Exp Mar Biol Ecol 399: 188–200.
55. Krauss KW, Lovelock CE, McKee KL, López-Hoffman L, Ewe SML, et al. (2008) Environmental drivers in mangrove establishment and early development: a review. Aquat Bot 89: 105–127.
56. Bosire JO, Dahdouh-Guebas F, Kairo JG, Koedam N (2003) Colonization of non-planted mangrove species into restored mangrove stands in Gazi Bay, Kenya. Aquat Bot 76: 267–279.
57. Bosire JO, Dahdouh-Guebas F, Kairo JG, Wartel S, Kazungu J, et al. (2006) Success rates of recruited tree species and their contribution to the structural development of reforested mangrove stands. Mar Ecol Prog Ser 325: 85–91.
58. Kairo JG, Lang'at JKS, Dahdouh-Guebas F, Bosire JO, Karachi M (2008) Structural development and productivity of replanted mangrove plantations in Kenya. For Ecol Manage 255: 2670–2677.
59. Chapin SF III, Kofinas GP, Folke C, Chapin MC (2009) Principles of ecosystem stewardship: resilience-based natural resource management in a changing world. The Netherlands: Springer Science, Dordrecht. 402 p.
60. Alfaro AC (2006) Benthic macro-invertebrate community composition within a mangrove/seagrass estuary in northern New Zealand. Estuar Coast Shelf Sci 66: 97–110.
61. Aung TT, Than MM, Katsuhiro O, Yukira M (2011) Assessing the status of three mangrove species restored by the local community in the cyclone-affected area of the Ayeyarwady Delta, Myanmar. Wetl Ecol Manage 19: 195–208.
62. Donato DC, Boone Kauffman J, Murdiyarso D, Kurnianto S, Stidham M, et al. (2011) Mangroves among the most carbon-rich forests in the tropics. Nat Geosci 4: 293–297.
63. Cohen R, Kain J, Okello JA, Bosire JO, Kairo JG, et al. (2013) Propagating uncertainty to estimates of above-ground biomass for Kenyan mangroves: A scaling procedure from tree to landscape level. For Ecol Manage 310: 968–982.
64. Komiyama A, Ong JE, Poungparn S (2008) Allometry, biomass, and productivity of mangrove forests: A review. Aquat Bot 89: 128–137.
65. FAO (2007) The world's mangroves 1980–2005. Rome: Forestry Paper 153, FAO.
66. Alongi DM (2002) Present state and future of the world's mangrove forests. Environ Conserv 29: 331–349.
67. UNEP (2014) Community involvment. In: van Bochove J, Sullivan E, Nakamura T, editors. The importance of mangroves to people - a call to action. UNEP (in press).
68. Kairo JG, Wanjiru C, Ochiewo J (2009) Net pay: economic analysis of a replanted mangrove plantation in Kenya. Journal of Sustainable Forestry 28:395–414.
69. Quispe Zuniga M, Satyanarayana B, Piou C, Berger U, Mohd-Lokman H, et al. (2014) Building an ecologic-econopmic model to improve silvicultural management practices in Matang (Malaysia). VLIZ Special publication 67: 87–88.

70. Ahmed I, Jangsawang W, Gupta AK (2012) Energy recovery from pyrolysis and gasification of mangrove. Applied Energy 91: 173–179.
71. Choudhury JK (1997) Sustainable management of coastal mangrove forest development and social needs. Proceedings of the XI World Forestry Congress 6: 266–286.
72. Ellison J (2000) How South Pacific mangroves may respond to predicted climate change and sea level rise. In: Gillespie A, Burns W, eidtors. Climate change in the South Pacific: Impacts and responses in Australia, New Zealand, and small islands states. The Netherlands: Kluwer Academic Publishers. pp. 289–301.
73. Dahdouh-Guebas F, Koedam N (2002) A synthesis of existent and potential mangrove vegetation structure dynamics from Kenyan, Sri Lankan and Mauritanian case-studies. Meded. Zitt. K. Acad. overzeese Wet./Bull. Séanc. Acad. r. Sci. Outre-Mer 48: 487–511.
74. Komiyama A, Moriya H, Prawiroatmodjo S, Toma T, Ogino K (1988) Forest primary productivity. In: Ogino K, Chihara M, editors. Biological System of Mangrove. Japan: Ehime University. pp. 97–117.
75. Komiyama A, Ogino K, Aksornkoae S, Sabhasri S (1987) Root biomass of a mangrove forest in southern Thailand. 1. Estimation by the trench method and the zonal structure of root biomass. J Trop Ecol 3: 97–108.
76. Wang G, Guan D, Peart MR, Chen Y, Peng Y (2013) Ecosystem carbon stocks of mangrove forest in Yingluo Bay, Guangdong Province of South China. For Ecol Manage 310: 539–546.
77. Mall LP, Singh VP, Garge A (1991) Study of biomass, litter fall, litter decomposition and soil respiration in monogeneric mangrove and mixed mangrove forests of Andaman Islands. Trop Ecol 32: 144–152.
78. Amarasinghe MD, Balasubramaniam S (1992) Net primary productivity of two mangrove forest stands on the northwest coast of Sri Lanka. Hydrobiol 247: 37–47.
79. Kairo JG, Bosire JO, Langat J, Kirui B, Koedam N (2009) Allometry and biomass distribution in replanted mangrove plantations at Gazi Bay, Kenya. Aquatic Conserv: Mar Freshw Ecosyst 19: S63–S69.
80. Coronado-Molina C, Day JW, Reyes E, Prez BC (2004) Standing crop and aboveground partitioning of a dwarf mangrove forest in Taylor River Slough, Florida. Wetlands Ecol Manage 12: 157–164.

14

Variation in Indigenous Forest Resource Use in Central Guyana

Claire M. P. Ozanne[1]*, Christie Cabral[2], Peter J. Shaw[1]

1 Centre for Research in Ecology, Department of Life Sciences, University of Roehampton, London, United Kingdom, **2** School for Social & Community Medicine, University of Bristol, Bristol, United Kingdom

Abstract

Sustainable forest conservation strategies should be based on local as well as landscape-scale forest resource use data. Using ecological and sociological techniques, we test the hypotheses that (1) forest resource use differs between ethnic and socioeconomic indigenous groups and (2) that this difference results in differing spatial patterns of resource use, with implications for forest diversity and for conservation planning. In the North Rupununi Guyana, three adjacent indigenous communities (differing in their indigenous/immigrant balance) were recorded using 73 animal and 164 plant species (plus several unidentified ethno-species). Farm sites formed important foci for most forest based activities and ex-farm sites supported similar floristic diversity to surrounding forest. Resource usage differences between communities could be attributed to socio-cultural drivers, e.g. mammal meat consumption and the use of the fruits from the palm tree *A. maripa* were higher in more traditional households. When extracting household construction timber, lower income groups created small scattered felling sites akin to tree fall gaps whereas higher income groups created larger gaps. Lower income (indigenous) households tended to clear larger but more contained sites for farming while mixed or non-Amerindian household tended to clear smaller but more widely dispersed farm sites. These variations resulted in different patterns of forest disturbance originating from agriculture and timber extraction.

Editor: John P. Hart, New York State Museum, United States of America

Funding: This work was supported by the University of Roehampton PhD Bursary Botanical Research Fund. The funders had no role in study design, data collection and analysis, decision to publish, or preparation of the manuscript.

Competing Interests: The authors have declared that no competing interests exist.

* Email: c.ozanne@roehampton.ac.uk

Introduction

Global strategies to reduce the rate of forest loss include setting up Protected Areas (PAs) of various types (including those which prohibit exploitation and those which allow it), the designation of Indigenous Reserves (IRs) and of Community Managed Forests. Recent evaluation of the effectiveness of IRs in the Amazon indicates that they play a significant role in reducing the number of deforestation fires [1]. In the Colombian Guyana Shield both Protected Areas and Indigenous Reserves act as buffers to deforestation [2]), particularly where large scale agriculture is a key agent of change [3] and recent global analyses of CMFs indicate they can be effective at reducing deforestation in the tropics [4]. Whilst gazetting land clearly has a positive impact overall [4] there is considerable variation in the effectiveness at a local level. Differences in the degree of protection afforded by the designated areas are dependent on factors such as road proximity and pattern and practices of colonisation [5], with density of habitation and approaches to management and resource use being key [1,2,3,6]. Much tropical forest lies outside protected areas and there remains a lack of information about anthropogenic effects on biodiversity patterns in these areas which is needed to inform regional conservation plans [8]). Although higher order analysis of land use change is of importance in the development of conservation strategies and in predicting impact of climate change, if models are to respond usefully to the complexity of this process, local "place based" information is required [9,10,11]. Such inputs should encompass both historically traditional and emergent patterns of management and resource use and must involve data collection at both ecologically and sociologically relevant scales.

Research in the Brazilian Amazon [12], Mexico [13], South Korea [14] and Tanzania [10] has documented the interdependence of indigenous people and forest habitat, as well as the influence of this interaction on the spatial arrangement of a wide range of timber and non-timber forest (NTFPs) products and resources. Recent studies have focussed on spatial patterns of hunting [7,15] and the effects on vertebrate conservation work. It is apparent from these studies that a good understanding of the effects of resource use requires disentangling the differing patterns across particular ethnic and socioeconomic groups within communities [6].

The Guiana shield, encompassing parts of northern Brazil, French Guiana, Suriname, Venezuela and eastern Colombia, supports the largest unfragmented extent of 'frontier forest' in the world [16]. In Guyana approximately 80 percent of the land remains forested, with high levels of endemism [17,18]. Of this forest, approximately 80 percent is categorised as State owned and 13.9 percent held by the indigenous Amerindian communities [19]. The main threats to the forest are logging, gold mining and the Georgetown to Lethem road. However, the deforestation rate remains relatively low, standing at approximately 0.3 percent per annum (543 ha/annum) [20] (0.29% in 2009) [21].

Local communities are integral to the protection of these areas as well as those owned by the state and the forest that does not lie within the protected zones and yet many remain dependent on forest resources for their livelihoods.

The objectives of this study were to test the hypotheses that: (1) forest resource use (animal and plant) varies between ethnic and socioeconomic indigenous groups and (2) that this variation results in differing spatial patterns of resources use.

Study area

Our study was conducted in central Guyana (4° 20′N, 59°10′W, see Figure 1) where the North Rupununi Savanah meets the southern border of the Iwokrama Wilderness Reserve. We worked with three Amerindian communities: Fair View (4°39′ N; 58° 41′ W), the only village located within the Iwokrama Reserve, and Surama (4° 8′ N; 59° 4′ W) and Wowetta (4° 1′ N; 59° 4′ W), two savannah based villages within the Annai Amerindian District which were closest to the Iwokrama Reserve boundary (Figure 2).

The study area is roughly encompassed by the Essequibo, Rupununi and Burroburro rivers [22]. There are two distinct rainy seasons: a long one from May to mid-August (which creates seasonal swamps; [23] and a short one from November to January. In the north, Fair View is surrounded by mixed forest with some nearby patches of 'muri-scrub', a degraded form of a fire prone forest type that occurs when fires are very regular [24,25]. Wowetta is just within the northern boundary of the savannah, with mixed forest all around, while the Surama is in an area touched by several types of mixed forest with a very small area of savannah.

Materials and Methods

Informed consent was obtained from all participants. Where possible written consent was obtained; however, many of the participants were not sufficiently literate to be able to provide written consent, in these cases verbal consent was obtained and recorded by the researcher. All procedures were approved by the University of Roehampton Committee on Ethics.

Permits to carry out this work were obtained from the Ministry of Amerindian Affairs and from the Environmental Protection Agency in Guyana. All necessary permits were obtained for the described study, which complied with all relevant regulations.

Participants

The majority of Amerindians living in the North Rupununi belong to the Makushi tribe [26]. At the time of our research, 94 percent of the population of the study communities were Amerindian, mainly of the Makushi tribe (62%), with some from the Arawak (3%) and Wapishana (7%) tribes, and some mixed-tribe households (22%). The residents of Surama and Wowetta were both predominantly Makushi, while in Fair View these three tribal groups were fairly equally represented and there was a higher proportion of mixed-tribe households. A further 5 percent were of mixed-ethnicity (one non-Amerindian parent) and there were two (<1%) residents of non-Amerindian ethnicity both living with Amerindian partners (one in Surama and one in Fair View). We grouped all 84 households (defined as a group of people living in a discrete compound who communally gather, manage and use forest resources) into three classes: Makushi in which both household heads (male and female) were Makushi; Mixed tribe in which both household heads were Amerindian but not both Makushi; Mixed ethnicity in which at least one head is either of non-Amerindian ethnicity or of mixed ethnicity.

All households engaged to varying degrees in both traditional subsistence farming, hunting and gathering practices and in trading or work for cash payment. Study participants ranked household income and we confirmed ranks by visual observation of household possessions; the rankings were then combined and simplified into four roughly equally spaced income groups, 1 being the lowest income group and 4 being the highest.

Data collection methods

In researching resource use practices, we used a combination of participant observation [27,28] semi-structured interviews [29,30] and elicitation techniques from participatory methodologies such as free listing, ranking (used during interviews) and transect walking and mapping (used during participant observation) [27,28,29,31]. We carried out interviews in English or Makushi, the latter using local translators.

Data were collected for two periods; five months in 1998 and 13 months from early 1999–2000. Data on normal patterns of resource use were collected during the first period. Semi-structured interviews with 48 key informants elicited descriptions of hunting, fishing, harvesting of wild plant products and farming practices covering topics including seasonality, locations and reasons behind resource use decisions. Descriptions of these activities were confirmed and further explored through participant observations (following the methods of [32,33]) during which the researcher joined participants during resource use activities. During the second period, a sample of 43 households (roughly half the households in each village, attempting to sample equally across income and ethnicity) was used to quantify resource use. In Surama, householders were asked to keep resource use diaries recording all plant and animal resources used during the month and the researcher visited each household monthly to collect the diaries and to conduct semi-structured interviews about their resource use practices. Topics covered in the semi-structured interviews included: what resources had been collected, in what quantities, from which locations, to which household resources had been distributed and why those resource use decisions had been taken. Where possible, quantities were measured using hand held scales (for weights) or visual counts (for individual items). However, where part of the resources had already been consumed, participants were asked to estimate the quantities of the original amount harvested. The accuracy of participant weight estimation was measured and average error rates were found to be <5%, with no bias toward either over or under estimation. Sample households recorded resources used for 52 consecutive weeks from March 1999. In Wowetta and Fair View, the researcher visited monthly and participating households were asked to recall and report resource use from the previous week. These two villages were visited ten times at 4–6 week intervals during the same period from March 1999. The quantities of resources recorded for each household were scaled according to the number and composition of each household using standard consumption units (C.U). Standard consumption units are scaled relative to adult male consumption (which scores 1) and are calculated for each gender in several age bands [34]. Seasonal differences were compared by looking at consumption during the two wet seasons (June to Sept; Dec) and the two dry seasons (Oct-Nov; Jan-May).

Forest composition surveys involved locating quadrats (0.25ha) in a stratified random design, sampling 4 different forest types (palm forest: 12 plots, mixed forest: 27 plots, Mora forest: 15 plots, mixed mora: 7 plots) and ex-farm sites: 15 plots aged approximately 5, 10, 15, 20 and 30 years (3 replicates per age). The sampling method follow that of Johnston and Gillman [35] and Johnston [36] for surveying locally defined forest types: local

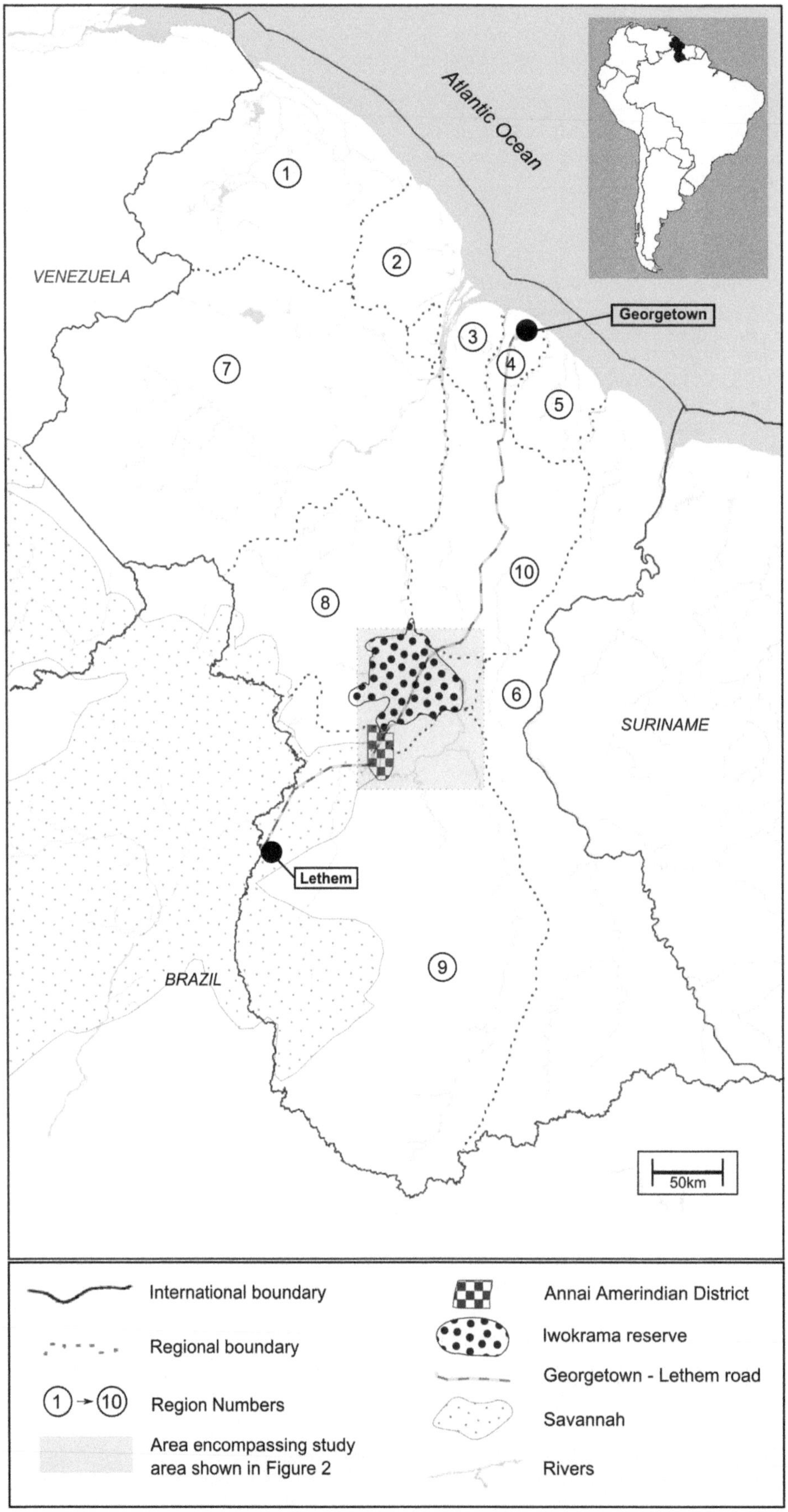

Figure 1. Map showing location of study site in central Guyana, South America [67].

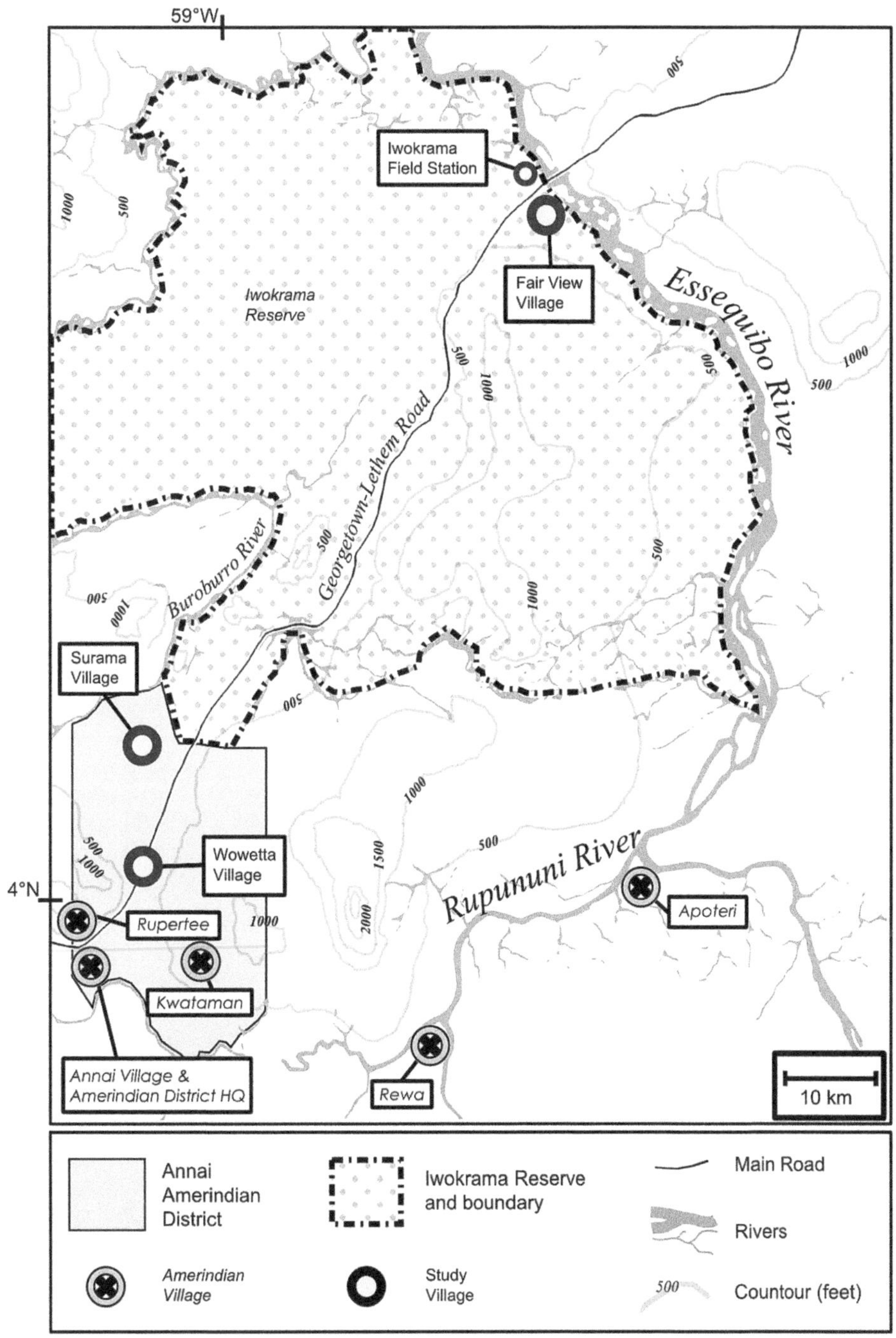

Figure 2. Map showing location of study villages in the Rupununi, Guyana [68].

guides located areas they deemed typical of a local forest type and identified the ethno-species within them. This process was repeated in the major forested areas around the community so plots for a certain forest type were drawn from several locations, all within 5 km of the community. Trees over 10 cm dbh were identified to species using [37–45]. A herbarium reference collection was prepared in duplicate and one set of specimens were deposited at the University of Guyana Herbarium (BRG< IH>) where field identification were confirmed by the curator of the botanical collection. (See S1 for list of deposited specimens).

Statistical analyses used ANOVA where data met the assumptions or could be transformed to meet them; otherwise non-parametric tests (Kruskal-Wallis) were used [46]. When multiple significance tests were carried out on the same data, a new α value, appropriate for the number of tests carried out, was calculated using the Dunn-Šidák method as described in [47]. Diversity was calculated using Simpson's index: $D = 1 - \Sigma p_i^2$ [48] (using proportional composition of all 78 tree species), and ordinated using PCA on log(X+1)-transformed data for species whose mean density per plot exceeded 1.0 (the commonest 15 species) [48].

Results

Household profiling

A breakdown of the villages by income level and ethnicity is given in Table 1. The demographic profiles of these communities are similar to other Amerindian communities in Amazonia, although having more in common with relatively acculturated groups.

Forest types

A total of 78 tree species were recorded in the forest plots, of which the dominant 15 (mean value >1 tree per plot) were used in floristic analyses and are listed in Table 2. The ex-farm plots had slightly lower total numbers of species than "Palm" and "Mixed" forest types, but were dominated by scarcer species leading to comparable diversity (Figure 3). PCA ordination of the tree data showed clear clustering (Figure 4). Mora and Mora-mix stands clustered together (and were associated with *Mora excelsa* Benth, unsurprisingly). Palm forest contained the greatest number of common species (notably *Attalea maripa* (Aubl.) Mart, *Astrocaryum aculeatum* G.Mey and *Couratari multiflorum* Eyma). Ex-farm plots also contained a distinct community dominated by *Byrsonima stipulacea* Juss, along with *Loxopterygium sagotti* Hook. When all non-farm sites were excluded, there were no significant differences between the tree communities in different ages of ex-farm plots (Kruskal-Wallis test with 4df; all p>0.05).

Forest resource use

During the study period, utilisation was recorded by the sample households for 73 named animal species (plus at least six ethno-species of fish) and 164 named plant species (plus at least 19 unidentified ethno-species). For ease of comparison with other studies these resources are grouped into five use categories: wild animals killed for consumption; wild fruits gathered for consumption; firewood; construction materials; and craft materials (Table 3).

Patterns of use of animal resources

Of the vertebrate species taken from the wild, fish contributed both the highest number of individuals and the largest proportion (by weight) of wild caught vertebrates consumed (53%). Fish comprised over 98% of the individuals taken; mammals were the most commonly taken terrestrial vertebrates (55%), followed by reptiles (30%) and then birds (15%).

The fish species most commonly caught in Surama and Wowetta are those that habitually live in seasonal waterways and smaller creeks and ponds, while those most commonly caught in Fair View are those that generally live in large rivers or associated habitats (Figure 5).

The seasonal ($F_{1,\ 79} = 25.7$; $P<0.01$) and inter-community ($F_{2,\ 79} = 36.7$; $P<0.01$) differences in quantities of fish consumed were significant (Figure 6), but the relationship with income was non-significant ($rs = -0.18$, $N = 86$, $p>0.05$).

There was a significant difference in mammal meat consumption between ethnicity classes (K-W H = 10.7; df = 2; P = 0.005), (Figure 7), but not in the number of mammals caught (K-W H = 4.4; df = 2; P = 0.111). In contrast, there was a significant difference in the number of mammals caught between income classes (K-W H = 14.1, df = 2, P = 0.003), but not in consumption of mammal meat (K-W H = 2.3; df = 2; P = 0.522). A slight trend of decreasing capture rate with increasing income class was apparent between income class 2 and 4 (Figure 7).

Patterns of use of plant resources

By weight, 81 percent of the wild fruits consumed were from the two palm species, *Astrocaryum aculiatum* (51%) and *Attalea maripa* (30%). Consumption of *A. maripa* was greatest in households with Makushi traditional origins and least in households with some Mixed-ethnicity, (Figure 8) (K-W H = 14.4; df = 2; P = 0.001). There were no other significant differences in consumption of wild fruits overall or of *A. aculiatum* separately (ANOVA, F = 1.585; df = 3, 39; P = 0.209).

Table 1. Number of households in each village, income and ethnicity class in a study of the North Rupununi, Guyana.

	Income class	Ethnicity class			
Village		**1**	**2**	**3**	**Total**
Surama	1	0 [1]	[0]	[0]	**0 [1]**
	2	3 [8]	0 [1]	[0]	**3 [9]**
	3	4 [6]	2 [5]	[0]	**6 [11]**
	4	0 [1]	6 [9]	2 [3]	**8 [13]**
	Total	**7 [16]**	**8 [15]**	**2[3]**	**17 [34]**
Fair View	1	[0]	[1]	1 [1]	**1 [2]**
	2	[0]	6 [9]	[1]	**6 [10]**
	3	[0]	2 [4]	1 [2]	**3 [6]**
	4	[0]	[0]	[0]	
	Total		**8 [14]**	**2 [4]**	**10 [18]**
Wowetta	1	2 [5]	1 [1]	[0]	**3 [6]**
	2	7 [13]	3 [8]	[0]	**10 [21]**
	3	2 [2]	1 [3]	[0]	**3 [5]**
	4	[0]	[0]	[0]	
	Total	**11 [20]**	**5 [12]**	**[0]**	**16 [32]**

(Total number of households in each class is given in square brackets): Ethnicity classes: 1 = makushi, 2 = Mixed amerindian, 3 = Mixed ethnicity.

Table 2. Interquartile range for the dominant 15 tree species) for 4 forest types and farm plots in the North Rupununi, Guyana.

Species	Forest type				
	Palm	Mora	Mora/mixed	Mixed	Farm plots
Attalea maripa (Aubl.) Mart.	15–33	0–10	8–12	4–17	3–12
Mora excels Benth.	0	17–26	21–25	0–3	0
Astrocaryum gynacanthum Mart	6–13	0–5	0–5	0–11	0
Astrocaryum aculiatum G.Mey	3–8	0	0	0–7	0
Oenocarpus bacaba Mart.	0–3	0	0	0–10	0
Eschweilera subglandulosa (Steud. ex O.Berg) Miers	0–4	0–6	0–1	0–4	0
Heliconia spathocircinata Aristeg.	0–3	0	0	0–2	0
Catostemma commune Sandwith	1–6	0–1	0–1	1–6	0
Euterpe precatoria Mart.	0–4	0	0–2	0–5	0
Blackseed	0–1	0–2	1–3	0	0–9
Couratari multiflora Eyma	0–7	0–1	0	0–1	0
Byrsonima stipulacea Juss	0	0	0	0	1–9
Loxopterygium sagotii Hook	0–1	0	0	0–1	0–5
Sandwood	0–1	0	0	0–2	0–2
Brosimum guianensis (Aubl.) Huber	0–2	0	0	0–1	0–4
All species	16–20	7–12	1–8	12–21	9–12

Dominant trees classified as those with a mean value >1 tree per plot. Data are trunks >10dbh per 2500 m^2.

The most preferred firewood species overall was *Swartzia leiocalycina* Benth, (20% of all firewood collected and used in all three communities). The next most commonly used was *B. spicata*, comprising 19% and only used in Surama and Wowetta. We found no significant differences in species diversity or quantity of firewood used between villages, income groups or ethnic groups. The number of individual timber poles in the 50–150 cm dbh size class (used for house construction) did differ between villages (K-W $H = 10.5$; $df = 2$; $P = 0.005$) with use greatest in Surama, very rare in Wowetta and intermediate in Fair View.

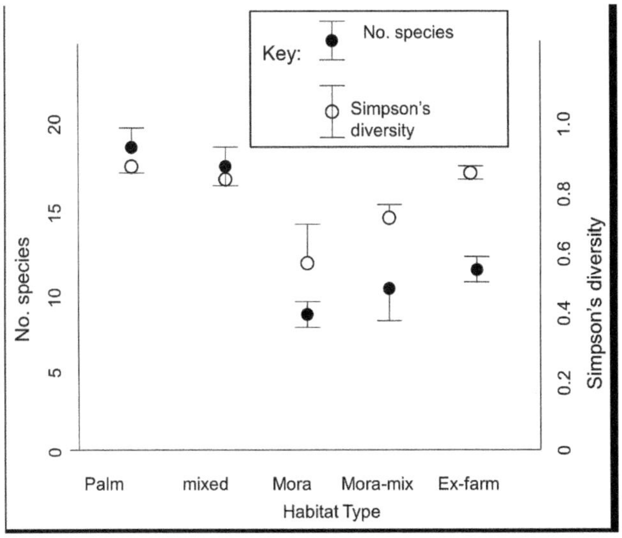

Figure 3. Floristic diversity and species richness in 4 forest types including ex-farm plots, North Rupununi, Guyana. (Representing the 15 dominant species).

A. maripa was the most commonly used palm species. The average number of *A. maripa* leaves used per household was significantly greater in Wowetta than in Fair View and tended towards being significantly greater in Wowetta than in Surama (ANOVA, $F = 11.848$; $df = 2,40$; $P<0.001$; Tukey's post hoc test Wowetta cf. Fair View $P<0.001$ and Wowetta cf. Surama $P<0.019$). *A. maripa* leaf use was very low in Fair View where most houses relied upon recycled zinc sheets for roofing. There were no significant differences in use of *A. maripa* leaves for thatching between groups defined according to traditions or income.

The most commonly used species for craft production was *Ischnosiphon arouma (Aubl.)Körn*, used to manufacture traditional household items. The average number of plants used per household differed significantly between communities (K-W $H = 77.4$; $df = 2$; $P = 0.024$), Wowetta using the highest number on average, followed by Surama, while use in Fair View was very rare.

Forest management

The main forest management activities included agriculture, hunting, fishing, harvesting of wild plant products such as edible fruits, construction materials, craft materials, those with medicinal and/or supernatural purposes, and the cutting of trails to facilitate activities in the forest. All participants very uniformly described the general form of activities and their seasonal cycles.

Agriculture provided a central focus for plant related forest management activities. Farm sites were selected, usually by the male head of household each year, with decisions based on soil qualities, landform and proximity to sites of relatives. Non-inundated, predominantly sandy soils were preferred and 84 percent of farms were in sand or sand-loam soils. The area under cultivation varied between villages from only 9.31 ha in Fairview, where all farms were in close proximity to the village (average distance of 0.39 km), to 29.73 ha in Wowetta and 36.72 ha in Surama, where farms were further away (average distances of

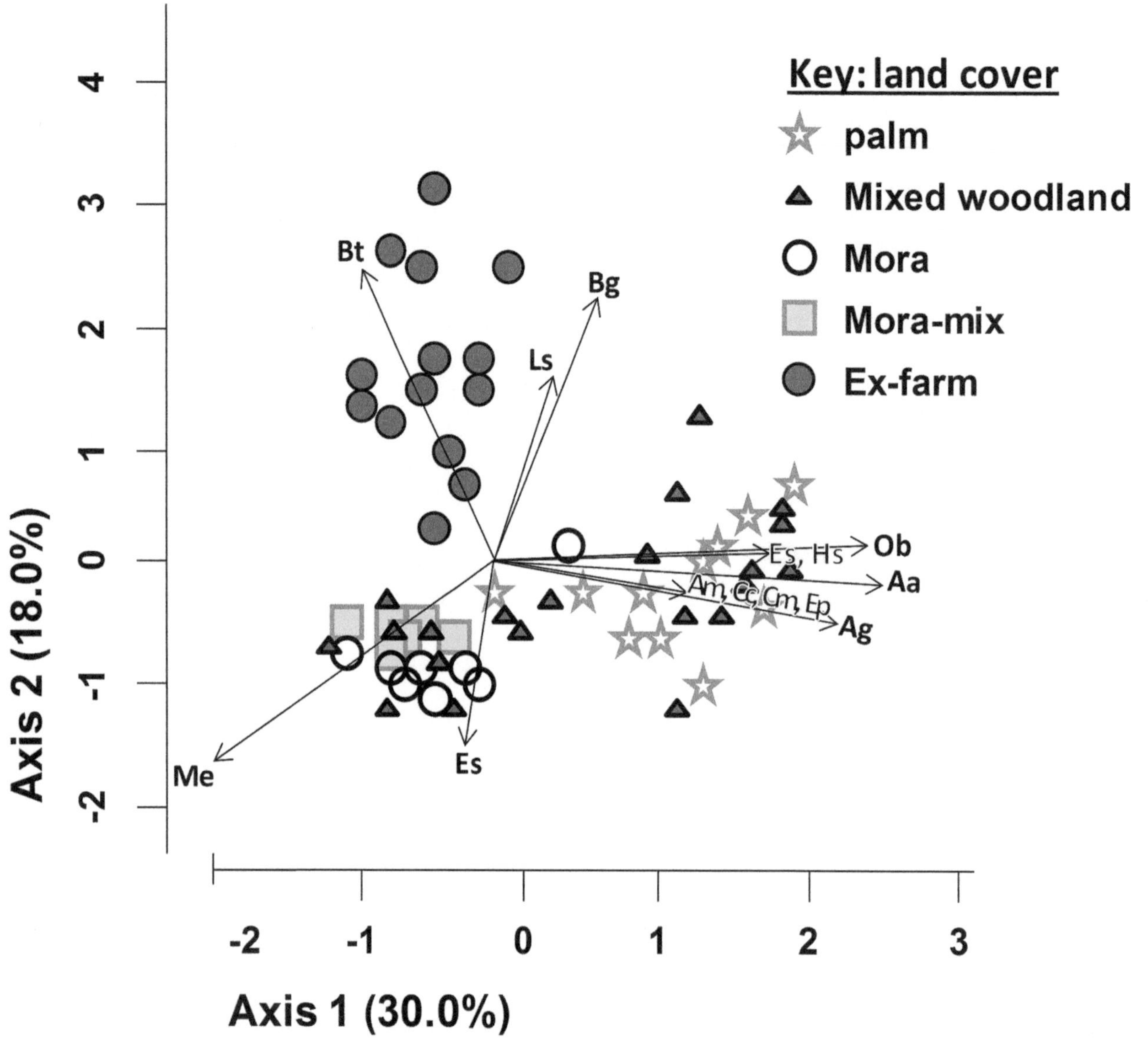

Figure 4. PCA ordination of floristic data (15 dominant tree species) for 4 forest types including ex-farm plots, North Rupununi Guyana. Key: Aa: *A. aculeatum*, Ag: *A. gynacanthum*, Am: *A. maripa*, Bg: *B. guianensis*, Bt: *B. stipulacea*, Cc: *C. commune*, Cm: *C. multiflora*, Es: *E. subglandulosa*, Es: *E. subglandulosa*, Hs: *H. spathocircinata*, Ls: *L. sapotis*, Me: *M. excels*, Ob: *O. bacaba*.

2.94 km and 2.25 km respectively). The average area per household under cultivation was similar for Surama (1.08 ha, S.E. = 0.18) and Wowetta (0.97 ha, S.E. = 0.18) but roughly half that area in Fair View (0.4 ha, S.E. = 0.06).

Opportunism played a key role in the extraction of resources, particularly wild fruits and tortoises, which were usually collected if encountered in the forest but rarely sought out. For example, only 13 percent of fruit collection events were obtained on a trip for which fruit collection was the primary or one of the primary purposes; 65 percent of households did not report seeking out wild fruits at all despite two of the most widely used edible species (the palms *A.maripa* and *A.aculiatum*) being dominant in the "palm" woodland type.

Fishing and hunting trips were made almost entirely to supply household needs rather than primarily to obtain fish or meat to sell (<2%). Men reported that they covered up to 300 km^2 of the forest during their hunting activities, whilst women usually reported and were observed to visit only more proximate forest locations e.g. the forest on the farm road. A factor universally identified as affecting the decisions to fish or hunt was the amount of mammal meat or fish received as an offering from another household, a practice known as "sharing". Although a commonly given reason for sharing was to achieve an even distribution of the resource across the community, meat was much more likely to be shared to family than non family (paired sample T-test $t = 4.8$; $df = 42$; $P < 0.001$).

Spatial patterns in extraction of forest resources

The majority of plant resources were collected from the forest around the farm or wild species growing up in the farm itself and it was also an important hunting area. 56 percent (by weight) of firewood was collected from farming grounds or the forest immediately around, 71 percent (by weight) of wild fruits, 31 percent (by number of individuals cut) of 'round-wood' (i.e. not cut

Table 3. Number of species used by participants in the North Rupununi, Guyana for each resource use category defined in the study.

Resource category	Number of species used	Comment
Fish	47 plus 6 unidentified ethnospecies	Represents 64% of species record by Eigenmann [66] and 40% of those recorded by Forte [26]
Mammals	10	
Birds	12	Represents 22% of species recorded as 'generally eaten' Forte [26]
Reptiles	4 (1 iguana, 2 forest tortoise, 1 water turtle)	
Wild fruit	16 plus 8 unidentified ethnospecies	Forte [26] lists 20 species and a further 40 ethnospecies of which 5 species and 9 ethnospecies were recorded here
Firewood	27 plus 13 unidentified ethnospecies of wild growing trees and 6 domesticated fruit trees	
Construction wood	23 plus 8 unidentified	
Construction material	3 palm species; bark from 1 identified and 1 unidentified tree; stems of 1 identified and 1 unidentified liana	
Craft materials	13 (1 herb, 4 liana, 2 palms and 6 hard wood trees)	

into boards with a chainsaw), 30 percent of thatching palm leaf, and 28 percent of *I. arouma* for craft. Hunting in the farm provided 11 percent of the mammal and 99 percent of the bird meat collected. Participants often referred to the association between some of the useful species and the secondary forest around the farming areas.

This study recorded no traditionally prohibited areas within the normal ranges of any of the study villages, although specific areas within the range of neighbouring villages were referred to. Hunting trips did tend to visit areas at some distance from the village centres (up to 50 km), as game was scarce in areas close to the population centres. Firewood not collected from the farm areas, was usually obtained from the forest edge as close to the household as possible, although in Wowetta, this still entailed trips of over 2 km from some households.

Socio cultural differentiation in management strategies

Participants perceived ethnicity and the related notion of traditionalism (both of which were inversely related to household income), as being the main sources of differences in forest management strategies. The number of different locations from which household timber had been obtained varied with income class (ANOVA, $F = 4.282$; $df = 3, 48$; $P = 0.009$) with a division between the two lowest and the two highest income classes (Figure 9). Households in the lowest income group sourced timber from significantly more locations than those in the two highest income classes. While the majority of hunting was carried out with bow and arrow, three households (all in upper two income groups) in Surama owned guns and relatives from other households regularly borrowed them.

The described norm for traditional farming was to have large clusters of farms belonging to households of relatives. Households in the Makushi ethnic group had more contiguous neighbours than those in the Amerindian mixed group, which in turn had more contiguous neighbours than those in the mixed ethnicity group (Figure 10) this difference was significant (K-W $H = 12.6$; $df = 2$; $P = 0.002$). There was also a significant difference in the number of contiguous neighbouring farms between households in income class 2 and the two highest, 3 and 4 (ANOVA, $F = 4.952$; $df = 3,60$; $P = 0.004$) (Figure 10).

Discussion

Forest management

The norms of forest resource use and management found in the North Rupununi were similar to those described for many other forest-based Amerindian groups living a subsistence lifestyle (e.g. [49–52]). Although early researchers sought to classify Amerindian communities into those that were 'hunters', 'fishers', 'gatherers' or 'cultivators', this has long been recognised as an over-simplification [53,54]. Here we observed that trips to the forest could not be definitively classified as one type of activity, there was an interlinked strategy of activities focussed around supplying household needs. Explicit management activities were rare and although some traditional management activities may have been lost, we suggest that a system of prohibited areas may never have been important in a region where there are large tracts of forest which have not been settled in living memory and resources are not limited. The distances between households and farm sites were very short compared to the estimated 6–10 km average distance between households and farm sites for savannah communities throughout the Rupununi [55]. While distance was certainly a factor in limiting resources extraction from areas around the villages, there were no definite limits except those imposed by difficult to cross topographical features. The forest management observed here is the result of activities that have management effects, but which are incidental to the main agricultural and subsistence objectives of those activities.

Variation in resource use and forest management strategy with ethnicity and socioeconomic grouping

We found that whilst the use of wild fish is ubiquitous, driven by availability, the amount and type of other sources of dietary protein differed between ethnic groups. In particular, mammal meat consumption and the use of the palm fruits from *A. maripa* were higher in the more traditional Makushi households (Figures 7 and 8). In interviews, participants attributed this pattern of protein consumption to the knowledge and skills base of the people in this group and to ethnically based food preferences. Participants perceived Makushi households as being more "traditional", a view partly based on their style of house construction, and as more reliant on forest resources. This is the first time such significant

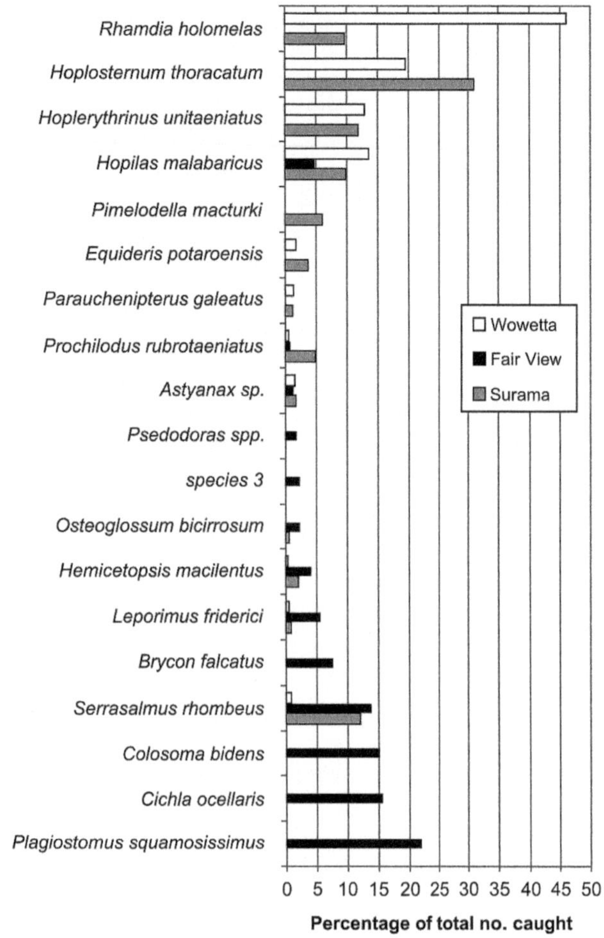

Figure 5. Most commonly caught fish species in each of three villages, North Rupununi Guyana. (Representing 1 percent or more of fish caught in all three communities during the sample year).

differentiation in wild resource use has been recorded at an inter-household level within communities.

Household socioeconomic status also influenced resource use patterns. For example, middle income groups hunted significantly more mammal meat than other groups, but were not consuming more, suggesting they were selling or sharing their catches with members of other income groups. This relationship formed part of a complex pattern of traditional deferred exchange, acting to promote community cohesion and increase equality of access to resources. The patterns of harvesting of construction materials also differed between income groups with more localised extraction and greater number of larger trees being harvested by villagers in the higher income groups. This latter pattern could lead to localised resource depletion and has clear impacts on plant spatial distribution.We found significant within community variations in tree extraction patterns, with lower income groups creating scattered felling sites more akin to tree fall gaps. Such felling gaps have been shown to return to similar levels of diversity and species complement to pristine forest within five years of disturbance in Africa, Bolivia and Borneo [56–58]. Higher income groups were more likely to create larger gaps from which greater numbers of trees are extracted which would take longer to recover.

Much of the forest resource use in these communities (with the exception of fish) was concentrated around current and 'old' or

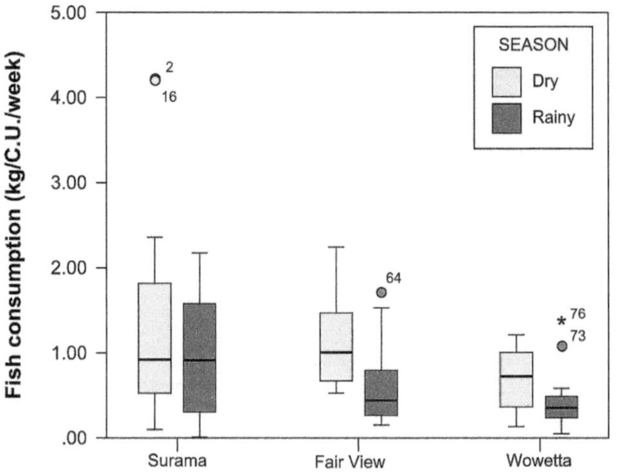

Figure 6. Seasonal and inter-community differences in fish consumption (kg/C.U./week) for three villages, North Rupununi, Guyana.

fallow farm sites. Whilst hunting trips did take participants significant distances from the villages, between 7 and 11 percent of these resources were also derived from the farm sites. Thus these locations form foci for forest management activities. Farms were generally set close to the villages (within 3 km on average), but there were significant differences between villages and between ethnic groups in terms of the spatial patterns of farm distribution and therefore forest resource use, reinforcing the potential for local "placed based" effects on biodiversity patterns.

At an intermediate landscape scale, forest clearance for farming shows different patterns, with more traditional Makushi households clearing contiguous areas resulting in larger gaps and less traditional households clearing more disparate sites. Evidence from other areas in the tropics suggests smaller more disparate sites may return to similar levels of plant diversity within 50–60 years, but that rare species may not return easily [57,58].

The spatial pattern of the farms within the North Rupununi forest lies on a scale suggested to increase biodiversity [59,60]

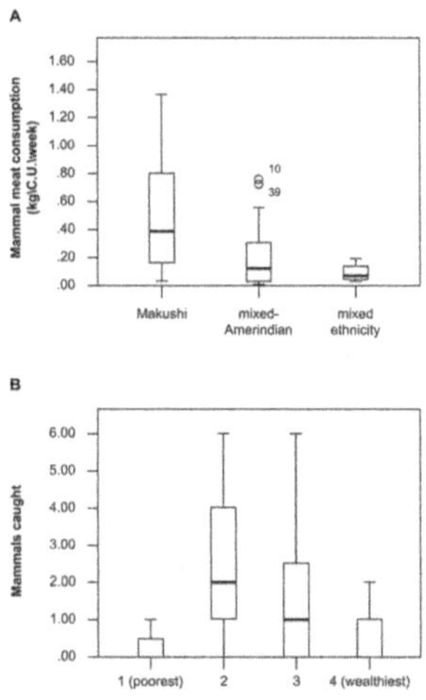

Figure 7. Patterns of animal resource use amongst Amerindian groups, North Rupununi, Guyana. A) Household consumption of mammal meat in the three classes of traditional origins, B) Capture of mammals in four income classes in the North Rupununi, Guyana.

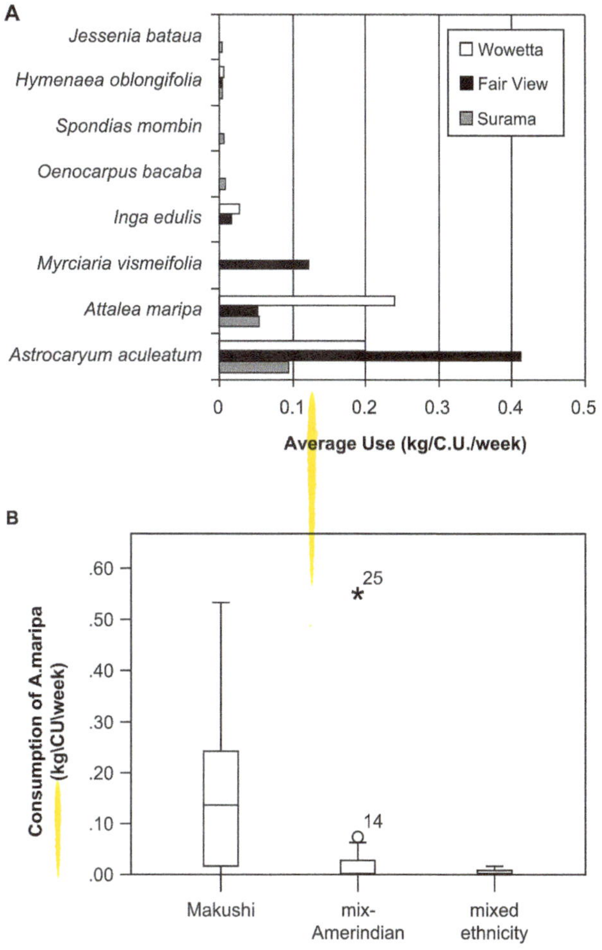

Figure 8. Patterns of wildfruit and palm use by Amerindian groups, North Rupununi, Guyana. A) Wild fruits consumed in three villages (representing more than 1% of total consumption, B) Consumption of *Attalea maripa* by ethnicity.

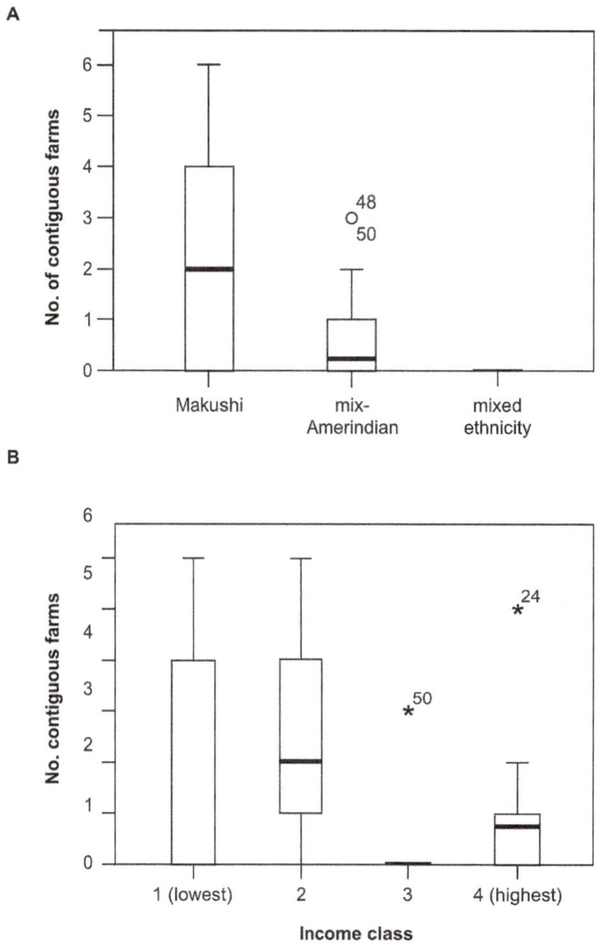

Figure 10. Farm contiguity in the North Rupununi Guyana, by ethnicity and income class. A) Number of contiguous neighbouring farms by ethnicity class of owning household, B) Number of contiguous neighbouring farms by income class of owning household in the North Rupununi, Guyana (Differences significant by Tukey's post hoc, 2>3 $P=0.013$, 2>4 $P=0.028$).

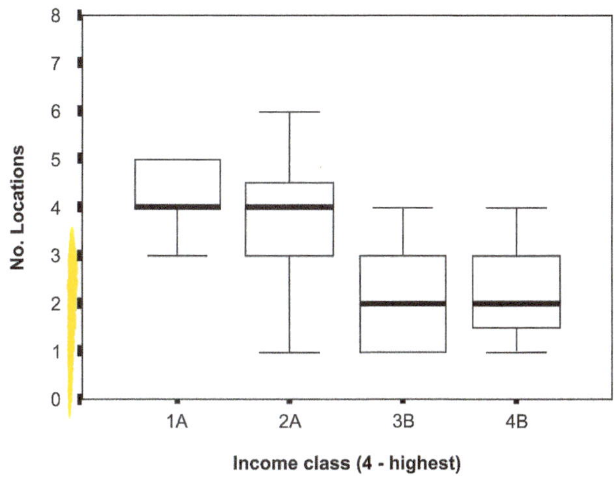

Figure 9. Number of locations for timber collection by income class for Amerindians, North Rupununi, Guyana. Columns with the same letter do not differ according to Tukey's multiple range test.

under the intermediate disturbance model [61]. Shifting cultivation and small scale timber extraction have been shown to increase floristic diversity in Northern Guyana [62], also in Africa [63] and Asia [64], marked by an increase in the number of pioneer species. The floristic data here indicate that ex-farm plots do contribute to supporting species that are less common in the surrounding forest therefore increasing diversity reinforcing the conclusions of others about the influence of humans on tropical forest diversity in other regions [8].

We conclude that there are ecologically significant differences in resource use between adjacent villages in the North Rupununi, both within communities (driven by relative affluence) and between community variations (driven by differences in ethnicity). Spatial patterns related to agriculture differed between villages, while differences between timber extraction patterns were driven more by financial status. Our findings that there is a significant role for both income and ethnicity in driving resource use and hence forest biodiversity patterns in the North Rupununi, have implications for models of diversity which underlie conservation planning and can contribute to assessments used in monitoring for REDD+ [65] which has particular importance in Guyana.

Acknowledgments

This research was made possible by the collaboration of the residents of Surama, Wowetta and Fairview. Advice on the project was received from C. Cotton, C. Ross and G. Marvin. Field work was supported by the Iwokrama International Centre for Conservation and Development. Preparation of figures was assisted by M. Evans.

Author Contributions

Conceived and designed the experiments: CO CC PS. Performed the experiments: CC. Analyzed the data: CO CC PS. Wrote the paper: CO CC PS.

References

1. Adeney JM, Christensen NL Jr, Pimm SL (2009) Reserves protect against deforestation fires in the Amazon. PLoS ONE 4(4): e5014. DOI:10.1371/journal.pone.0005014.
2. Armenteras D, Rodriguez N, Retana J (2009) Are conservation strategies effective in avoiding the deforestation of the Colombian Guyana Shield? Biol Conserv 142: 1411–1419.
3. Nepstad DC, Schwartzman S, Bamberger B, Santili M, Ray D, et al. (2006) Inhibition of Amazon deforestation and fire by parks and indigenous reserves. Conserv Biol 20: 65–73.
4. Porter-Bolland L, Ellis EA, Guariguata MR, Ruiz-Mallén I, Negrete-Yankelevich S, et al. (2012) Community managed forests and forest protected areas: An assessment of their conservation effectiveness across the tropics. For Ecol Manage 268: 6–17.
5. Walker R, Moorea NJ, Arimab Perzc EP, Simmonsa C, Vergaraa MCD, et al. (2009) Protecting the Amazon with protected areas. Proc Natl Acad Sci U S A 106 (26): 10582–10586.
6. Lu F, Gray C, Bilsborrow RE, Mena CF, Erlien CM, et al. (2011) Contrasting Colonist and Indigenous Impacts on Amazonian Forests. Conserv Biol 24 (3): 881–885.
7. Smith DA (2008) The spatial patterns of indigenous wildlife use in western Panama: Implications for conservation management. Biol Conserv 141: 925–937.
8. Chazdon RL, Harvey CA, Komar O, Griffith DM, Ferguson BG, et al. (2009) Beyond reserves: a research agenda for conserving biodiversity in human modified tropical landscapes. Biotropica 41: 142–153.
9. Lambin EF, Geist HJ, Lepers E (2003) Dynamics of land-use and land-cover Change in tropical regions. Annu Rev Environ Resour 28: 205–41.
10. Käyhkö N, Fagerholma N, Asseidb BS, Mzee AJ (2011) Dynamic land use and land cover changes and their effect on forest resources in a coastal village of Matemwe Zanzibar Tanzania. Land Use Policy 28: 26–37.
11. Brooks JS, Waylen KA, Borgerhoff Mulder M (2012) How national context, project design, and local community characteristics influence success in community-based conservation projects. Proc Natl Acad Sci U S A 109: 21265–70.
12. Posey DA (1985) Indigenous management of tropical forest ecosystems: the case of the Kayapo' Indians of the Brazilian Amazon. Agrofor Syst 3: 139–158.
13. Toledo VM, Ortiz-Espejel B, Cortés L, Moguel P, Ordoñez MDJ (2003) The multiple use of tropical forests by indigenous peoples in Mexico: a case of adaptive management. Conserv Ecol 7(3): 9.
14. Youn Y-C (2009) Use of forest resources traditional forest-related knowledge and livelihood of forest dependent communities: Cases in South Korea. For Ecol Manage 257 (10): 2027–203.
15. Peres CA, Nascimento HS (2006) Impact of game hunting by the Kayapo' of south-eastern Amazonia: implications for wildlife conservation in tropical forest indigenous reserves. Biodivers Conserv 15: 2627–2653.
16. Haden P (1999)Forestry issues in the Guiana shield region: a perspective on Guyana and Suriname. European Union Tropical Forestry Paper 3 Overseas Development Institute. 19 p.
17. Ecosecurities (2002) A baseline analysis for the Guiana Shield. Region Baseline report for NC-IUCN April 2002.
18. Ellenbroek A (1996) Rainforests of the Guyana Shield: Evergreen or forever gone? BOS NiEuWSLETTER. 15(2), no. 34. pp. 7–12.
19. Conservation International Climate Focus Terracarbon LLC SarVision. Development and Policy Management Consultants (2009) Reducing Deforestation and Forest Degradation while Promoting Sustainable Development. South American Regional Infrastructure Development Forests and REDD: Implications for Guyana. IDB (ATN/NP-10749-GY – ATN/SF-10750-GY).
20. FAO (2001) Forest Resource Assessment 2000. Main Report. FAO Forestry Paper 140. Rome. 479 p.
21. Guyana Forestry Commission (2009) Quick assessment paper. Deforestation and forest degradation in Guyana. Guyana Forestry Commission Georgetown Guyana.
22. Department of Lands and Surveys (1973) Guyana. Surama: sheet 55 SE. Government of Guyana, Georgetown.
23. Salisbury RF, Dummett MJ, Hills TL, Cook D (1968) Ethnographic Notes on Amerindian Agriculture. Department of Geography, McGill University, Montreal.
24. Ter Steege H (2001) National vegetation map of Guyana. Guyana Forestry Commission Forest Resources Information Unit, Georgetown.
25. Hammond DS (2005) Tropical forests of the Guiana shield: ancient forests in a modern world. CABI. 528 p.
26. Forte J (1996) About Guyanese Amerindians. Janette Forte. Georgetown. 129 p.
27. Ellen RF (1984) Ethnographic research: a guide to general conduct. Research Methods Social Anthropology 1. Academic Press London. 403 p.
28. Silverman D (1997) Qualitative research: Theory method and practice. SAGE Publications London. 390 p.
29. Chambers R (1992) Rural appraisal: rapid relaxed and participatory. IDS Discussion Paper No. 311. IIED London.
30. Chambers R (1994) The origins of practice of participatory rural appraisal. World Dev 22(7): 953–969.
31. Messerschmidt DA (1995) Rapid Appraisal for community forestry: The RA process and rapid diagnostic tools. IIED London. 132p.
32. Wolcott H (1990) Writing-up Qualitative Research. SAGE, London
33. Silverman D (1993) Interpreting qualitative data: Methods for analysing talk, text and interaction. SAGE Publications, London.
34. Strickland SS, Tuffrey VR (1997) Form and Function. A study of nutrition adaptation and social inequality in three Gurung villages of the Nepal Himalayas. London. Smith-Gordon. 304 p.
35. Johnston M, Gillman MP (1995) Tree population studies in low-diversity forests Guyana. I. Floristic composition and stand structure. Biodivers Conserv 4: 339–362.
36. Johnston M (1998) Tree population studies in low-diversity forests, Guyana. II. Assessments on the distribution and abundance of non-timber forest products. Biodivers Conserv 7: 73–86.
37. Maguire B (1981) Memoirs of The New York Botanical Garden Vol. 32: The botany of the Guyana Highland - Part XI. New York Botanical Garden, New York.
38. van Roosmalen MGM (1985) Fruits of the Guianan Flora. Institute of Systematic Botany, Utrecht.
39. Mennega EA, Tammens-de Rooij WCM, Jansen-Jacobs MJ (1988) Check-list of the woody plants of Guyana. The Tropenbos Foundation, Ede, The Netherlands. 281p.
40. Polak AM (1992) Major Timber Trees of Guyana: A Field Guide. The Tropenbos Foundation, Wageningen, N.L.
41. Brunner M, Kucera LJ, Zürcher E, Hirzel A (1994) Major timber trees of Guyana: A lens key. The Tropenbos Foundation, Wageningen, The Netherlands.
42. Henderson A, Galeano G, Bernal R (1995)Field guide to the palms of the Americas. Princetown University Press, Princetown. 64 p.
43. Gérard J, Miller RB, ter Welle BJH (1996) Major timber trees of Guyana: Timber characteristics and utilization. The Tropenbos Foundation, Wageningen, The Netherlands. 183 p.
44. van Andel TR (2000) Non-Timber Forest Products of the Northwest District of Guyana. Part I & II. Unpublished Doctoral Dissertation. Utrecht University. Tropenbos-Guyana Series 8A & 8B.
45. Rose SA (2000) Seeds, seedlings and gaps - size matters. A study in the tropical rainforest of Guyana. PhD thesis, Utrecht University, Tropenbos Guyana Series 9. The Tropenbos Foundation, Wageningen, the Netherlands.
46. Siegal S, Castellan NJ Jr (1988) Non-parametric statistics for the behavioural sciences. McGraw-Hill Book Company New York. 2nd Edition. 399 p.
47. Sokal RR, Rohlf FJ (2012) Biometry: the principles and practice of statistics in biological research. 4th edition. W. H. Freeman and Co.: New York. 937 p.
48. Shaw PJA (2003) Multivariate Statistics for the Environmental Sciences. Hodder Arnold. 248 p.
49. Balée W (1989) The culture of Amazonian forests. Adv Econ Bot 7: 1–21.
50. Milliken W, Miller RP, Pollard SR, Wandelli EV (1992) Ethnobotany of the Waimiri Atroari Indians of Brazil. Royal Botanic Gardens, Kew London. 146 p.
51. Beckerman S (1993) Major patterns in indigenous Amazonian subsistence. In: Hladik CM, Hladik A, Linares OF, Pagezy H, Semple A, Hadley M, editors. Tropical forests people and food - biocultural interactions and applications to development. UNESCO Paris. pp.52–78.
52. de Souza-Mazurek RR, Pedrinho T, Feliciano X, Hilario W, Geroncio S, et al.(2000) Subsistence hunting among the Waimiri Atroari Indians in central Amazonia Brazil. Biodivers Conserv 9: 579–596.

53. Moran F (1989) Models of native and folk adaptation in the Amazon. Adv Econ Bot 7: 22–29.
54. Meggers BJ (1996) Amazonia: Man and culture in a counterfeit paradise. Smithsonian Institution Press. Revised Edition. Washington & London.
55. Hills TL (1968) Amerindian Agriculture. Dept. Geography, McGill University, Montreal, Canada.
56. Cannon CH, Peart DR, Leighton M (1998) Tree species diversity in commercially logged Bornean rainforest. Science 281: 1366–1368.
57. Peña-Claros M (2001) Secondary forest succession: processes affecting the regeneration of Bolivian tree species. PROMAB Scientific Series 3. PROMAB Riberalta Bolivia.
58. van Gemerden BS, Shu GN, Olff H (2003) Recovery of conservation values in central African rainforest after logging and shifting cultivation. Biodivers Conserv 12: 1553–1570.
59. Holling CS (1992) Cross-scale morphology geometry and dynamics of ecosystems. Ecol Monogr 62: 447–502.
60. Risser PG (1995) Biodiversity and ecosystem function. Conserv Biol 9(4): 742–746.
61. Connell JH (1978) Diversity in tropical rain forests and coral reefs. High diversity of trees and corals is maintained only in a non-equilibrium state. Science 119: 1302–1310.
62. van Andel T (2001) Floristic composition and diversity of mixed primary and secondary forests in northwest Guyana. Biodivers Cons 10: 1645–1682.
63. Mwampamba TH, Schwartz MW (2011) The effects of cultivation history on forest recovery in fallows in the Eastern Arc Mountain Tanzania. For Ecol Manage 261: 1042–1052.
64. Wangpakapattanawonga P, Kavinchana N, Vaidhayakarna C, Schmidt-Vogt D, Elliott S, (2010) Fallow to forest: applying indigenous and scientific knowledge of swidden cultivation to tropical forest restoration. For Ecol Manage 260: 1399–1406.
65. Dickson B, Kapos V (2012) Biodiversity monitoring for REDD+. Current Opinion in Environ Sustain 4: 717–725.
66. Eigenmann CH (1912) Some results from an ichthyological reconnaissance of Colombia South America. Part I. (Contrib. Zool. Lab. Ind. Univ. No. 127.). Indiana University Studies No. 16 [sic No. 8]: 1–27.
67. Government of Guyana (1974) Gazetteer of Guyana. Ministry of Lands and Surveys, Government of Guyana, Georgetown.
68. Hawkes MD, Wall JRD (1993) The Commonwealth and Government of Guyana Rain Forest Programme, phase I, site resource survey, main report. Natural Resources Institute, Chatham, UK.

The Effect of Leaf Litter Cover on Surface Runoff and Soil Erosion in Northern China

Xiang Li, Jianzhi Niu*, Baoyuan Xie

Key Laboratory of Soil and Water Conservation and Desertification Combating of Education Ministry, Beijing Forestry University, Beijing, China

Abstract

The role of leaf litter in hydrological processes and soil erosion of forest ecosystems is poorly understood. A field experiment was conducted under simulated rainfall in runoff plots with a slope of 10%. Two common types of litter in North China (from *Quercus variabilis*, representing broadleaf litter, and *Pinus tabulaeformis*, representing needle leaf litter), four amounts of litter, and five rainfall intensities were tested. Results revealed that the litter reduced runoff and delayed the beginning of runoff, but significantly reduced soil loss ($p<0.05$). Average runoff yield was 29.5% and 31.3% less than bare-soil plot, and for *Q. variabilis* and *P. tabulaeformis*, respectively, and average sediment yield was 85.1% and 79.9% lower. Rainfall intensity significantly affected runoff ($R = 0.99$, $p<0.05$), and the efficiency in runoff reduction by litter decreased considerably. Runoff yield and the runoff coefficient increased dramatically by 72.9 and 5.4 times, respectively. The period of time before runoff appeared decreased approximately 96.7% when rainfall intensity increased from 5.7 to 75.6 mm h^{-1}. Broadleaf and needle leaf litter showed similarly relevant effects on runoff and soil erosion control, since no significant differences ($p\leq0.05$) were observed in runoff and sediment variables between two litter-covered plots. In contrast, litter mass was probably not a main factor in determining runoff and sediment because a significant correlation was found only with sediment in *Q. variabilis* litter plot. Finally, runoff yield was significantly correlated ($p<0.05$) with sediment yield. These results suggest that the protective role of leaf litter in runoff and erosion processes was crucial, and both rainfall intensity and litter characteristics had an impact on these processes.

Editor: Ben Bond-Lamberty, DOE Pacific Northwest National Laboratory, United States of America

Funding: This study was supported by the Fundamental Research Funds for the Central Universities (No.BLYJ201406); National Natural Science Fund of China (41171028); National Advanced Project of the 12th Plan, China (2011BAD38B05); and National Forestry Public Welfare Industry Research Project (201104005). The funders had no role in study design, data collection and analysis, decision to publish, or preparation of the manuscript.

Competing Interests: The authors have declared that no competing interests exist.

* Email: nexk@bjfu.edu.cn

Introduction

Soil erosion has become a serious problem worldwide, causing to decrease productivity of agricultural and forest land, environmental and ecological degradations, and natural disasters such as mudflow that threaten to human safety and infrastructure [1]. Approximately five billion tons of soil is lost annually in China [2,3]. Soil erosion often damages forest ecosystems, including reduction of soil organic matter content and water-holding capacity [4], loss of valuable soil nutrients, and biota, also declines in biodiversity, and collectively lead to ecosystem instability [1,3]. While vegetation litter is often considered as an effective cover above soil surface that prevents soil erosion, it is often burned or removed for fuel by local population in many forests of Northern China. These activities may increase detachment of soil aggregates, so raising the sediment generation and transportation via runoff [5].

Although it is widely recognized that vegetation canopy is important in hindering soil erosion [6], by comparison, the role of vegetation litter layers in modulating surface runoff and soil erosion remains poorly understood. Litter layers are known to protect soil from raindrop splashes by intercepting rainfall, preventing surface sealing and crusting of soil, extend the time of soil infiltration, and enhance sediment deposition by increasing soil surface roughness [7–9]. Previous studies have mainly evaluated the effectiveness of various surface covers in reducing surface runoff and soil loss, including rock fragments [10–13], crop residues [14,15], grass [16–18], geo-textiles [19], post-fire ash and needle cover [20–22], and combined cover such as rock and litter [23]. Nonetheless, few leaf litter materials have been tested [24,25], with variable findings [26–34]. The effects of litter layers on surface runoff and soil loss were not consistent across different types of litter, soil, or different environments. In general, runoff volume and sediment yield were reduced in the presence of litter cover [26–30]. However, some disputable results have shown that runoff and soil erosion were accelerated by impervious plastic mulch covers [31–34].

Moreover, the early studies have mainly examined the percentage of litter cover as a dominant influential factor in surface runoff and soil erosion [20,23,28]. Rainfall intensity and litter characteristics such as litter type and areal litter mass (i.e., litter mass per unit area, kg m^{-2}), have rarely been regarded as key factors in hydrologic and erosion processes in forest ecosystems. These factors were proved to be effective in rainfall interception [35–38], thus possibly exerting strong effects on soil infiltration, surface runoff, and soil erosion.

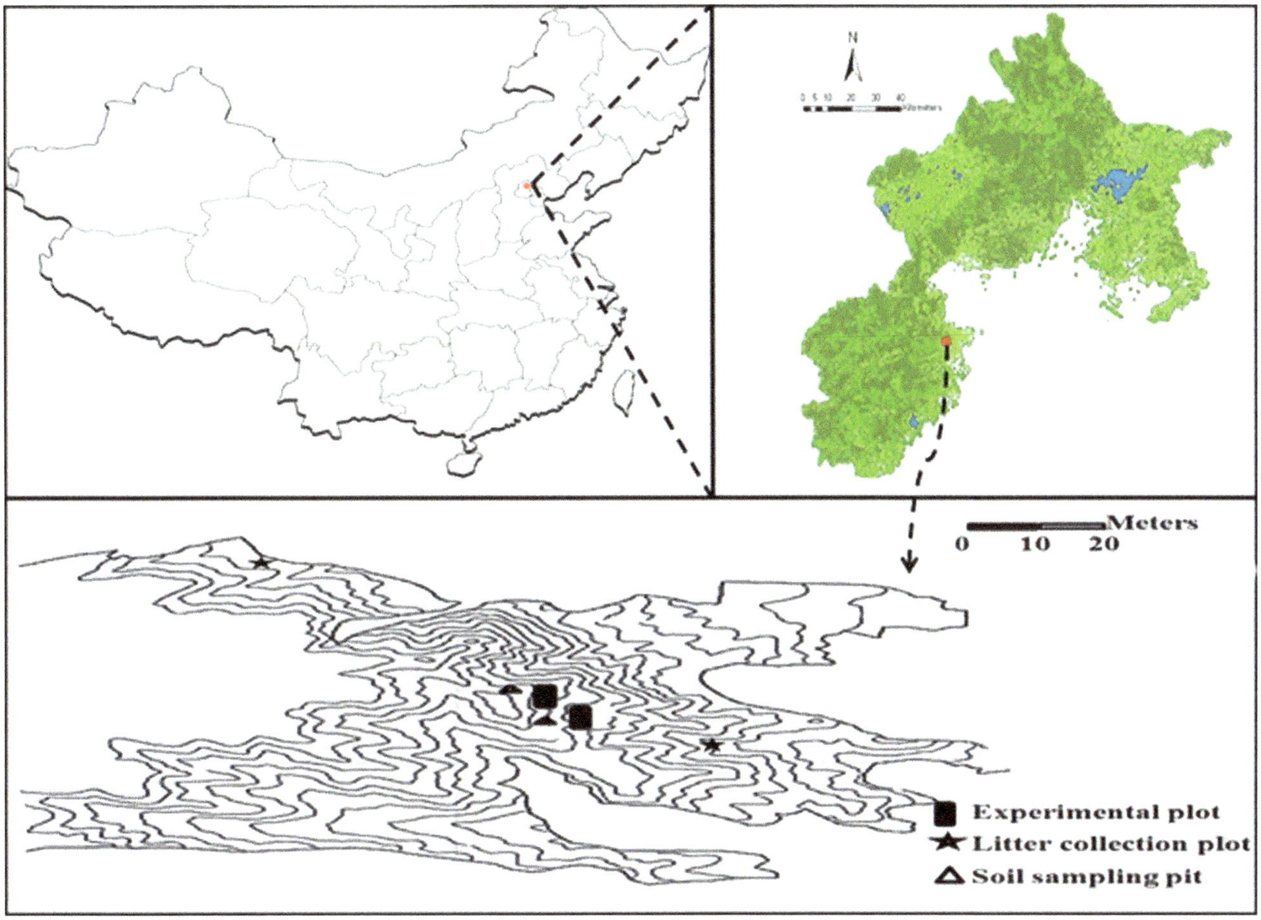

Figure 1. The location of the study area.

In this study, the effects of litter cover (i.e., little type and areal litter mass) and rainfall intensity on surface runoff and soil erosion were quantified through rainfall simulation experiments on sloped field plots with the presence or absence of litter covers. The purpose of the present work was to clarify the role of litter layers on surface runoff and soil erosion, which have caused serious environmental and ecological problems and sometimes even catastrophic events such as debris or mudslides in Northern China [39,40].

Materials and Methods

Ethics Statement

The experimental site, Jiu Feng National Forestry Park is managed by the Forestry Committee of Beijing Forestry University and is available for teaching and research of the university. This field study did not involve any endangered or protected species, and the tree species we selected were common in Northern China.

Study site

This study was performed at Jiu Feng National Forest Park, which is overseen by the Forestry Committee of Beijing Forestry University and available for the university's research and teaching activities. The park is located northwest of Beijing, China (116°28′E, 39°34′N) (Fig. 1). In its warm temperate climate, summers are hot and wet, and winters are cold and dry. Mean daily temperatures are in the range of 23–28°C from May to mid-October, and −5–15°C throughout the winter, resulting in mean annual temperature of 12°C. Mean annual precipitation was approximately 630 mm, most of which occurs as rainfall between June and September, i.e., the period with very intense and erosive rainfall events, usually following the summer drought period. In

Table 1. Forest stands and litter characteristics [38].

Litter species	Plot area (m^2)	Density (trees ha^{-1})	DBH* (cm)	Height (m)	Leaf litter length (cm)	Leaf litter width (cm)
Q. variabilis	25×15	1225	11.2	9.8	11.3–13.7	3.5–4.3
P. tabulaeformis	25×15	1748	7.6	5.2	9.5–12.5	3.3–5.1 (twigs + needles)

*Diameter at breast height.

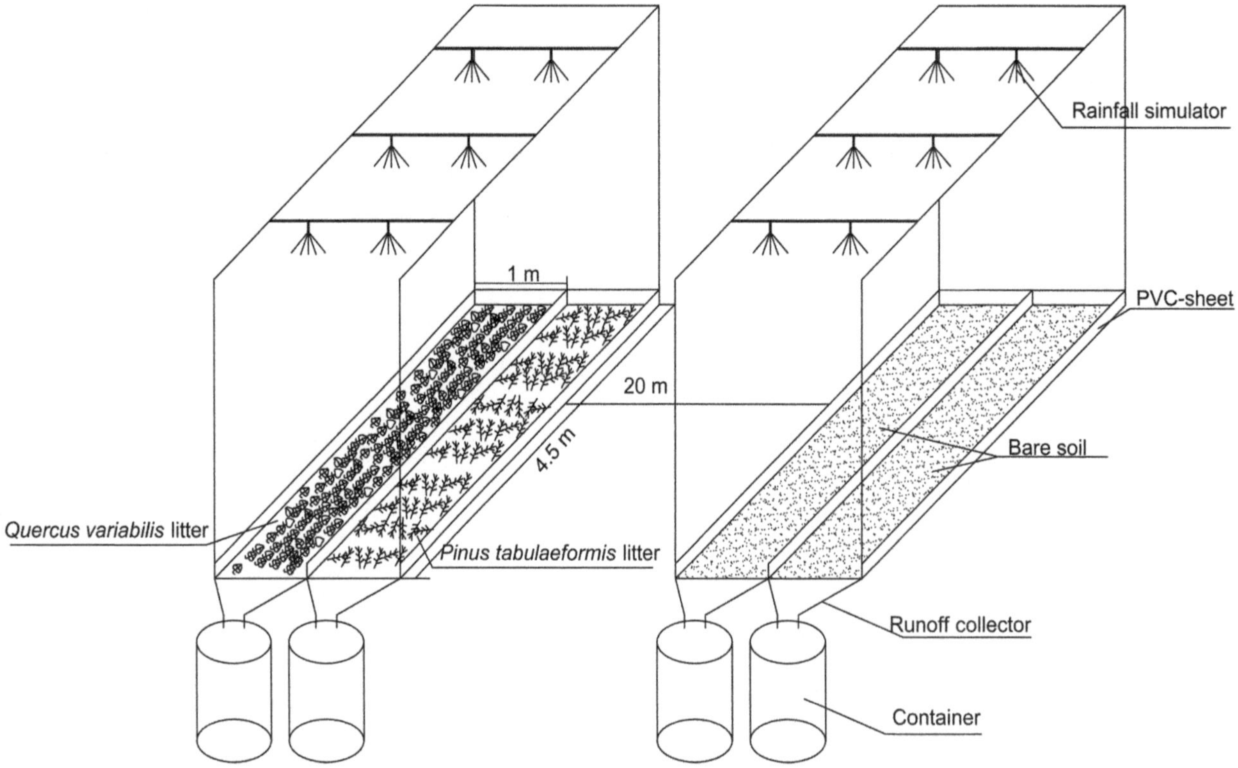

Figure 2. A single set of schematic diagram of experimental runoff plots.

the park Pinaceae *Pinus tabulaeformis*, Fagaceae *Quercus variabilis*, and Cupressaceae *Platycladus orientalis* are dominant tree species, most of which were planted in the 1950s and 1960s. At an elevation of 70–900 m, soils are mainly consisted of cinnamon soils, changing to brown soils at elevations above 900 m. The average slope is approximately 10% with a northeastern aspect.

Leaf litter

Collection and characterization of leaf litter used in this study have been described in detail in Li *et al.* [38]. Briefly, because the tree species of *Pinus tabulaeformis* and *Quercus variabilis* are broadly planted not only in the park but also across the Northern China to prevent wind and water erosion, two dominant and typical types of leaf litter was collected and tested: *Q. variabilis* litter (*QVL*), which here represents broadleaf litter, and *P. tabulaeformis* litter (*PTL*), which here represents needle-leaf litter. Regulators and cleaners remove and burn the litter for fuel almost every year. Prior to this study, the most recent removal and burning occurred in April 2010. In order to investigate the natural litter layer distribution, a 25×15 m^2 experimental stand for each type of litter, respectively. These were further divided into 1×1 m^2 sub-plots, where human activities were not allowed during the collection period from May 2010 to April 2013. Afterwards, in order to consistently investigate the variation of litter mass to ensure that the measured mass values were representative of field conditions, 10 sub-plots were randomly chosen and litter mass was measured and recorded every four months to cover seasonal variation. A 2-cm-high aluminum rectangular frame was placed over the ground to facilitate observation and collection of fallen leaves in each sub-plot. No cover was set in the sub-plots so that the litter was not isolated from the topsoil, also the biological decomposition of litter was not affected. Because the litter was moved around by wind, the mass of collected litter varied from one

Table 2. Soil properties (0–40 cm deep) in experimental and control plots.

	Bare-soil plot	Litter-covered plot
Soil moisture (%)	21.2	17.5
Porosity (%)	57.4	55.4
Bulk density (g cm^{-3})	1.57	1.43
pH	5.6	5.7
Sand (%)	79.5	73.6
Silt (%)	10.9	12.5
Clay (%)	9.6	13.9

Table 3. Runoff yield and runoff coefficient data corresponding to each treatment and simulated rainfall event.

Type of cover	Litter mass (kg m^{-2})	Rainfall intensity (mm h^{-1})				
		5.8	11.8	25.2	49.8	75.6
		Runoff yield (mm), runoff coefficient (%)				
Bare soil	0	2.09, 36.7	6.58, 56.2	21.3, 84.5	43.98, 88.3	68.42, 90.5
Q. variabilis	0.3	0.81, 14.2	4.64, 39.7	13.64, 54.1	26.71, 53.6	51.05, 67.5
	0.5	1.28, 22.5	3.66, 31.3	13.24, 52.5	26.53, 53.3	54.50, 72.1
	0.8	0.98, 17.2	3.66, 31.3	13.43, 53.3	33.26, 66.8	52.10, 68.9
	1.0	0.28, 4.9	2.75, 23.5	13.79, 54.7	33.14, 66.6	51.90, 68.7
P. tabulaeformis	0.3	0.71, 12.5	2.98, 25.5	12.31, 48.9	35.74, 71.8	52.06, 68.9
	0.5	0.65, 11.4	3.06, 26.2	11.58, 46.0	29.10, 58.4	43.54, 57.8
	0.8	0.69, 12.1	3.29, 28.1	9.60, 38.1	28.82, 57.9	54.24, 71.8
	1.0	0.35, 6.1	3.16, 27.0	12.76, 50.6	36.35, 73.0	50.30, 66.5

sub-plot to another. Because the interface between the litter and soil was clearly distinguishable, the uncomposed litter was selected as the experimental material instead of the fresh leaves, twigs and decomposed litter, only litter with intact shapes were collected. These samples were manually transferred into plastic bags, and transported to the laboratory, where they were allowed to air-dry.

The lengths and widths of the collected needle-leaf litter and broadleaf litter were measured (Table 1). Because the twigs and needles were difficult to separate, they were weighed together, and decomposed needles and broadleaves were excluded from the measurements. The areal litter mass in kg per m^2 was then calculated by dividing the weighed litter mass over the sub-plot areas. The undecomposed litter was chosen for two main reasons: first, it was the dominant component of the litter (approximately 85%) and was easy to identify; second, the undecomposed litter was located in the upper layer where it would directly intercept the raindrops. For this reason, that the undecomposed litter was assumed to play a more significant role in modulating surface runoff and soil erosion processes than decomposed litter and half-decomposed litter. Accordingly, the mass of this component of litter ranged from 0.33 to 1.24 kg m^{-2} for *QVL* and from 0.18 to 0.77 kg m^{-2} for *PTL* in the sub-plots (1×1 m^2). Based on the measurements mentioned above, to represent the litter variation characteristics and erosion in the slope accurately and to effectively illustrate and compare the hydrological and erosive responses of the two types of litter, four areal litter masses of 0.3, 0.5, 0.8, and 1.0 kg m^{-2} were selected for testing in rainfall simulation studies. This may explain the role of litter in runoff and sediment depletion.

Experimental plot design

Based on initial site survey, two northeastern-facing open runoff plots (4.5×2 m^2 each) with an approximate slope of 10% were selected for rainfall simulation experiments. These served as the bare plot and the litter-covered plot, respectively. The plots were approximately 20 m apart, the soil moisture and soil texture composition were almost the same, so they were not decisive factors in the infiltration and runoff generation in further analysis of runoff yields. All large stones were removed from both plots, and the remaining decomposed litter was removed from the litter-covered plot. No trees or other shrubs were present in either plot. Both plots were divided in half with each part measuring 4.5×1 m^2. The two halves of the divided plots were then covered with *Q. variabilis* litter (*QVL*) and *P. tabulaeformis* litter (*PTL*). These served as litter-covered plots in comparisons of the effects of the two different types of litter on runoff and erosion control (Fig. 2). Before each test run, leaf litter was brought out from plastic bags, then carefully distributed by hand in the litter-covered plot at litter masses of 0.3, 0.5, 0.8, and 1.0 kg m^{-2}. Litter thickness was measured in the upper, middle, and lower part of the plot to test whether the litter covered soil surface uniformly. Several adjustments were made if the litter thickness varied by more than 0.5 cm. The remaining two plots (4.5×1 m^2 each) were regarded as the control (or bare-soil) plot with no litter cover. Within almost 30 min of the cessation of rainfall, leaf litter was re-collected in plastic bags and then oven-dried at 80°C, and evenly returned to the plots before the next rainfall simulation.

Soil properties

The soil properties of each plot are presented in Table 2 (A common Chinese Soil Taxonomy was reported by CRGCST [41] and Shi *et al.* [42]). To comprehensively represent the soil characteristics of the experimental plot, bulk soil samples were collected (at a depth of 0–40 cm) from three different points outside each plot (2–5 m away from the plot) to determine soil properties from May to June, 2013. At each point, soil samples were separated into four sub-samples based on soil depth of 10 cm (0–10 cm, 10–20 cm, 20–30 cm, 30–40 cm) and stored in aluminum specimen boxes and trays. In order to determine soil bulk density (D_b) a weight basis of a 100 cm^3 soil core sample was collected under field-moisture conditions, oven-dried at 105°C for 24 h, and measured for the mass of oven-dried solids, V_t refer to the bulk volume of the soil, which includes the volume of the soil (m_s). Then D_b was calculated as m_s/V_t and the pore space between the soil particles [43], here $V_t = 100$ cm^3. Total porosity S_t can be calculated from the particle density (D_p) and bulk density (D_b) as follows:

$$S_t = 1 - \frac{D_b}{D_P}$$

Here, particle density ($D_p = \frac{m_s}{V_s}$) refers to the mass (m_s) of a unit volume of solid soil particles (V_s) [44]. Gravimetric soil water content (water mass/dry soil mass) was also measured after the

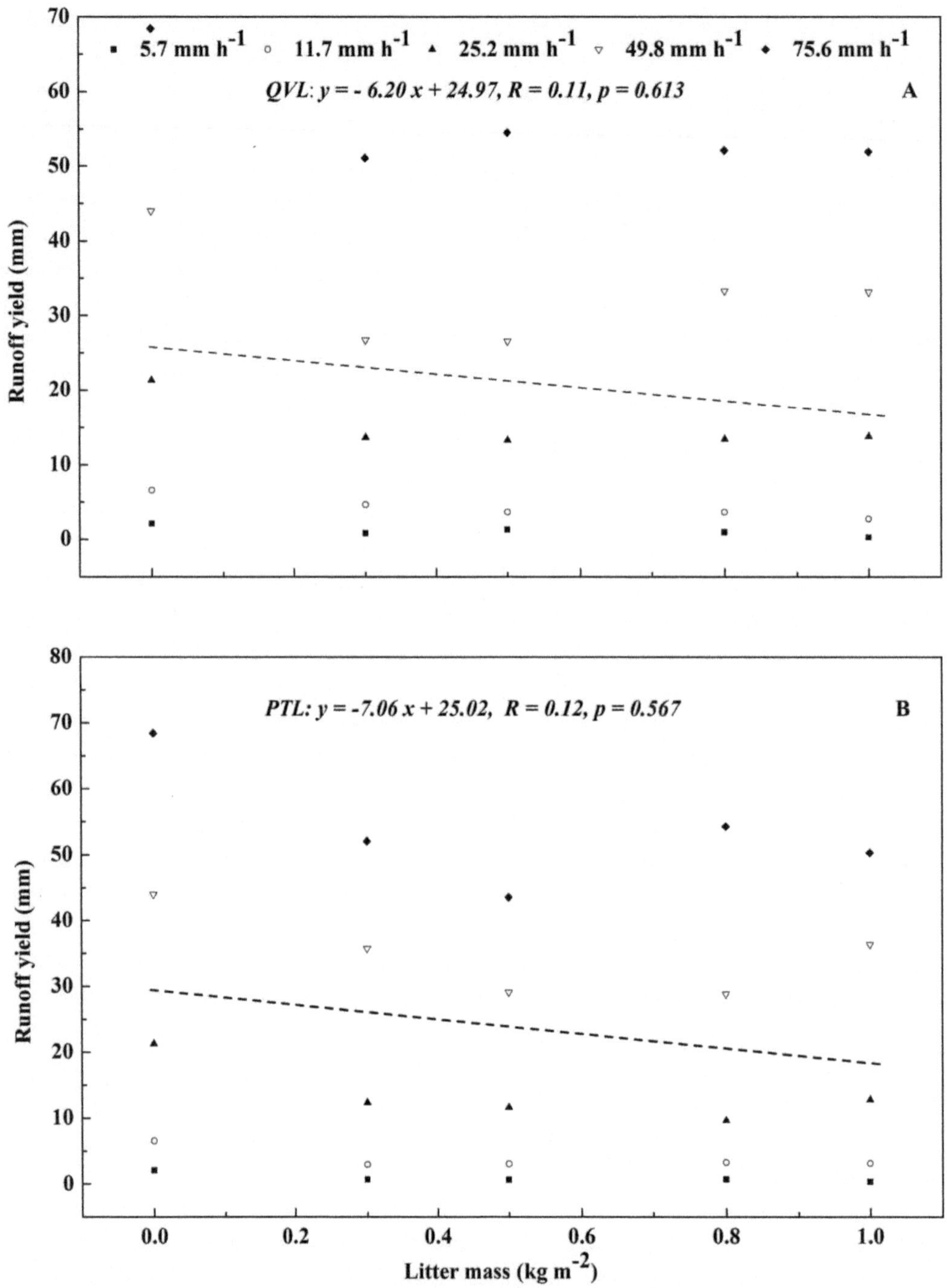

Figure 3. Runoff yield versus litter mass for five rainfall intensities for bare-soil plots (Litter mass equals zero represents 'bare'), *Quercus variabilis* litter (A), and *Pinus tabulaeformis* litter (B).

samples were oven dried [45]. Soil particle size distribution was measured after the soil samples were treated with H_2O_2 and dispersed in sodium hexametaphosphate solution using the Bouyoucos densimeter method [46,47]. Soil pH was determined with an electrode pH-meter on the saturated soil paste of 1:2.5 soil to distilled water ratio [48,49].

Rainfall simulation

The relatively high variability of the rainfall intensity in Northern China makes it difficult to examine its role in rainfall-runoff and soil erosion processes that take place under litter cover [50,51]. Here, simulated rainfall produced by an artificial rainfall simulator jointly developed by Beijing Normal University and Beijing Jiaotong University in 2006 was used [52,53]. A detailed description on the rainfall simulator was provided in Li *et al* [38]. The simulator delivered water at a height of 4.5 m and at varying speeds, spraying an area 2.2 m long ×1.5 m wide under a range of rainfall intensities. This rainfall simulator was designed to allow raindrops to reach terminal velocity and allow the flow rates to be easily controlled, resulting in simulated stable rainfall events with

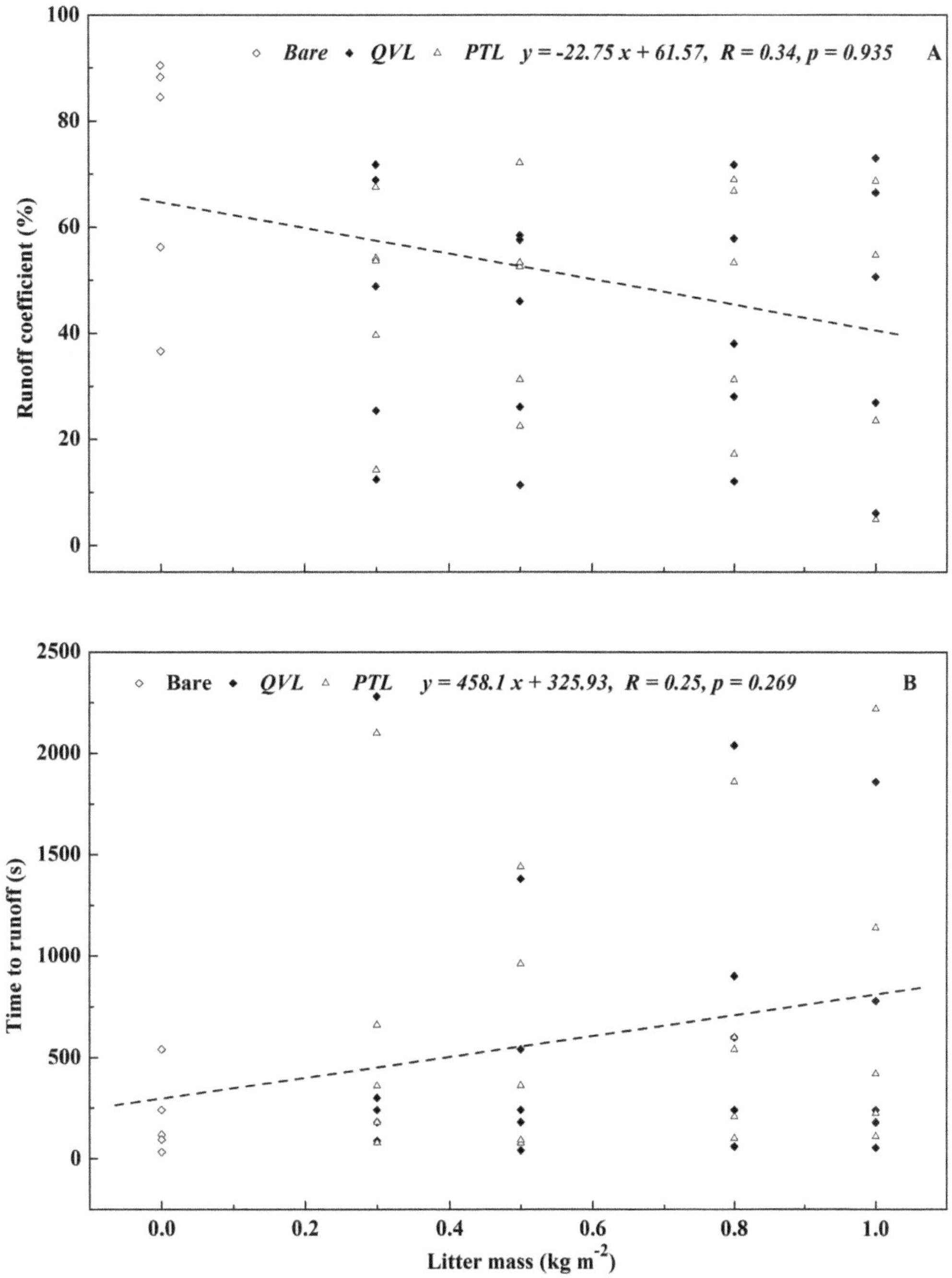

Figure 4. Runoff coefficient and time to runoff versus litter mass for bare-soil plots (open diamonds), *Quercus variabilis* litter (closed diamonds), and *Pinus tabulaeformis* litter (open triangles) for five rainfall intensities.

80% uniformity and raindrop sizes of 2.3±0.3 mm. Simulated rainfall chemistry is also an important factor to be considered in the simulation experiments. Water with high electrical conductivity tends to flocculate soil particles, and water with low electrical conductivity typical of natural rainfall may disperse and readily erode the same soil particles [54,55]. Distilled water was modified using NH_4Cl and NaCl to produce a solution with a pH of 5.5 and electrical conductivity ranged from 12.2 to 20.7 μS cm^{-1} (a detailed description of the electrical conductivity measurement was presented in the National Standard of China [56]), which was similar to the reference value of 14.8 μS cm^{-1} [57].

According to the precipitation data in the study region over 53 years from 1956 to 2008, the maximum rainfall intensity between June and September ranged from 2.6 to 82 mm h^{-1} [58], each plot was subjected to rainfall at intensities of 5.7, 11.7, 25.2, 49.8, and 75.6 mm h^{-1} for 1 h each to represent natural rainfall conditions and to measure the hydrological and erosive response to rainfall accurately. The duration of simulated rainfall was 1 h, which was consistent with the duration of the stable natural rainfall in the study region since 1984 [59,60]. Runoff and sediment samples were collected using a metal runoff collector, which was placed at the bottom of each plot to capture the runoff during the

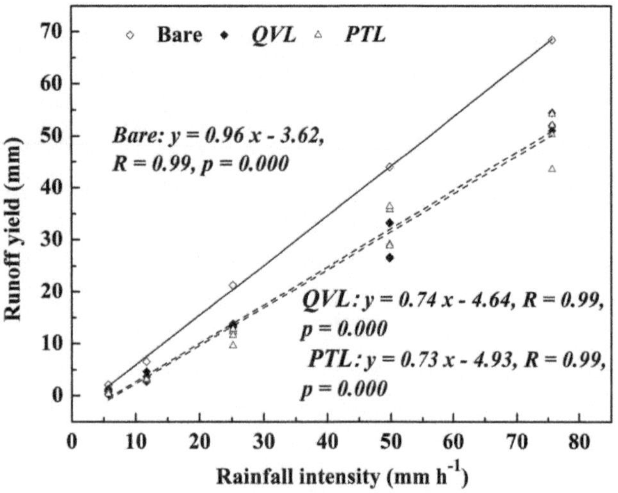

Figure 5. Runoff yield versus rainfall intensity for bare-soil plots (open diamonds), *Quercus variabilis* litter (closed diamonds), and *Pinus tabulaeformis* litter (open triangles) for four litter masses.

test run at 1-minute intervals, and the runoff sample was weighed using an electronic balance. Sediment was settled, separated from the water, and oven-dried at 105°C for 24 h, after which it was weighed to determine the sediment yield. Time to runoff (T_r) was measured as the time when runoff started to develop on the plot surface after the rainfall began [21,61]. Within 30 min after the cessation of rainfall, by which the gravitational water had drained out of the litter, the litter was removed from the soil manually and collected in two plastic bags.

Runoff coefficient (R_c) defined as the proportion of total rainfall that becomes runoff during a storm event, is used to describe the variation of runoff and water resource development [62]. R_c was calculated for each rainfall event. Sediment concentration (*SC*) was determined by the equation $SC = \frac{Y}{Q}$, where *Y (g)* represents sediment yield, and *Q (L)* represents runoff volume. In addition, infiltration and the antecedent soil water content affected runoff process very visibly [63]. Given the wide range of simulated rainfall intensities, the simulated rainfall intensity (*RI*) order 5.7, 25.2, 49.8, 11.7, and 75.6 mm h^{-1} was set to avoid that two high-intensity of rainfalls were simulated consecutively. To lessen the impact of infiltration on runoff and sediment generation, changes in soil water content in the soil profiles were measured by inserting time-domain reflectometry probes to a depth of 20 cm before and after the experiment at 2-day (48-h) intervals to determine whether the soil water content had reached 20%±2% (Table 2). If the soil water content reached or exceeded 25%, the rainfall simulation would be delayed for another 24 h until the initial water content before every experimental run was 20%±2%. All test runs were carried out from August to September 2013, during which no intensive rain or significant wind occurred. Each test was repeated once.

Statistical analysis

Linear, polynomial and non-linear regressions were used to determine the relationship between the litter mass (or *RI*) and runoff parameters (or sediment parameters). Correlation analysis among runoff (e.g. runoff yield, R_c, T_r), and sediment (e.g. sediment yield, *SC*) as dependent variables and litter mass and of *RI* as an independent variable was performed to evaluate possible relationships among them, and to facilitate understanding of the hydrological and erosion processes. Parameters were considered to be significantly correlated when they were at or above the 95% confidence level ($p \leq 0.05$). A test of normality was conducted by comparing the Sig. value in Kolmogorov-Smirnov and Shapiro-Wilk. Homoscedasticity was carried out to determine the whether the data was homogeneous. Sig>0.05 was defined as homoscedasticity in the Levene Statistic. One-way analysis of variance (ANOVA) was carried out to determine if there were any differences in runoff and sediment yield among three plot treatments (one bare-soil, two litter-covered plots), specifically differences between the broadleaf litter-covered plot and needle-leaf litter covered plot. The Fisher LSD (Least Significant Difference) test at $p \leq 0.05$ was used to test for significant differences. Data were grouped by treatment (bare, *Q. variabilis* litter cover and *P. tabulaeformis* litter cover). All statistical analyses were performed using IBM SPSS Statistics 20.0 software.

Results and Discussion

Runoff

Runoff yield and runoff coefficient (R_c) data are presented in Table 3. Generally, the most runoff took place in the control plot (bare soil). The mean amount of runoff in the *QVL* and *PTL* cover plots was lowered by 29.5% and 31.3% compared to the bare plot, respectively. In agreement, similar results were reported by Singer and Blackard [28], Pannkuk and Robichaud [20], Findeling *et al.* [15], and Cerdà and Doerr [21], with a reduction that ranged from 0.12–55.6% in contrast with the bare soil. This was likely due to the protection provided by the litter cover, which absorbed the energy of raindrops and also increased the roughness of soil surface to increase infiltration rate, and delayed and reduced runoff [49]. However, the opposite result was observed for straw mulch cover: runoff yield was slightly higher (by 0.09%) than that in a bare-soil plot [28,33]. Similar results were obtained for rock fragments cover [12,13,64] and plastic mulch cover [34], presumably because the physical shapes or impervious nature of these surface covers concentrated the flow of water to decrease the infiltration rate and increase the flow velocity [25,65]. However, results of LSD test in ANOVA revealed no significant differences ($p \leq 0.05$) in runoff yield between the control and two litter-covered plots ($p = 0.417$ and 0.390 respectively). In contrast, significant differences ($p \leq 0.05$) were found in R_c ($p = 0.000$ and 0.001) and T_r ($p = 0.028$ and 0.012) between bare and litter-covered plots. The results implied that litter cover had an impact on runoff, but other factors, such as precipitation and rainfall intensity should be taken into account as well.

The manner in which litter type (broadleaf *vs* needle-leaf), specially differences in physical leaf shape, might affect runoff was studied. However, no significant differences in runoff yield ($p = 0.696$), R_c ($p = 0.677$) and T_r ($p = 0.681$) were found between *QVL* and *PTL* plots, indicating that litter type was a the major influencing factor in controlling runoff in the present study as it was supposed to be. Very few studies concentrated on the role of litter type in runoff reduction, particularly on the physical differences in leaf shape. In a previous work, Neris *et al.* argued that runoff from the pine needle litter cover was twice of the rainforest litter cover, which does not match the current results [47]. The discrepancy suggests that further research is required.

Litter mass was treated as an influencing element in runoff process by covering the topsoil, but its effect on runoff remains unclear. In general, no significant linear correlations were found between litter mass and runoff yield for *QVL* ($R = 0.11$, $p = 0.923$)

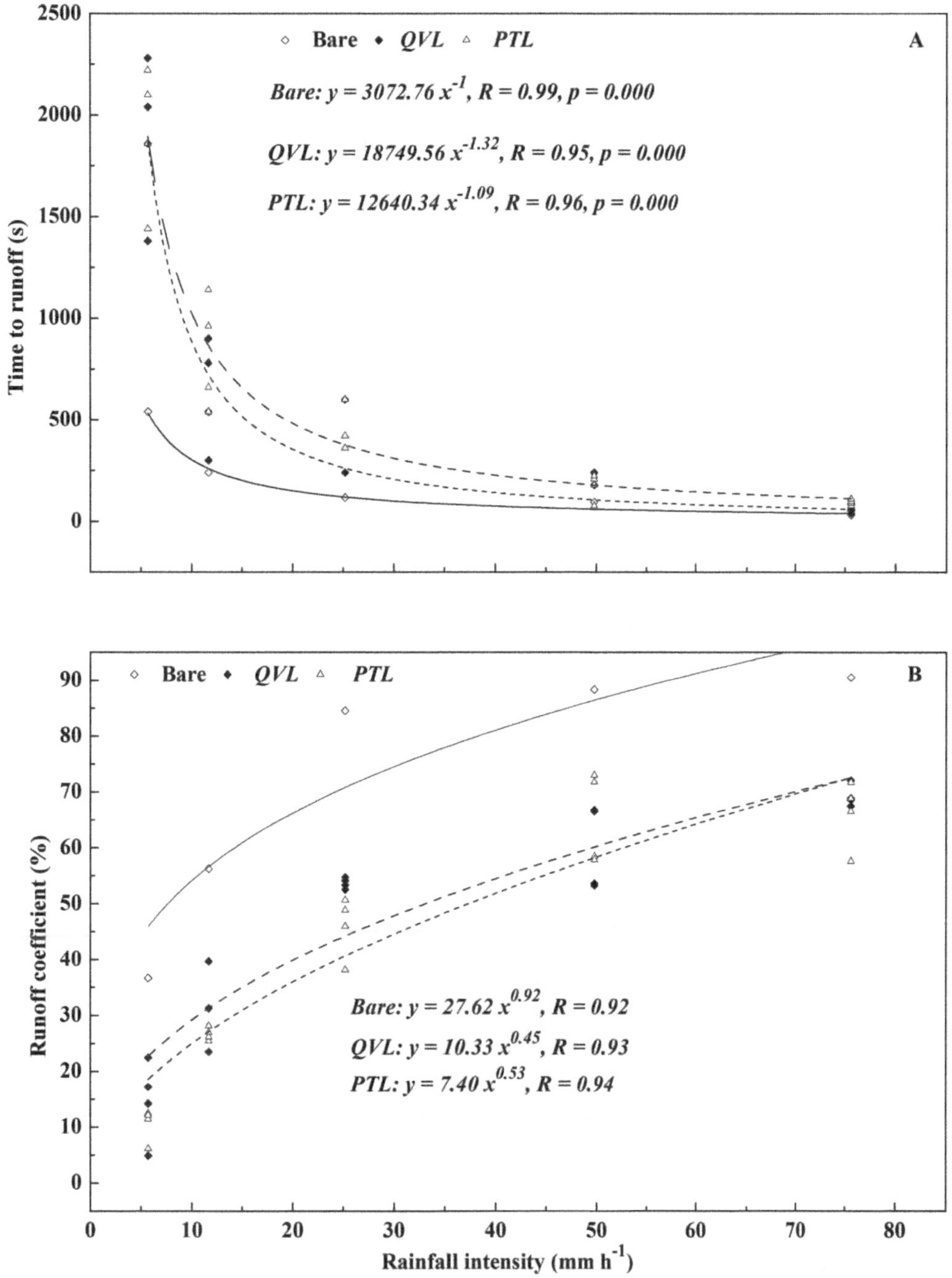

Figure 6. Rainfall intensity versus time to runoff and runoff coefficient for bare-soil plots (open diamonds), *Quercus variabilis* litter (closed diamonds), and *Pinus tabulaeformis* litter (open triangles) for four litter masses.

and *PTL* ($R = 0.12$, $p = 0.964$) (Fig. 3), and no non-linear correlations were observed. Runoff yield decreased dramatically when litter mass increased from 0 to 0.3 kg m^{-2} but then remained steady regardless of the increasing litter mass afterwards, which proved that litter cover had a strong impact on runoff reduction as stated above. R_c ($R = 0.34$, $p = 0.114$) and T_r ($R = 0.25$, $p = 0.078$) showed a similar correlations with litter mass (Fig. 4). One likely explanation was that the increased litter mass might intercept and store more rainwater. However, on the other hand, it may provide a flow channel for runoff by covering more bare topsoil so that runoff flowed on rather than through the leaf litter, particularly under rainfall intensities of 49.8 and 75.6 mm h^{-1}. For this reason, more runoff was observed with higher litter masses of 0.8 and 1.0 kg m^{-2}. While Findeling *et al.* compared the efficiency of different corn residue masses (0, 0.15 and 0.45 kg m^{-2}) in cutting runoff down, and reported that mean values of R_c were 0.44, 0.16, and 0.05, respectively [15]. They confirmed that surface cover contributed notably to the decrease of runoff in contrast with the bare soil, but because they tested only a narrow range of masses,, the slight increase in mass was not sufficient to facilitate extrapolation that runoff would decrease as the ground

Table 4. Sediment yield and sediment concentration data corresponding to each treatment and simulated rainfall event.

Type of cover	Litter mass (kg m^{-2})	Rainfall intensity (mm h^{-1})				
		5.8	11.8	25.2	49.8	75.6
		Sediment yield (kg), sediment concentration (g L^{-1})				
Bare soil	0	0.17, 18.29	0.23, 7.77	1.65, 17.26	4.70, 23.73	7.02, 22.79
Q. variabilis	0.3	0.07, 17.89	0.13, 6.19	0.79, 12.91	0.58, 4.79	2.30, 10.02
	0.5	0.07, 12.00	0.14, 8.42	0.51, 8.53	0.55, 4.64	1.24, 5.06
	0.8	0.05, 10.88	0.08, 4.98	0.18, 2.96	0.31, 2.09	0.42, 1.80
	1.0	0.06, 49.37	0.14, 11.39	0.16, 2.63	0.16, 1.10	0.22, 0.96
P. tabulaeformis	0.3	0.10, 30.14	0.12, 8.61	0.21, 3.77	0.85, 5.29	3.38, 14.44
	0.5	0.06, 19.97	0.15, 10.65	0.32, 6.14	0.68, 5.21	2.94, 15.03
	0.8	0.06, 17.97	0.09, 5.89	0.17, 3.92	0.50, 3.86	0.52, 2.11
	1.0	0.04, 24.44	0.13, 9.17	0.18, 3.08	0.26, 1.58	0.32, 1.40

cover mass increased, which suggested that a broader range of mass should be studied in further works.

Significant linear correlations were found between rainfall intensity (*RI*) and runoff yield for all the plots in Figure 5 ($R = 0.99$, $p < 0.05$). In all plots, runoff yield increased with increasing *RI*. Runoff yield at the maximum *RI* of 75.6 mm h^{-1} was 32.7, 62.4, and 83.4 times larger than that at the minimum *RI* of 5.7 mm h^{-1} in bare, *QVL* and *PTL* plots respectively. Based on regression analysis, R_c and time to runoff (T_r) showed a significant power ($R > 0.9$, $p < 0.05$) and reverse-power ($R > 0.95$, $p < 0.05$) relationships with *RI* (Figure 6A and 6B). As expected, mean R_c increased from 12.6% ($RI = 5.7$ mm h^{-1}) to 67.8% ($RI = 75.6$ mm h^{-1}) in litter-covered plots (Figure 6B), but T_r decreased apparently from nearly 30 min to 1~1.5 min with increasing *RI*. The findings were in accordance with those reported by Gholami *et al.* [65], who noted that mulch cover was more effective in controlling runoff at a low *RI* of 30 mm h^{-1} ($R_c = 49.9\%$, $T_r = 81.83$ s) than at a high *RI* of 90 mm h^{-1} ($R_c = 66.6\%$, $T_r = 55.22$ s). The phenomena in the present study also proved that *RI* (rather than litter mass) was a dominant factor in determining runoff irrespective of litter cover. This was probably because when *RI* was high, leaf litter was pressed close to the soil surface, blocking the channels between leaves. The accumulative precipitation was assumed to surpass infiltration rate, which resulted in the generation of more runoff.

Sediment

The magnitude of the sediment yield and sediment concentration (*SC*) are shown in Table 4. Generally, bare-soil plot generated 2.75 kg sediment, which was almost 6.7 and 5 times larger than that in *QVL* (0.41 kg) and *PTL* (0.55 kg) plots, respectively. Further, ANOVA analysis indicated there were significant differences ($p = 0.000$ and $p = 0.001$) between bare plot and two litter-covered plots. Similar results were obtained in terms of *SC* in bare and litter-covered plots ($p = 0.001$ and 0.003). In this way, litter cover was able to retard erosion. The observed outcomes were similar to those presented in previous studies. For example, Pannkuk and Robichaud reported that rill and inter-rill erosion were reduced by 70% when plots were covered with Douglas fir needles and 30% when they were covered with Ponderosa pine needles, as compared with the erosion in a bare-soil plot [20]. Cerdà and Doerr reached similar conclusions in their study on soil erosion from the plots covered with ash and pine needles [21]. Benkobi *et al.* suggested that soil loss was reduced by about 87% in 100% litter cover relative to bare soil [23]. Miyata *et al.* reported that annual soil erosion in uncovered plots was 3.7 times greater than in covered plots [30]. However, Singer and Blackard found that more soil loss was observed when mulch covers <40% in contrast with bare soil [28], this was mainly because they excluded the splash erosion off the inter-rill plot, and overall sediment was reduced under mulch cover.

In the present study, because the litter cover prevented raindrops from hitting the soil surface directly by intercepting and storing the rainfall, the detachment of soil aggregates and splash erosion was lessened [66,67]. Less soil was lost in the litter-covered plots than in the bare plot.

Additionally, mean sediment yield and *SC* in the *QVL* plot were slightly lower than in the *PTL* plot, but no statistical differences were found in sediment yield ($p = 0.544$) and *SC* ($p = 0.810$) between the two plots. The findings indicated that litter type (broadleaf or needle-leaf) was may not be a defining predictor in reducing soil loss. This was probably because even though more topsoil was exposed and soil aggregations were detached under needle-leaf litter cover compared with broadleaf litter, needle-leaf litter tended to form mini-debris dams to trap down the soil particles in the runoff and return them to the soil surface [20]. The results were consistent with Neris *et al.*, who also reported that no significant differences were observed between pine forest floor (including needle leaf litter) and a rainforest floor (including broadleaf litter) [47].

Linear relationships between litter mass and sediment yield in *QVL* ($R = 0.50$, $p = 0.05$) and *PTL* plot ($R = 0.52$, $p = 0.112$) are shown in Figure 7. In general, total sediment yield decreased from 13.77 to 0.75 kg (*QVL*) and 0.92 kg (*PTL*) when the litter mass increased from 0 (bare) to 1 kg m^{-2}, but the largest decreases of 71.9% and 66.2% were revealed when litter mass increased from 0 to 0.3 kg m^{-2}. Similar relationships are presented in Figure 8 in terms of litter mass and *SC*. *SC* decreased as litter mass increased. Although no significant correlations were observed, the effect of litter mass on reducing sediment was found to be relevant, as the increasing litter mass covered the remaining bare soil, which indirectly protected soil from raindrop impact, and reduced the velocity of runoff [15,67]. Similarly, Lal reported that soil loss decreased from 52.4 to 0.04 kg m^{-2} when the maize grain yields increased from 0.06 to 0.68 kg m^{-2} [68].

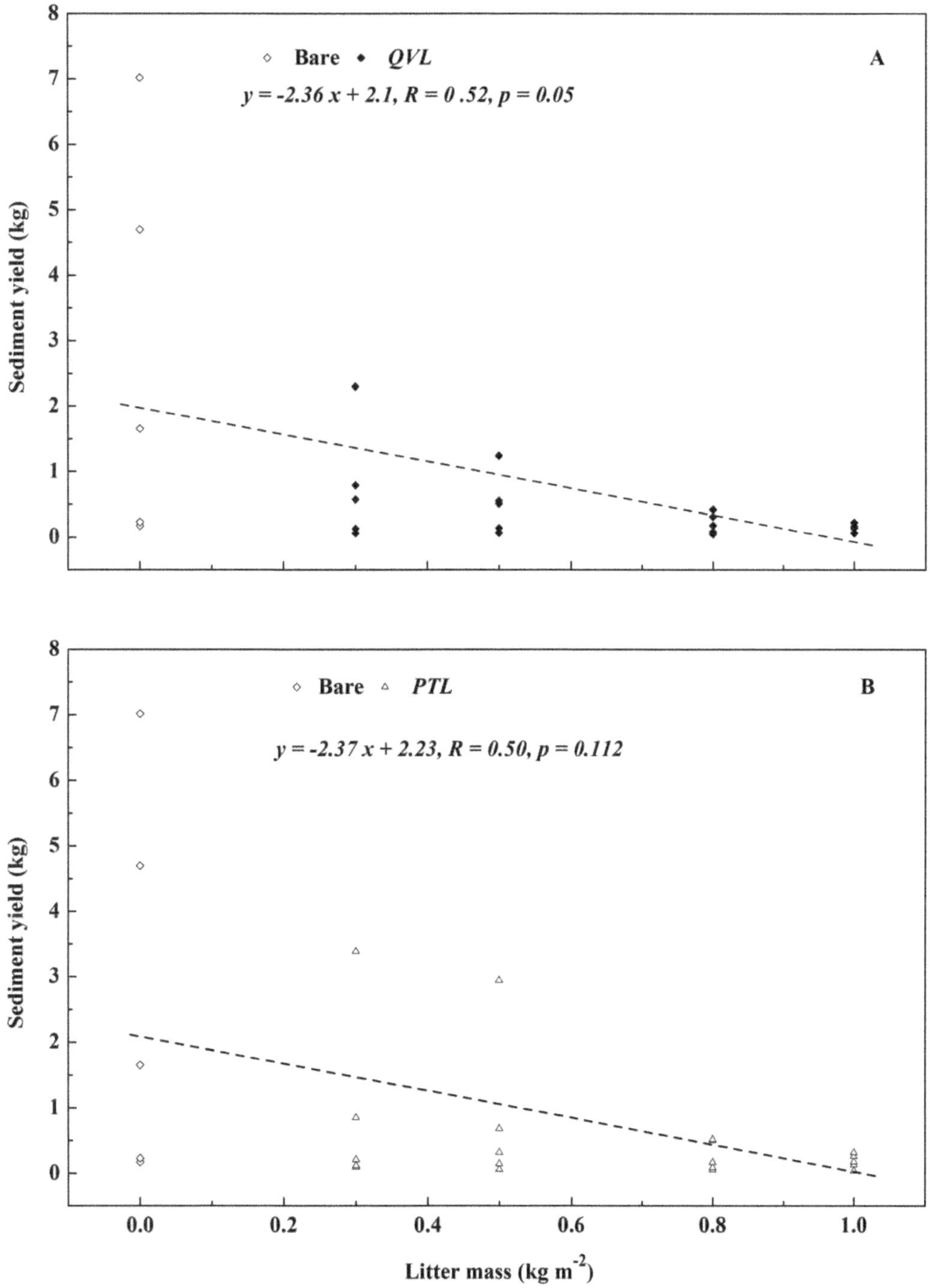

Figure 7. Relationship between litter mass and sediment yield for bare-soil plots (open diamonds), *Quercus variabilis* litter (closed diamonds), and *Pinus tabulaeformis* litter (open triangles) for five rainfall intensities.

RI also had an impact on sediment generation. Both linear (bare soil: $R = 0.99$) and polynomial relationships (*QVL*: $R = 0.63$, $p = 0.055$; *PTL*: $R = 0.71$, $p = 0.022$) between *RI* and sediment yield are presented in Figure 9A. When *RI* increased from 5.7 to 75.6 mm h^{-1}, sediment yield increased 41.3, 17.5, and 29.8 times in bare, *QVL* and *PTL* plots, respectively. *SC* showed a reverse-power decrease trend with *RI* in *QVL* and *PTL* but increased in bare plots. These results verified the protective role of the leaf litter layer, which was in line with the previous studies [24,25,46]. This was mainly because on one hand, leaf litter reduced the detachment of soil aggregation by covering on the topsoil, particularly, high-intensity rainfall (49.8 and 75.6 mm h^{-1}) with intense striking force pressed leaf litter closer to the soil surface [17]; on the other hand, the coverage by leaf litter increased the surface roughness, which facilitated sediment deposition in the runoff [30,67].

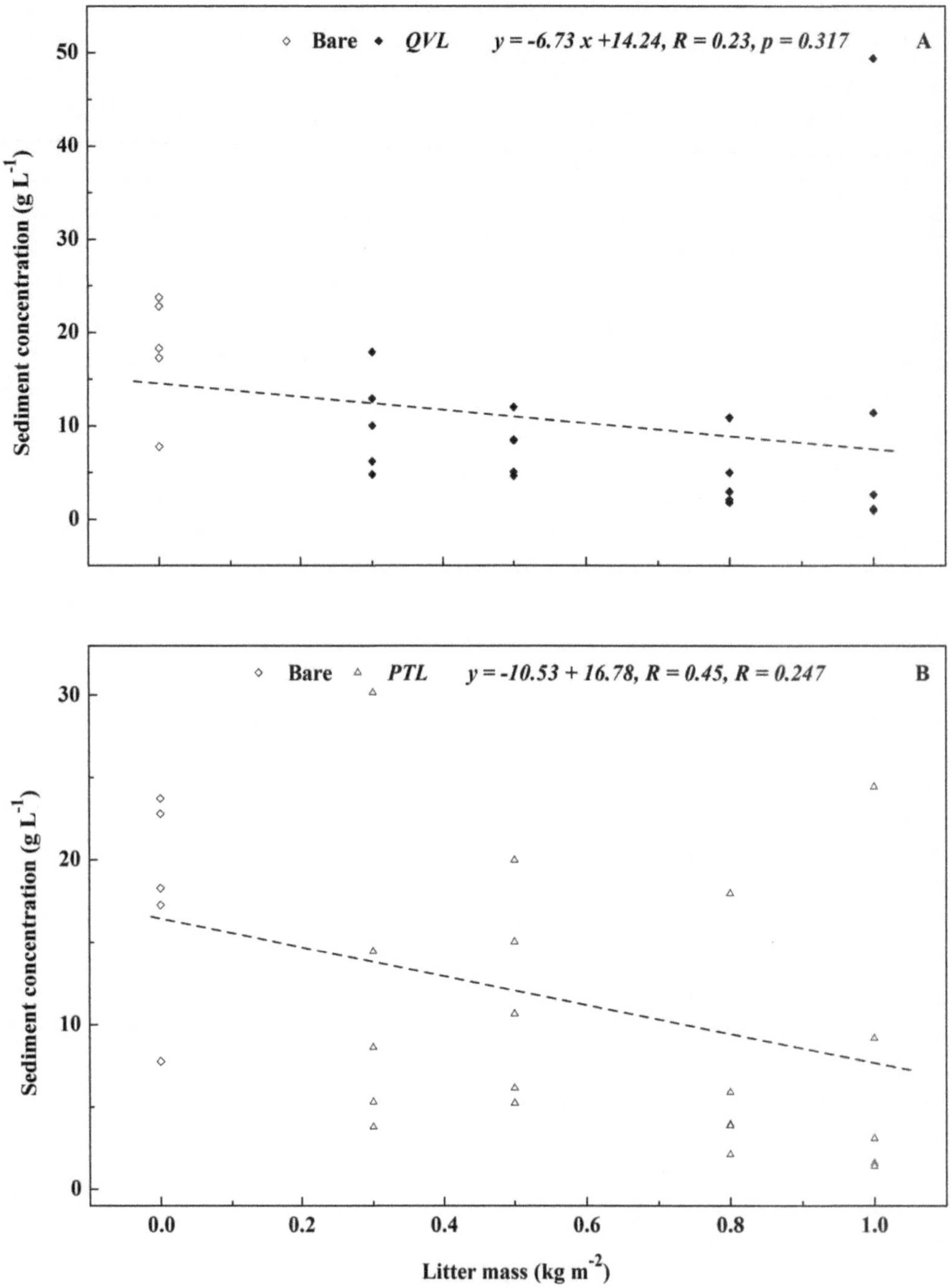

Figure 8. Relationship between litter mass and sediment concentration for bare-soil plots (open diamonds), *Quercus variabilis* litter (closed diamonds), and *Pinus tabulaeformis* litter (open triangles) for four litter masses.

Relationship between runoff and sediment

As analyzed above, runoff yield in the litter-covered plots was not significantly different ($p = 0.226$ and 0.220) from that in the bare plot, but significant differences ($p = 0.000$ and 0.000) were shown in sediment between the bare plot and litter-covered plots. This can be explained briefly by the different forms of interaction between litter type and runoff-erosion process. For broadleaf litter, runoff generated and flowed on the leaf surface, while the presence of needle-leaf litter increased the soil surface roughness and trapped down soil particles in the flow as a barrier [15,21,67,69]. Significant correlations ($p \leq 0.05$) between runoff and sediment were observed for plots in Figure 10A and 10B. Apparently, sediment yield increased with the increasing runoff. This is similar with the previous studies (e.g. [30,65,70]). In addition, an excellent linear regression was observed in bare plot ($R = 0.99$). The greatest runoff, which showed large amounts of energy, was capable of detaching soil aggregation and moving the soil particles. However, leaf litter reduced runoff and sediment by covering soil to prevent splash erosion and largely increasing surface roughness which in turn increased infiltration. The phenomenon confirmed again that litter cover played an irreplaceable role in runoff and soil erosion control.

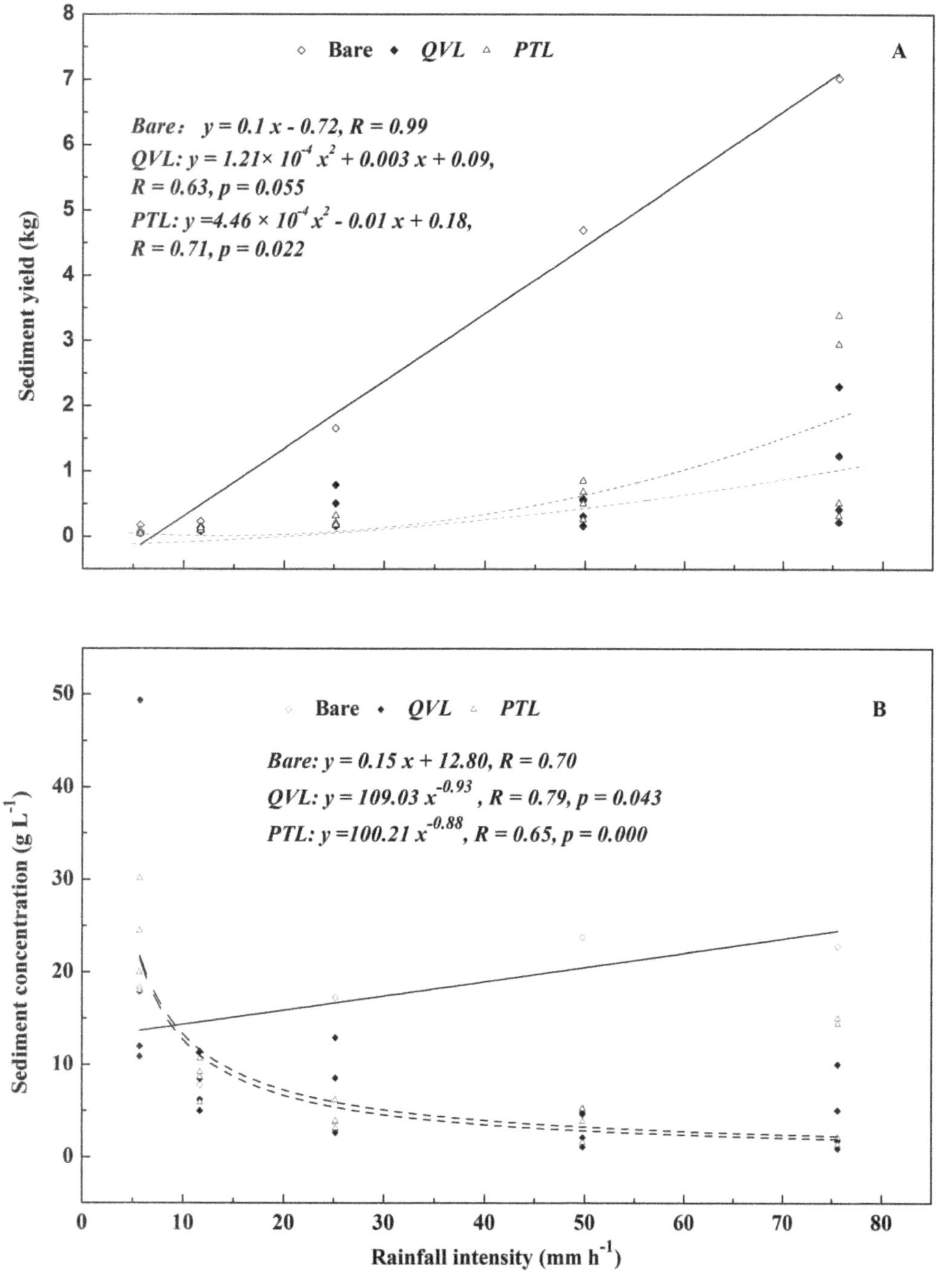

Figure 9. Relationships between sediment yield (A) and sediment concentration (B) versus rainfall intensity for bare-soil plots (open diamonds), *Quercus variabilis* litter (closed diamonds), and *Pinus tabulaeformis* litter (open triangles) for four litter masses.

Conclusion

The research effort was conducted to examine the role of leaf litter in runoff and soil erosion. Results showed that runoff and sediment yield were lower in litter-covered plots than in bare plot, litter cover had a relevant impact on runoff and sediment control. A detailed analysis of how runoff and sediment interacted with litter type, litter mass and rainfall intensity showed that rainfall intensity rather than litter type (broadleaf or needle-leaf) or litter mass was the decisive factor in generating runoff. The correlation between runoff yield and sediment yield was significantly positive ($p<0.05$) in all the plots, which confirmed that increasing runoff with larger energy was able to detach soil aggregation and cause severe erosion afterwards.

These results improved our understanding of litter as a protective layer and rainfall re-distributor. The data may be used to establish hydrological models to predict changes in runoff and soil erosion, and to provide scientific support for managing water resources in Northern China. Future studies should focus on the effects of slope, plot length, other litter types, and hydraulic characteristics (e.g., flow speed, erosive power, and shearing force)

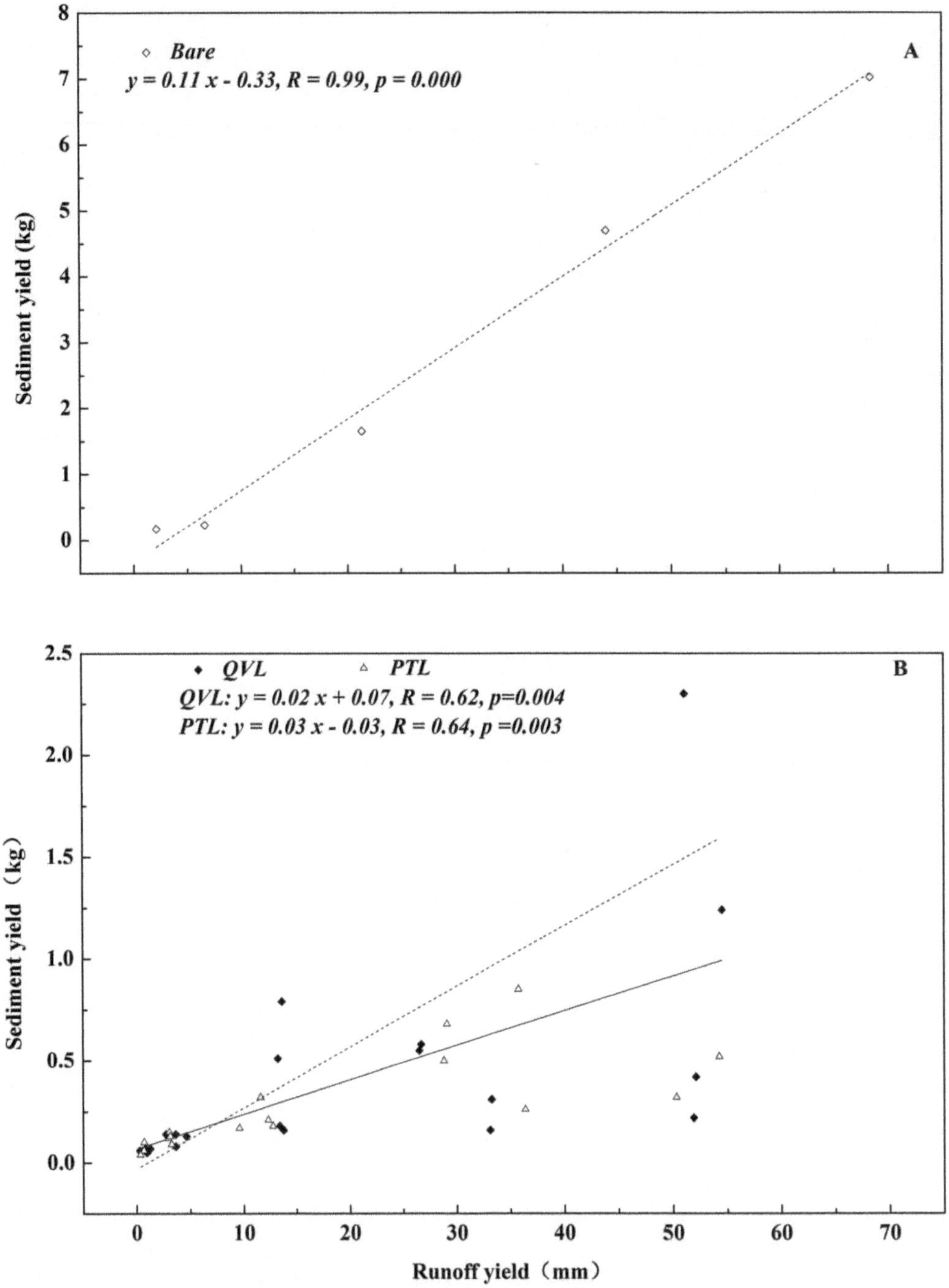

Figure 10. Relationships between runoff yield and sediment yield for bare-soil plots (open diamonds), *Quercus variabilis* litter (closed diamonds), and *Pinus tabulaeformis* litter (open triangles).

on runoff and soil erosion under litter cover to draw comprehensive conclusions.

Acknowledgments

We would like to thank the Jiu Feng National Forest Park Administration for their help and support. Further, we deeply appreciate the efficient and effective assistance provided by graduate students Shangjie Chen, Xiaoqing Du, Mingxing Wang, Bingpeng Qu, and Zhentai Pan during the field experiment.

Author Contributions

Conceived and designed the experiments: XL JN BX. Performed the experiments: XL. Analyzed the data: XL JN BX. Contributed reagents/materials/analysis tools: JN. Wrote the paper: XL JN.

References

1. Morgan RPC (2005) Soil and Conservation, 3rd edition. London: Blackwell Publishing Ltd. 304 p.
2. Hugo V, Pleguezuelo C (2008) Soil-erosion and runoff prevention by plant covers. A review. Agron J 28: 65–86. doi: 10.1007/978-90-481-2666-8_48.
3. Li ZG, Cao Y, Liu BZ, Luo ZD (2008) Current status and developing trend of soil erosion in China. Science of Soil and Water Conservation 6(1): 57–62. (In Chinese with English abstract)
4. Zhang HJ (2008) Soil erosion theory, 2nd edition. Beijing: China Forestry Publishing House. 355 p.
5. Kimoto A, Uchida T, Mizuyama T, Li CH (2002) Influences of human activities on sediment discharge from devastated weathered granite hills of southern China: effects of 4-year elimination of human activities. Catena 48: 217–233. doi: 10.1016/S0341-8162(02)00029-2.
6. Bochet E, Poesen J, Rubio JL (2006) Runoff and soil loss under individual plants of a semi-arid Mediterranean shrubland: influence of plant morphology and rainfall intensity. Earth Surf Proc Land 31: 536–549. doi: 10.1002/esp.1351.
7. Walsh RPD, Voight PJ (1977) Vegetation litter: an underestimated variable in hydrology and geomorphology. J Bio 4: 253–277.
8. Geddes N, Dunkerley D (1999) The influence of organic litter on the erosive effects of raindrops and of gravity drops released from desert shrubs. Catena 36: 303–313. Doi: 10.1016/S0341-8162(99)00050-8.
9. Sayer EJ (2006) Using experimental manipulation to assess the roles of leaf litter in the functioning of forest ecosystems. Biol Rev 81: 1–31. Doi: 10.1017/S1464793105006846.
10. Abraham AD, Parsons AJ (1991) Relation between infiltration and stone cover on a semiarid hillslope Arizona. J Hydrol 122: 45–59. doi: 10.1016/0022-1694(91)90171-D.
11. Valentin C, Casenave A (1992) Infiltration into sealed soils as influenced by gravel cover. Soil Sci Soc Am J 56: 1667–1673. doi:10.2136/sssaj1992.03615995005600060002x.
12. Peosen JW, Torri D, Bunte K (1994) Effects of rock fragments on soil erosion by water at different spatial scales: a review. Catena 23: 141–166. Doi: 10.1016/0341-8162(94)90058-2.
13. Jean JS, Ai KF, Shih K, Hung CC (2000) Stone cover and slope factors influencing hillside surface runoff and infiltration: laboratory in investigation. Hydrol Process 14: 1829–1849. doi: 10.1002/1099-1085(200007)14:10<1829::AID-HYP66>3.0.CO;2-#.
14. Dickey EC, Shelton DP, Jasa PJ, Perterson TR (1985) Soil erosion from tillage systems used in soybean and corn residues. T ASAE 28: 1124–1129.
15. Findering A, Ruy S, Scopel E (2003) Modeling the effects of a partial residue mulch on runoff using a physically based approach. J Hydrol 275: 49–66. doi: 10.1016/S0022-1694(03)00021-0.
16. Pan CZ, Shangguan ZP (2006) Runoff hydraulic characteristics and sediment generation in sloped grassplots under simulated rainfall conditions. J Hydrol 331: 178–185. doi: 10.1016/j.jhydrol.2006.05.011.
17. Pan CZ, Ma L, Shangguan ZP (2010) Effectiveness of grass strips in trapping suspended sediments from runoff. Earth Surf Proc Land 35: 1006–1013. doi: 10.1002/esp.1997.
18. Adekalu KO, Olorunfemi IA, Osunbitan JA (2007) Grass mulching effect on infiltration, surface runoff and soil loss of three agricultural soils in Nigeria. Bioresource Technol 98(4): 912–917. doi: 10.1016/j.biortech.2006.02.044.
19. Bhattacharyya R, Smets T, Fullen MA, Poesen J, Booth CA (2010) Effectiveness of geotextiles in reducing runoff and soil loss. A ynthesis. Catena 81: 184–195. doi: 10.1016/j.catena.2010.03.003.
20. Pannkuk CD, Robichaud PR (2003) Effectiveness of needle cast at reducing erosion after forest fires. Water Resour Res 39 (12): 1–9. doi: 10.1029/2003WR002318.
21. Cerdà A, Doerr SH (2008) The effect of ash and needle cover on surface runoff and erosion in the immediate post-fire period. Catena 74: 256–263. doi: 10.1016/j.catena.2008.03.010.
22. Robichaud PR, Lewis SA, Wagenbrenner JW, Ashmun EL, Brown RE (2013) Post-fire mulching for runoff and erosion mitigation Part 1: Effectiveness at reducing hillslope erosion rates. Catena 105: 75–92. doi: 10.1016/j.catena.2012.11.015.
23. Benkobi L, Trlica MJ, Smith JL (1993) Soil loss as affected by different combinations of surface litter and rock. J Environ Qual 4: 657–661. doi:10.2134/jeq1993.00472425002200040003x.
24. Smets T, Poesen J, Knapen (2008) Spatial scale effects on the effectiveness of organic mulches in reducing soil erosion by water. Earth-Sci Rev 89: 1–12. doi: 10.1016/j.earscirev.2008.04.001.
25. Smets T, Poesen J, Bochet E (2008) Impact of plot length on the effectiveness of different soil-surface covers in reducing runoff and soil loss by water. Prog Phys Geog 32 (6): 654–677.
26. Lowdermilk WC (1930) Influence of forest litter on runoff, percolation, and erosion. J Forest 28: 474–491.
27. Adams JE (1966) Influence of mulches on runoff, erosion, and soil moisture depletion. Soil Sci Soc Am J 30 (1): 110–114. doi:10.2136/sssaj1966.03615995003000010036x.
28. Singer MJ, Blackard J (1978) Effect of mulching on sediment in runoff from simulated rainfall. Soil Sci Soc Am J 42 (3): 481–486. doi: 10.2136/sssaj1978.03615995004200030022x.
29. Netto ALC (1987) Overlandflow production in a tropical rainforest catchment: The role of litter cover. Catena 14: 213–231. doi: 10.1016/S0341-8162(87)80019-x.
30. Miyata S, Kosugi K, Gomi T, Mizuyama T (2009) Effect of forest floor coverage on overland flow and soil erosion on hillslopes in Japanese cypress plantation forests. Water Resour Res 45: 1–17. doi: 10.1029/2008WR007270.
31. Pierce SP (1967) Evidence of overland flow on forest watersheds. Inter Symp Forest Hydrol. 25: 247–253.
32. Ruxton BP (1967) Slopewash under primary rainforest in Northern Papua. In: Jennings JN, Mabbutt JA, editors. Landform Studies from Australia and New Guinea. Cambridge University Press. pp. 85–94.
33. McGregor KC, Bengtson RL, Muchler CK (1988) Effects of surface straw on interrill runoff and erosion of Grenada Silt Loam soil. T ASAE 31: 111–116.
34. Y . Wan, SA . El-Swaify (1999) Runoff and soil erosion as affected by plastic mulch in a Hawaiian pineapple field. Soil Till Res 52: 29–35. doi: 10.1016/S0167-1987(99)00055-0.
35. Putuhena W, Cordery I (1996) Estimation of interception capacity of the forest floor. J Hydrol 180: 283–299. doi: 10.1016/0022-1694(95)02883-8.
36. Sato Y, Kumagai T, Kume A, Otsuki K, Ogawa S (2004) Experimental analysis of moisture dynamics of litter layers – the effect of rainfall conditions and leaf shapes. Hydrol Process18: 3007–3018. doi: 10.1002/hyp.5746.
37. Guevara-Escobar, Gonzalez-Sosa, Ramos-Salinas, Hernandez-Delgado (2007) Experimental analysis of drainage and water storage of litter layers. Hydro Earth Syst Sci 11 (5): 1703–1716. doi:10.5194/hess-11-1703-2007.
38. Li X, Niu JZ, Xie BY (2013) Study on Hydrological Functions of Litter Layers in North China. Plos One 8 (7): 1–11. doi: 10.1371/journal.pone.0070328.
39. Wu ZH (2001) The mud-rock flow disaster and their touch off condition by rainfall in Beijing Area. Research Soil Water Conserv 8 (1): 67–72. (In Chinese with English abstract).
40. Wang HZ (2008) A study of the critical rainfall volume for mudflows based on historical data in the mountainous area of Beijing. Analysis Research 3 (1): 18–21. (In Chinese with English abstract).
41. Cooperative Research Group on Chinese Soil Taxonomy (CRGCST) (2001) Chinese Soil Taxonomy, Beijing: Beijing Science Press.
42. Shi XZ, Yu DS, Yang GX, Yang GX, Wang HJ, et al. (2006) Cross-Reference Benchmarks for Translating the Genetic Soil Classification of China into the Chinese Soil Taxonomy. Pedosphere 16(2): 147–153.
43. Blake GR, Hartge KH (1986) Bulk density. In: Klute AK, editor. Methods of soil analysis. Part I: Physical and mineralogical method. American Society of Agronomy-Soil Science Society of America, Madison, pp. 363–375.
44. Carter MC, Gregorich EG (2008) Soil Samping and Methods of Analysis. Canadian Society of Soil Science, pp. 743–760.
45. Carter MC, Gregorich EG (2008) Soil Samping and Methods of Analysis. Canadian Society of Soil Science, pp. 911–938.
46. Gee GW, Bauder JW (1986) In: Klute AK, editor. Methods of soil analysis. Part I: Physical and mineralogical method. American Society of Agronomy-Soil Science Society of America, Madison, pp. 383–411.
47. Neris J, Tejedor M, Rodriguez M, Fuentes J, Jimenez C (2013) Effect of forest floor characteristics on water repellency, infiltration, runoff and soil loss in Andisols of Tenerife (Canary Islands, Spain). Catena 108: 50–57. doi: 10.1016/j.catena.2012.04.011.
48. Carter MC, Gregorich EG (2008) Soil Sampling and Methods of Analysis. Canadian Society of Soil Science, pp. 173–175.
49. Ayed GM, Mohammad AA (2010) The impact of vegetative cover type on runoff and soil erosion under different land uses. Catena 81: 97–103. doi: 10.1016/j.catena.2010.01.008.
50. Jiali W, Renhe Z, Yingchun W (2012) Characteristics of precipitation in Beijing and the precipitation representativeness of Beijing Weather Observatory. Journal of Applied Meteo Sci 23: 265–273. (In Chinese with English abstract).
51. Ma JJ, Li SY, Wang J (2012) The study of Beijing's rainstorm division and its returning periods. Meteo Monthly 38: 569–576. (In Chinese with English abstract).
52. Zhang GH, Liu BY, Li PK (2007) Principles and properties of artificial trough rainfall simulator. Bulletin of Soil and Water Conservation 27 (6): 56–60. (In Chinese with English abstract).
53. Xie Y, Lin XJ, Liu YN, Zhen YZ, Liu BY, et al. (2008) Calibration of simulated rainfall intensity and its special distribution for trough rainfall simulator. Bulletin of Soil and Water Conservation 28 (4): 1–6. (In Chinese with English abstract).
54. Battany MC, Grismer ME (2000) Rainfall runoff and erosion in Napa Vally vineyards: effects of slope, cover and surface roughness. Hydrol Process 14: 1289–1304. doi: 10.1002/(SICI)1099-1085(200005)14:7<1289::AID-HYP43>3.0.CO;2-R.
55. Kim KH, Miller WP (1996) Effect of rainfall electrolyte concentration and slope on infiltration and erosion. J Soil Tech 9: 173–185. doi: 10.1016/S0933-3630(96)00011-6.
56. Ministry of Environmental Protection of the P.R. China (1993) Determination of specific conductance in the wet precipitation. http://www.mep.gov.cn/image20010518/1881.pdf. Accessed 2014 Aug 29.
57. Tang J, Xue HS, Yu XL Cheng HB, Xu XB, et al. (2000) The preliminary study on chemical characteristics of precipitation at Mt. Waliguan. Acta Sci Circums 20(4): 420–425. (In Chinese with English abstract).

58. Zhong YD, Jia YW, Li ZW (2013) Spatial and Temporal Changes of Maximum 1h Precipitation Intensity in Beijing Region in Last 53 Years. J China Hydrol 33 (1): 32–37. (In Chinese with English Abstract).
59. Jiali W, Renhe Z, Yingchun W (2012) Characteristics of precipitation in Beijing and the precipitation representativeness of Beijing Weather Observatory. J Applied Meteo Sci 23: 265–273. (In Chinese with English abstract).
60. Ma JJ, Li SY, Wang J (2012) The study of Beijing's rainstorm division and its returning periods. Meteo Monthly 38: 569–576. (In Chinese with English abstract).
61. Imeson AC (1983) Studies of erosion thresholds in semi-arid areas:field measurements of soil loss and infiltration in northern Morocco. Catena Sup 4: 79–89.
62. Savenije HHG (1996) The runoff coefficient as the key to moisture recycling. J Hydrol 176: 219–225. doi: 10.1016/0022-1694(95)02776-9.
63. Uttam KM, Rao KV, Mishra PK, Vittal KPR, Sharma KL, et al. (2005) Soil infiltration, runoff and sediment yield from a shallow soil with varied stone cover and intensity of rain. Eur J Soil Sci 56: 435–443. doi: 10.1111/j.1365-2389.2004.00687.x.
64. Abrahams AD, Parsons AJ (1991) Relation between infiltration and stone cover on a semiarid hillslope, southern Arizona. J Hydrol 122: 45–59. doi: 10.1016/0022-1694(91)90171-D.
65. Gholami L, Sadeghi SH, Homaee M (2013) Straw mulching effect on splash erosion, runoff, and sediment yield from eroded plots. Soil Sci Soc Am J 77: 268–278. doi:10.2136/sssaj2012.0271.
66. Ghahramani A, Ishikawa Y, Gomi T (2011) Downslope soil detachment–transport on steep slopes via rain splash. Hydrol. Processes 25: 2471–2480. doi: 10.1002/hyp.8086.
67. Ghahramani A, Ishikawa Y, Gomi T (2011) Effect of ground cover on splash and sheetwash erosion over a steep forested hillslope: A plot-scale study. Catena 85: 34–47. doi: 10.1016/j.catena.2010.11.005.
68. Lal R (1981) Soil erosion problems on alfisols in Western Nigeria, VI. Effects of erosion on experimental plots. Geoderma 25: 215–230. doi: 10.1016/0016-7061(81)90037-9.
69. Miura S, Yoshinaga S, Yamada T (2003) Protective effect of floor cover against soil erosion on steep slopes forested with Chamaecyparis obtusa (hinoki) and other species. J Forest Res-Japan 8: 27–35. doi: 10.1007/s103100300003.
70. Dunjó G, Pardini G, Gispert M (2004) The role of land use–land cover on runoff generation and sediment yield at a microplot scale, in a small Mediterranean catchment. Journal of Arid Environments 57: 99–116. doi: 10.1016/S0140-1963(03)00097-1.

16

The Influence of Vegetation Height Heterogeneity on Forest and Woodland Bird Species Richness across the United States

Qiongyu Huang[1]*, Anu Swatantran[1], Ralph Dubayah[1], Scott J. Goetz[2]

1 Department of Geographical Sciences, University of Maryland, College Park, Maryland, United States of America, 2 Woods Hole Research Center, Falmouth, Massachusetts, United States of America

Abstract

Avian diversity is under increasing pressures. It is thus critical to understand the ecological variables that contribute to large scale spatial distribution of avian species diversity. Traditionally, studies have relied primarily on two-dimensional habitat structure to model broad scale species richness. Vegetation vertical structure is increasingly used at local scales. However, the spatial arrangement of vegetation height has never been taken into consideration. Our goal was to examine the efficacies of three-dimensional forest structure, particularly the spatial heterogeneity of vegetation height in improving avian richness models across forested ecoregions in the U.S. We developed novel habitat metrics to characterize the spatial arrangement of vegetation height using the National Biomass and Carbon Dataset for the year 2000 (NBCD). The height-structured metrics were compared with other habitat metrics for statistical association with richness of three forest breeding bird guilds across Breeding Bird Survey (BBS) routes: a broadly grouped woodland guild, and two forest breeding guilds with preferences for forest edge and for interior forest. Parametric and non-parametric models were built to examine the improvement of predictability. Height-structured metrics had the strongest associations with species richness, yielding improved predictive ability for the woodland guild richness models ($r^2 = \sim 0.53$ for the parametric models, 0.63 the non-parametric models) and the forest edge guild models ($r^2 = \sim 0.34$ for the parametric models, 0.47 the non-parametric models). All but one of the linear models incorporating height-structured metrics showed significantly higher adjusted-r^2 values than their counterparts without additional metrics. The interior forest guild richness showed a consistent low association with height-structured metrics. Our results suggest that height heterogeneity, beyond canopy height alone, supplements habitat characterization and richness models of forest bird species. The metrics and models derived in this study demonstrate practical examples of utilizing three-dimensional vegetation data for improved characterization of spatial patterns in species richness.

Editor: Ricardo Bomfim Machado, University of Brasilia, Brazil

Funding: The research received support from the NASA Applied Sciences programs on Ecological Forecasting and Biological Diversity (http://www.nasa.gov/applied-sciences/eco-forecasting.html) (grant # NNX09AK20G to SJG and RD). The funders had no role in study design, data collection and analysis, decision to publish, or preparation of the manuscript.

Competing Interests: The authors have declared that no competing interests exist.

* Email: qhuang@umd.edu

Introduction

Avian diversity has been under increasing pressure from anthropogenic disturbances such as habitat loss and fragmentation [1]. Successful conservation planning relies upon understanding how the distribution of avian richness responds to existing and potential changes in environmental conditions which influence their distributions. Discovering the drivers of large-scale spatial variation of species richness has been a central debate in ecology [2–5], and many hypotheses have been proposed to address this issue [6–11]. One major hypothesis suggests that habitat heterogeneity is a key factor because it leads to greater spatial variability of habitat physical conditions, and therefore permits greater niche specialization resulting in more species richness [12–15]. Particularly in North America, habitat heterogeneity theory predicted the richness of some faunas significantly better than the species-energy theory [7,14,15]. This latter theory also has widespread support, and hypothesizes that productive energy through food webs or species physiological constraints to ambient energy determines species richness [4,6,8,16].

Traditionally large scale habitat heterogeneity has been quantified mostly as topographical variability [7,14,17] or two dimensional habitat characteristics derived from remote sensing products [18,19]. Vertical habitat structure may also lead to niche generalization, and as such be an important element of habitat heterogeneity affecting biodiversity [20]. However, it has rarely been used to explain species richness at broad scales. The incorporation of vertical heterogeneity is especially important for avian richness models where vertical habitat structure at local scales has long been recognized as a critical factor influencing bird life history [21–23] and abundance [24,25].

Until recently, there have been relatively few studies utilizing three-dimensional habitat information due to difficulties of acquiring measurements of vertical vegetation structure beyond the plot scale over extended geographical areas [20]. This has changed significantly since the emergence of active remote sensing systems such as Light Detection and Ranging (lidar) and Radio Detection and Ranging (radar) which provide capability to map the vertical dimension of vegetation at local to regional scales [20,26]. There is an increasing number of studies using lidar and radar derived three-dimensional vegetation structure to model biodiversity, many of which have revealed significant association between vegetation vertical structure, habitat quality, species richness and abundance [27–32]. However, none of the existing habitat metrics sufficiently characterize the spatial arrangement of vegetation height (i.e. the heterogeneity of height), nor its potential for predicting avian richness distributions over large geographical extents.

Related advances have been made in the development of statistical fusion models that provide a means to effectively combine remotely sensed data from radar, lidar, optical remote sensing systems and forest inventory data, yielding wall-to-wall high resolution vegetation structure maps at the continental scale [33–37]. The production of these maps not only enables the creation of habitat metrics that capture rich vegetation height heterogeneity, but also the comparison of the predictive abilities in various forms of these metrics. Our study is designed to embrace these opportunities by examining the relationship between forest bird richness, height-structured habitat metrics and avian richness models involving various degrees of forest height heterogeneity.

The overall goal of our study is to examine the potential of three-dimensional habitat structure in improving avian richness models at broad geographical scales. In doing so we hope to expand our understanding of the relationship between habitat structure and the spatial distribution of avian species richness, and to lay the foundation for constructing habitat metrics that better utilize increasingly available three-dimensional habitat data. Specifically we address the following questions:

1. How do the height-structured metrics compare with traditional habitat metrics in their ability to associate and predict forest bird richness in the forested ecoregion of the U.S.? Does incorporating the height-structured metrics improve the explanatory ability of avian richness models that use traditional habitat metrics?
2. How do the predictive abilities of richness models vary among forest bird guilds with contrasting preferences to habitat edges?

First, we introduce the conceptual similarities and differences between traditional habitat metrics and two types of height-structured metrics. Next, we describe the data and the methods we used to create the habitat metrics in this study. We then use correlation analysis and multivariate regression models to examine the relationships between different combinations of metrics and the species richness of three forest breeding bird guilds. Lastly, we examine the models' explanatory abilities and the importance of individual metrics in predicting the richness of the three guilds.

Background

Traditional habitat metrics are based primarily on two-dimensional habitat structure, such as land cover types, patch size and shape statistics. Developing such metrics generally depends on two steps: a) classifying scene space into binary habitat and non-habitat land cover types; b) delineating habitat patches based on the rule of contiguity (Figure 1) [38]. There have been numerous studies using habitat patch metrics and derivative habitat edge and contrast metrics to associate with ecological attributes such as species richness, reproductive success and individual fitness of birds [39–41]. However vegetation height information generally plays little role in the process of delineating habitat patches and characterizing their properties.

Some studies have applied three-dimensional habitat information in habitat quality and species diversity models [28,30,31,42–44]. Usually, these applications rely on simple summary statistics such as mean, maximum, minimum and standard deviation to characterize three-dimensional vegetation structure. Summary statistics are straightforward and easy to obtain, but they cannot fully capture the heterogeneity of vegetation vertical structure. To give an example, one can have two forested landscapes with the same mean, maximum, minimum and standard deviation of tree height but with greatly different spatial arrangements of trees (e.g. tall trees can cluster in a few locations or can randomly distribute over the landscape which would have very different ecological implications for bird communities).

To account for more height heterogeneity, we created two groups of height-structured habitat metrics, the first of which integrates vegetation height information into the habitat patch framework while the second one characterizes canopy height distribution directly using second-order texture algorithms.

At the canopy level, vertical differences in vegetation create boundaries that segment contiguous habitats into smaller patches, each with similar height values (Figure 1). We first classified height pixels into a few height classes to characterize vertical edges and patches. Next, we grouped adjacent pixels from the same height class into patches. We treated the boundaries dividing those vertical patches as vertical edges (Figure 1). We also weighted the vertical edges by their depth (the height difference between two sides of a vertical edge) to capture the contrast of the height values of neighboring patches. By doing so, we could adapt a wide range of conventional habitat patch and edge metrics to account for complex spatial variability of canopy height.

Besides utilizing habitat patch and edge metrics to capture vegetation height heterogeneity, the second approach we introduce here involves calculation of the second-order (co-occurrence) texture statistics [45] directly from the gridded vegetation height maps. Second-order texture measures indicate the probabilities of each combination of pixel values co-occurring in a specific direction and distance [45]. These metrics can quantify spatial heterogeneity in terms of the spatial distribution and dependencies of height values [46] through grey level co-occurrence matrix. Texture measures are conventionally extracted from individual band of remotely sensed imagery and aerial photographs to assist object or land cover type discriminations [46,47]. Normally a small moving window is used to calculate the grey level co-occurrence matrix in specified neighborhoods. Texture measures extracted from optical remote sensing imageries have been used to infer broadly defined habitat heterogeneity that includes various environmental factors (e.g. land cover type, vegetation type, soil condition as well as vertical structure). This type of habitat structural information has been linked to avian species richness in many studies [48–50]. Here we derived the second-order texture metrics from gridded canopy height maps to directly characterize habitat height structure and to associate them with variation in avian richness.

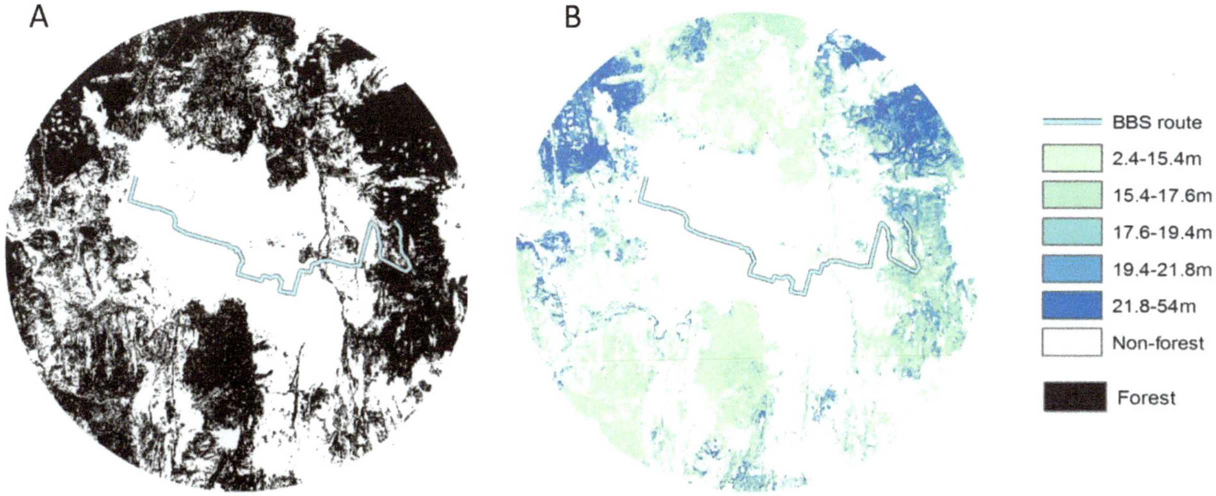

Figure 1. An example of the delineation of habitat patches at one BBS location. A two-dimensional vegetation map (A) and a vegetation map segmented by height structure (B) are shown. The pixel-based segmentation method (supporting information S1) is used to segment two dimensional habitat maps by using height thresholds.

Data Sets and Methods

Avian Data

The study area includes 21 predominately forested ecological regions (provinces) [51] across the conterminous U.S. (Table S1) (Figure 2). We used avian records from the Breeding Bird Survey (BBS) to model species richness over the entire study range. BBS is an annual road side survey organized by U.S. Geological Survey (USGS) [40,52]. Initiated in 1966, BBS has over 4000 survey routes located on secondary roads across the continental U.S. and Canada. Each survey route is 39.4 km long. Every year, during the avian breeding season, surveys are conducted by competent volunteers using the protocol of three-minute point count at 50 stops at 0.8 km intervals. All birds seen or heard within 0.37 km radius are recorded [53]. We removed the records whose survey procedures or associated data are not acceptable by BBS standard. We also removed the records surveyed by first year observers to minimize observer bias [54]. We selected 134 broadly grouped woodland breeding birds species (here after "woodland guild") based on the USGS species groupings [55]. We also selected 26 and 49 bird species as the forest breeding guilds with preference for interior forest habitat and forest edge habitat respectively (here after "interior forest guild" and "forest edge guild") based on the classification of Boulinier et al.1998 [56]. A complete list of birds involved in this study and their guild assignment are given in Table S2. Because most of the interior forest and forest edge bird species are distributed in the Eastern U.S., we limited our analysis on these two guilds to the 10 forested ecoregions in the east (Figure 2).

Adjustments were made to take into account the detection probability bias [57]. We used the "fossil" package [58] in the R statistical program [59] to calculate the adjusted species richness using a first-order jackknife estimator [60,61]. This estimator is based on multiple recapture studies in closed populations, which allows detection probability to vary among species. It is also the basic estimator underlying the species richness adjustments used by a USGS-developed BBS pre-processing program called COMDYN [62]. We averaged the available first-order jackknife richness within the five year period between 1998 and 2002 to temporally approximate the acquisition time of the radar data which played a key role in developing the vegetation height maps as discussed in the following section. The resulting mean avian richness is the richness we refer to in the rest of the study.

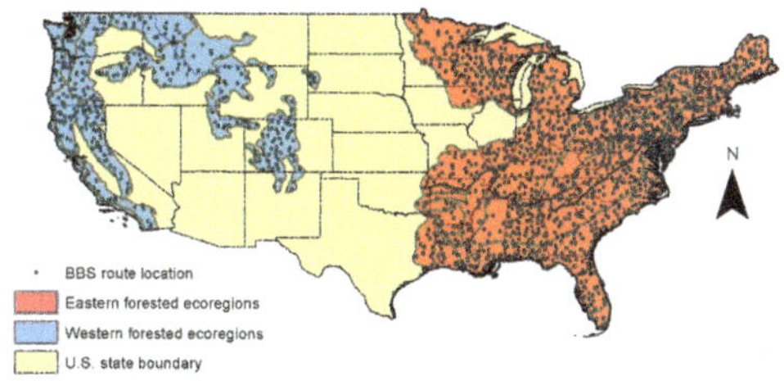

Figure 2. Distribution of BBS routes through the primarily forested ecoregions in the U.S. The richness models for the woodland guild were built using data from both eastern and western forested ecoregions. The forest edge and interior forest bird richness was modeled in the eastern forested ecoregions only.

Forest height data and habitat metrics

The National Biomass and Carbon Database of the year 2000 (NBCD) [33] provides an estimate of vegetation height distribution and variation at fine resolution for the conterminous U.S. The dataset is based on combined information from U.S. Department of Agriculture's Forest Service Forest Inventory and Analysis data, high-resolution Interferometric Synthetic Aperture Radar data acquired from 2000 Shuttle Radar Topography Mission and optical remote sensing data from the Landsat ETM+ sensor. Products from the USGS' National Land Cover Dataset 2001 and the Landscape Fire and Resource Management Planning Tools Project were also used during the process as input to build the empirical model for tree height estimation. The basal area weighted tree height (hereafter, "tree height") maps produced by the model gives spatially explicit vegetation vertical structure maps over the conterminous U.S. of 30 m-resolution.

We adapted a method to use 19 km (~half the length of a BBS route) radius buffers placed on the centroid of each BBS route, encompassing ~1100 km^2 areas to characterize the surrounding habitat around BBS locations [31,63,64]. We created habitat metrics on 1751 such circular landscapes where there are available BBS species richness data. (A), (B), (C), and (D), four metric sets incorporating a total of 26 metrics were calculated for each landscape (Table 1). The methods to produce each set of metrics are described in more details in the supporting information S1. The first two metric sets, embedded with little to no vegetation height heterogeneity, included (A) summary height statistics (hereafter "summary statistics") and (B) traditional patch-based metrics. The other two metric sets incorporated height heterogeneity: (C) patch metrics characterizing vertical patches and edges (hereafter, "height-structured patch-based metrics"), and (D) second-order texture metrics capturing vertical heterogeneity of height distributions (Table 1). The metric sets (A) and (B) were created as baselines to compare with the height-structured metric sets (C) and (D).

All the metrics created are listed in Table 1, and the detailed formula and descriptions for each metric are presented in Table S3. In order to differentiate the metrics with the same name from metrics set (B) and (C), capital letter "B" or "C" were given as prefixes to acronyms to indicate metric set membership (Table 1).

Species Richness Models

We first explored the statistical correlation between richness of the three avian guilds and the habitat metrics to evaluate the association between individual habitat metrics and the richness of different guilds. The woodland species richness models were based on data of all 21 forested ecoregions, and the interior forest and forest edge guild models were limited to data of the 10 forested ecoregions from Eastern U.S. as noted earlier (Figure 2).

We selected 2 metrics from each of metric set (C) and (D) that on average had the best association with the richness of the three guilds as the best performing height-structured metrics (BPHMs). These four BPHMs were later combined with the traditional habitat metrics in multivariable models for comparisons of improvement. We limited our choice to only the four best metrics to avoid subsequent overfitting of our multivariate models while still maintaining enough representativeness.

We next constructed 6 multivariate linear models to explain each guild's richness. The first 4 models were created using the complete list of metrics from set (A), (B), (C), and (D) respectively. They served to compare the explanatory abilities of models that characterize habitat condition with very different approaches. The two other models combined metric set (A) and (B) individually with the 4 BPHMs. We created the combined models to examine the impacts of adding spatial arrangement of height in richness models characterizing habitat in traditional ways.

We used a bootstrapping technique to provide the mean value and confidence intervals for the richness models' adjusted-r^2 values and AIC values to assess models' explanatory ability and goodness

Table 1. List of all metrics developed in the study.

Data	Metrics type (Set)	Metric name (Abbreviation)
NBCD vegetation height map	Summary height statistics (A)	Mean height (MEAN)
		Standard deviation of height (SD)
		Minimum height (Min)
		Maximum height (Max)
Two-dimensional vegetation cover map	Traditional patch-based metrics (B)	Number of patches(B.NP)
		Mean patch area (B.Area.MN)
		Standard deviation of patch area (B.Area.SD)
		Edge density (B. ED)
		Total edge (B.TE)
		Mean fractal dimension index (B.FRAC.MN)
		Standard deviation of fractal dimension index (B.FRAC.SD)
Vegetation cover map segmented by height structure	Height-structured patch-based metrics (C)	Number of patches(C.NP)
		Mean patch area (C.Area.MN)
		Standard deviation of patch area (C.Area.SD)
		Total edge (C.TE)
		Mean fractal dimension index (C.FRAC.MN)
		Standard deviation of fractal dimension index (C.FRAC.SD)
		Contrast weighted edge density (C.CWED)
		Mean of edge contrast index(C.ECON.MN)
		Standard deviation of edge contrast index (C.ECON.SD)
		Shannon's diversity index (C.SHDI)
NBCD vegetation height map	Second-order texture metrics (D)	Entropy
		Contrast
		Angular second moment (ASM)
		Homogeneity
		Dissimilarity

of fit, as well as the variability of these measures. The bootstrap resampling was repeated 3000 times for each model. To examine the significance level of model improvements the 95% confidence interval of adjusted-r^2 values and AIC values were obtained with the bias-corrected and accelerated (BCA) bootstrap algorithm [65] to make the interval's median unbiased and adjusted for skewness.

Lastly we explored the effect of combining the 26 metrics from all four metric sets using a non-parametric Random Forest (RF) model. The RF model [66] is known for being able to handle large number of input variables without overfitting [67]. It is also well-suited for our study because the model allows for covariance between predictor variables, which commonly exists between different habitat metrics. The RF model also provides a mechanism for assessing predictor variable importance using a measure of cross-validated mean square error (out of bag mean square error (OOB MSE)). The higher the increase of OOB MSE (IncMSE) is, the more important a specific metric is. More detailed introduction of RF model is described in the supporting information S1. We also ran 6 RF models on the same combinations of metrics used by the linear models to compare the differences between linear and RF models. We set the number of trees to be 2000 for all models to allow for the mean residual error to converge. In our study the RF models were built with Random Forests package [68] in the R statistical program [59].

Results

Predictor metric correlation

The predictor metrics that correlate best with bird species richness varied among guilds. For woodland species richness, (D) the second order texture metrics generally had the greatest predicative ability, followed by (B) the traditional patch-based metrics, and (C) the height-structured patch metrics. For forest edge bird richness, the metrics with the strongest correlation were (C) the height-structured patch-based metrics followed by (D) the second order texture metrics and (A) the summary height statistics (Figure 3). Interior forest bird species richness in general had consistently low correlation with any metrics. Among all the metrics developed (Table 1, Table S3), the metric with the greatest predictive capability for interior guild richness was mean vegetation height. ASM had the strongest average predictive capability over the richness models for three guilds, followed by entropy, C.TE, homogeneity, and C.CWED (Figure 3), all of which are height-structured metrics. We selected ASM, entropy, C.TE and C.CWED as the four BPHMs to be combined with models relied on the traditional metrics.

The direction of the correlation between metrics and the bird richness was generally consistent across three guild types except for metrics with weak correlation (Table S4). Among the variables with highest average correlation, ASM and homogeneity both had negative correlation with the richness of all three guilds. Conversely, entropy, C.TE, C.CWED all showed strong positive correlation for each guild's richness (Figure 4, Table S4).

After incorporating vegetation height heterogeneity in patch-based metrics, the metrics characterizing patch number and area (AREA.MN, AREA.SD, and NP) showed a decreased correlation with the woodland guild richness. Conversely, the strength of the correlation between edge metrics (ED, TE) and the woodland guild richness increased. For the forest edge species both the patch and edge related metrics showed a prominent increase of correlation after incorporating vegetation height heterogeneity. The direction of the correlation for some patch-based metrics also changed. The NP metric showed an exceptionally large change for the woodland guild richness: from −0.45 to 0.25 after incorporating vertical patches (Table S4, Figure 3).

Predictive models

The non-parametric RF models combining all 26 metrics (all-inclusive models) from the four metric sets were the ones with greatest ability to predict species richness for each guild (Figure 4, Table S5). Among those models the lowest species richness variability was explained for the interior forest guild ($r^2 = 0.11$), but the forest edge guild richness was predicted moderately well ($r^2 = 0.47$) and the predictive model was strong for the woodland guild ($r^2 = 0.63$) (Figure 5). The most important variable for predicting the woodland guild richness were two traditional patch-based metrics (B.AREA.MN and B.AREA.SD) followed by two second order texture metrics (entropy and ASM). The forest edge species richness model was most dependent on two height-structured patch metrics (C.CWED and C.NP) followed by two summary height statistics (MAX and MEAN). The most important predictive metrics for the interior forest guild model were MEAN followed by B.AREA.MN and B.AREA.SD (Figure 5).

For the RF models, our results consistently showed that adding height-structured metrics improved the model predictive ability. Specifically, the explained variance of the all-inclusive RF models for woodland and forest edge guild were up to 0.27 and 0.21 higher respectively than the RF models with only traditional habitat metrics. In addition for these two guilds, when the RF models were combined with the four BPHMs, the improvement for explained variance were up to 0.21 (woodland guild) and 0.13 (forest edge guild). The interior forest guild however showed only minor improvements when combined with any height-structured metrics. In general for woodland and forest edge guild, RF models' predictabilities were higher than the comparable linear models by a prominent margin. (Figure 4 and Table S5).

The linear models had a lower explanatory ability than their RF counterparts. For a specific combination of habitat metrics, the linear models explained the most amount of variation in the woodland guild richness and the least in the interior forest guild richness. The one exception was the model using summary statistics of height (set A), which showed the highest predictability for forest edge guild richness, followed by woodland guild richness, and then the interior forest guild richness (Figure 4, Table S5). In every guild, the models incorporating the four BPHMs showed consistently higher predictability than the models without (Figure 4).

Combining the BPHMs with the summary height statistics resulted in significantly higher adjusted-r^2 values in the woodland and forest edge models (Figure 4). The AIC value for the woodland richness model also improved significantly. In comparison, when combined with traditional patch-based metrics, the BPHMs significantly increased the adjusted-r^2 for the forest edge guild model, while significantly improving the AIC values for both the forest edge and woodland guild models (Table S5).

Discussion

A large number of hypotheses have been proposed to explain the spatial patterns of species richness over broad geographical scales [2,4,6,69–71]. While it is unlikely that one single mechanism can explain species richness patterns completely, a large portion of the literature testing habitat heterogeneity hypothesis has focused on the association between species richness and two dimensional habitat structure, often combined with land cover type composition and distribution [63,64,72,73]. On the other hand other studies testing species-energy hypothesis have relied on covariates

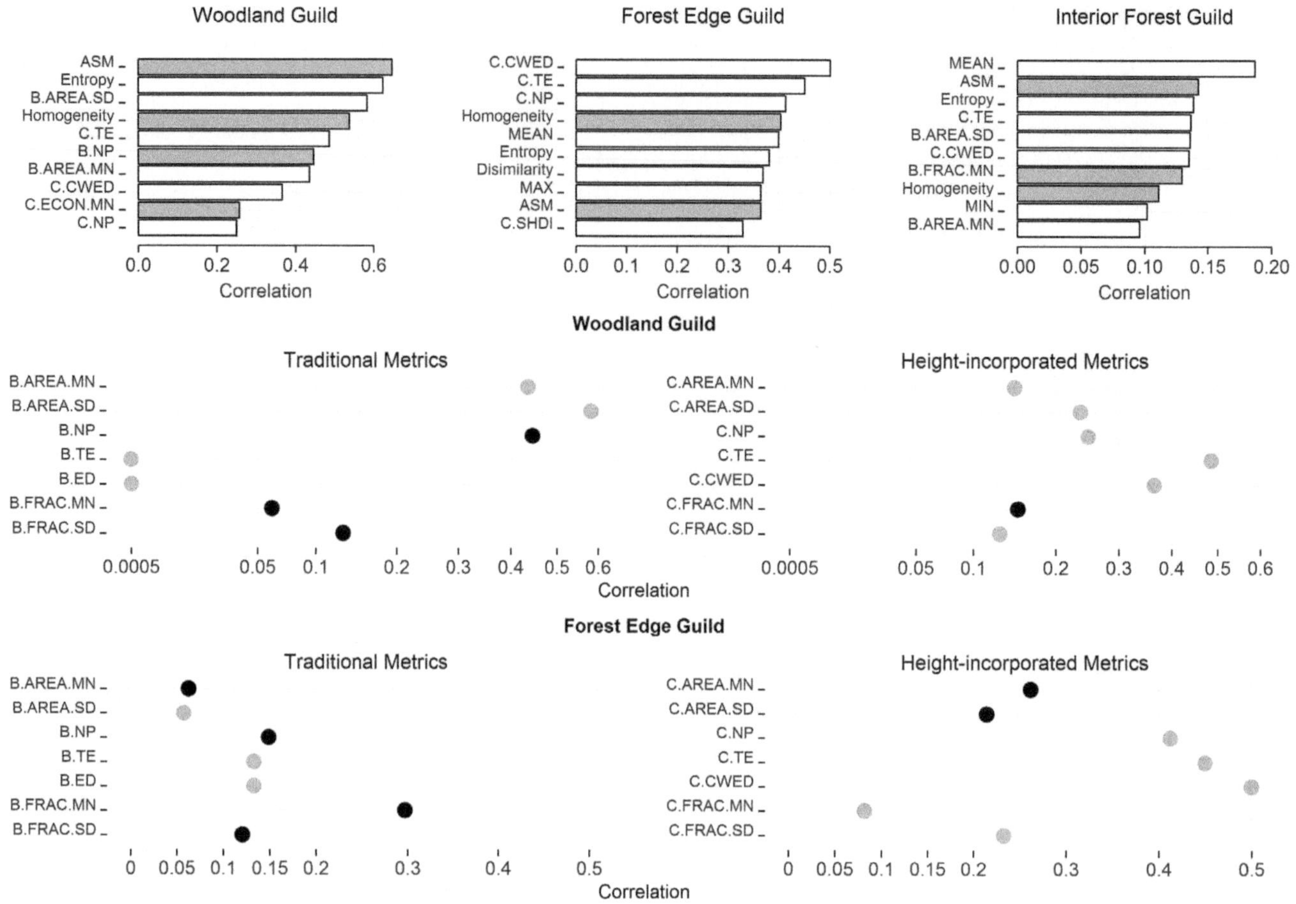

Figure 3. Guild richness associations with various metrics. (Top row): correlation bar plots of the most predictive metrics of species richness by guild. White bars represent a positive correlation and grey indicate a negative correlation. (Bottom rows): correlation comparisons between comparable patch-based metrics with and without considering the vertical patches and edges for the woodland and forest edge guild. The left panels show traditional metrics without accounting for height-heterogeneity; the right panels are height-structured counterparts. The black dots indicate a negative correlation and the grey ones indicate a positive correlation.

related to ecosystem productivity and energy such as evapotranspiration and photosynthetic capacity indices like the normalized difference vegetation index (NDVI) [74–76] to explain large scale species richness patterns. Studies to associate habitat vertical structure with species richness are, however, often focused at local scale [31], which limited the efficacies of habitat heterogeneity models to explain species richness at broad scale.

Only recently was vegetation height information assessed as a predictor of avian species richness across the conterminous U.S. in two studies [31,77]. One of these [31] used the same NBCD data we employed here, but they explored only summary statistics of vegetation height and biomass combined with land cover type composition and distribution. The other used sparsely sampled height metrics from a satellite lidar system that is no longer operating, and included climatic data as predictive variables [77]. Although our models employed only the distribution vegetation structures, with no input from other land cover type or climatic data, their explanatory ability for the woodland guild was comparable to these recent results ($r^2 = 0.70$ for the forest guild model [31], and $r^2 = 0.60$ for the open woodland model [77]). We found that models combining only vegetation vertical and horizontal structure can explain a significant amount of species richness for the broadly grouped woodland guild and the forest breeding guild with preferences for the forest edge habitat. More importantly, our results showed that incorporating vegetation vertical heterogeneity, and not just mean and standard deviation of height, greatly improves the ability to explain variability in avian richness for the two guilds. The spatial arrangement of vegetation height plays an important role in associating the quality of habitat condition and diversity of ecological niches for bird species within the two groups.

Traditionally habitat edges are thought to affect species movement, interaction, mortality and community dynamics [78]. The summary height statistics are considered indicators of habitat diversity and forest successional stage [20,79,80]. The traditional way of characterizing habitat through two-dimensional habitat patch distribution and summary height statistics still play important roles in our multivariate richness models. The large pool of traditional patch-based metrics provides a well-known framework to readily incorporate vertical height distribution once habitat patches are segmented by height. Both traditional and our height-structured metrics contribute to explanation of the variance of avian richness, although the importance of individual metrics in the models varies from guild to guild. Thus, our study shows that for the woodland avian guild and forest edge guild, the species richness is highly sensitive to the vegetation height heterogeneity, and the addition of the spatial arrangement of vegetation height provides significantly improved estimates of species richness for the

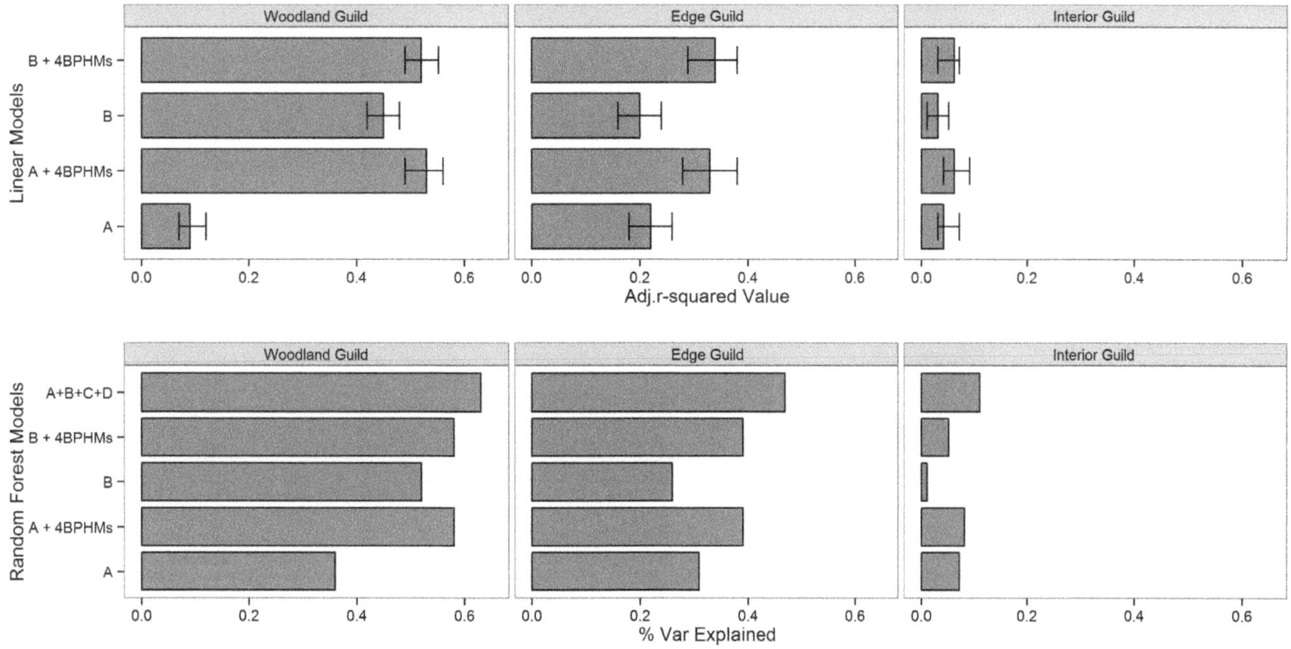

Figure 4. Predictive ability of multivariable models. A, B, C, and D are the four habitat metric sets, and 4BPHMs are the four best predictive height-structured metrics. Each of the top panels shows four linear models with whiskers giving 95% confidence interval of adjusted-r^2 values. The length of the bar represents the mean adjusted-r^2 for these models. The lower panels show the explained variance of the comparable random forest (RF) models. Uniquely the top bars at lower pannels are the results from the models employed all metrics from the four metric sets.

two guilds. The patch-based height-structured metrics and the second order texture metrics thereby supplement and extend common methods of characterizing habitat condition and predicting avian species richness.

We note that the BBS data is collected along roadways where volunteers can easily and regularly traverse, thus the areas along the survey routes could be subject to disturbances such as motor vehicle traffic or habitat conversion [81,82]; i.e. they may not be representative samplings of forest spatial and vertical variability. This characteristic of the data set could pose a challenge for systematical sampling of interior forest bird species in the surrounding areas and is likely one of the contributing factors for the consistently low species richness and weak correlations with our metrics and models in the case of the interior forest guild. Alternatively, forest edge habitats are relatively more exposed to stressors such as wind damages and human disturbances. They can exhibit higher vertical structure diversity than the interior forest areas [83]. It may be that interior birds are less adapted to habitat structure heterogeneity, and thus exhibit limited sensibility to habitat structure metrics. Lastly the results could also be attributed to the different ways members of avian guilds utilize habitat. Forest edge and majority of woodland bird species tend to use a wide range of habitat, and their degree of co-existence can vary in a broad spectrum over space. In comparison the interior forest guild, composed mostly of forest specialists that avoid other habitat types [84] with overlapping ecological niches, are more likely to face greater interspecific competition which limits species richness despite diverse height structure across landscapes [85]. However while the models see low association between height heterogeneity metrics and interior forest guild richness, there are still likely more specific vertical structure preferences associated with individual bird species [30,86].

While four BPHMs highlighted in our study showed a good ability to associate with species richness and to improve broad scale avian richness modeling, it is reasonable to assume that height-structured metrics have potential to be improved further given the large number of options that remained unexplored. First, the pixel-based segmentation method used in our study (supporting information S1) is one of the simplest algorithms to delineate vertical patches and edges. The method is based on a set of global threshold values while not considering neighboring heterogeneity [87]. The process of setting up the threshold values and weight matrix (for contrast metrics) inevitably involves somewhat arbitrary decisions. More complex segmentation methods such as edge and region-based methods can be performed readily with commercial and open-source software packages that potentially may produce more efficacious vertical patches and be less arbitrary [88]. Secondly, there are many untested texture measures [45]. The relationship between texture metrics and the avian richness varies as the size of moving window changes [48]. More work is needed to understand the impact of those methodological options for further improving species richness models.

Conclusions

As active remote sensing technologies like radar and lidar mature and become more widely available, data sets characterizing vegetation vertical structure should become increasingly useful for biodiversity applications and management. Our study showed that vegetation height heterogeneity is associated with habitat diversity and species richness for some forest avian guilds. Thus, while recognizing the advances conveyed by incorporating height information, there is an imperative to explore in more depth the role of such heterogeneity. Furthermore we suggest not just height, but vertical canopy heterogeneity, e.g. foliar profiles and layering, will provide an even richer source of information from which to develop new metrics and models [30,83].

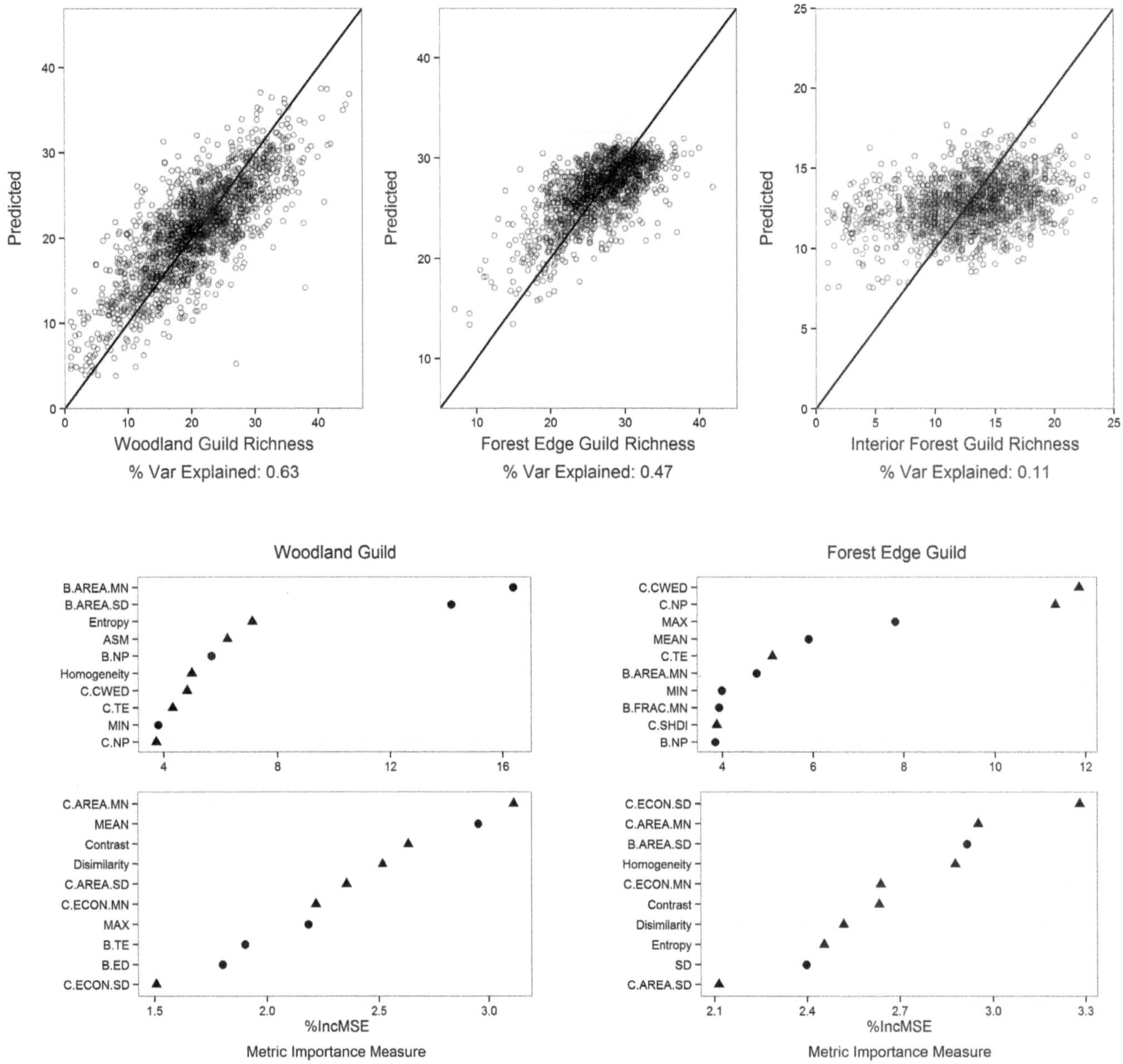

Figure 5. Random Forest model results. (Top row): Modeled vs. actual species richness for three guilds using all-inclusive random forest models. (Below the scatter plots): variable importance plots show the percent increase in mean square error (%IncMSE) of the top 20 most influential metrics in the woodland guild richness model and the forest edge guild richness model (note different scales on X-axes). The metrics characterizing vegetation height heterogeneity are plotted with triangles and the rest of the metrics are circles.

Incorporating such information will require data on not only canopy height but canopy vertical structure, the latter of which is unavailable at continental scales. Nonetheless, the metrics and models used in our analyses provide a means to incorporate and utilize three-dimensional habitat information, with the goal of better understanding the controls on avian species richness and habitat use.

Acknowledgments

We thank Dr. John Sauer and Dr. James Kellner for valuable discussions on the project, as well as the academic editor and the peer reviewers for their input in helping to improve the manuscript. Lastly we also thank the BBS staff and volunteers for their effort in conducting the surveys and making their data widely available.

Author Contributions

Conceived and designed the experiments: QH RD AS. Performed the experiments: QH. Analyzed the data: QH. Contributed to the writing of the manuscript: QH AS RD SG.

References

1. Gaston KJ, Blackburn TM, Goldewijk KK (2003) Habitat conversion and global avian biodiversity loss. Proc R Soc Lond B Biol Sci 270: 1293–1300. doi:10.1098/rspb.2002.2303.
2. Palmer MW (1994) Variation in species richness: Towards a unification of hypotheses. Folia Geobot Phytotaxon 29: 511–530. doi:10.1007/BF02883148.
3. Rosenzweig ML (1995) Species Diversity in Space and Time. Cambridge; New York: Cambridge University Press. 460 p.
4. Gaston KJ (2000) Global patterns in biodiversity. Nature 405: 220–227. doi:10.1038/35012228.
5. Gaston KJ, Spicer JI (2004) Biodiversity: An Introduction. 2 edition. Malden, MA: Wiley-Blackwell. 208 p.
6. Waide RB, Willig MR, Steiner CF, Mittelbach G, Gough L, et al. (1999) The Relationship Between Productivity and Species Richness. Annu Rev Ecol Syst 30: 257–300. doi:10.1146/annurev.ecolsys.30.1.257.
7. Rahbek C, Graves GR (2001) Multiscale assessment of patterns of avian species richness. Proc Natl Acad Sci 98: 4534–4539. doi:10.1073/pnas.071034898.
8. Hawkins BA, Field R, Cornell HV, Currie DJ, Guégan JF, et al. (2003) Energy, Water, and Broad-Scale Geographic Patterns of Species Richness. Ecology 84: 3105–3117. doi:10.1890/03-8006.
9. Willig MR, Kaufman DM, Stevens RD (2003) Latitudinal gradients of biodiversity: pattern, process, scale, and synthesis. Annu Rev Ecol Evol Syst: 273–309.
10. Currie DJ, Mittelbach GG, Cornell HV, Field R, Guégan JF, et al. (2004) Predictions and tests of climate-based hypotheses of broad-scale variation in taxonomic richness. Ecol Lett 7: 1121–1134. doi:10.1111/j.1461-0248.2004.00671.x.
11. Colwell RK, Rahbek C, Gotelli NJ (2004) The mid-domain effect and species richness patterns:what have we learned so far? Am Nat 163: E1-23. doi:10.1086/382056.
12. Davies RG, Orme CDL, Storch D, Olson VA, Thomas GH, et al. (2007) Topography, energy and the global distribution of bird species richness. Proc R Soc B Biol Sci 274: 1189–1197. doi:10.1098/rspb.2006.0061.
13. Koh CN, Lee PF, Lin RS (2006) Bird species richness patterns of northern Taiwan: primary productivity, human population density, and habitat heterogeneity. Divers Distrib 12: 546–554. doi:10.1111/j.1366-9516.2006.00238.x.
14. Kerr JT, Packer L (1997) Habitat heterogeneity as a determinant of mammal species richness in high-energy regions. Nature 385: 252–254.
15. Kerr JT, Southwood TRE, Cihlar J (2001) Remotely sensed habitat diversity predicts butterfly species richness and community similarity in Canada. Proc Natl Acad Sci 98: 11365–11370. doi:10.1073/pnas.201398398.
16. Evans KL, Warren PH, Gaston KJ (2005) Species-energy relationships at the macroecological scale: a review of the mechanisms. Biol Rev Camb Philos Soc 80: 1–25.
17. Richerson PJ, Lum K (1980) Patterns of plant species diversity in California: relation to weather and topography. Am Nat: 504–536.
18. Turner JRG, Lennon JJ, Lawrenson JA (1988) British bird species distributions and the energy theory. Nature 335: 539–541. doi:10.1038/335539a0.
19. Duro DC, Coops NC, Wulder MA, Han T (2007) Development of a large area biodiversity monitoring system driven by remote sensing. Prog Phys Geogr 31: 235–260. doi:10.1177/0309133307079054.
20. Bergen KM, Goetz SJ, Dubayah RO, Henebry GM, Hunsaker CT, et al. (2009) Remote sensing of vegetation 3-D structure for biodiversity and habitat: Review and implications for lidar and radar spaceborne missions. J Geophys Res 114:GE00E06. doi:10.1029/2008JG000883.
21. Halaj J, Ross D, Moldenke A (2000) Importance of habitat structure to the arthropod food-web in Douglas-fir canopies. Oikos 90: 139–152. doi:10.1034/j.1600-0706.2000.900114.x.
22. Kelly J (1993) The Effect of Nest Predation on Habitat Selection by Dusky Flycatchers. Condor 95: 83–93. doi:10.2307/1369389.
23. Robinson SK, Holmes RT (1984) Effects of plant species and foliage structure on the foraging behavior of forest birds. The Auk 101: 672–684.
24. MacArthur RH, MacArthur JW (1961) On Bird Species Diversity. Ecology 42: 594–598. doi:10.2307/1932254.
25. Whittaker RJ, Willis KJ, Field R (2001) Scale and Species Richness: Towards a General, Hierarchical Theory of Species Diversity. J Biogeogr 28: 453–470.
26. Lefsky MA, Cohen WB, Parker GG, Harding DJ (2002) Lidar Remote Sensing for Ecosystem Studies. BioScience 52: 19. doi:10.1641/0006-3568(2002)052[0019:LRSFES]2.0.CO;2.
27. Bergen KM, Gilboy AM, Brown DG (2007) Multi-dimensional vegetation structure in modeling avian habitat. Ecol Inform 2: 9–22. doi:10.1016/j.ecoinf.2007.01.001.
28. Goetz S, Steinberg D, Dubayah R, Blair B (2007) Laser remote sensing of canopy habitat heterogeneity as a predictor of bird species richness in an eastern temperate forest, USA. Remote Sens Environ 108: 254–263. doi:10.1016/j.rse.2006.11.016.
29. Imhoff ML, Sisk TD, Milne A, Morgan G, Orr T (1997) Remotely sensed indicators of habitat heterogeneity: Use of synthetic aperture radar in mapping vegetation structure and bird habitat. Remote Sens Environ 60: 217–227. doi:10.1016/S0034-4257(96)00116-2.
30. Swatantran A, Dubayah R, Goetz S, Hofton M, Betts MG, et al. (2012) Mapping Migratory Bird Prevalence Using Remote Sensing Data Fusion. PLoS ONE 7: e28922. doi:10.1371/journal.pone.0028922.
31. Culbert PD, Radeloff VC, Flather CH, Kellndorfer JM, Rittenhouse CD, et al. (2013) The Influence of Vertical and Horizontal Habitat Structure on Nationwide Patterns of Avian Biodiversity. The Auk 130: 656–665. doi:10.1525/auk.2013.13007.
32. Zellweger F, Braunisch V, Baltensweiler A, Bollmann K (2013) Remotely sensed forest structural complexity predicts multi species occurrence at the landscape scale. For Ecol Manag 307: 303–312. doi:10.1016/j.foreco.2013.07.023.
33. Kellndorfer J, Walker W, Kirsch K, Fiske G, Bishop J, et al. (2011) National Biomass and Carbon Dataset for the Year 2000. Available: http://www.whrc.org/mapping/nbcd.
34. Kellndorfer JM, Walker WS, LaPoint E, Kirsch K, Bishop J, et al. (2010) Statistical fusion of lidar, InSAR, and optical remote sensing data for forest stand height characterization: A regional-scale method based on LVIS, SRTM, Landsat ETM+, and ancillary data sets. J Geophys Res 115. doi:10.1029/2009JG000997.
35. Saatchi SS, Harris NL, Brown S, Lefsky M, Mitchard ETA, et al. (2011) Benchmark map of forest carbon stocks in tropical regions across three continents. Proc Natl Acad Sci 108: 9899–9904. doi:10.1073/pnas.1019576108.
36. Walker WS, Kellndorfer JM, LaPoint E, Hoppus M, Westfall J (2007) An empirical InSAR-optical fusion approach to mapping vegetation canopy height. Remote Sens Environ 109: 482–499. doi:10.1016/j.rse.2007.02.001.
37. Baccini A, Goetz SJ, Walker WS, Laporte NT, Sun M, et al. (2012) Estimated carbon dioxide emissions from tropical deforestation improved by carbon-density maps. Nat Clim Change 2: 182–185. doi:10.1038/nclimate1354.
38. Girvetz EH, Greco SE (2007) How to define a patch: a spatial model for hierarchically delineating organism-specific habitat patches. Landsc Ecol 22: 1131–1142. doi:10.1007/s10980-007-9104-8.
39. Helzer CJ, Jelinski DE (1999) The Relative Importance of Patch Area and Perimeter-Area Ratio to Grassland Breeding Birds. Ecol Appl 9: 1448–1458. doi:10.1890/1051-0761(1999)009[1448:TRIOPA]2.0.CO;2.
40. Robbins CS, Dawson DK, Dowell BA (1989) Habitat Area Requirements of Breeding Forest Birds of the Middle Atlantic States. Wildl Monogr: 3–34.
41. Strelke WK, Dickson JG (1980) Effect of Forest Clear-Cut Edge on Breeding Birds in East Texas. J Wildl Manag 44: 559–567. doi:10.2307/3808003.
42. Broughton RK, Hinsley SA, Bellamy PE, Hill RA, Rothery P (2006) Marsh Tit Poecile palustris territories in a British broad-leaved wood. Ibis 148: 744–752. doi:10.1111/j.1474-919X.2006.00583.x.
43. Hill RA, Hinsley SA, Gaveau DLA, Bellamy PE (2004) Predicting habitat quality for Great Tits (Parus major) with airborne laser scanning data. Int J Remote Sens 25: 4851–4855. doi:10.1080/0143116031000139962.
44. Hinsley SA, Hill RA, Fuller RJ, Bellamy PE, Rothery P (2009) Bird species distributions across woodland canopy structure gradients. Community Ecol 10: 99–110. doi:10.1556/ComEc.10.2009.1.12.
45. Haralick RM, Shanmugam K, Dinstein I (1973) Textural Features for Image Classification. Syst Man Cybern IEEE Trans On 3: 610–621. doi:10.1109/TSMC.1973.4309314.
46. Coburn CA, Roberts ACB (2004) A multiscale texture analysis procedure for improved forest stand classification. Int J Remote Sens 25: 4287–4308. doi:10.1080/0143116042000192367.
47. Franklin SE, Hall RJ, Moskal LM, Maudie AJ, Lavigne MB (2000) Incorporating texture into classification of forest species composition from airborne multispectral images. Int J Remote Sens 21: 61–79. doi:10.1080/014311600210993.
48. St-Louis V, Pidgeon AM, Radeloff VC, Hawbaker TJ, Clayton MK (2006) High-resolution image texture as a predictor of bird species richness. Remote Sens Environ 105: 299–312. doi:10.1016/j.rse.2006.07.003.
49. St-Louis V, Pidgeon AM, Clayton MK, Locke BA, Bash D, et al. (2009) Satellite image texture and a vegetation index predict avian biodiversity in the Chihuahuan Desert of New Mexico. Ecography 32: 468–480. doi:10.1111/j.1600-0587.2008.05512.x.
50. Culbert PD, Radeloff VC, St-Louis V, Flather CH, Rittenhouse CD, et al. (2012) Modeling broad-scale patterns of avian species richness across the

Midwestern United States with measures of satellite image texture. Remote Sens Environ 118: 140–150. doi:10.1016/j.rse.2011.11.004.
51. Bailey RG (1995) Description of the Ecoregions of the United States: Miscellaneous Publication.
52. Robbins CS, Bystrak D, Geissler PH (1986) The breeding bird survey: its first fifteen years, 1965–1979. U.S. Dept. of the Interior, Fish and Wildlife Service. 198 p.
53. Sauer JR, Hines JE, Fallon JE, Pardieck KL, Ziolkowski DJ Jr, et al. (2011) The North American Breeding Bird Survey, Results and Analysis 1966–2010. USGS Patuxent Wildlife Research Center. Available: http://www.mbr-pwrc.usgs.gov/bbs/.
54. Kendall WL, Peterjohn BG, Sauer JR (1996) First-Time Observer Effects in the North American Breeding Bird Survey. The Auk 113: 823–829.
55. U.S. Geological Survey (2012) List of Species Groupings. Available: http://www.mbr-pwrc.usgs.gov/bbs/guild/guildlst.html.
56. Boulinier T, Nichols JD, Hines JE, Sauer JR, Flather CH, et al. (1998) Higher temporal variability of forest breeding bird communities in fragmented landscapes. Proc Natl Acad Sci U S A 95: 7497–7501.
57. Kéry M, Schmid H (2004) Monitoring programs need to take into account imperfect species detectability. Basic Appl Ecol 5: 65–73. doi:10.1078/1439-1791-00194.
58. Vavrek MJ (2011) fossil: palaeoecological and palaeogeographical analysis tools. Palaeontol Electron 14: 1T.
59. R Development Core Team (2011) R: A Language and Environment for Statistical Computing. Vienna, Austria: R Foundation for Statistical Computing. Available: http://www.R-project.org.
60. Burnham KP, Overton WS (1978) Estimation of the Size of a Closed Population When Capture Probabilities Vary Among Animals. Biometrika 65: 625–633. doi:10.1093/biomet/65.3.625.
61. Burnham KP, Overton WS (1979) Robust Estimation of Population Size When Capture Probabilities Vary Among Animals. Ecology 60: 927–936. doi:10.2307/1936861.
62. Hines JE, Boulinier T, Nichols JD, Sauer JR, Pollock KH (1999) COMDYN: software to study the dynamics of animal communities using a capture-recapture approach. Bird Study 46: 209–217. doi:10.1080/00063659909477247.
63. Rittenhouse CD, Pidgeon AM, Albright TP, Culbert PD, Clayton MK, et al. (2012) Land-cover change and avian diversity in the conterminous United States. Conserv Biol J Soc Conserv Biol 26: 821–829. doi:10.1111/j.1523-1739.2012.01867.x.
64. Pidgeon AM, Radeloff VC, Flather CH, Lepczyk CA, Clayton MK, et al. (2007) Associations of Forest Bird Species Richness with Housing and Landscape Patterns across the USA. Ecol Appl 17: 1989–2010.
65. Efron B (1987) Better Bootstrap Confidence Intervals. J Am Stat Assoc 82: 171–185. doi:10.1080/01621459.1987.10478410.
66. Breiman L (2001) Random Forests. Mach Learn 45: 5–23.
67. Biau G (2012) Analysis of a Random Forests Model. J Mach Learn Res 98888: 1063–1095.
68. Liaw A, Wiener M (2002) Classification and Regression by randomForest. R News 2: 18–22.
69. Hawkins BA, Diniz-Filho JAF, Jaramillo CA, Soeller SA (2007) Climate, niche conservatism, and the global bird diversity gradient. Am Nat 170: S16–S27.
70. Rahbek C, Gotelli NJ, Colwell RK, Entsminger GL, Rangel TFLVB, et al. (2007) Predicting continental-scale patterns of bird species richness with spatially explicit models. Proc R Soc B Biol Sci 274: 165–174. doi:10.1098/rspb.2006.3700.
71. Guégan JF, Lek S, Oberdorff T (1998) Energy availability and habitat heterogeneity predict global riverine fish diversity. Nature 391: 382–384. doi:10.1038/34899.
72. Donovan TM, Flather CH (2002) Relationships among North American Songbird Trends, Habitat Fragmentation, and Landscape Occupancy. Ecol Appl 12: 364–374. doi:10.2307/3060948.
73. Griffiths GH, Lee J (2000) Landscape pattern and species richness; regional scale analysis from remote sensing. Int J Remote Sens 21: 2685–2704. doi:10.1080/01431160050110232.
74. Phillips LB, Hansen AJ, Flather CH (2008) Evaluating the species energy relationship with the newest measures of ecosystem energy: NDVI versus MODIS primary production. Remote Sens Environ 112: 4381–4392. doi:10.1016/j.rse.2008.08.002.
75. Hurlbert AH, Haskell JP (2003) The Effect of Energy and Seasonality on Avian Species Richness and Community Composition. Am Nat 161: 83–97. doi:10.1086/345459.
76. Seto KC, Fleishman E, Fay JP, Betrus CJ (2004) Linking spatial patterns of bird and butterfly species richness with Landsat TM derived NDVI. Int J Remote Sens 25: 4309–4324. doi:10.1080/0143116042000192358.
77. Goetz SJ, Sun M, Zolkos S, Hansen A, Dubayah R (2014) The relative importance of climate and vegetation properties on patterns of North American breeding bird species richness. Environ Res Lett 9: 034013. doi:10.1088/1748-9326/9/3/034013.
78. Fagan WE, Cantrell RS, Cosner C (1999) How habitat edges change species interactions. Am Nat 153: 165–182. doi:10.1086/303162.
79. Morgan K, Freedman B (1985) Breeding Bird Communities in a Hardwood Forest Succession in Nova Scotia Canada. Can Field Nat 100: 506–519.
80. North MP, Franklin JF, Carey AB, Forsman ED, Hamer T (1999) Forest Stand Structure of the Northern Spotted Owl's Foraging Habitat. For Sci 45: 520–527.
81. Griffith EH, Sauer JR, Royle JA (2010) Traffic Effects on Bird Counts on North American Breeding Bird Survey Routes. The Auk 127: 387–393. doi:10.1525/auk.2009.09056.
82. Keller CME, Scallan JT (1999) Potential Roadside Biases Due to Habitat Changes along Breeding Bird Survey Routes. The Condor 101: 50–57. doi:10.2307/1370445.
83. Whitehurst AS, Swatantran A, Blair JB, Hofton MA, Dubayah R (2013) Characterization of Canopy Layering in Forested Ecosystems Using Full Waveform Lidar. Remote Sens 5: 2014–2036. doi:10.3390/rs5042014.
84. Hagan JM, Vander Haegen WM, McKinley PS (1996) The Early Development of Forest Fragmentation Effects on Birds. Conserv Biol 10: 188–202. doi:10.1046/j.1523-1739.1996.10010188.x.
85. Cody ML (1974) Competition and the Structure of Bird Communities. Princeton University Press.
86. Goetz SJ, Steinberg D, Betts MG, Holmes RT, Doran PJ, et al. (2010) Lidar remote sensing variables predict breeding habitat of a Neotropical migrant bird. Ecology 91: 1569–1576.
87. Schiewe J (2002) Segmentation of high-resolution remotely sensed data – concepts, applications and problems. Symposium on Geospatial theory, Processing and Applications.
88. Baatz M, Benz U, Dehghani S, Heynen M, Höltje A, et al. (2003) eCognition user guide. Defin Imaging GmbH.

17

Large Impacts of Climatic Warming on Growth of Boreal Forests since 1960

Pekka E. Kauppi[1,2]*, Maximilian Posch[3], Pentti Pirinen[4]

1 Department of Environmental Sciences, University of Helsinki, Helsinki, Finland, 2 Finnish Forest Research Institute (METLA), Vantaa, Finland, 3 Coordination Centre for Effects (CCE), National Institute for Public Health and the Environment (RIVM), Bilthoven, The Netherlands, 4 Climate Service Centre, Finnish Meteorological Institute, Helsinki, Finland

Abstract

Boreal forests are sensitive to climatic warming, because low temperatures hold back ecosystem processes, such as the mobilization of nitrogen in soils. A greening of the boreal landscape has been observed using remote sensing, and the seasonal amplitude of CO_2 in the northern hemisphere has increased, indicating warming effects on ecosystem productivity. However, field observations on responses of ecosystem productivity have been lacking on a large sub-biome scale. Here we report a significant increase in the annual growth of boreal forests in Finland in response to climatic warming, especially since 1990. This finding is obtained by linking meteorological records and forest inventory data on an area between 60° and 70° northern latitude. An additional increase in growth has occurred in response to changes in other drivers, such as forest management, nitrogen deposition and/or CO_2 concentration. A similar warming impact can be expected in the entire boreal zone, where warming takes place. Given the large size of the boreal biome – more than ten million km^2– important climate feedbacks are at stake, such as the future carbon balance, transpiration and albedo.

Editor: Olivier Boucher, Université Pierre et Marie Curie, France

Funding: The authors have no support or funding to report.

Competing Interests: The authors have declared that no competing interests exist.

* Email: pekka.kauppi@helsinki.fi

Introduction

Variations of ambient temperature in boreal forests strongly affect the ecosystem processes [1,2,3]. Increased greenness and amplified seasonal changes of CO_2 have been observed [4,5], but the improvement of productivity has not been correlated with warming observations. If climatic warming has impacts on forest growth, this may significantly affect the carbon budget, transpiration and forest albedo [6,7].

Growing degree-days (GDD) are the annual sum of daily temperatures above a pre-defined threshold. GDD combines the effects of favourable temperature on the duration and intensity of growth. Two articles published in the 1980 s projected an increase in the growth of boreal forests under future climate warming [8,9]. A regression was reported, showing how forest growth is positively correlated with GDD ranging from 700 degree-days and low increment in northernmost boreal forests to 1,350 degree-days and high increment in southern boreal forests [8]. Moreover, the regression was used to explore hypothetical warming scenarios [9]. The eco-physiological mechanisms of this correlation are not fully understood. However, process-based ecosystem models suggest that an accelerating nitrogen cycle is an important, though conceivably not the only mechanism [3].

And indeed, a considerable warming has occurred in northern latitudes since the 1980 s [10]. In Canada, a positive and significant growth response to warming has been reported, based on inventory measurements on 1,267 sites, combining observational data with models, which relate stand growth to stand age [11]. Here we analyse a large amount of data from Finland, collected and compiled by the Finnish Meteorological Institute (FMI) and the Finnish Forest Research Institute (METLA), and apply the empirical methodology used earlier [8]. All boreal forests growing on a land area of 302,000 km^2 in Finland were covered in the monitoring system. Forest data from Finland are very well documented and have been made available in reference [12]. We describe the study region, derive spatially representative climate trends in terms of Growing Degree Days (GDD), estimate a regression between forest growth and GDD from recent data drawing on the geographic co-variation of the two variables, and finally compare the results with the findings and projections from the 1980 s.

Materials and Methods

The method of this research is based on the empirical finding that spatial and temporal variations of forest growth (annual increment) correlate with variations of GDD [8,9,13]. The study area is located in Finland, stretching from the southern to the northern limit of the boreal biome and is divided in 15 regions (Fig. 1). The Ahvenanmaa region (Åland Islands) is excluded as it has a temperate character and the forested area is small (about 1,000 km^2). Region 1 surrounding Helsinki is at the southern edge of the boreal zone, where agriculture and settlements are widespread but, nevertheless, 54% of the land is covered by

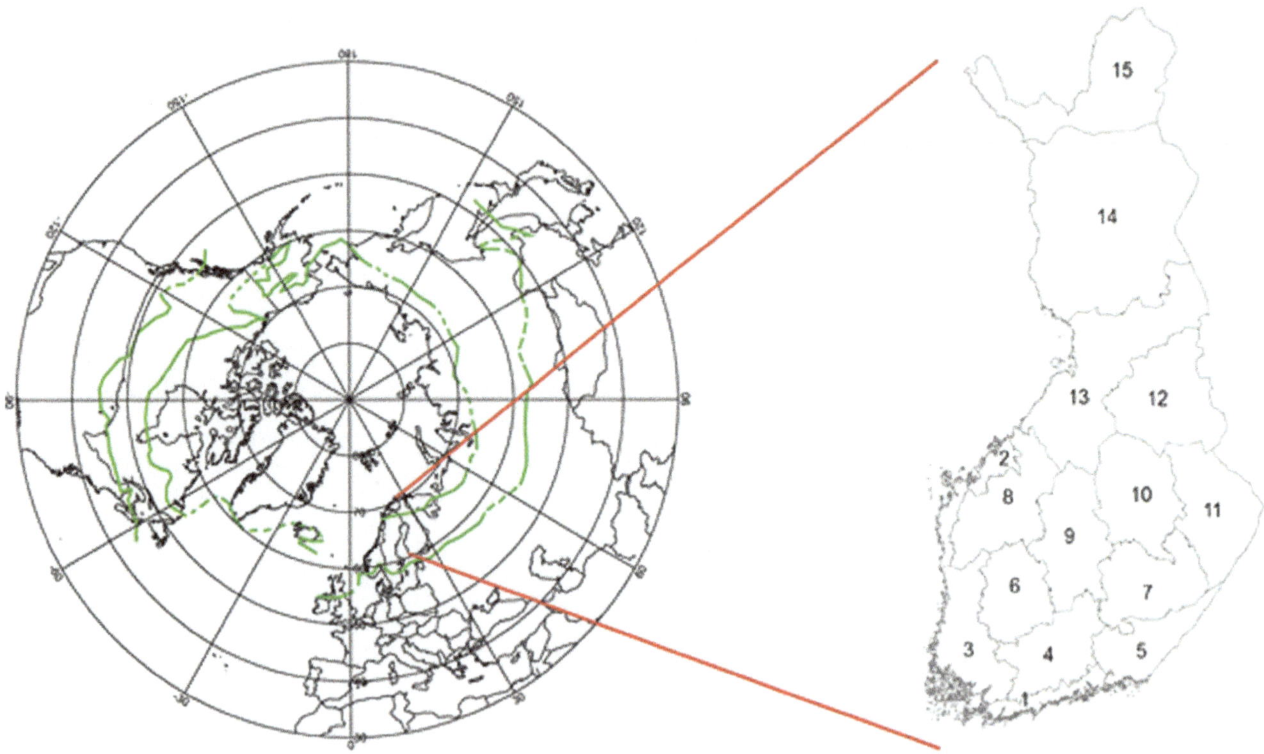

Figure 1. The boreal zone as presented in [9] and the 15 regions of this study located within the boreal biome in Finland.

forests. Region 15 between 68° and 70°N is partly located beyond the Arctic timber line and is the only one where forest land does not predominate as the main land cover type. The distance between the most northern and southern regions is about 900 km.

During 1961–2013 the FMI has recorded daily mean temperature at 120–210 stations across Finland. Station data on the daily mean temperature and the beginning and end of the growing season have been interpolated onto a 10 km ×10 km grid using a kriging method [14,15]. GDD for each grid point and each year has been calculated using a threshold temperature of +5°C. From these estimates we computed the spatial average GDD for each region and each year. Linear regressions were fitted to these time series.

The 11th National Forest Inventory (NFI) of Finland was carried out in 2009–2012. The sample plots of this inventory provide a rigorous network of measurements, which was distributed over all regions. Forest growth (or annual increment) refers to the accumulation of woody tissue on tree stems during one growing season. A national (and Nordic) convention of the concept 'forest growth' was applied, which does not take into account the contribution of the trees lost in natural mortality. However, the growth of recently harvested trees is included into the growth estimate. The concept is similar to 'gross growth' as referred to in [11].

The methods of measuring forest increment are well tested, because there has been a persistent economic incentive to obtain accurate information on this natural resource [16]. The number of sample plots measured in the 11th NFI varied between regions from 567 to 4,377 (Table 1). While the total number of plots was large (39,276), an even more important characteristic of the forest inventory system is that the plots were located using a well-tested statistical algorithm [16]. Therefore, the sample represents all forests in each region. Forest growth is directly measured from sample trees observing the width of the five most recent complete tree rings and the stem elongation of the leader shoot during a five-year period preceding the observation. As the sample trees for growth measurements are statistically chosen on each plot, the results represent the growth of all trees growing on forest land in each region. A linear regression was fitted to the data for describing the dependence of growth on GDD. The data refer to the year 2008, which coincides with the average time of tissue formation of the retrospective growth observations as taken in 11th NFI during 2009–2012.

The approach of this study is strictly empirical and is based on high-quality observations on daily temperature and forest growth, which in Finland are unique within the boreal biome. Measurements from the period 1961 to 2012 (to 2013 for GDD) were chosen, thus obtaining a sufficiently long period of time for trend analyses. Earlier records and reports were available referring to the period 1931–1960 and were used as a reference.

The earlier work [8,9] addressed the same area in Finland and applied growth observations of Finland's 3rd NFI, which was conducted in 1951–1953. Noting these sampling years and the time lag of retrospective monitoring, the analysis referred to the forest increment during the period 1946–1952. In the earlier papers [8,9] GDD data of the climate normal period 1931–1960 were applied. The method was improved in this new analysis as follows. The kriging method, which was not yet available for [8], provides a better spatial representation of the temperature data. As the forested area varies between regions, the growth observations in regression analysis were weighed by the area of forest land (see Table 1). In determining the growth-GDD regression, the timing of the temperature and growth data was better matched (the observations referring to 2006–2011).

Table 1. Land cover and forest sample size by regions.

Region	Land area	Forestry land	Forest land	Number of measured forest plots in 2009–2012
	1000 km^2			
1	6.7	4.3	3.6	918
2	7.2	5.1	4.7	1066
3	16.9	10.9	9.8	2513
4	14.3	9.6	9.3	2380
5	10.5	7.7	7.4	1879
6	12.6	9.7	9.2	2300
7	14.3	12.6	12.2	3095
8	19	14.3	12.6	3034
9	16.7	14.4	13.8	3258
10	16.8	14	13.5	3197
11	17.8	15.8	14.6	3430
12	21.5	20.4	17	3264
13	35.5	31.3	24.8	4377
14	64.5	62.5	42.2	3998
15	28.1	27.9	7.8	567
Total	302.4	260.5	202.5	39276

'Forestry land' is a land use concept referring to lands with no other priority assigned except forestry. 'Forest land' is a land cover concept referring to lands, which are estimated to produce at least 1 m^3 of wood per hectare and year as a long-term average. Data are from [12] and from METLA (A. Ihalainen, pers. comm.). While the location of the study area is the same as in [8] the geographical borders of regions slightly differ and, in particular, region 15 in the very north was not reported in [8].

Results

Warming trends

Climatic warming was quantified from the GDD regressions, comparing their estimates for the growing seasons of 1961 and 2008 (Fig. 2). The latter year coincided with the forest growth as measured in the 11th NFI. Warming was clearly observed as the GDD was higher for 2008 than for 1961 in all regions. The GDD for 2008 (and 1961) in each region refer to the estimates as obtained from the time series regression, not GDD as measured in that particular year, nor the five-year moving average (see Fig. 2). The warming trends were statistically significant ($p > 95\%$ in all regions). In absolute terms, the GDD warming was largest in the southernmost region 1 (+233 degree days) and smallest in the northernmost region 15 (+124 dd). However, in line with projections [9], the largest relative change was in the northernmost region 15 (+23.8%); while the smallest relative change occurred in the south-eastern region 5 (+15.2%). Comparing the moving averages with the regressions it appears that the warming in all regions was particularly significant since 1990. In all regions, the five-year average for the latest observed year 2011 is higher than the regression estimate for 2011 and for 2008 (Fig. 2).

Impact of warming on growth

The observations of this study can be compared with the earlier results [8]. Relating the most recent growth observations of the 11th NFI to the average GDD of 2006–11, a new regression was obtained with almost the same slope as earlier (Fig. 3). Differences between the results were as follows: (i) The new region 15 in the very north widened the range to lower values; (ii) The warming impact moved the data points to the right; and (iii) Other drivers affecting growth were reflected in the elevated position of the new regression above the old one.

Reproducing successfully the growth-to-GDD regression and obtaining an improved fit lends new support to using GDD in analyses of spatial and inter-annual variations of forest growth in the boreal zone. The new data were temporally better matched and spatially more representative than those reported earlier [8], and in the new analysis the growth estimates were weighted by the area of forest land. These improvements of the method and the inclusion of region 15 contributed to the better fit ($r^2 = 0.94$).

GDD above 1,350 degree-days were common in many regions in 2006–11, unlike in the mid-20th century (Figs. 2 and 3). However, warming was not the only driver affecting growth. Forest management, a gradual change in the species composition and age structure of forests, CO_2 fertilization, improving water-use efficiency, nitrogen deposition and other (unknown) factors can have played a role. In the earlier projections a maximum growth of 6 m^3 ha^{-1} yr^{-1} was postulated, expecting drought limitations to come into play [9]. This turned out to be incorrect, as growth above 6 m^3 ha^{-1} yr^{-1} was now observed in the majority of regions, even though the area-weighted country average did not exceed 5 m^3 ha^{-1} yr^{-1}. We cannot predict whether drought limitations will come into play in the future, especially as future climate may trigger changes in precipitation [17].

What would have been the growth of these boreal forests had the warming not occurred? This question was addressed in the following way: For every region we calculated the GDD for the years 1961 and 2008 from the regional regressions given in Fig. 2. Inserting the pairs of GDD values into the new GDD-growth regression (Fig. 3), we computed the warming impact on growth per hectare in every region as the difference of the two obtained growth values. This difference was multiplied by the forest area of each region (see Table 1), resulting in estimates of regional growth due to warming; and the difference to the overall growth was assigned to other factors (Table 2).

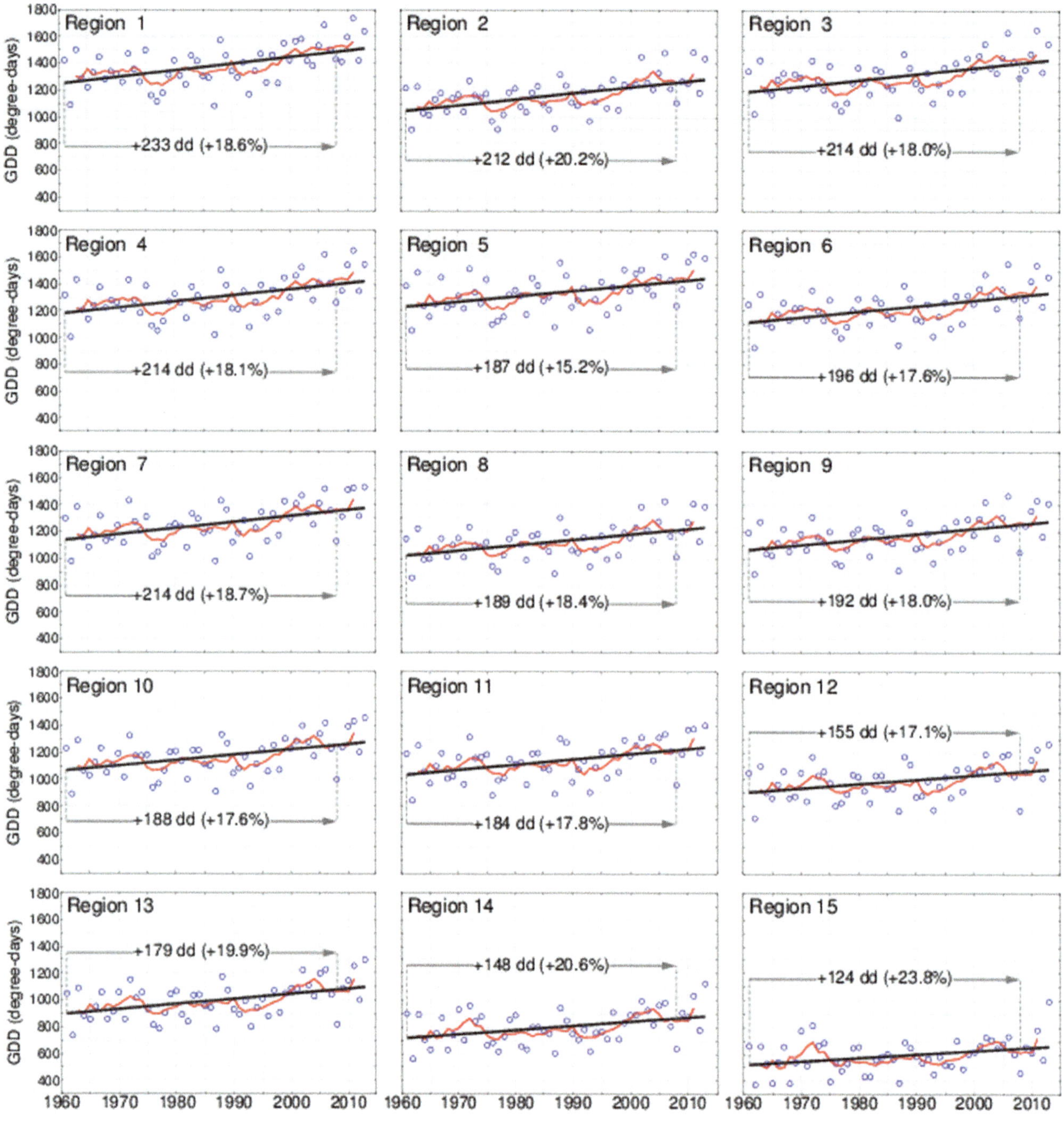

Figure 2. Annual Growing Degree Days (GDD) 1961–2013 for the regions (blue circles). Also shown are the linear regression lines (black) and the 5-year moving averages from 1963 to 2011 (red lines).

A growth of 103.9 million m^3 yr^{-1} was observed in the boreal forests of Finland referring to approximately 2008 as described in the 11th NFI (Table 1, based on [12], excluding Ahvenanmaa (Åland)). This can be compared to the observed growth of 49.5 million m^3 yr^{-1} as measure in the 4th NFI in 1960–63 [18]. Warming-induced increment is estimated at 28.4 million m^3 yr^{-1} (Table 2). An additional change of +26 million m^3 yr^{-1} would hence be the response to other drivers, such as improved forest management, CO_2 fertilization and other (unknown) mechanisms. Given uncertainties, we estimate that the growth of boreal forests of Finland would have been about 75 million m^3 in 2008, had the growing season temperatures remained at the level of the 1960's.

Discussion and Conclusions

More than half of the significant growth improvement in the boreal forests of Finland has been a response to climatic warming. Such a large change in productivity has greatly affected the ecological and economic performance of these forests. The long interval of about 60 years and the large geographic scale from the northern to the southern edge of the boreal zone give confidence

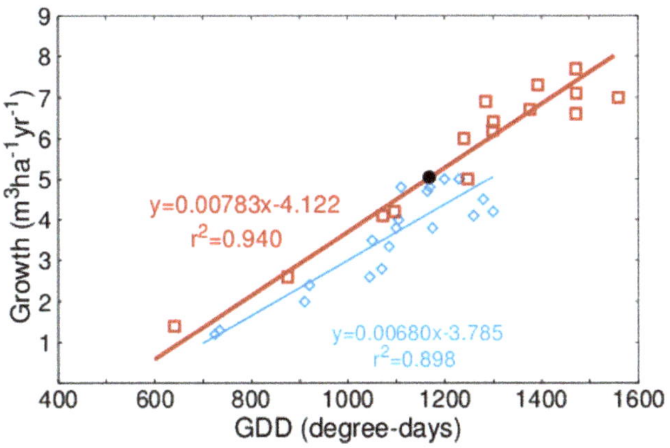

Figure 3. Regressions relating forest growth to Growing Degree Days. The new regression (red) referring to 2006–11 is based on data from the 15 regions shown in Figure 1. The black dot on the red line shows the area-weighted average of all 15 regions. The old regression (blue) as published in [8] was based on 19 data points as recorded in the mid-20th century.

for obtaining conclusive results. The regression analyses were applied as they are simple and transparent and enable comparison with earlier results [8,9].

The analysis in this paper addressed the statistical correlation between forest growth and growing degree days. There could be causal mechanisms affecting forest growth, which vary in time and space and are unrelated to temperature and, nevertheless, trigger similar spatial patterns as those depicted in Table 2. However, it has been well documented that the inter-annual variation of growth in the boreal zone correlates with changes in temperature. Moreover, the slope of the regression is reproduced using two independent data sets (see Fig. 3). Therefore, we conclude that the spatial and temporal variation of growing season temperature is the main causal factor affecting variations of forest growth in this region.

New research is needed to elaborate on the ecological and eco-physiological mechanisms by which increasing temperatures stimulate forest growth. An empirical analysis like this cannot characterize the mechanisms of such large impacts. It is possible that the various drivers reinforce one another based on synergistic mechanisms, thus creating an aggregated warming impact. Analogous attribution problems are met in plant sciences in assessing the relative effect of plant genotype and of the environment in affecting plant phenotype and yield [19]. Finland-specific land management, nitrogen deposition, or disturbance history may have affected the warming response. However, warming has been widely observed [10], CO_2 concentration is spatially invariant and similar warming responses have been reported recently from Canada [20] and in Russia [21]. Indirect observations reinforce this view [4,5]. Therefore, a significant warming impact on productivity has likely occurred more broadly in forest ecosystems of the boreal biome.

The growth concept that we applied is similar to "gross growth" as described in [11]. The fraction of gross growth that decomposes in the forest has been relatively small in Finland (national total <5 million m^3 annually). Forest fires have been virtually non-existent; and trees, which have been damaged by storms or insects, are often harvested for industrial or energy use. The growth concept in the Finnish NFI has not changed since the 1960's and, therefore, the time series are directly comparable.

Ecological and economic assets are at stake in Finnish forestry. For example, a major expansion plan of forest industries in Finland was recently announced, though not yet confirmed, drawing on the ample timber resources. The new plant would start operating in 2017 and would be located in region 6. If confirmed, the new plant is the largest single investment project in forest industry to date in Finland. The plant would process about 4

Table 2. Measured growth by regions, the warming impact, and the growth unrelated to warming.

Region	Mesured growth	Warming impact	Growth unrelated to warming
		Million m^3 yr^{-1}	
1	2.6	0.7	1.9
2	2.9	0.8	2,1
3	6.6	1.7	4.9
4	7.2	1.6	5.6
5	5.3	1.1	4.2
6	6.2	1.4	4.8
7	8.9	2.0	6.8
8	6.4	1.9	4.5
9	8.8	2.1	6.8
10	9.4	2.0	7.4
11	8.8	2.1	6.7
12	7.2	2.1	5.1
13	10.7	3.5	7.2
14	11.7	4.9	6.8
15	1.5	0.8	0.7
Total	103.9	28.4	75.6

All results refer to the year 2008, with estimates for 1961 used as reference.

million m^3 of round wood annually. Timber shortage would restrict such initiatives in absence of the warming impacts. More broadly, the implications of climatic warming are immensely diverse. The increment of woody biomass feeds into forest carbon sinks [22]. Research should address the climate feedbacks and assess how hydrology or forest albedo will change as forest growth and standing biomass gradually increase in the circumpolar boreal forests [23–26].

Papers in the 1980 s suggested that boreal forests are sensitive to climatic warming, should a warming occur [8,9]. This study reports that a warming has occurred by 2013 and boreal forests have responded with accelerating growth. Although the data cover only Finland, similar responses have been reported based on research in other parts of northern hemisphere forests [21]. It is important to continue and expand the monitoring programs of both climatic changes and ecosystem responses, and to further improve their scientific quality.

Acknowledgments

The authors thank Antti Ihalainen, Pekka Nöjd, Pauline Stenberg, Heikki Tuomenvirta and two anonymous reviewers for valuable comments, and the Finnish Forest Research Institute (METLA) and Finnish Meteorological Institute(FMI) for collecting the big data, processing them and making them publicly available. The authors thank University of Helsinki, METLA, FMI, and RIVM for their research infrastructure.

Author Contributions

Analyzed the data: PK MP. Contributed to the writing of the manuscript: PK MP PP. Initiated the research: PK. Processed the meteorological observations: PP.

References

1. Goulden ML, Wofsy SC, Harden JW, Trumbore SE, Crill PM, et al. (1998) Sensitivity of boreal forest carbon balance to soil thaw. Science 279: 214–217.
2. Hyvönen R, Ågren GI, Linder S, Persson T, Cotrufo MF, et al. (2007) The likely impact of elevated [CO2], nitrogen deposition, increased temperature and management on carbon sequestration in temperate and boreal forest ecosystems: a literature review. New Phytologist 173: 463–480.
3. Melillo JM, McGuire AD, Kicklighter DW, Moore III B, Vorosmarty CJ, et al. (1993) Global climate change and terrestrial net primary production. Nature 363: 234–240.
4. Myneni RB, Keeling CD, Tucker CJ, Asrar G, Nemani RR (1997) Increased plant growth in the northern high latitudes from 1981 to 1991. Nature 386: 698–702.
5. Graven HD, Keeling RF, Piper SC, Patra PK, Stephens BB, et al. (2013) Enhanced seasonal exchange of CO_2 by northern ecosystems since 1960. Science 341: 1085–1089.
6. Foley JA, Kutzbach JE, Coe MT, Levis S (1994) Feedbacks between climate and boreal forests during the Holocene epoch. Nature 371: 52–54.
7. Bonan GB (2008) Forests and climate change: forcings, feedbacks, and the climate benefits of forests. Science 320: 1444–1449.
8. Kauppi P, Posch M (1985) Sensitivity of boreal forests to possible climatic warming. Climatic Change 7: 45–54.
9. Kauppi P, Posch M (1988) A case study of the effects of CO_2-induced climatic warming on forest growth and the forest sector: A. Productivity reactions of northern boreal forests. In: Parry ML, Carter TR, Konijn N, editors. The Impact of Climatic Variations on Agriculture, Volume 1: Assessments in Cool Temperate and Cold Regions. Kluwer, Dordrecht, Netherlands. 183–195.
10. Hartman DL, Klein Tank AMG, Rusticucci M, Alexander LV, Broenniman B, et al. (2013) Observations: atmosphere and surface. In: Stocker TF, Qin D, et al., editors. Climate Change 2013: The Physical Science Basis, Cambridge University Press. 159–254.
11. Hember RA, Kurz WA, Metsaranta JM, Black TA, Guy RD, et al. (2012). Accelerating regrowth of temperate-maritime forests due to environmental change. Global Change Biology, 18: 2026–2040.
12. Ylitalo E, editor (2013) Finnish Statistical Yearbook of Forestry. Finnish Forest Research Institute.
13. Mikola P (1950) On variations in tree growth and their significance to growth studies. Communicationes Instituti Forestalis Fenniae 38: 1–131.
14. Henttonen H (1991) Kriging in interpolating July mean temperatures and precipitation sums. Report 12, Dept. of Statistics, University of Jyväskylä, Jyväskylä, Finland. 41 pp.
15. Venäläinen A, Heikinheimo M (2002) Meteorological data for agricultural applications. Physics and Chemistry of the Earth 27: 1045–1050.
16. Tomppo E, Heikkinen J, Henttonen HM, Ihalainen A, Katila M, et al. (2011) Designing and Conducting a Forest Inventory - case: 9th National Forest Inventory of Finland. Managing Forest Ecosystems Vol.22, Springer. 272 pp.
17. Bintanja R, Selten FM (2014) Future increases in Arctic precipitation linked to local evaporation and sea-ice retreat. Nature 509: 479–482.
18. Huikari O, Seppälä R, editors (1984) Yearbook of forest statistics 1983. Folia Forestalia 590: 76–81.
19. Finlay KW, Wilkinson GN (1963) The analysis of adaptation in a plant-breeding programme. Crop and Pasture Science 14: 742–754.
20. Peng Y, Arora VK, Kurz WA, Hember RA, Hawkins BJ, et al. (2014) Climate and atmospheric drivers of historical terrestrial carbon uptake in the province of British Columbia, Canada. Biogeosciences 11: 635–649.
21. Frost GV, Epstein HE (2014) Tall shrub and tree expansion in Siberian tundra ecotones since the 1960 s. Global Change Biology 20: 1264–1277.
22. Pan Y, Birdsey RA, Fang J, Houghton R, Kauppi PE, et al. (2011) A large and persistent carbon sink in the world's forests. Science 333: 988–993.
23. Swann AL, Fung IY, Levis S, Bonan GB, Doney SC (2010) Changes in Arctic vegetation amplify high-latitude warming through the greenhouse effect. Proc Natl Acad Sci USA 107: 1295–1300.
24. Betts RA (2000) Offset of the potential carbon sink from boreal forestation by decreases in surface albedo. Nature 408: 187–190.
25. Luyssaert S, Jammet M, Stoy PC, Estel S, Pongratz J, et al. (2014) Land management and land-cover change have impacts of similar magnitude on surface temperature. Nature Climate Change 4: 389–393.
26. Lukeš P, Rautiainen M, Manninen T, Stenberg P, Mõttus M (2014) Geographical gradients in boreal forest albedo and structure in Finland. Remote Sensing of Environment 152: 526–535.

18

Accumulation and Connectivity of Coarse Woody Debris in Partial Harvest and Unmanaged Relict Forests

Robert C. Morrissey[1,2]*, Michael A. Jenkins[2], Michael R. Saunders[2]

1 Department of Forest Ecology, Faculty of Forestry and Wood Sciences, Czech University of Life Sciences, Prague, Czech Republic, 2 Hardwood Tree Improvement and Regeneration Center, Department of Forestry and Natural Resources, Purdue University, West Lafayette, Indiana, United States of America

Abstract

When a tree dies, it continues to play an important ecological role within forests. Coarse woody debris (CWD), including standing deadwood (SDW) and downed deadwood (DDW), is an important functional component of forest ecosystems, particularly for many dispersal-limited saproxylic taxa and for metapopulation dynamics across landscapes. Processes, such as natural disturbance or management, modify forest composition and structure, thereby influencing CWD abundance and distribution. Many studies have compared older forests to forests managed with even-aged silvicultural systems and observed a prolonged period of low CWD occurrence after harvesting. With fine-scale spatial data, our study compares the long-term impacts of light partial harvesting on the CWD structure of eastern deciduous hardwood forests. We mapped and inventoried DDW and SDW using variable radius plots based on a 10 m×10 m grid throughout an unmanaged, structurally-complex relict forest and two nearby forests that were partially harvested over 46 years ago. The relict stand had significantly larger individual pieces and higher accumulations of DDW and SDW than both of the partially harvested stands. Connectivity of CWD was much higher in the relict stand, which had fewer, larger patches. Larger pieces and higher proportion of decay-resistant species (e.g. *Quercus* spp.) in the relict forest resulted in slower decomposition, greater accumulation and increased connectivity of CWD. Partial harvests, such that occur with selection forestry, are generally considered less disruptive of ecosystem services, but this study highlights the long-term impacts of even light partial harvests on CWD stocks and distribution. When planning harvesting events, forest managers should also consider alternative methods to ensure the sustainability of deadwood resources and function.

Editor: Christopher Carcaillet, Ecole Pratique des Hautes Etudes, France

Funding: This work was funded by the Hardwood Tree Improvement and Regeneration Center, and the Department of Forestry and Natural Resources at Purdue University. The funders had no role in study design, data collection and analysis, decision to publish, or preparation of the manuscript.

Competing Interests: The authors have declared that no competing interests exist.

* Email: robcmorrissey@gmail.com

Introduction

Even after death, trees continue to influence ecological processes and biodiversity of forest ecosystems [1]. Coarse woody debris (CWD), both standing dead trees (SDW) and downed deadwood (DDW), provides regeneration substrate for some plants, serves as habitat for many insect and wildlife taxa, mitigates runoff and erosion from slopes, contributes to soil development, and provides long-term sequestration of carbon and other nutrients [2]. Because CWD decays slowly, it can potentially influence these ecological processes at local, stand, and landscape levels for many decades. As forest managers have attempted to integrate a broader array of ecological objectives into their planning, provision of CWD to preserve structural attributes and ecological processes has become increasingly emphasized.

The importance of CWD as habitat has been well documented for many organisms including birds [3], arthropods [4], fungi [5], and small mammals and amphibians [6]. Given the abundance of CWD in mature forests, it is conceivable that many species evolved to rely on a suite of habitats based on abundance, species, size, decay stages, and position (standing vs. fallen) of CWD [1], [2]. In these forests, CWD-dependent species with limited dispersal should predominate, and the connectivity and spatial arrangement of CWD should be critical to the abundance of species and individuals [7]. Availability, connectivity, patch size and abundance of CWD are critical to supporting and maintaining diverse natural communities [8]. Connectivity, an emergent property across multiple ecological scales, strongly influences the population and metapopulation dynamics of multiple taxa.

Not surprisingly, the literature is replete with differences among species regarding the importance of connectivity of CWD at the stand scale [5], [7], [9]. Organisms that are highly mobile are likely unaffected at stand-level scales [10], but there may be population fitness benefits associated with high connectivity for these species, e.g. prey avoidance, perch sites [11]. For saproxylic taxa that may be dispersal-limited, including many insects, bryophytes, lichens, amphibians, and reptiles, the scale of connectivity or, conversely, fragmentation, may influence an individual's ability to disperse and occupy new habitat [12], [13], [14]. Thus, changes to spatiotemporal patterns of habitat continuity in managed and fragmented forests likely influence species dispersal, success, and fitness.

Understanding the spatiotemporal dynamics of CWD is essential to framing strategies for forest and wildlife management

Table 1. Coarse woody debris piece summaries.

a) Downed deadwood				
Tract	n	LED (cm)	Length (m)	Volume (m^3)
Relict	924	34.3±0.7 A	11.2±0.2 A	1.19±0.08 A
Single-harvest	288	26.0±0.9 B	8.7±0.2 B	0.48±0.06 B
Double-harvest	153	31.3±2.2 B	7.6±0.3 C	0.84±0.21 B
b) Standing deadwood				
Tract	**n**	**DBH (cm)**	**Height (m)**	**Volume (m^3)**
Relict	182	46.8±2.5 A	13.2±0.7 A	2.27±0.26 A
Single-harvest	96	36.5±2.4 A	10.9±0.8 A	1.24±0.28 A
Double-harvest	57	28.7±3.7 B	7.0±0.5 B	0.56±0.16 B

Total number of pieces (n) and mean (± standard error) of individual piece sizes of sampled a) downed deadwood (minimum diameter of 10 cm) and b) standing deadwood in relict, single-harvest, and double-harvest stands. Large end diameter (LED), length, and volume of downed deadwood, and diameter at 1.37 m (DBH), total height, and volume of standing deadwood are presented. We tested for differences ($p<0.05$) of log-transformed values using a generalized linear model and post-hoc multiple comparisons with Tukey's honestly significant difference. Different letters indicate significant differences among stands.

[15], and is thus an essential element of biological conservation in forest ecosystems. Spatiotemporal CWD dynamics are largely influenced by tree mortality and decay rates. Tree mortality is not evenly distributed over time or space [1] and is related to disturbance history, stand structure, and site factors [2], all of which influence patterns of CWD dispersion. Several studies of old-growth forests have observed high abundance and aggregated patterns of DDW [5], [7], [10], but patterns of SDW varied across many studies [16], [17]. Light partial harvests, which include thinnings, selection cut, and some timber stand improvement treatments, generally disproportionally remove low vigor and individual or small groups of trees, in order to favor more vigorous, higher quality trees throughout the stand. Therefore, these harvests may impact future spatiotemporal inputs of CWD by influencing CWD size, species, volume, input rate, and spatial distribution, which thereby can both reduce the abundance of CWD over time and either homogenize or concentrate CWD distribution, depending on the management regime. Light partial harvests have typically been considered to have a short-term and minor ecological impact, but studies of CWD impacts related to known management events is critical to understand long-term dynamics of ecological processes.

We examined the quantity, quality, and spatial arrangement of deadwood in a relict forest and two managed forests that were partially harvested more than 46 years ago. All stands are isolated forests within a fragmented agricultural matrix; thus, connectivity at a stand level may be especially important to sustain dispersal-limited taxa under the assumption that highly connected populations are less likely to succumb to local random extinction events. We chose not to consider individual species or taxa in this study, but rather focus on structural connectivity of the CWD within stands. Based on differences in disturbance histories of these stands, we hypothesized: 1) the overall volume of CWD would be less in managed stands; 2) average piece size and abundance of CWD would be less in managed stands than in the relict, unmanaged stand; 3) the distribution of CWD volume among decay stages would shift towards less decayed material in managed stands relative to the relict stand; and 4) distribution of CWD would exhibit higher levels of connectivity in less disturbed stands.

Methods

Study sites

Study sites were located on the Davis Purdue Agricultural Center property; permissions were granted by the Woodlands Management Committee through the Department of Forestry and Natural Resources of Purdue University. They included three *Quercus*-dominated deciduous forest stands within a fragmented, agricultural landscape in east-central Indiana, U.S.A. (40°15′26″, 85°09′16″). These stands have gently rolling, low relief topography with elevation changes <5 m across the entire area. Since the middle of the 20th century, all stands have been impacted by Dutch elm disease and elm phloem necrosis, both of which resulted in the death of most elm trees before they recruit into the overstory canopy. The relict forest is the largest stand (20.6 ha), of which we sampled the 8.5 ha interior. Records indicate that this stand has not been managed, except for the removal of some dead and dying trees in 1941, 1948, and 1955 (volume removed of 5.3 m^3 ha^{-1}). The smallest stand (2.8 ha) was sampled in its entirety; this stand, referred to as the double-harvest stand henceforth, was selectively harvested in 1951 (volume removed of 9.8 m^3 ha^{-1}) and again in 1964 (volume removed of 7.8 m^3 ha^{-1}). The third stand (2.9 ha) was also selectively harvested in 1951 (volume removed of 8.1 m^3 ha^{-1}); this stand will henceforth be referred to as the single-harvest stand. According to historical records, all trees removed from these stands were overstory individuals ≥40 cm at 1.37 m (DBH). For a more thorough description of these sites, see [18].

Data collection

We sampled CWD on a 10 m by 10 m grid using probability proportional to size (PPS) sampling methods. At each grid point, DDW was sampled using point relascope sampling [19] with an angle of 60°, and SDW was sampled using a 10-factor prism. For each individual DDW piece with a small end diameter (SED) ≥ 10 cm diameter, we measured the large end diameter (LED), SED, and the length of the piece. When pieces were long enough to still have intact multiple branches within a partially intact crown, we chose the single most linear piece to measure the SED and length. In rare cases, branches of intact crowns had sufficiently large diameters and were sampled as independent pieces. For

SDW, pieces had to be ≥1.3 m in height, but no minimum diameter limit was used; height was recorded to either top of crown, if still intact, or top of the standing stem.

To estimate individual DDW piece volumes, we assumed a conic-paraboloid shape to minimize bias of estimates and field measures required [20]. We calculated volume (V; m^3) using:

$$V = \frac{L}{12} \times \left(5A_l + 5A_s + 2\sqrt{A_l A_s}\right) \qquad (5.1)$$

where L is DDW length (m), A_l is cross-sectional area (cm) at large end, and A_s is cross-sectional area (cm) at small end. Because we were only interested in historical inputs of DDW as a function of forest dynamics, we did not account for decreased volumes of collapsed pieces in advanced states of decay, as recommended by [20]. Volume for SDW with intact or partially intact crowns was based on unpublished local volume tables using a 10 cm top diameter (Table S1). For sheared stems, volume was estimated with equation 5.1 using height for length, breast height diameter to calculate A_l, and a form class of 78 was used to estimate SED and calculate A_s; form class mathematically describes the shape of the tree bole and provides an estimate of the bole diameter at any given length (or height in this case) based on the taper of the bole and DBH [21].

We classified each piece according to its decay stage; when a piece is sufficiently large, it often had different sections of the piece in differing decay stages, so we assigned decay stage based on the section most representative of the piece based on volume. For DDW, we classified decay stage as: 'sound' if recently dead with wood still hard and bark largely present, either loose or tight; 'intermediate-decay' as bark mostly gone, bole periphery softened whereby a blade can penetrate the outer layer; and 'high-decay' as little to no bark remains, bole periphery is very soft and extends in to the core, often partially incorporated into forest floor or vegetation has begun to colonize. For SDW, we had only two categories: 'sound' and 'intermediate-decay', as described above.

Data analyses

We compared mean length, LED, and volume of individual DDW among all three stands using a generalized linear model (GLM) to account for the unbalanced number of pieces sampled in each stand. We used a log transformation of the response variable and a Gaussian distribution of errors. The same procedure was used for individual SDW pieces to examine DBH, height, and volume. To test for differences of DDW and SDW abundance among stands, we used a subset of the sampling grid (30 m by 30 m) distribution to limit correlation among plots related to sampling of individual pieces from multiple different plots. We used a GLM procedure on the square root transformed response variable and a Gaussian distribution of errors. Post-hoc tests of comparison were conducted using Tukey's honest significant differences method.

We tested for differences between stands based on volume ha^{-1} by size classes (≤30 cm, >30 cm and ≤60 cm, >60 cm) and decay stages using a chi-square test. For density, we could only test for differences in the number of pieces of DDW by decay stages; DDW by size classes and SDW for decay and size classes had small bin values that could not be tested formally, although we do present them graphically.

Estimates of connectivity were developed to characterize the potential for dispersal-limited organisms to traverse the stand based on their affinity for CWD of a certain form or size. We assumed organisms could disperse 10 m, or the distance between two grid cells, if CWD was present in adjacent cells; thus, data included SDW and DDW pieces that were tallied from at least two gridpoints. Connectivity was defined by the number of gridpoints that tallied a CWD piece divided by the total number of gridpoints, and reported as a percentage by diameter size classes (10 cm width). We standardized patch area and patch density using per hectare values; patches of CWD were defined either by the edge of the sample area, adjacent plots that tallied no CWD, or some combination of the two. Connectivity interactions between DDW and SDW were not substantiated because the PPS methods differ both in angle used to select sample pieces (i.e., DDW is sampled with 60° and SDW is sampled with 1° 48′) and also the size criterion used for each sample procedure (i.e., DDW uses length of piece and SDW uses diameter).

Given the time since harvest in these stands and the lack of stumps, we assumed that all CWD was related to natural processes rather than the harvest events. All tests were conducted with an alpha level of 0.05, and all analyses were done using R [22].

Table 2. Coarse woody debris stand-level summaries.

a) Downed deadwood			
Tract	**n**	**Pieces ha^{-1}**	**Volume ha^{-1} (m^3 ha^{-1})**
Relict	817	242.8±17.0 A	130.8±4.2 A
Single-harvest	392	154.8±14.8 B	39.9±2.5 B
Double-harvest	234	199.3±28.1 AB	59.8±6.4 B
b) Standing deadwood			
Tract	**n**	**Pieces ha^{-1}**	**Volume ha^{-1} (m^3 ha^{-1})**
Relict	817	21.7±1.9 A	24.3±1.0 A
Single-harvest	392	18.4±2.1 A	11.2±0.7 B
Double-harvest	234	41.2±9.1 A	7.6±0.7 B

Total plots (n), and mean (± standard error) number of pieces ha^{-1} and volume ha^{-1} of a) downed deadwood (minimum diameter of 10 cm) and standing deadwood in relict, single-harvest, and double-harvest stands. To minimize correlation between plots, we used a subset of plots based on 30 m×30 m spacing in each stand. We tested for differences ($p<0.05$) of square root-transformed values using a generalized linear model and post-hoc multiple comparisons with Tukey's honestly significant difference. Different letters indicate significant differences among stands.

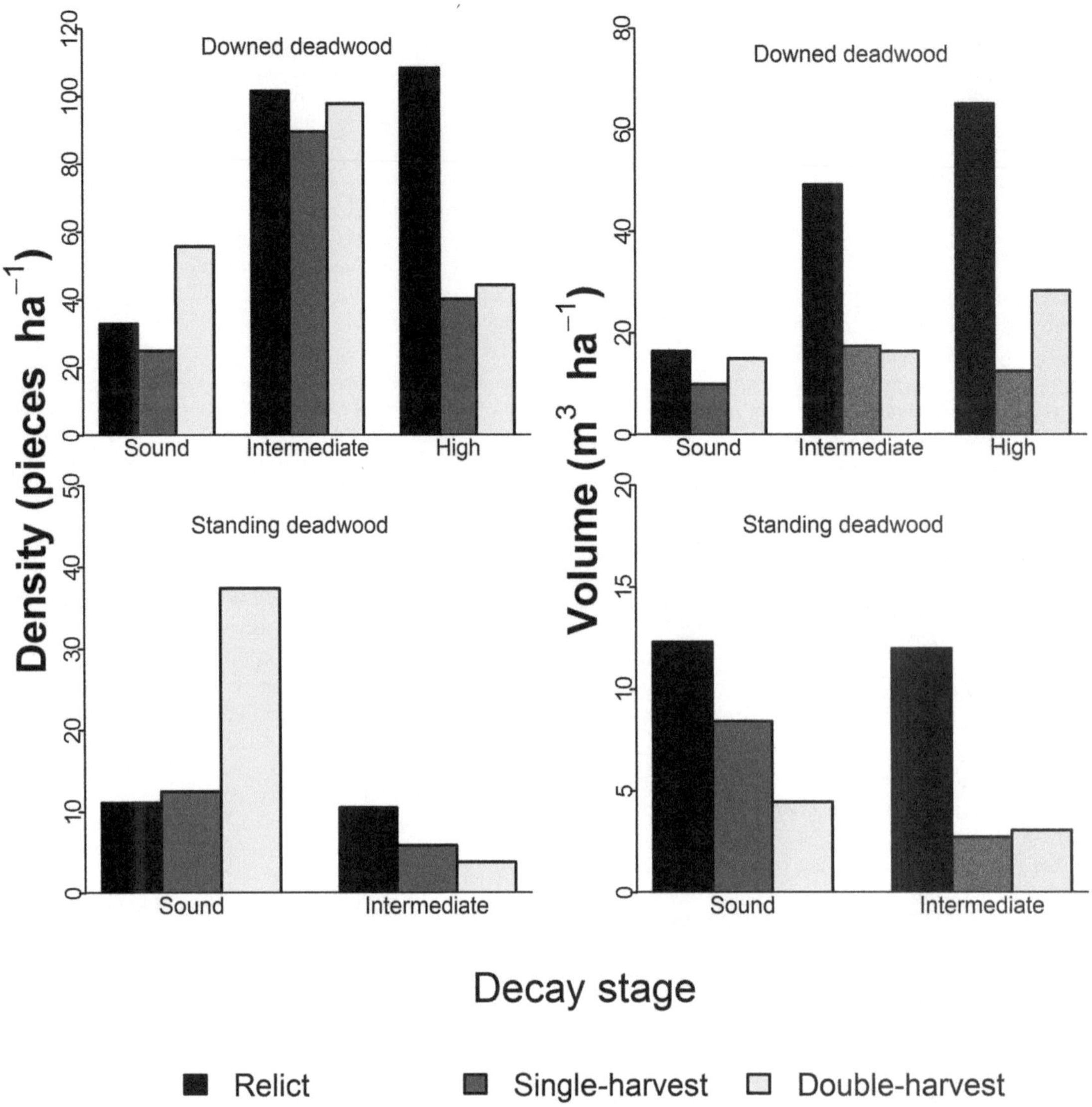

Figure 1. Coarse woody debris decay stage summaries. Downed deadwood (DDW; top row) and standing deadwood (SDW; bottom row) number of pieces ha^{-1} and volume ha^{-1} by decay stage for relict, single-harvest, and double-harvests stands. Decay stages were assigned as: sound, intermediate, or high; for SDW, sound and intermediate categories were used. For a more thorough description, see 2.2 Data Collection.

Results

Individual pieces of DDW in the relict stand were significantly larger in LED, length, and volume compared to the partial-harvest stands (Table 1; all $p \leq 0.003$). Mean volume of individual DDW pieces within the relict stand was 142% larger compared to the double-harvest stand and 248% larger than those of the single-harvest stand. Individual SDW pieces of the relict and single-harvest stands were not significantly different (all $p \geq 0.202$); however, individual piece means of DBH, height, and volume of both stands differed from those of the double-harvest stand (all $p \leq 0.02$). Within the double-harvest stand, the population of SDW was dominated by smaller pieces with correspondingly low volumes.

Mean number of DDW pieces ha^{-1} within the relict stand was significantly higher compared to the single-harvest stand ($p = 0.033$), but did not differ from the double-harvest stand ($p = 0.134$; Table 2); the partial-harvest stands were not significantly different from one another ($p = 0.975$). Significant differences between the relict and partial-harvest stands were evident for volume ha^{-1} of DDW (all $p < 0.001$). The relict stand had 328% and 219% more volume compared to the single-harvest and double-harvest stands, respectively; the partial-harvest stands were not significantly different from one another ($p = 0.918$). Mean density of SDW pieces ha^{-1} did not differ among all three stands (all $p \geq 0.764$), but SDW volume ha^{-1} of the relict stand was significantly higher (both $p \leq 0.008$) than that of both partial-harvest stands, which were not different from one another ($p = 0.773$). Although density of SDW pieces ha^{-1} did not significantly differ among stands (all $p < 0.245$), the double-harvest stand had 41.2±9.1 (mean ± SD) pieces ha^{-1} on average, almost double the relict stand. The SDW volume ha^{-1} of the relict stand was 217% greater than that of the single-harvest stand and 320% greater than that of the double-harvest stand.

We observed a significant difference in the number of DDW pieces ha^{-1} ($\chi^2 = 15.2$, $p = 0.004$) and volume ha^{-1} ($\chi^2 = 131.0$, $p < 0.001$) by decay stages among stands (Figure 1). In the relict stand, density of high-decay DDW pieces in the relict stand (108.4

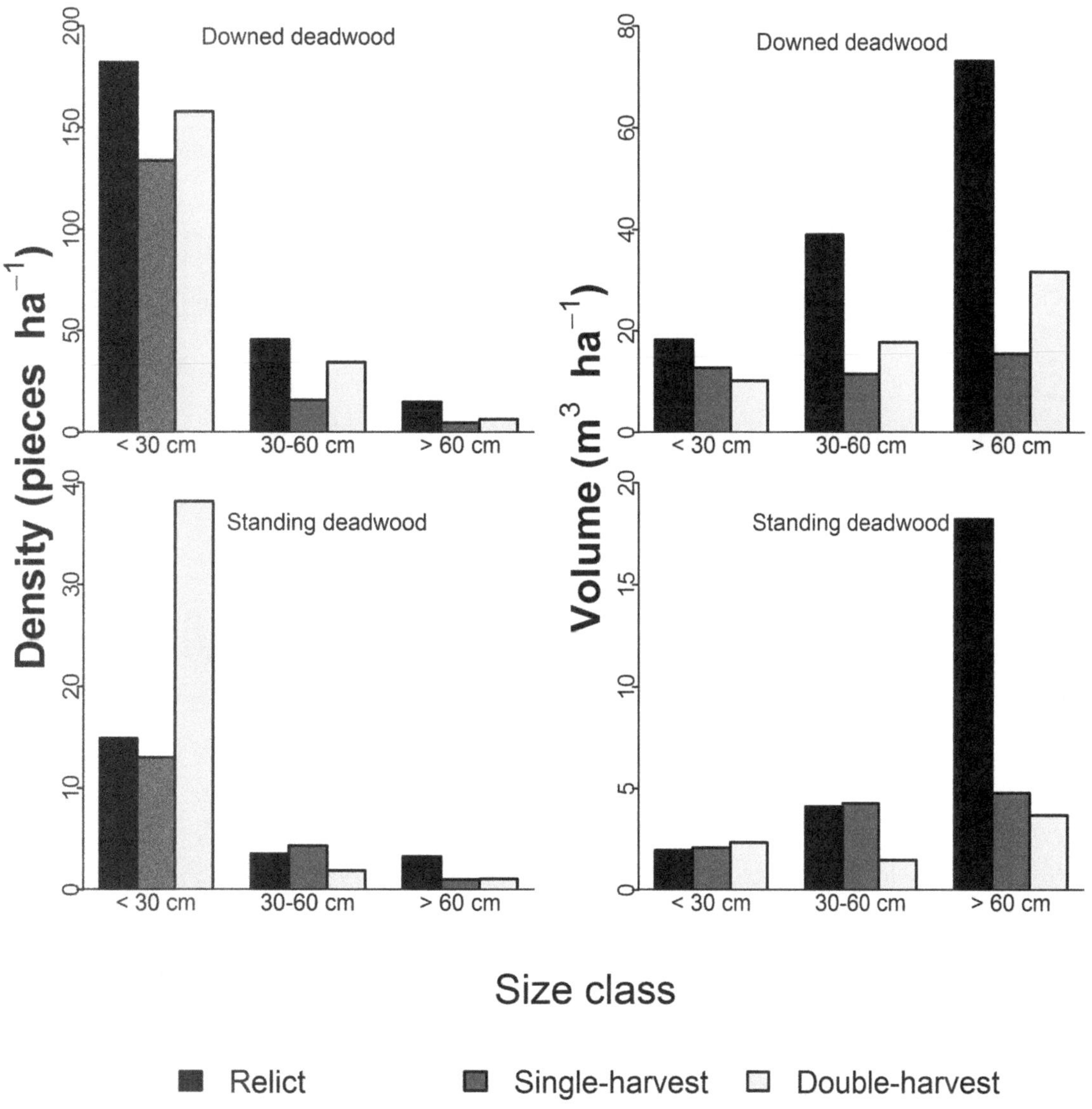

Figure 2. Coarse woody debris size class summaries. Downed deadwood (DDW; top row) and standing deadwood (SDW; bottom row) number of pieces ha^{-1} and volume ha^{-1} by size class for relict, single-harvest, and double-harvests stands. Size classes were defined as diameters of ≤30 cm, >30 cm and ≤60 cm, and >60 cm; DDW diameter was based on large end diameter, and we used diameter at 1.37 m (DBH) for SDW.

pieces ha^{-1}) was much higher than either of the partial-harvest stands (≤44.4 pieces ha^{-1}), and volumes in intermediate- and high-decay, 49.4 m^3 ha^{-1} and 65.1 m^3 ha^{-1}, respectively, were more than twice the amount in either of the partial-harvest stands. Density of SDW pieces by decay stage could not be tested because there were too few pieces in the intermediate-decay stage, but there was more than three times the number of sound SDW pieces (37.4 pieces ha^{-1}) in the double-harvest stand compared to the other stands (Figure 1). Volume ha^{-1} of SDW by decay stage was significantly different among stands ($\chi^2 = 177.6$, $p<0.001$); the relict stand had 12 m^3 ha^{-1} in the intermediate decay stage compared to <5 m^3 ha^{-1} in each of the partial-harvest stands.

For size class distributions, we could not test for differences in the number of DDW or SDW pieces ha^{-1} among stands because the largest size classes had very low values (Figure 2). All stands exhibited high densities of DDW <30 cm, similar to the living tree distribution [18]. The double-harvest stand had more than double the density of SDW pieces, 38.2 pieces ha^{-1}, in the <30 cm size class compared to the other two stands. Volume of DDW ha^{-1} across size classes was different across stands ($\chi^2 = 103.8$, $p<0.001$), with the relict stand having larger volumes in all size classes, most notably in classes >30 cm. Volume of SDW ha^{-1} across size classes differed across stands ($\chi^2 = 76.3$, $p<0.001$), attributable to the large volume of large diameter (≥60 cm) SDW, 18.2 m^3 ha^{-1}, compared to 4.8 m^3 ha^{-1} and 3.7 m^3 ha^{-1} for the single- and double-harvest stands, respectively.

Connectivity of DDW and SDW declined as minimum diameter increased in a similar pattern for all stands, but the relict stand always displayed much higher connectivity (Figures 3, 4, and 5). Mean patch area ha^{-1} of DDW generally decreased as minimum diameter increased, but the partial-harvest stands area did not vary as much by minimum diameter class compared to the relict stand (Figure 3). Mean patch area ha^{-1} of the relict stand DDW exhibited two abrupt decreases at minimum LEDs of 40 cm and 60 cm; beyond 60 cm, all three stands had similar mean patch area ha^{-1}. For SDW, patch sizes show a similar flat profile for partial-harvest stands, and mean patch area ha^{-1} of the relict stand abruptly decreases beyond a diameter of 40 cm. Number of

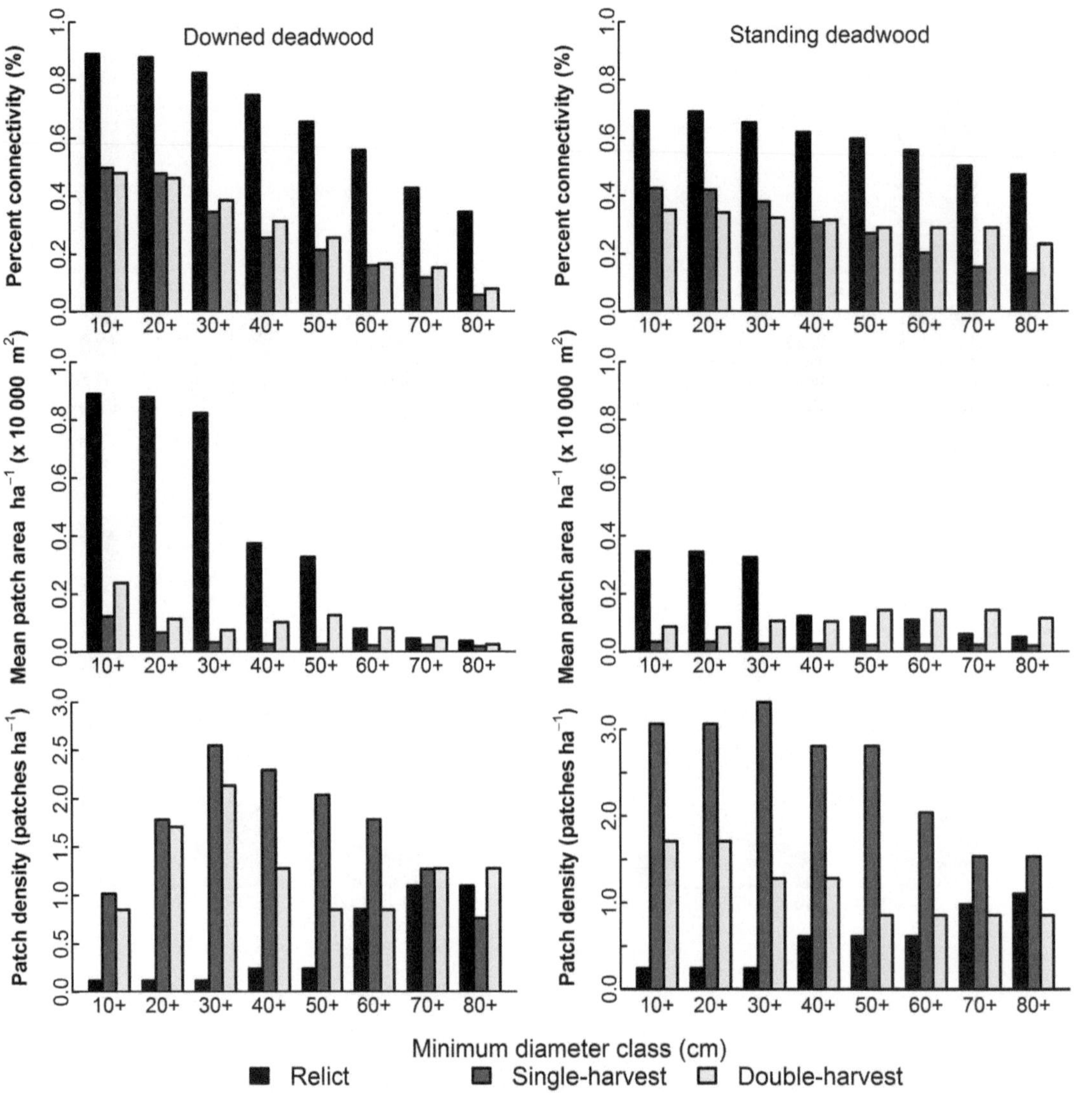

Figure 3. Coarse woody debris connectivity and patch-level summaries. Percent of all plots connected based upon presence of CWD (top row), mean patch area ha^{-1} (middle row), and number of patches ha^{-1} (bottom row) by 10 cm diameter classes for relict, single-harvest, and double-harvests stands. Downed deadwood (left column) diameter was based on large end diameter, and we used diameter at 1.37 m (DBH) for standing deadwood (right column); each diameter class indicates pieces of indicated size and larger were included in the analysis.

patches ha^{-1} generally shows an inverse relationship to mean patch area ha^{-1}, and the partial-harvest stands tended to have a greater number of smaller patches (Figures 4 and 5). Beyond a diameter of 60 cm for DDW and SDW, number of patches was more similar across all stands, although the single-harvest stand generally exhibited a greater number of patches than the other two stands. Although we used no metrics to compare overlay of DDW and SDW sample volume $plot^{-1}$, differences between stands were apparent (Figure 6). In the relict stand, plots with large SDW volumes tend to occur in areas with little or no DDW; we observed a similar pattern in the partial-harvest stands, but volumes of DDW and SDW throughout the stands were much lower. There were very few plots with no DDW or SDW sampled in the relict stand, but there are numerous patches throughout the partial-harvest stands with little or no DDW or SDW sampled.

Discussion

Almost half a century later, the effects of light partial harvesting on the type, amount, and spatial distribution of CWD remains evident. The relict forest had significantly larger pieces, greater volume, and higher connectivity of DDW and SDW compared to both partially harvested stands. Partial harvesting has altered species composition and structure relative to the relict stand [18], and the abundance and spatial distribution of CWD are a function of the resulting mortality and decomposition rates [23], [24].

Density-independent mortality related to windthrow, senescence, disease, insects, or some combination of these agents occurred in all stands, but they may affect species and/or different size trees in differing capacities [25]. For example, Dutch elm disease has been present in all stands since about the 1950s [26]. *Ulmus* trees, most of which were <20 cm DBH, were more abundant in the double-harvest stand (86 trees ha^{-1}), but were also common in the relict (41 trees ha^{-1}) and single-harvest stands (25 trees ha^{-1}). Given the relatively small size of *Ulmus* trees in the

Figure 4. Downed deadwood connectivity examples. Display of connected plots for downed deadwood (DDW) in relict, single-harvest, and double-harvests stands for pieces ≥30 cm (top row) and ≥60 cm (bottom row) based on large end diameter. Large dots indicate plots that sampled at least one DDW piece of the stated size, and small dots indicate plots with no DDW pieces of designated size sampled.

understory, they are a relatively ephemeral source of CWD throughout all stands with limited impact on long-term volumes and connectivity. In contrast, windthrow and senescence were likely the primary disturbance types in the relict stand given the higher density of larger trees compared to the partially harvested stands [18]. Relict stand CWD volume ha^{-1} was significantly higher than both partially harvested stands similar to several other studies [27], [28], [29]. This observation is presumably related to the changes in the size distribution of living trees, and the associated change in potential exogenous natural disturbances, particularly wind, related to tree size.

Removal of individual or small groups of large canopy trees allows for the release, growth and development of advance regeneration, or the establishment of new regeneration. These areas tend to have higher stem densities of smaller stems that will experience endogenous disturbances, particularly density-dependent mortality, as the stand develops rather than exogenous natural disturbances. Although these areas may contribute numerous CWD pieces, their cumulative low volume dictates a more ephemeral source of CWD because of faster rates of decomposition [2], [29]. Using a chronosequence approach, one study [30] indicated that CWD dynamics of managed stands were more unidirectional related to succession in contrast to virgin forests that exhibited a more cyclical pattern. In boreal mesic forests of southern Finland, researchers observed larger average diameter pieces of DDW and SDW in old-growth stands versus managed stands, with considerably more large-diameter (≥40 cm) SDW in old-growth stands and twice as many CWD stems < 10 cm in the managed stands compared to old-growth stands [27]. In boreal *Picea* forests of Sweden, significant differences in the availability of large diameter, highly-decayed CWD logs between selectively managed and semi-natural forests were observed, even after about 100 years [31]. In contrast, no differences were evident in DDW piece sizes in old-growth and selection cut stands of *A. saccharum*-dominated stands in Quebec, Canada, although SDW average diameter was lower in diameter-limit cuts [32]. Intermediate harvests can also influence species composition.

The overstory composition of the relict and partially harvested stands differed, which likely influenced decay rates and stand-level abundance of CWD. *Quercus* species dominated the overstory of all stands, but the relict stand *Quercus* basal area was 17.1 m^2 ha^{-1} compared to only 7.9 m^2 ha^{-1} and 8.6 m^2 ha^{-1} in the single-harvest and double-harvest stands, respectively [18]. The partially harvested stands had a higher abundance of *Acer*, *Carya*, and *Fraxinus* individuals, which replaced the harvested overstory *Quercus* trees. At an old-growth deciduous forest in Indiana,

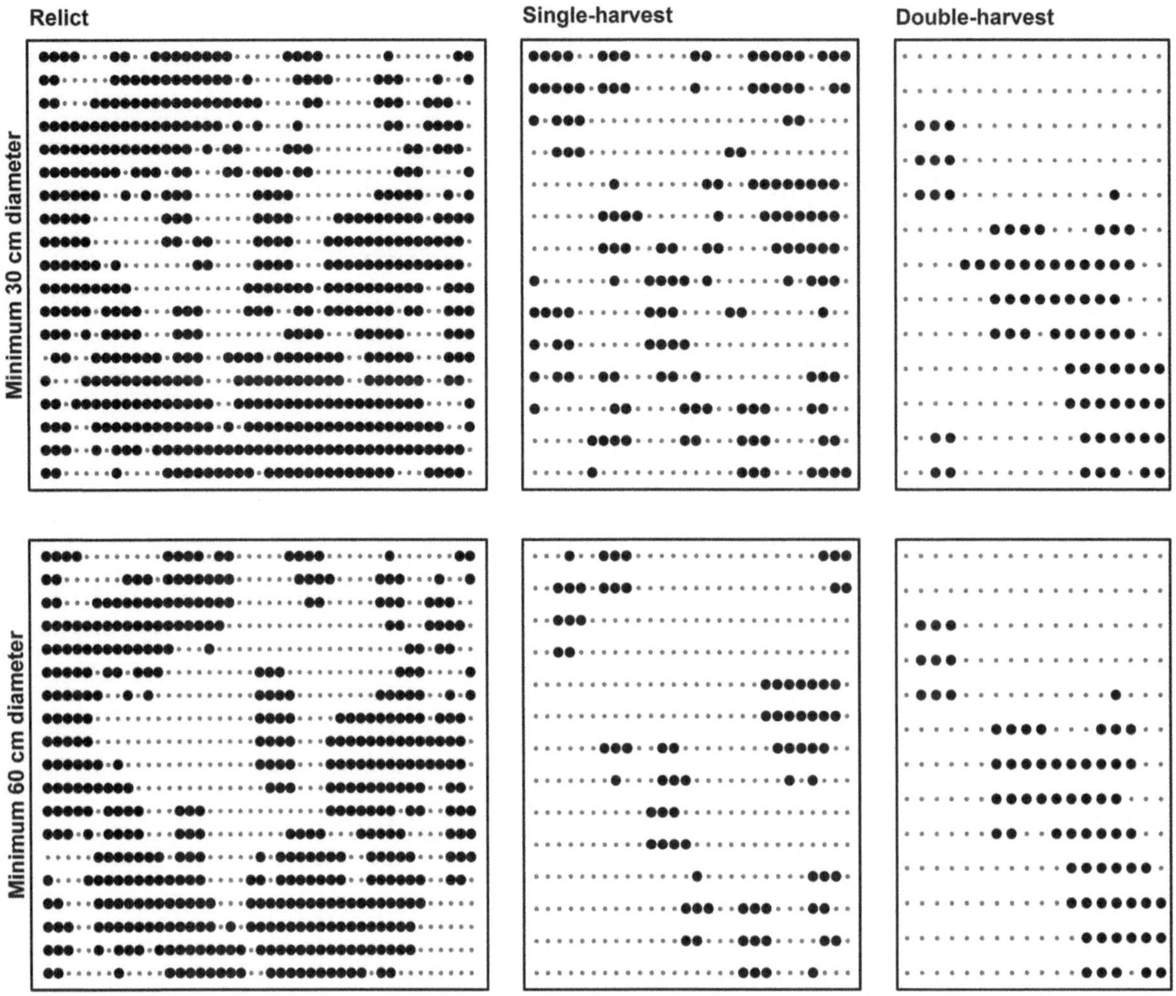

Figure 5. Standing deadwood connectivity examples. Display of connected plots for standing deadwood (SDW) in relict, single-harvest, and double-harvests stands for pieces ≥30 cm (top row) and ≥60 cm (bottom row) based on diameter at 1.37 m height. Large dots indicate plots that sampled at least one SDW piece of the stated size, and small dots indicate plots with no SDW pieces of designated size sampled.

U.S.A., it was predicted that 95% loss of CWD density of *Quercus*, *Carya*, and *Acer* species would occur at 171, 86, and 66 years, respectively [29]. Although we were unable to consistently identify CWD pieces to genus, the higher volumes of CWD within intermediate and high-decay stages likely reflects of the interaction of piece size and species composition within the relict stand; larger, more decay-resistant pieces will remain in situ compared to smaller pieces that decay quicker.

Rates of inputs (mortality events) and outputs (decomposition) are modified by management decisions, thus, CWD stores within a stand will also be modified. Assuming random mortality events with a 1% coefficient of variation over time [33], combined with slow decomposition rates related to large piece sizes and species composition, CWD can accumulate over time. This phenomenon may explain the aggregated patterns of CWD observed in one study [34] despite randomly dispersed mortality, as well as aggregated patterns of CWD reported for other old-growth forests [7], [10], [35]. Harvesting often alters rates of inputs and outputs, thus influencing potential CWD stores. One study observed that accumulation of CWD was slower after selective harvests, and that CWD levels had not returned to primeval forest levels even after 100 years [28]. Similarly, it has been proposed that selective logging operations interrupted the continuity of CWD inputs over time, thus influencing proportions of decaying logs and large CWD pieces [31]. Because there are fewer slow-decaying large pieces and a greater number of smaller CWD pieces derived from aggregated mortality events due to density-dependent competition within newly-formed harvest gaps, CWD is more ephemeral in selection harvests due to faster rates of decomposition of smaller pieces. Faster decomposition of CWD stores will result in lower volumes and decreased connectivity.

Decreased connectivity of DDW and SDW resulted in a larger number of smaller patches in the partially harvested stands, notably in diameter classes <60 cm. As minimum diameter class size increases, differences between all stands become minimized around 60 cm. Although this is partly an artifact of the methodology used whereby larger pieces are sampled more frequently across larger areas, connectivity, patch size, and number of patches were regulated by large diameter CWD. For taxa that can utilize smaller pieces, connectivity was still maintained in the in the partially harvested stands, being more than half of that observed in the relict stand, although the patch size and density were not comparable. Abundance of DDW and SDW sampled at each plot (Figure 6) indicated an inverse spatial relationship between the two pools of CWD; areas with abundant DDW tended to have low volumes of SDW and vice-versa. This

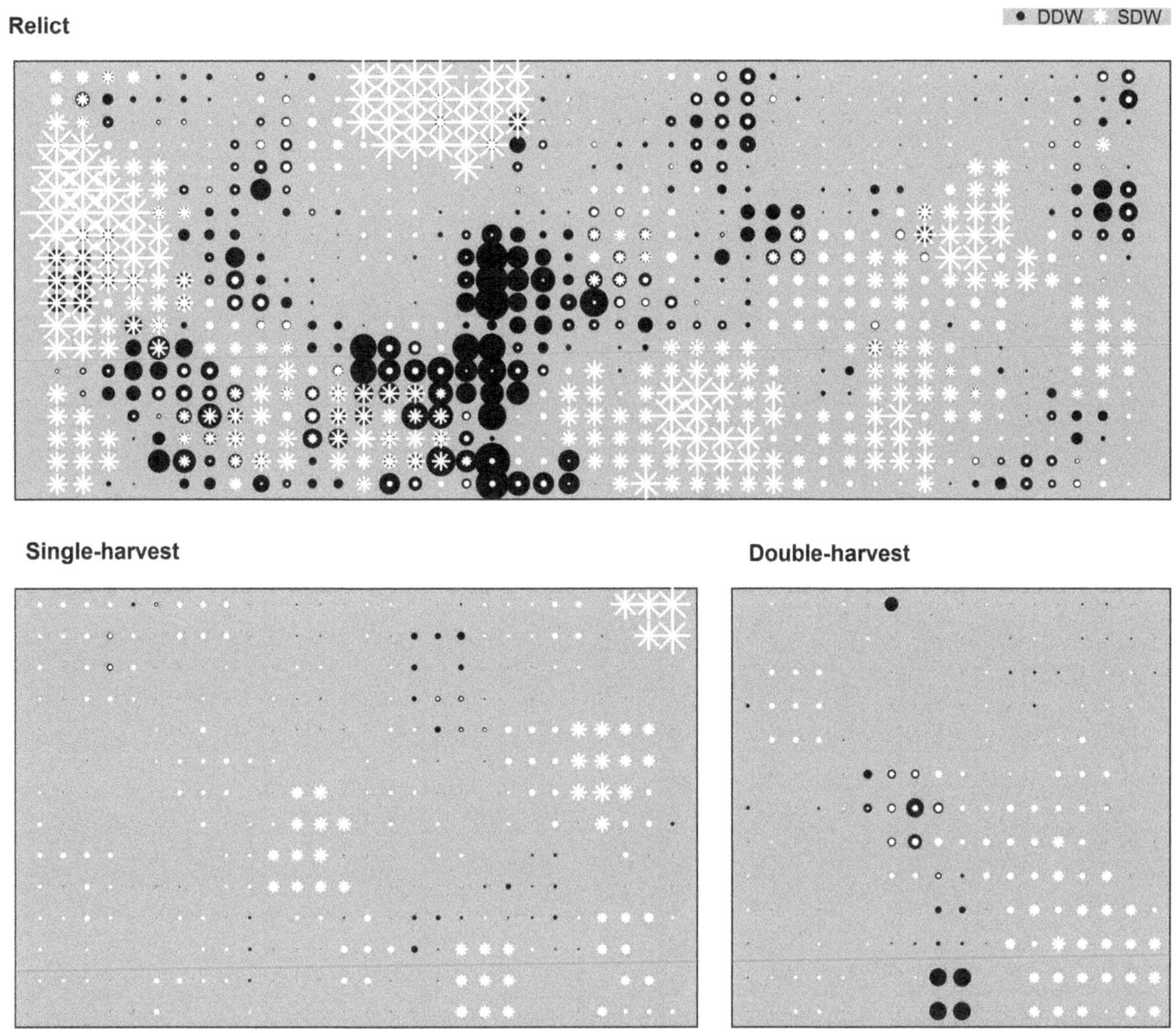

Figure 6. Coarse woody debris spatial distributions by volume. Each panel displays the volume (m^3) of downed deadwood (DDW; black circles) and standing deadwood (SDW; white stars) sampled at each plot location for relict, single-harvest, and double-harvests stands; size of symbol within each pane is standardized.

may suggest a dynamic relationship between the two pools whereby SDW eventually contribute to DDW pools and enhances connectivity. In the partially harvested stands, large patches with little or no CWD are indicative of areas that were previously harvested but now contain vigorous trees; many of these areas will be devoid of large CWD for the foreseeable future.

Traditional forest management practices can reduce CWD stores in forest stands for extended periods of time; differences in CWD connectivity and abundance among the three stands remained evident 46+ years after harvesting. Because the partially harvested stands have received no management since the initial harvest, this scenario differs from conventional uneven-aged, selection management in the region where most trees larger than 60 cm are harvested, stands are entered at short intervals, often 10–15 years, and removals tend to include declining individuals. This management approach reduces CWD inputs and drives reduced accumulation and connectivity across the stand. Extended rotation and entry times are often cited as ecologically viable harvest practices that promote greater heterogeneity in forest structure. However, these options may not be economically feasible or sufficient to preserve CWD pools in many management regimes. Several other studies have observed low levels of CWD long after active management operations ceased [28], [30], [31], and it may take a century or more for CWD abundance to return to the range of natural levels without adequate planning. Retaining individual and groups of trees within harvested stands can help preserve current and future CWD associated dispersal-limited taxa, and enhance biodiversity [36]. Using an "island" approach (*sensu* [37]), within stands, especially in fragmented forest landscapes, may be more appropriate to preserve dispersal-limited, smaller populations and metapopulation dynamics across the landscape akin to the "lifeboat hypothesis". We propose that differences in abundance and connectivity are attributable to decreased CWD stores in the managed stands. Although most studies of CWD focus on abundance within forests, understanding of CWD spatiotemporal dynamics may prove more critical to understand sustainability of CWD and its associated functions within forests.

Acknowledgments

We appreciate co-operation and logistical support from the staff of the Davis Purdue Agricultural Center. J. Dyer, K. Leffel, D. Luchik, A. Meier, M. Schuessler, J. Stark and C. Sullivan assisted with field data collection. Also thank you to B. Graham, J. Shields, and C. Zellers for critique and discussion in project development. We would also like to thank two anonymous reviewers for their comments and critiques to improve the manuscript.

Author Contributions

Conceived and designed the experiments: RCM MAJ MRS. Performed the experiments: RCM. Analyzed the data: RCM. Contributed reagents/materials/analysis tools: RCM MAJ MRS. Contributed to the writing of the manuscript: RCM MAJ MRS.

References

1. Franklin JF, Shugart HH, Harmon ME (1987) Tree death as an ecological process. Bioscience 37: 550–556.
2. Harmon ME, Franklin JF, Swanson FJ, Sollins P, Gregory SV, et al. (1986). Ecology of coarse woody debris in temperate ecosystems. Adv Ecol Res 15: 133–302.
3. Linden DW, Roloff GJ, Kroll AJ (2012) Conserving avian richness through structure retention in managed forests of the Pacific Northwest, USA. For Ecol Manage 284: 174–184.
4. Grove SJ (2002) Saproxylic insect ecology and the sustainable management of forests. Annu Rev Ecol Syst 33: 1–23.
5. Ylisirniö AL, Berglund H, Aakala T, Kuuluvainen T, Kuparinen AM, et al. (2009) Spatial distribution of deadwood and the occurrence of five saproxylic fungi in old-growth timberline spruce forests in northern Finland. Scand J For Res 24: 527–540.
6. Butts SR, McComb WC (2000) Associations of forest-floor vertebrates with coarse woody debris in managed forests of western Oregon. J Wildl Manage 64: 95–104.
7. Edman M, Jonsson BG (2001) Spatial Pattern of Downed Logs and Wood-Decaying Fungi in an Old-growth *Picea abies* Forest. J Veg Sci 12: 609–620.
8. Turner MG, Gardner RH, O'Neill RV (2001) Landscape ecology in theory and practice. New York: Springer-Verlag. 406 p.
9. Jönsson MT, Edman M, Jonsson BG (2008) Colonization and extinction patterns of wood-decaying fungi in a boreal old-growth *Picea abies* forest. J Ecol 96: 1065–1075.
10. Komonen A (2005) Local spatial pattern in the occurrence of two congeneric wood-decaying fungi in an old-growth boreal forest. Scand J For Res 20: 393–399.
11. McComb WC (2003) Ecology of coarse woody debris and its role as habitat for mammals. In: Zabel CJ, Anthony RG, editors. Mammal Community Dynamics: Management and Conservation in the Coniferous Forests of Western North America. New York: Cambridge University Press. 374–404.
12. Schiegg K (2000) Effects of deadwood volume and connectivity on saproxylic insect species diversity. Ecoscience 7: 290–298.
13. Laaka-Lindberg S, Korpelainen H, Pohjamo M (2006) Spatial distribution of epixylic hepatics in relation to substrate in a boreal old-growth forest. J Hattori Bot Lab 100: 311–323.
14. Manning AD, Cunningham RB, Lindenmayer DB (2013) Bringing forward the benefits of coarse woody debris in ecosystem recovery under different levels of grazing and vegetation density. Biol Conserv 157: 204–214.
15. Riffell S, Verschuyl J, Miller D, Wigley TB (2011) Biofuel harvests, coarse woody debris, and biodiversity – A meta-analysis. For Ecol Manage 261: 878–887.
16. Szwagrzyk J, Czerwczak M (1993) Spatial patterns of trees in natural forests of East-Central Europe. J Veg Sci 4: 469–476.
17. Aakala T, Kuuluvainen T, Grandpré LD, Gauthier S (2007) Trees dying standing in the northeastern boreal old-growth forests of Quebec: spatial patterns, rates, and temporal variation. Can J For Res 37: 50–61.
18. Morrissey RC (2013) Long-term disturbance histories of managed and natural mixed-hardwood forests of Central Indiana. Ph.D. Dissertation, Purdue University.
19. Gove JH, Ducey MJ, Ståhl G, Ringvall A (2001) Point relascope sampling: a new way to assess downed coarse woody debris. J For 99: 4–11.
20. Fraver S, Ringvall A, Jonsson BG (2007) Refining volume estimates of down woody debris. Can J For Res 37: 627–633.
21. Mesavage C, Girard JW (1946) Tables for estimating board foot volume of timber. USDA Forest Service, Washington, DC. 94.
22. R Core Team (2014) R: A language and environment for statistical computing. R Foundation for Statistical Computing, Vienna: Austria. ISBN 3-900051-07-0, URL: http://www.R-project.org/.
23. Siitonen J (2001) Forest Management, Coarse Woody Debris and Saproxylic Organisms: Fennoscandian Boreal Forests as an Example. Ecol Bull 49: 11–41.
24. Harmon ME (2009) Woody detritus mass and its contribution to carbon dynamics of old-growth forests: the temporal context. IN: Wirth C, Gleixner G, Heimann M, editors. Old-growth forests: function, fate and value – an overview old-growth forests (Ecological Studies). Berlin- Heidelberg: Springer. 159–190.
25. Turcotte RM, Elliott TR, Fajvan MA, Park YL, Snider DA, et al. (2012) Effects of ice storm damage on hardwood survival and growth in Ohio. North J Appl For 29: 53–59.
26. Parker GR, Leopold DJ (1983) Replacement of *Ulmus americana* L. in a mature east-central Indiana woods. Bull Torrey Bot Club 110: 482–488.
27. Siitonen J, Martikainen P, Punttila P, Rauh J (2000) Coarse woody debris and stand characteristics in mature managed and old-growth boreal mesic forests in southern Finland. For Ecol Manage 128: 211–225.
28. Sippola AL, Lehesvirta T, Renvall P (2001) Effects of Selective Logging on Coarse Woody Debris and Diversity of Wood-Decaying Polypores in Eastern Finland. Ecol Bull 49: 243–254.
29. MacMillan PC (1988) Decomposition of coarse woody debris in an old-growth Indiana forest. Can J For Res 18: 1353–1362.
30. Vandekerkhove K, De Keersmaeker L, Menke N, Meyer P, Verschelde P (2009) When nature takes over from man: Dead wood accumulation in previously managed oak and beech woodlands in North-western and Central Europe. For Ecol Manage 258: 425–435.
31. Bader P, Jansson S, Jonsson BG (1995) Wood-inhabiting fungi and substratum decline in selectively logged boreal spruce forests. Biol Conserv 72: 355–362.
32. Angers VA, Messier C, Beaudet M, Leduc A (2005) Comparing composition and structure in old-growth and harvested (selection and diameter-limit cuts) northern hardwood stands in Quebec. For Ecol Manage 217: 275–293.
33. Runkle JR (2000) Canopy tree turnover in old-growth mesic forests of eastern North America. Ecology 81: 554–567.
34. Aakala T, Fraver S, Palik BJ, D'Amato AW (2012) Spatially random mortality in old-growth red pine forests of northern Minnesota Can J For Res 42: 899–907.
35. Ylisirniö AL, Berglund H, Aakala T, Kuuluvainen T, Kuparinen AM, et al. (2009) Spatial distribution of deadwood and the occurrence of five saproxylic fungi in old-growth timberline spruce forests in northern Finland. Scand J For Res 24: 527–540.
36. Gustafsson L, Baker SC, Bauhus J, Beese WJ, Brodie A, et al. (2012). Retention Forestry to Maintain Multifunctional Forests: A World Perspective. Bioscience 62: 633–645.
37. MacArthur RH, Wilson EO (1967) The Theory of Island Biogeography. Princeton: Princeton University Press. 203.

Spatial Patterns and Associations between Species Belonging to Four Genera of the Lauraceae Family

Lin Li[1], Wan Hui Ye[2]*, Shi Guang Wei[1]*, Ju Yu Lian[2], Zhong Liang Huang[2]

1 Guilin University of Electronic Technology, Guilin, China, 2 South China Botanical Garden, Chinese Academy of Science, Guangzhou, China

Abstract

Spatial distribution pattern of biological related species present unique opportunities and challenges to explain species coexistence. In this study, we explored the spatial distributions and associations among congeneric species at both the species and genus levels to explain their coexistence through examining the similarities and differences at these two levels. We first used DNA and cluster analysis to confirmed the relative relationship of eight species within a 20 ha subtropical forest in southern China. We compared Diameter at breast height (DBH) classes, aggregation intensities and spatial patterns, associations, and distributions of four closely related species pairs to reveal similarities and differences at the species and genus levels. These comparisons provided insight into the mechanisms of coexistence of these congeners. O-ring statistics were used to measure spatial patterns of species. Ω_{0-10}, the mean conspecific density within 10 m of a tree, was used as a measure of the intensity of aggregation of a species, and g-function was used to analyze spatial associations. Our results suggested that spatial aggregations were common, but the differences between spatial patterns were reduced at the genus level. Aggregation intensity clearly reduced at the genus level. Negative association frequencies decreased at the genus level, such that independent association was commonplace among all four genera. Relationships between more closely related species appeared to be more competitive at both the species and genus levels. The importance of competition on interactions is most likely influenced by similarity in lifestyle, and the habitat diversity within the species' distribution areas. Relatives with different lifestyles likely produce different distribution patterns through different interaction process. In order to fully understand the mechanisms generating spatial distributions of coexisting siblings, further research is required to determine the spatial patterns and associations at other classification levels.

Editor: Helge T. Lumbsch, Field Museum of Natural History, United States of America

Funding: This work was funded by the National Nature Science Foundation of China (31200326, 31200412, 31460155), The Project of Scientific Research and Technological Development of GuangXi (No.gkg1140002-2-4), and the Knowledge Innovation Project of the Chinese Academy of Sciences (KSCX2-EW-Z). The funders had no role in study design, data collection and analysis, decision to publish, or preparation of the manuscript.

Competing Interests: The authors have declared that no competing interests exist.

* Email: why@scbg.ac.cn (WHY); weishig@mails.ucas.ac.cn (SGW)

Introduction

The spatial distribution of species and inter-species interactions provide fundamental information for understanding species structures and coexistence in communities. One major focus of ecological research is to reveal the outcome of the interactions of biological and ecological processes by analyzing spatial distribution patterns and associations [1–5]. For instance, a recent study [6], which simulated spatial processes of seed dispersal and habitat association found that niche- and neutral-based interactive operations may have important roles in generating spatial patterns. Furthermore, spatial distribution of species can essentially be used to understand and model biodiversity patterns over space [7–13]. For example, Pofessor He and Legendre [11] and Green and Ostling [14] interpreted relationships based on the spatial patterns of individuals within their distribution and endemic-range, respectively. Most studies have utilized statistical analyses of species data to reveal processes and phenomena explaining species coexistence mechanisms. Gotelli *et al.* [15] used a general simulation model to explain the patterns and causes of species richness. The simulation approach offers new insights into the origin and maintenance of species richness patterns, combined with contemporary climate, evolutionary history and geometric constraints on global biodiversity gradients [15]. Rahbek *et al.* [16] found correlative climatic models substantially underestimate the importance of historical factors and small-scale niche-driven assembly processes in shaping contemporary species-richness patterns.

Analyses of the spatial patterns of sympatric congeneric species present unique opportunities and challenges to explain species coexistence [17–19]. Congeneric species, and even species within the same families stemmed from a common ancestor. Sympatric siblings usually possess many phenotypic and ecological trait similarities, and therefore, utilize analogical resources similarly. Survival under conditions of limited resources, thus requires repulsion among sibling species preventing their coexistence [20]. However, congeneric trees are commonly found to coexist within high-diversity tropical communities and low-diversity temperate communities; Inter-specific differences and ontogenetic shifts in light requirements with life-form differences may contribute to the coexistence of the Acer species in old-growth forests [21]. Partitioning of the topographic and light environments was verified may double or treble the number of species able to coexist, but no evidence that partitioning of physical habitats can explain the coexistence of these closely related species [22].

Many ecologist and evolutionary biologists devote themselves to addressing the following questions: One is what factors drive the relative distribution of congeneric trees within the same community. Competitive ability is known to influence distribution patterns of species. Species with stronger competitive ability may retain more individuals than others species, a species' relative abundance is positively related to its competitive ability modeled by Du [23]. Also competition and facilitation can structure plant distribution and assemblages [24]. Additionally, the distributions of two strongly repulsed species within the same community usually differ. A second question is why does the abundance of coexisting sibling trees tend to differ? The literature on plant rarity often speculates that rare species are poor competitors [25,26], resulting in their reduced abundance. Though some rare species may have low competitive ability, it should not be assumed as the cause of rarity for all rare species [27]. In fact, other studies have attributed superior competitive ability to rare species [28,29]. The conclusions of these studies were based on a consensus that competition exists between congeneric trees and that competitive ability and action differ with spatial distribution patterns and abundance. Furthermore, habitat is most likely another important factor influencing congener distributions [30]. Large plots provide the basic element for the study of congener coexistence. A study of a 52 ha tropical rain forest plot in Borneo found habitat heterogeneity (light, established micro sites, and soil textural properties) lead to differences in tree distribution patterns, and played important roles in their species coexistence [31]. However, another 50 ha tropical plot revealed no strong evidence of a relationship between partitioning of physical habitat and 16 coexisting sibling species belonging to the family Myristicaceae [22].

Most previous research on spatial patterns and associations of related species has been performed at the species level, but biological function is influenced at multiple levels. Therefore, studies on species interactions at different levels are required. While there are little is known about species interactions at multiple levels so far. Study in tropical of Myristicaceae tree seedlings in two separate taxonomic level (species and genus) analyses were hinted that different mechanisms of coexistence among tropical tree taxa may function at different taxonomic or phylogenetic scales [32]. Here, we studied the family Lauraceae which is made up of a large number of genetically-related species. Lauraceae is also a family with a long history in East Asia. We compared the population structure, spatial patterns and associations of eight congeneric species within a large (20 ha) subtropical forest. Our objectives involved answering the following questions: (1) Do genetic relationships at the species or genus level influence the population structure of coexisting species? (2) What mechanisms control spatial distributions and associations and how do they change with scale, genetic relationship or habitat heterogeneity? (3) Do patterns in aggregation intensity response to Diameter at Breast Height (DBH) differ at the species and genus level? Finally we discuss mechanisms of coexistence among genetically related species at different relationship levels, and in different forests. This analysis will contribute to the understanding of congeneric species coexistence and diversity maintenance in subtropical forests.

Methods

Ethics Statement

No specific permits were required for the described field studies. The study site Dinghusan plot (DHS plot) is owned by the Chinese government and the Chinese Forest Biodiversity Monitoring Network, DHS plot is managed by South China Botanical Garden, Chinese Academy of Sciences. We can do our research works freely in these plots under the Regulations of the People's Republic of China on Nature Reserves. Our field studies did not involve endangered or protected species.

Study Site

The study area was located in the Dinghushan Mountain (112°30′39″–112°33′41″E, 23°09′21″–23°11′30″N) in Guangdong Province. Dinghushan was the first Nature Reserve established in China in 1956 and has been significantly important to the conservation of forest ecosystems over the past 50 years [33]. The reserve is covered by tropical-subtropical forests and is comprised of low mountains and hilly landscapes. Its total area is 1155 ha, with an altitude of 14.1–1000.3 m. Dinghushan has a south subtropical monsoon climate with a mean annual temperature of 20.9°C, and a mean monthly temperature of 12.6°C in January and 28.0°C in July. Average annual precipitation is 1929 mm, with most of the precipitation occurring between April and September. Annual evaporation is 1115 mm and relative humidity 82%.

A permanent 20 ha (400×500 m) plot called the Dinghushan plot (DHS plot) was established in the Dinghushan reserve in November 2004. Mapping of the plot mainly took place from January to March, but was completed in October 2005. Following the field protocols of the Center for Tropical Forest Science (CTFS), to identify, measure and map the trees, including all free standing trees and shrubs of at least 1 cm in DBH, we used the same methodology as in [34],The plot features rough terrain with a steep hillside in the southeast corner. Topography varies with ridge and valley in the plot and the elevation ranges from 240 to 470 meters. There were 56 families, 119 genera, 210 species and 71617 individuals counted in the first census. Thirty of the identified species were composed of solitary individuals and 110 species were made up of fewer than 20 individuals. The most abundant species *Aidia canthioides* owned 5996 individuals. Mean stand density was 3580.85 living trees and shrubs per hectare. Mean basal area was 28.24 cm^2 per ha [33].

Study Species

The Lauraceae family is made up of about 45 genera and 2000 to 2500 species. Almost all of these species are native to tropical and subtropical regions, with the distribution centers in Southeast Asia and Brazil. As a typical subtropical forest plot, the DHS plot has 19 species of Lauraceae belonging to 5 genera. Considering the species abundance and genetic relationship distance, eight focus species (abundance >200 each species) belonging to 4 genera, were chosen for comparisons of population structure and spatial patterns in this study. The selected species were *Cryptocarya concinna* (CRCO), *Cryptocarya chinensis* (CRCH), *Lindera metcalfiana* (LIME), *Lindera chunii* (LICH), *Machilus breviflora* (MABR), *Machilus chinensis* (MACH), *Neolitsea umbrosa* (NEUM) and *Neolitsea membranaceum* (NEME).

Data Analysis

We used the relative neighborhood density index Ω_r to characterize the distribution of eight Lauraceae species found in the plot. The Ω_r is the abundance-based O-ring scaled for the species evaluated, and is calculated as $\Omega_r = D_r/\lambda$ [8], λ is the mean density of a given species across the whole plot. Where $D_r = \sum N_r / \sum A_r$, A_r is the area of annulus at distance r and N_r is the number of conspecifics within the annulus. In this study the annulus width is 10 m. Therefore, D_r is the density of conspecifics as a function of distance r, λ is the mean density of a given species across the whole plot. A Monte Carlo simulation (999 iterations) was used to test the hypothesis that a distribution is not significantly different from a random distribution, i.e., $\Omega_r = 1$. If the observed Ω_r fell within the 2.5th (bottom dash-line) and 97.5th (top dash-line) quartiles, the null hypothesis was not rejected, and the species distribution within the DHS plot was concluded to be significant randomly distribution [35]. If the observed Ω_r located above the top dash-line, it indicates the significant aggregation distribution. If the observed Ω_r located below the bottom dash-line, it indicates the significant regular distribution.

We used DNA test results to perform cluster analysis, to quantify the genetic relationships among the species examined and accordingly group the species. DNA sequences were generated for 1–2 tagged individuals located within the DHS plot. Genomic DNA was extracted from leaf and/or bark tissue using a standard CTAB protocol [36]. We used the relative neighborhood density Ω_r to assess spatial patterns and $\Omega_{0\text{–}10}$, the mean conspecific density within 10 m of a tree, as a measure of the intensity of aggregation of a species [8]. And then we compared spatial intensity of aggregations of trees of different DBH classes at species and genus levels. The DBH classes were binned at 2 cm intervals. Growth type, shade tolerance and life history traits rooted in empirical data from field investigation personnel with long-term experience.

Finally we used bivariate pair-correlation function ($g_{12}(r)$) to analyze spatial associations among and between species (within the four genera). The bivariate pair-correlation function $g_{12}(r)$ is the analogy to Ripley's $K_{12}(r)$ [37], but it replaced the circles of radius r by rings with radius r. The function $g_{12}(r)$ quantifies the type of spatial association between species 1 species 2:

$$g_{12}(r) = \frac{dK_{12}(r)}{2\pi r dr}$$

For a bivariate pattern of two objective species, $g_{12}(r) = 1$ indicates non-association (independence), at distance r, and $g_{12}(r) > 1$ indicates a positive association between the two species at given distance r, whereas $g_{12}(r) < 1$ indicates a negative spatial interaction (spatial repulsion or segregation) between the two species at distances r [38]. Here the annulus width r is 10 meters.

Results

Population Structure

The abundance of the eight study species ranged from 223 individuals of NEME to 4478 of CRCO (Table 1). Total numbers of individuals were 13362. Two *Cryptocarya* species were much more abundant than the *Lindera*, *Machilus* and *Neolitsea* species. The total abundance of *Cryptocarya* was also much greater than for the other three genera. The basal areas of CRCH and MACH were much larger than for the other species.

Cluster analysis for DNA test results showed the closest relative of each of the eight species was within the same genus (Fig. 1). However, the degree of within genus species relatedness differed among the four genera. Relatedness was greatest for *Lindera* and *Machilus*, then *Neolitsea*, followed by *Cryptocarya*.

Table 1. Population structure of the eight Lauraceae species studied in the DHS plot.

Species name	Sp. code	Genera	Growth type	Ind.	Base area (m^2/ha)	Base area/ind	Shade tolerance order
Cryptocarya concinna	CRCO	*Cryptocarya*	Understory	4478	1703.21	0.38	2
Cryptocarya chinensis	CRCH	*Cryptocarya*	Upperstory	2557	11238.98	4.40	1
Lindera metcalfiana	LIME	*Lindera*	Midstory	2118	1789.99	0.85	7
Lindera chunii	LICH	*Lindera*	Midstory	1302	1092.76	0.84	6
Machilus breviflora	MABR	*Machilus*	Upperstory	800	3261.17	4.08	4
Machilus chinensis	MACH	*Machilus*	Upperstory	532	8250.27	15.51	3
Neolitsea umbrosa	NEUM	*Neolitsea*	Understory	1352	163.85	0.12	8
Neolitsea membranaceum	NEME	*Neolitsea*	Understory	223	125.53	0.56	5

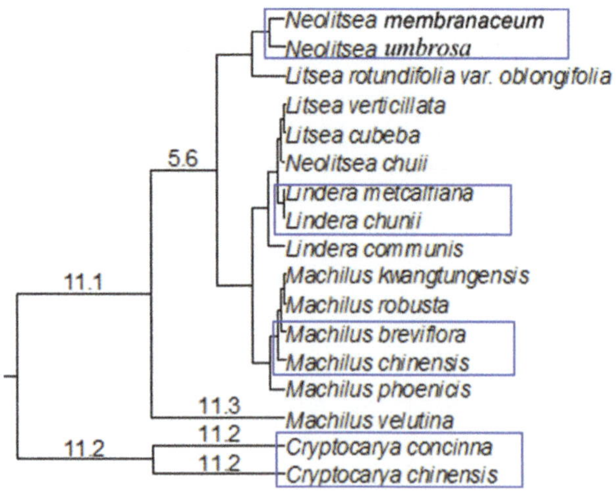

Figure 1. Genetic relationships of Lauraceae species. The numbers related clustering coefficient. The eight species enclosed within rectangles are the same species in Table 1.

At the species level, eight species showed different size class distributions (Fig. 2). Using a DBH of 2 cm to compare individual's structure of congeneric species, we found CRCO and CRCH were distinctly different. Individuals of CRCO tended to have DBH less than 4 cm and mostly were aggregated within the range of 1 to 2 cm DBH. CRCH, MABA and MACH were liked a reversed DBH J-shaped distribution. LIME, LICH, NEUM and NEME were liked a DBH L-shaped distribution. At the genera level, *Cryptocarya*, *Lindera* and *Neolitsea* had a distinct L-shaped DBH distribution (Fig. 3). *Machilus* had a larger number of big trees than the other three genera, and showed like a reversed J-shaped DBH distribution.

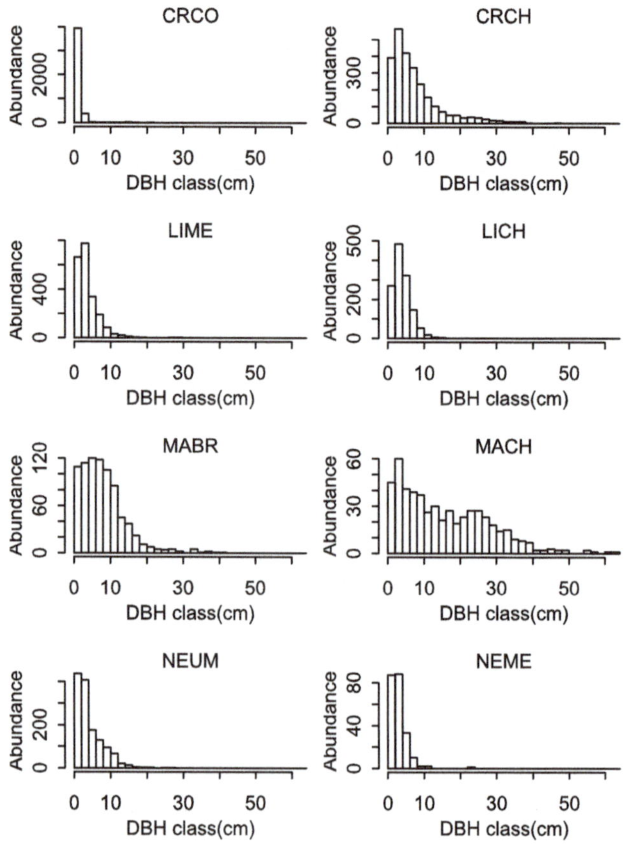

Figure 2. Size (DBH) class of the eight Lauraceae species belonging to four genera.

Spatial Patterns

The eight study species showed aggregation patterns in the DHS plot (Fig. 4), especially at small scales. CRCO, CRCH, LIME and LICH showed patterns of aggregation at scales 0–200 m, then became random, and then regular patterns as scale increased (Fig. 4). MABR and MACH showed significant aggregated patterns at scales of 0–150 m, which then became random, then regular with increasing spatial scale. NEUM and NEME were clumped at scales less than 175 m, then NEUM became random and quickly showed regular patterns at scales >180 m. NEME, on the other hand, had a random distribution at scales of 175–200 m, then became regular as scale increased (Fig. 4).

When examined at the genera level, the differences in spatial patterns were reduced. The four genera showed similar tendencies (Fig. 5). Aggregation patterns were at small and middle scales (*Cryptocarya*<200 m, *Lindera*<220 m, *Machilus*<160 m and *Neolitsea*<180 m), then random and then regular at larger scales, corresponding to their spatial pattern in the plot (Fig. 5).

Each species' spatial distribution corresponded with its spatial patterns within the plot (Fig. 6). CRCO and CRCH were relatively widely distributed throughout entire plot, and, though their distributions did not overlap at small spatial scale, the distribution of each complemented the other at middle to large spatial scales. LIME was located mainly in the western part of the plot, whereas LICH was in the eastern part, and thus spatially not overlapping. MABR and MACH were mosaicly distributed at small to middle spatial scales. MABR was abundant at the top of mountain, located in the southwest corner of plot, and small

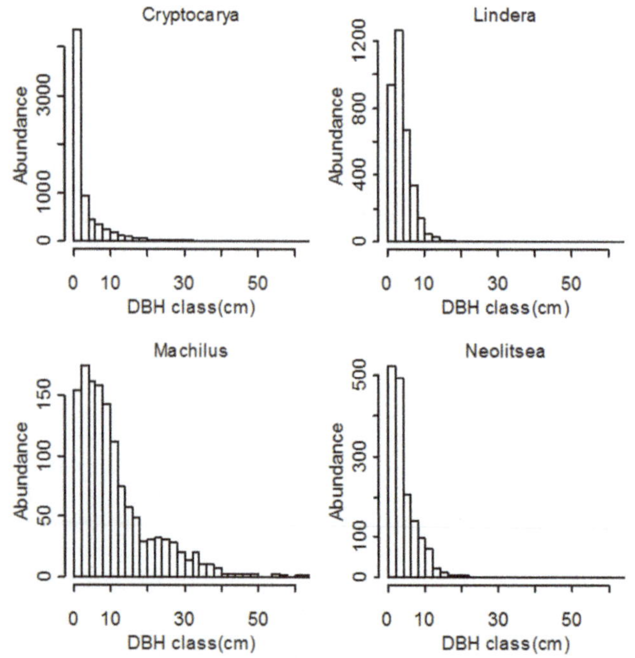

Figure 3. Size (DBH) class of the four genera including species from Table 1.

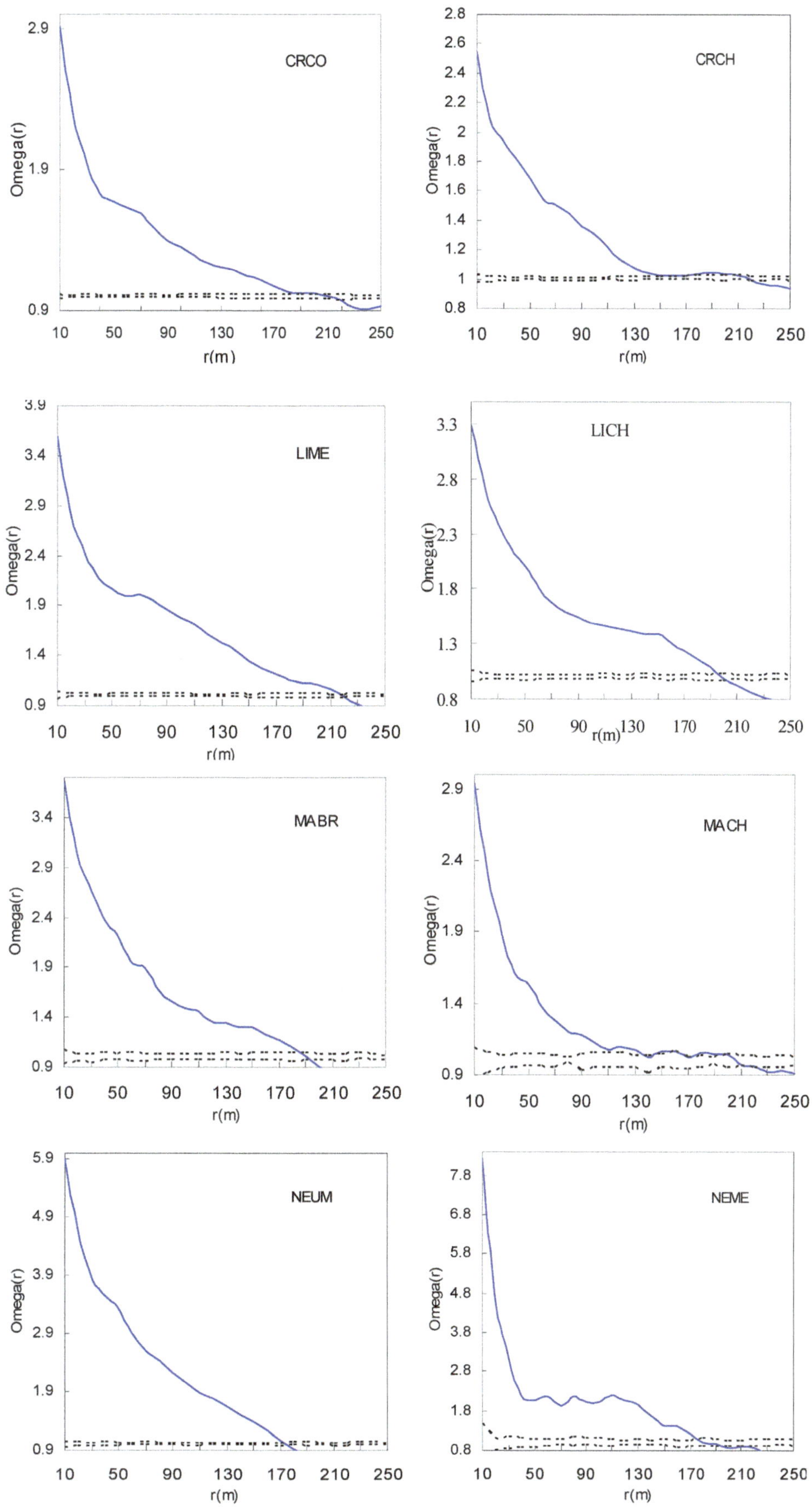
CRCO
CRCH
LIME
LICH
MABR
MACH
NEUM
NEME
Omega(r)
r(m)
10 50 90 130 170 210 250

Figure 4. Spatial patterns of eight Lauraceae species. The solid line curve is the Ω_r value. Dashed lines correspond to the confidence intervals generated from 999 Monte Carlo simulations under the null hypothesis of complete spatial randomness. See Table 1 for Species codes.

numbers of MACH existed in complement space with MABR. NEUM was concentrated in the southwest whereas NEME was in the southeast corner, and thus the distribution of both was nearly distinct. *Cryptocarya* and *Lindera* were distributed widely throughout the plot, whereas *Machilus* and *Neolitsea* were relatively concentrated in the southern part of the plot, *Neolitsea* was especially rare in the northern part of the plot.

Aggregation Intensity at Species and Genera Levels

The aggregation intensity as measured by $\Omega_{0\text{-}10}$ clearly decreased from the species to the genus level (Table 2). For example, *Cryptocarya* ($\Omega_{0\text{-}10} = 1.93$) was lower than that of species CRCO ($\Omega_{0\text{-}10} = 2.91$) and CRCH ($\Omega_{0\text{-}10} = 2.55$). The other three genera displayed similar trends. *Cryptocarya* had the lowest $\Omega_{0\text{-}10}$ and *Neolitsea* ($\Omega_{0\text{-}10} = 4.67$) had the highest $\Omega_{0\text{-}10}$ of the four genera. Species of all four genera showed tendencies of congeneric species aggregations within parts of the plot. Though the locations of the aggregates differed with species, individuals from different species inside congeners were fewer overlapped distribution (Fig. 6).

Relationship between Aggregation Intensity and DBH at the Genera Level

Where more than 50 individuals made up a DBH class, $\Omega_{0\text{-}10}$ was examined (Fig. 7). Results showed differences in the change in $\Omega_{0\text{-}10}$ with DBH group between *Cryptocarya* and *Machilus*. Although aggregation intensity of these two genera similarly tended to decrease from small to large DBH, the curves differed. *Cryptocarya* showed a wavelike decrease, compared to *Machilus*, appeared to linearly decrease with increasing DBH.

Spatial Associations

Spatial associations of the four pairs of closely relative species showed no positive associations (Fig. 8). CRCO and CRCH were exceptionally negatively associated until scales greater than 210 m. LIME and LICH showed a negative association at scales of

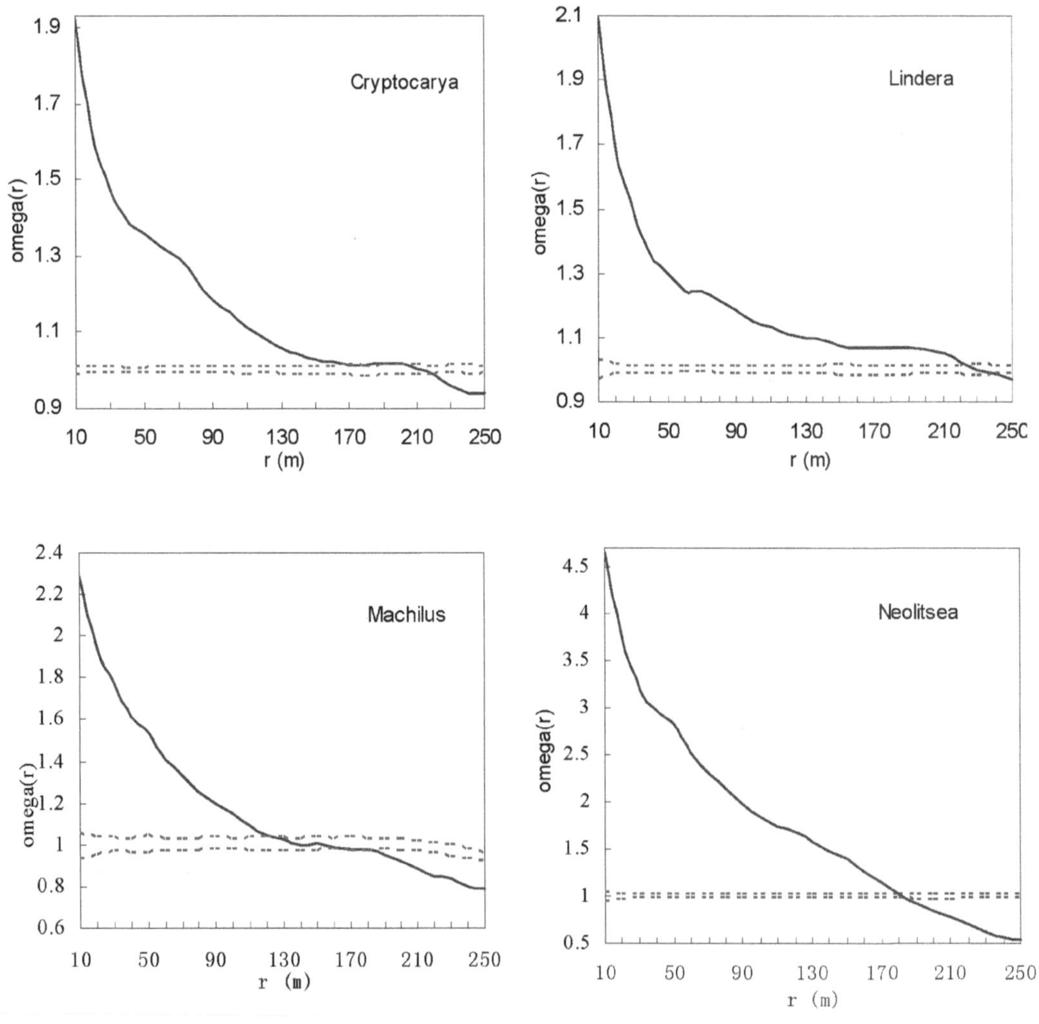

Figure 5. Spatial patterns of the four genera of Lauraceae species. The solid line curve is the Ω_r value. Dashed lines correspond to the confidence intervals generated from 999 Monte Carlo simulations under the null hypothesis of complete spatial randomness.

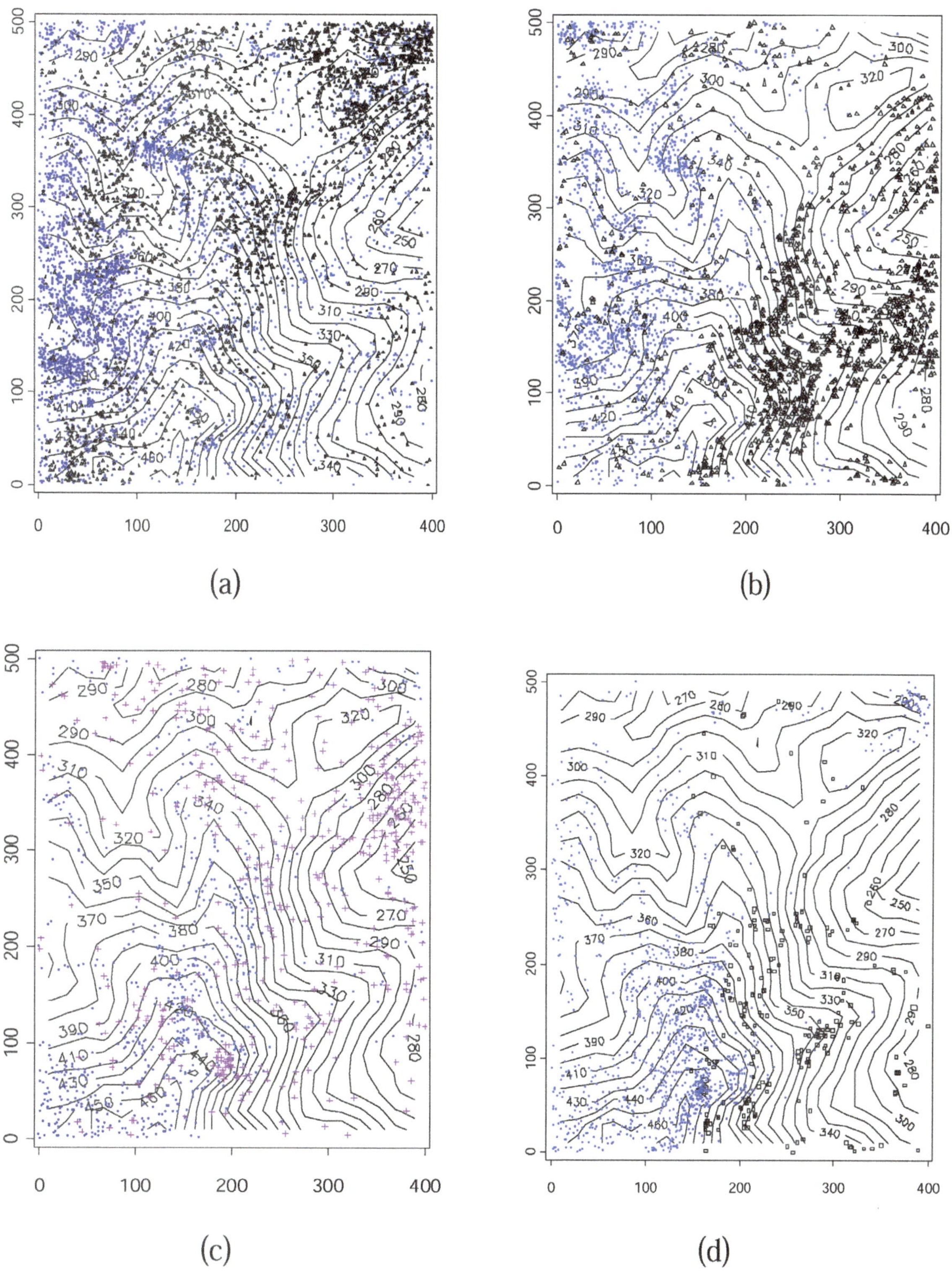

Figure 6. Spatial distribution of the eight species belonging to four genera in relation to the topography of the DHS plot. (a) *Cryptocarya*, dots represent CRCO and triangles represent CRCH; (b) *Lindera*, dots represent LIME and triangles represent LICH; (c) *Machilus*, circles represent MABR, crosses represent MACH. (d) *Neolitsea*, dots represent NEUM and triangles represent NEME.

20–140 m. The other two pairs were independently distributed within the plot at every spatial scale.

Independent association was commonplace among four genera (Fig. 9). Genera pairs were showed independent association, except for *Cryptocarya* and *Lindera* were positively distributed at scale <40 m, and *Machilus* and *Neolitsea* were showed positively association at scale <60 m.

Table 2. Statistic Ω_{0-10} of eight Lauraceae species at the species and genus level (Mixed congeneric species individuals).

Genera	Ω_{0-10}	Spcode	Ω_{0-10}
Cryptocarya	1.93	CRCO	2.91
		CRCH	2.55
Lindera	2.09	LIME	3.59
		LICH	3.29
Machilus	2.29	MABR	3.78
		MACH	2.96
Neolitsea	4.67	NEUM	5.93
		NEME	8.29

Discussion

The spatial distribution patterns of closely related congeneric trees in subtropical forests have important implications for how species interact with each other and their performance at different genetic levels. Closely related congeneric species originate from a common ancestor and, because they share many similar phenotypic and ecological traits, they tend to utilize a similar set of resources in similar ways. Therefore, they may exhibit relatively intense interspecific competition that limits their coexistence [23,29]. The spatial distributions of and associations between the four closely relative pairs studied here suggested that closer related species would repel each other when resources are limited. The degree of competition between species pairs, would be dependent on the similarities in their lifestyles [15], and also be subject to the diversity of habitat and small-scale niche-driven assembly processes in shaping contemporary species-richness patterns within their distribution area [16]. Relatives with different lifestyles would produce different distribution patterns through different interaction processes. CRCO and CRCH, and LIME and LICH were showed a negative association at most scales within the sampled plot (Fig. 8). These two closely relative pairs showed inter-specific competition in that they had different spatial distribution patterns (Fig. 6).

Long term natural selection may have driven relative trees to develop different, but mutually beneficial lifestyles resulting in mutual attraction. Congeneric trees are commonly found to coexist within different communities [21,22]. There is evidence that congeneric species can coexist if traits have diverged within the genus. When traits have diverged within the genus, the niche overlap is reduced and competition relaxed, thus allowing the coexistence of congeneric species [39]. Also, genetic differences at the subspecies level might lead to genetic subgroups taking advantage of environmental specialization, affecting spatial distribution above the species level [40]. Habitat is most likely another important factor influencing congener distributions [30]. Light, established micro sites, and soil textural properties played important roles in tree distribution patterns, and in their coexistence [31]. Habitat specialization plays an important role in maintaining the diversity of this species-rich subtropical forest. Former study found 83% of the species were related to topographic variables [41].

Inter-specific differences and ontogenetic shifts led to the law of spatial patterns and associations changed from genus level to species level. Aggregated distributions of species is a widespread pattern in nature, observed in both tropical plant communities with diverse species and temperate plant communities with relatively few species [10,42,43,44]. Four genera provided further evidence of clumped patterns both at species and genera levels in a subtropical forest. Eight species showed significant aggregation patterns, especially at small scales (Fig. 4 and Fig. 6). At the genus level, differences in spatial patterns were reduced, and significant aggregation patterns existed at small and middle spatial scales (Fig. 5). The aggregation intensity varied among species (or

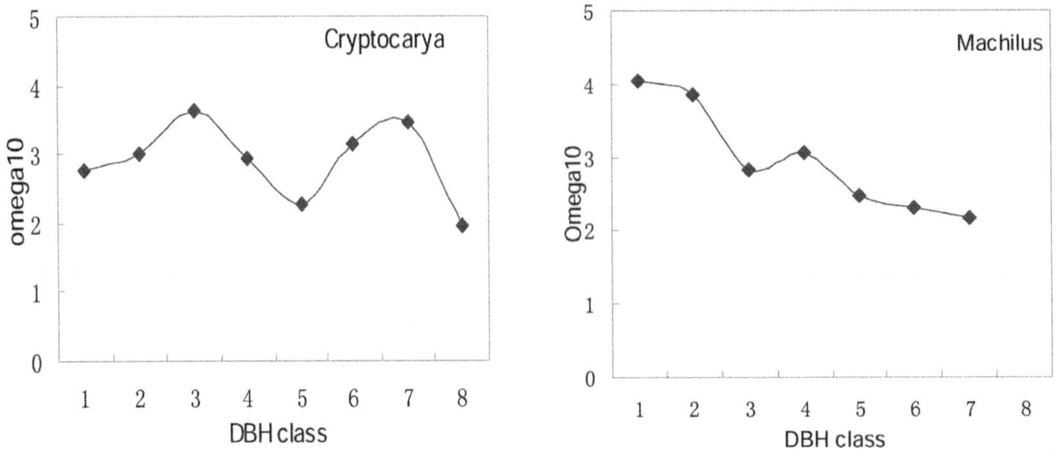

Figure 7. Relationships between Ω_{0-10} and DBH for *Cryptocarya* and *Machilus*. DBH classes were 3 cm increment bins beginning at 1 cm.

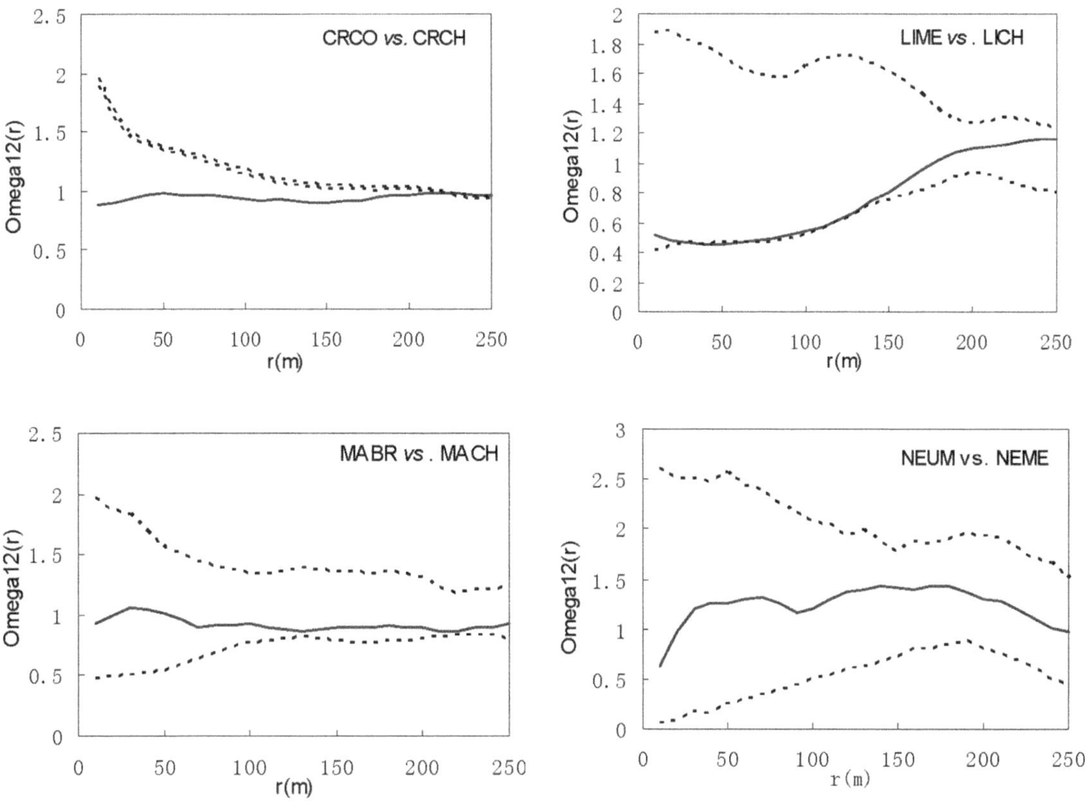

Figure 8. Spatial associations between Lauraceae species within genera. Solid line curve represents the Ω_r value. Dashed lines correspond to the confidence intervals generated from 999 Monte Carlo simulations under the null hypothesis of complete spatial randomness.

genera), and reduced at the genus level. Furthermore, aggregation intensity of congeneric individuals was much lower than for species within genera. Though the four congeneric species pairs displayed different biotic or abiotic processes to enhance or attenuate their drive to repulse each other, then formed current pattern. Repulsive interactions among relatives were seen through decreased aggregation intensity at the genera level (Table 2).

The mechanisms behind species coexistence has been explored at the species level and rarely above or below species level. Indeed the genetic influences at the species level produced a qualitative change in species distribution. However, the genetic influence was not restricted to the species level. From individual, to species, to genera and even to higher levels, genetic influences and variation from natural selection must affect species coexistence. Results differed at different levels. Therefore, information from more levels is required to understand species coexistence mechanisms. For example, a sub-species level analysis found genetic subgroups of *Castanopsis chinensis* that showed environmental specialization; a species level analysis would not have detected this association [40].

To our knowledge, the present study is the first to trace interactions between and distributions of congeneric species at both the species and genus levels, with the aim of explaining their coexistence through examining the similarities and difference at these two levels. In our study, this was specifically achieved through comparisons of DBH classes, aggregation intensities and spatial patterns, associations and distributions of eight Lauraceae species at both the species and genus levels. DBH structure was more stable at the genus than species level. Individuals at both levels were aggregated, while aggregation intensity at the genus level was lower than for species within genera. Previous studies have found aggregation intensity to clearly decrease with DBH in DHS plot [35]. Here we also found aggregation intensity at the genus level tended to decrease from small to large DBH. Spatial associations within genera showed no positive associations, with two pairs (CRCO and CRCH, LIME and LICH) possessing negative associations. Associations among genera were almost never negative, often independent, but sometimes positive (e.g., *Cryptocarya* and *Lindera* and *Machilus* and *Neolitsea*) at the same spatial scales. As the level of examination increases from species to genus level, the negative associations become less common due to increasingly distant relationships between the compared groups.

Former studies of congeneric species interactions focused on testing and verifying whether or not interspecific competition existed. The conditions, such as resource availability and number of potential competitors, required to produce competitive relationship have been neglected up to now. Perhaps competition appeared more frequently in high than low diversity communities because a large number competitors for single resource units lead to significant competition. Studies in tropical forest communities showed interspecific competition should be much stronger among congeneric species, especially at small scales [20,45]. In subtropical forest no positive spatial associations within genera and two pairs of negative associations were found. In the central Aegean archipelago strong evidence for widespread competition among congeneric species were not found, and most communities investigated show no significant patterns of species associations [46]. Temperate plant communities with few species showed slightly negative associations at small scales, indicating interspecific competition, few species pairs showed positive associations [47].

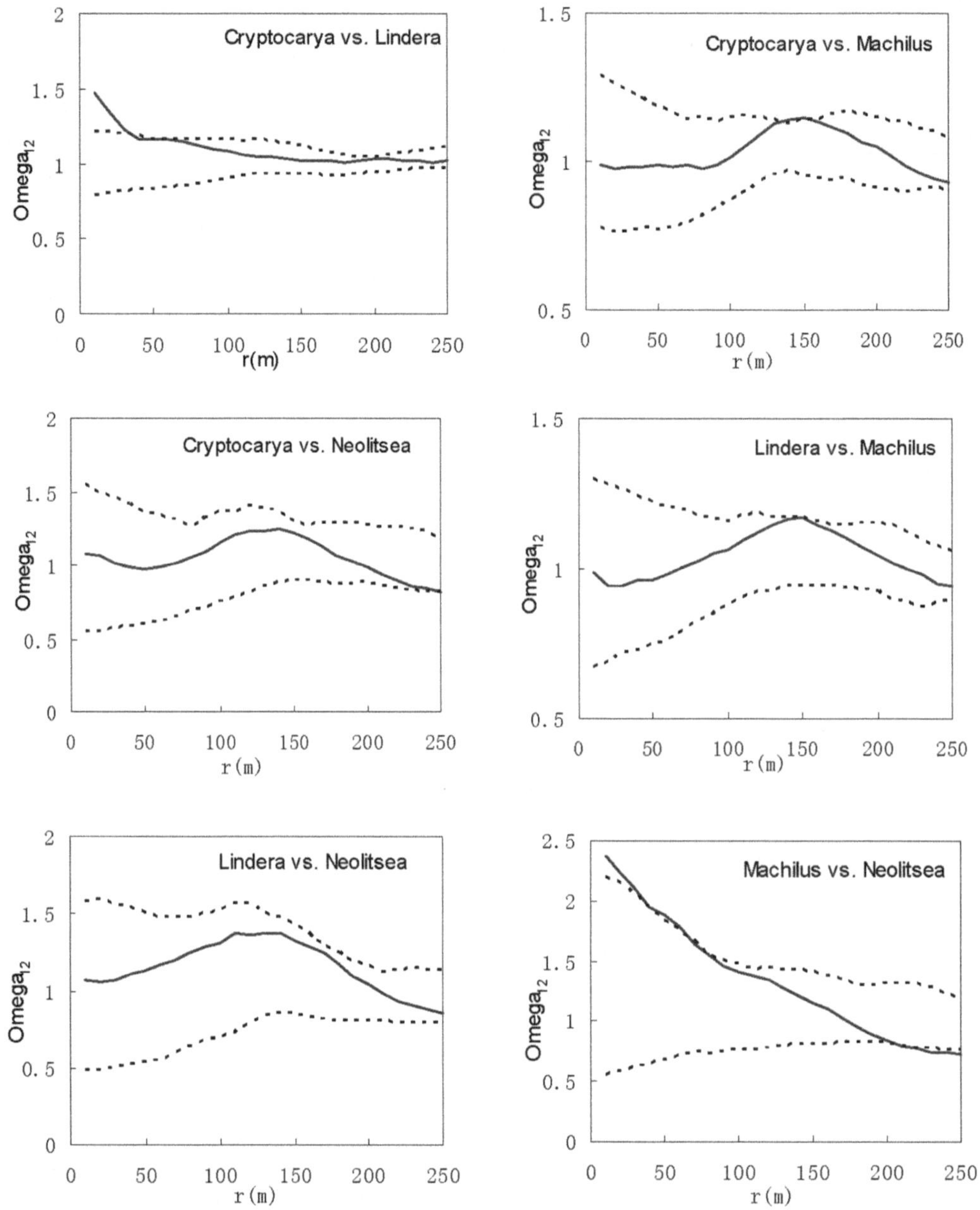

Figure 9. Spatial associations among genera of Lauraceae species. The Solid line curve represents the Ω_r value. Dashed lines correspond to the confidence intervals generated from 999 Monte Carlo simulations under the null hypothesis of complete spatial randomness.

There maybe sufficient resources for the coexistence of congeneric species in low diversity communities with relatively homogeneous environmental conditions. Among general congeneric species pairs no negative associations occurred, and even some positive correlations were reported [47].

Conclusion

From the study of eight Lauraceae species at genus and species level, we compared DBH classes, aggregation intensities and spatial patterns, associations and distributions of four pairs of these related species at both the species and genus level, to reveal similarities and differences at the two levels of relatedness. In conclusion, spatial aggregations were common, and the differences in spatial patterns were reduced at genus, relative to species level. The aggregation intensity clearly reduced at the genus level, the $\Omega_{0\text{-}10}$ value of all four genera was lower than the $\Omega_{0\text{-}10}$ value of the species within each genera. Aggregation intensity decreased with increasing DBH both at the species and genus levels, but due to the differences in DBH structure and growth type, the pattern of aggregation intensity change with DBH differed between individuals of two genera. Spatial associations between the four pairs of closely relative species showed no positive associations, and two pairs showed significant negative associations at several scales. The frequency of negative associations decreased at genus relative to species level, and independent associations were common among the four genera. Negative association illustrated competitive effect were existed in these negatively correlated relative species, however, coexistence mechanism of relative species at different levels was really a very complicated question. It's a good start, in order to fully understand the mechanisms generating distribution patterns of coexisting sibling trees, further investigation into interactions between coexisting relatives at scales ranging from subspecies to families is required, and even in combination with ecological and life-history data and experimental data.

Acknowledgments

We are grateful to many individuals who contributed to the field survey of the DHS plot.We thank the Chinese Forest Biodiversity Monitoring Network. We thank Professor Fangliang He of University of Alberta, Dr. I Fang Sun of National Dong Hwa University, Dr. Richard Condit of CTFS, Dr. Pierre Legendre of University of Montreal for training of data analysis. We would also like to thank Dr. Christine Verhille at the University of California for her assistance with English language and grammatical editing of the manuscript.

Author Contributions

Conceived and designed the experiments: WHY. Performed the experiments: ZLH JYL. Analyzed the data: SGW. Contributed to the writing of the manuscript: LL.

References

1. He FL, Duncan RP (2000) Density-dependent effects on tree survival in an old-growth Douglas fir forest. J Ecol 88.
2. Janzen DH (1970) Herbivores and the number of tree species in tropical forests. Am Nat 104: 501–528.
3. Connell JH (1971) On the role of natural enemies in preventing competitive exclusion in some marine animals and in rain forest trees. In: Den Boer PJ, Gradwell GR, editors. Dynamics of Numbers in Populations. Wageningen: Center for Agricultural Publication and Documentation. pp. 298–312.
4. Sterner RW, Ribic CA, Schatz GE (1986) Testing for life historical changes in spatial patterns of four tropical tree species. J Ecol 74: 621–633.
5. Kenkel NC (1988) Patten of self-thinning in jack pine: testing the random mortality hypothesis. Ecology 69: 1017–1024.
6. Lin YC, Chang LW, Yang KC, Wang HH, Sun IF (2011) Point patterns of tree distribution determined by habitat heterogeneity and dispersal limitation. Oecologia 165: 175–184.
7. Hubbell SP (1979) Tree dispersion, abundance, and diversity in a tropical dry forest. Science 203: 1299–1309.
8. Condit R, Ashton PS, Baker P, Bunyavejchewin S, Gunatilleke S, et al. (2000) Spatial patterns in the distribution of tropical tree species. Science 288: 1414–1418.
9. Condit R, Pitman N, Leigh EG, Chave J, Terborgh J, et al. (2002) Beta-diversity in tropical forest trees. Science 295: 666–669.
10. Plotkin JB, Potts MD, Manokaran N, Lafrankie J, Ashton PS (2000) Species-area curves, spatial aggregation, and habitat specialization in tropical forests. J Theor Biol 207: 81–99.
11. He FL, Legendre P (2002) Species diversity patterns derived from species-area models. Ecology 83: 1185–1198.
12. Wright SJ (2002) Plant diversity in tropical forests:a review of mechanisms of species coexistence. Oecologia 130: 1–14.
13. Wills C, Harms KE, Condit R, King D, Thompson J (2006) Nonrandom processes maintain diversity in tropical forests. Science 311: 527–531.
14. Green JL, Ostling A (2003) Endemics-area relationships: the influence of species dominance and spatial aggregation. Ecology 84: 3090–3097.
15. Chao A, Colwell RK, Lin C-W, Gotelli NJ (2009) Sufficient sampling for asymptotic minimum species richness estimators. Ecology 90: 1125–1133.
16. Rahbek C, Gotelli NJ, Colwell RK, Entsminger GL, Rangel TFLVB, et al. (2007) Predicting continental-scale patterns of bird species richness with spatially explicit models. P ROY SOC LOND B BIO 274: 165–174.
17. Valiente-Banuet A, Verdú M (2008) Temporal shifts from facilitation to competition occur between closely related taxa. J Ecol 96: 489–494.
18. Helmus MR, Savage K, Diebel MW, Maxted JT, Ives AR (2007) Separating the determinants of phylogenetic community structure. Ecol Lett 10: 917–925.
19. Swenson NG, Enquist BJ, Pither J, Thompson J, Zimmerman JK (2006) The problem and promise of scale dependency in community phylogenetics. Ecology 87: 2418–2424.
20. Mooney KA, Jones P, Agrawal AA (2008) Coexisting congeners: demography, competition, and interactions with cardenolides for two milkweed-feeding aphids. Oikos 117: 450–458.
21. Tanaka H, Shibata M, Masaki T, Iida S, Niiyama K, et al. (2008) Comparative demography of three coexisting Acer species in gaps and under closed canopy. J Veg Sci 19: 127–138.
22. Queenborough SA, Burslem DFRP, Garwood NC, Valencia R (2007) Habitat niche partitioning by 16 species of Myristicaceae in Amazonian Ecuador. Plant Ecol 192: 193–207.
23. Du X, Zhou S, Etienne RS (2011) Negative density dependence can offset the effect of species competitive asymmetry: a niche-based mechanism for neutral-like patterns. J Theor Biol 278: 127–134.
24. Pellissier L, Brathen KA, Pottier J, Randin CF, Vittoz P, et al. (2010) Species distribution models reveal apparent competitive and facilitative effects of a dominant species on the distribution of tundra plants. Ecography 33: 1004–1014.
25. Raven PH, Axelrod DI (1978) Origin and relationships of the California flora. Berkeley: University of California Press. 134 p.
26. Médail F, Verlaque R (1997) Ecological characteristics and rarity of endemic plants from southeast France and Corsica: implications for biodiversity conservation. Biol Conserv 80: 269–281.
27. Lloyd KM, Lee WG, Wilson JB (2002) Competitive abilities of rare and common plants: comparisons using Acaena (Rosaceae) and Chionochloa (Poaceae) from New Zealand. Conserv Biol 16: 975–985.
28. Snyder SW, Ladror US, Wade WS, Wang GT, Barrett LW, et al. (1994) Amyloid-beta aggregation: selective inhibition of aggregation in mixtures of amyloid with different chain lengths. Biophys J 67: 1216–1228.
29. Rabinowitz D, Rapp JK, Dixon PM (1984) Competitive abilities of sparse grass species: means of persistence or cause of abundance. Ecology: 1144–1154.
30. Debussche M, Thompson John D (2003) Habitat differentiation between two closely related Mediterranean plant species, the endemic Cyclamen balearicum and the widespread C. repandum. Acta Oecologica 24: 35–45.
31. Davies SJ, Palmiotto PA, Ashton PS, Lee HS, Lafrankie JV (1998) Comparative ecology of 11 sympatric species of Macaranga in Borneo: tree distribution in relation to horizontal and vertical resource heterogeneity. J Ecol 86: 662–673.
32. Queenborough SA, Burslem DFRP, Garwood NC, Valencia R (2009) Taxonomic scale-dependence of habitat niche partitioning and biotic neighbourhood on survival of tropical tree seedlings. 276: 4197–4205.
33. Li L, Wei SG, Huang ZL, Ye WH, Cao HL (2008) Spatial Patterns and Interspecific Associations of Three Canopy Species at Different Life Stages in a Subtropical Forest, China. J Integr Plant Biol 50: 1140–1150.
34. Harms KE, Condit R, Hubbell SP, Foster RB (2001) Habitat associations of trees and shrubs in a 50-ha neotropical forest plot. J Ecol 89: 947–959.
35. Li L, Huang Zl, Ye Wh, Cao Hl, Wei Sg, et al. (2009) Spatial distributions of tree species in a subtropical forest of China. Oikos 118: 495–502.
36. Pei NC, Lian JY, Erickson DL, Swenson NG, Kress WJ, et al. (2011) Exploring Tree-Habitat Associations in a Chinese Subtropical Forest Plot Using a Molecular Phylogeny Generated from DNA Barcode Loci. Plos One 6: e21273.
37. Ripley BD (1977) Modeling spatial patterns. J Roy Stat Soc 39: 172–212.
38. Stoyan D, Stoyan H (1994) Fractals, random shapes, and point fields: Methods of geometrical statistics. Chichester and New York: Wiley.
39. Beltrán E, Valiente-Banuet A, VerdúVerd M (2012) Trait divergence and indirect interactions allow facilitation of congeneric species. Ann Bot 110: 1369–1376.
40. Wang ZF, Lian JY, Huang GM, Ye WH, Cao HL, et al. (2012) Genetic groups in the common plant species Castanopsis chinensis and their associations with topographic habitats. Oikos 121: 2044–2051.
41. Wang ZG, Ye WH, Cao HL, Huang ZL, Lian JY, et al. (2009) Species-topography association in a species-rich subtropical forest of China. Basic Appl Ecol 10: 648–655.
42. Manabe T, Nishimura N, Miura M, Yamamoto S (2000) Population structure and spatial patterns for trees in a temperate old-growth evergreen broad-leaved forest in Japan. Plant Ecol 151: 181–197.
43. Mitsui H, Kimura MT (2000) Coexistence of drosophilid flies: Aggregation, patch size diversity and parasitism. Ecol Res 15: 93–100.
44. Parrish JK, Edelstein-Keshet L (1999) Complexity, pattern, and evolutionary trade-offs in animal aggregation. Science 284: 99–101.
45. Garcia-Serrano H, Sans FX, Escarré J (2007) Interspecific competition between alien and native congeneric species. Acta Oecologica 31: 69–78.
46. Sfenthourakis S, Tzanatos E, Giokas S (2006) Species co-occurrence: the case of congeneric species and a causal approach to patterns of species association. Global Ecol Biogeogr 15: 39–49.
47. Zhang J, Song B, Li B-H, Ye J, Wang X-G, et al. (2010) Spatial patterns and associations of six congeneric species in an old-growth temperate forest. Acta Oecologica 36: 29–38.

20

Ecosystem Services and Opportunity Costs Shift Spatial Priorities for Conserving Forest Biodiversity

Matthias Schröter[1,2*], Graciela M. Rusch[2], David N. Barton[2], Stefan Blumentrath[2], Björn Nordén[2]

1 Environmental Systems Analysis Group, Wageningen University, Wageningen, The Netherlands, **2** Norwegian Institute for Nature Research (NINA), Trondheim/Oslo, Norway

Abstract

Inclusion of spatially explicit information on ecosystem services in conservation planning is a fairly new practice. This study analyses how the incorporation of ecosystem services as conservation features can affect conservation of forest biodiversity and how different opportunity cost constraints can change spatial priorities for conservation. We created spatially explicit cost-effective conservation scenarios for 59 forest biodiversity features and five ecosystem services in the county of Telemark (Norway) with the help of the heuristic optimisation planning software, Marxan with Zones. We combined a mix of conservation instruments where forestry is either completely (non-use zone) or partially restricted (partial use zone). Opportunity costs were measured in terms of foregone timber harvest, an important provisioning service in Telemark. Including a number of ecosystem services shifted priority conservation sites compared to a case where only biodiversity was considered, and increased the area of both the partial (+36.2%) and the non-use zone (+3.2%). Furthermore, opportunity costs increased (+6.6%), which suggests that ecosystem services may not be a side-benefit of biodiversity conservation in this area. Opportunity cost levels were systematically changed to analyse their effect on spatial conservation priorities. Conservation of biodiversity and ecosystem services trades off against timber harvest. Currently designated nature reserves and landscape protection areas achieve a very low proportion (9.1%) of the conservation targets we set in our scenario, which illustrates the high importance given to timber production at present. A trade-off curve indicated that large marginal increases in conservation target achievement are possible when the budget for conservation is increased. Forty percent of the maximum hypothetical opportunity costs would yield an average conservation target achievement of 79%.

Editor: Runguo Zang, Chinese Academy of Forestry, China

Funding: MS received funding from the European Research Council under grant 263027 ('Ecospace'), and from the Research Council of Norway (Yggdrasil grant). GMR, DNB and SB received funding from the European Union's Seventh Programme for research, technological development and demonstration under grant agreement No. 244065 (POLICYMIX project (http://policymix.nina.no)). BN was supported by the project ECOSERVICE (Approaches for integrated assessment of forest ecosystem services under large scale bioenergy utilization) financed by the Research Council of Norway, RCN grant no 233641/E50. The funders had no role in study design, data collection and analysis, decision to publish, or preparation of the manuscript.

Competing Interests: The authors have declared that no competing interests exist.

* Email: matthias.schroter@wur.nl

Introduction

The ecosystem services (ES) concept comprises multiple contributions of ecosystems to human well-being [1], and has increasingly been used to raise awareness about the benefits that people derive from ecosystems [2,3]. Considering ES when making decisions about the use of ecosystems could provide additional, anthropocentric arguments to support either management aimed at sustainable use of ecosystems or biodiversity conservation [4]. However, there is a still unresolved debate about to what extent components of biodiversity correspond with ES provision [4–7] and about the extent to which considering ES in decision making matches with biodiversity conservation objectives. Furthermore, accounting for ES within conservation planning is a fairly new practice [8–11]. In a conservation decision-making context, ES can be seen as benefits of conservation (many cultural and regulating services), or in the case of extractive provisioning services as an opportunity cost of conservation since their use may become restricted [8]. Trade-offs between extractive provisioning services, such as clear-cutting timber harvest, and other ES [12] and biodiversity protection [11,13–15] require choices to be made on whether and where to protect an area. However, certain management systems restrict timber production and might thus allow for a synergy between an extractive provisioning service and other ecosystem services [16,17] as well as some aspects of biodiversity conservation [17–21]. This leads to the crucial question within cost-effective conservation planning on how multiple-use areas, in which extractive exploitation is restricted, can potentially contribute to biodiversity conservation [22–24]. Cost-effective conservation means minimizing opportunity costs in terms of foregone commodity production [25]. As some conservation targets are compatible with a certain level of use [26], and since the opportunity costs of setting aside areas can be potentially high, a mixture of fully protected areas and areas allowing for partial use is likely to render more cost-effective and less conflictive conservation solutions, and may open opportunities for overall higher levels of biodiversity protection.

Spatial considerations play an integral role in the assessment of cost-effectiveness of conservation as the spatial configurations of important habitats [27] and of opportunity costs of conservation do not necessarily coincide [28]. A 'policyscape' may be defined as the spatial configuration of a mix of policy instruments [29], which aims at conserving biodiversity and ES at an aggregated spatial level. This framing suggests that there is an optimal and complementary spatial allocation of different types of instruments across a space containing all possible combinations of conservation values and opportunity costs within a study area. The spatial configuration of the policyscape has important practical implications for decision-making. For instance, it opens opportunities to evaluate disproportionate economic burdens between administrative units.

In this study, we suggest ways of creating cost-effective policyscapes. We address a mix of instruments that combines non-use (strict protection) and partial use (forestry restricted) for the conservation of forest biodiversity and ES in the county of Telemark (Norway). Indicators of the state of forests in Norway show a decline of certain species populations, especially of species associated to old-growth forest and species whose habitats are threatened by current forestry practices [14,30]. There is a need to modify and adapt current conservation policies to help secure portions of unprotected biodiversity as well as to halt the processes that lead to forest biodiversity loss [14,30,31]. One approach is to increase protected forest areas in Norway, particularly within the ecological zones that are most favourable for forestry production [31]. Currently, new nature reserves in Norway are mostly implemented through voluntary forest conservation schemes that are based on a negotiation between forest owners and conservation authorities in Norway [32]. The exploration of different policyscapes for conservation of biodiversity and ES can give guidance to support such conservation efforts.

We used the conservation planning software Marxan with Zones [33] for near-optimal selection of areas for cost-effective policyscapes on a county level. Some experience has been developed in applying (earlier versions of) Marxan to conservation optimisation with ES [8,11,34–37]. However, to our knowledge integrated targeting of both biodiversity and multiple ES within a policyscape with different levels of protection has not been systematically studied before.

We addressed the following specific questions. We first analysed how optimal conservation outcomes differ between two scenarios that either take into account biodiversity only (scenario 1) or a set of ES next to biodiversity (scenario 2). The outcome of both scenarios was measured in terms of spatial configuration, area protected, conservation target achievement, and opportunity costs.

Second, we assessed the trade-off between biodiversity and ES conservation goals and timber production. We analysed this relationship by constructing a production possibility frontier (PPF) [25], while considering timber production as a private good and the sum of biodiversity features and other ES as public goods. These public goods are either spared from timber production in the case of full protection or jointly produced with the private good in the case of partial protection. We compared current instrument targeting, i.e. the effectiveness of current reserves to achieve conservation targets set in our scenario, to a 'benchmark' defined as the cost-effective policyscape traced by the PPF [38,39].

Third, we explored differences in conservation burden across administrative units. For this purpose, we calculated the expected opportunity costs of an optimal conservation outcome for each municipality in Telemark. Significant differences in conservation burden across municipalities would suggest potential efficiency gains with concomitant distributional consequences, which could justify considering the introduction of a conservation instrument such as ecological fiscal transfer schemes [40].

Methods

Study area

Telemark is a county in southern Norway with an area of 15,300 km^2 and a population of about 170,000 people [41], concentrated mainly in the south-eastern part of the county. The climate varies across the region with temperate conditions in the south-east (Skien, average temperature January −4.0°C, July 16.0°C, 855 mm annual precipitation) and alpine conditions in the north-west (Vinje, January −9.0°C, July 11.0°C, 1035 mm) [42]. The southern part of Telemark is mainly covered by forest exploited by forestry activities as well as by large inland lakes, with few towns and a small agricultural area (247 km^2, i.e. about 1.6% of the land area) [41]. The northern part is characterised by treeless alpine highland plateaus covered by bogs, fens and heathlands [43]. The forest landscape in Telemark is characterized by coniferous and boreal deciduous forest [43]. Important forest ecosystem services include moose hunting, free range sheep grazing and timber production [44]. In addition, forests of Telemark sequester and store considerable amounts of carbon, prevent snow slides and provide opportunities for recreational hiking and residential amenities [44]. In 2011, conservation areas comprised about 5.1% in national parks, 4.6% in landscape protection areas (both types cover mainly highland plateaus), as well as 1.7% in nature reserves [41]. As a result of forestry activities, the status of biodiversity in forests of Telemark shows relatively low values compared to other ecosystems and regions within Norway [14]. We conducted our analysis for the forest area within Telemark, however, as forest field mapping is lacking for a small south-eastern part of the county [45], this area was excluded from the analysis.

Principle of Marxan with Zones

Marxan with Zones [33] builds on a heuristic optimisation algorithm that incorporates key principles of systematic conservation planning, including comprehensiveness, cost-effectiveness and compactness of the reserve system [46]. Marxan with Zones enables to consider zones with different levels of protection and thus spatial differences in costs, thereby allowing for planning and evaluation of policyscapes that include full and partial protection. Marxan with Zones requires a series of inputs, which are specified below.

Data input Marxan with Zones

ES and biodiversity features and conservation targets. Depending on the scenario, a total of 59 (scenario 1, biodiversity) and 64 (scenario 2, biodiversity and ES) input features were used, respectively. Table 1 provides an overview of all features.

We included five key ES of importance within a Norwegian context for which spatial models have been developed (Table 1) [44]. We specifically included biodiversity features which are characteristic of old-growth, largely undisturbed forest and which are not maintained under current commercial forestry practices. We included 40 types of old-growth forest, to a large extent remnants of previously high-graded forests, occurring across a range of vegetation zones, climate zones and productivity conditions to represent the ecological variability across the county (Text S1 for details). Six proposed forest corridors of national importance that connect existing reserves [47] were included as a spatial indicator of conditions enabling species dispersal between habitats [48]. Forest habitats of particular conservation importance on a national level in Norway [49,50] were also included. Three classes of priority habitats for conservation (very important,

Table 1. Features, targets, fraction of targets to be achieved across the two zones (non-use and partial use), and contribution (effectiveness) of the partial zone in meeting respective targets.

Feature name	Feature target (%)	Fraction non-use (%)	Fraction partial (%) (contribution in %)
Wilderness-like areas (ES)	100	100	0 (0)
Recreational hiking (ES)	20	50	50 (100)
Carbon storage (ES)	10	50	50 (25)
Carbon sequestration (ES)	5.57	75	25 (25)
Snow slide protection (ES)	100	0	100 (100)
Old-growth forest types (40)	50	75	25 (50)
Corridors (6)	50	50	50 (50)
Priority habitats for conservation (very important)	100	100	0 (0)
Priority habitats for conservation (important)	100	100	0 (0)
Priority habitats for conservation (locally important)	50	100	0 (0)
Hollow deciduous trees	100	100	0 (0)
Late successional forests with deciduous trees	100	100	0 (0)
Logs	100	100	0 (0)
Old trees	100	100	0 (0)
Rich ground vegetation	100	100	0 (0)
Snags	100	100	0 (0)
Trees with nutrient-rich bark	100	100	0 (0)
Trees with pendant lichens	100	100	0 (0)
Recently burned forest	100	100	0 (0)
Stream gorges	100	100	0 (0)

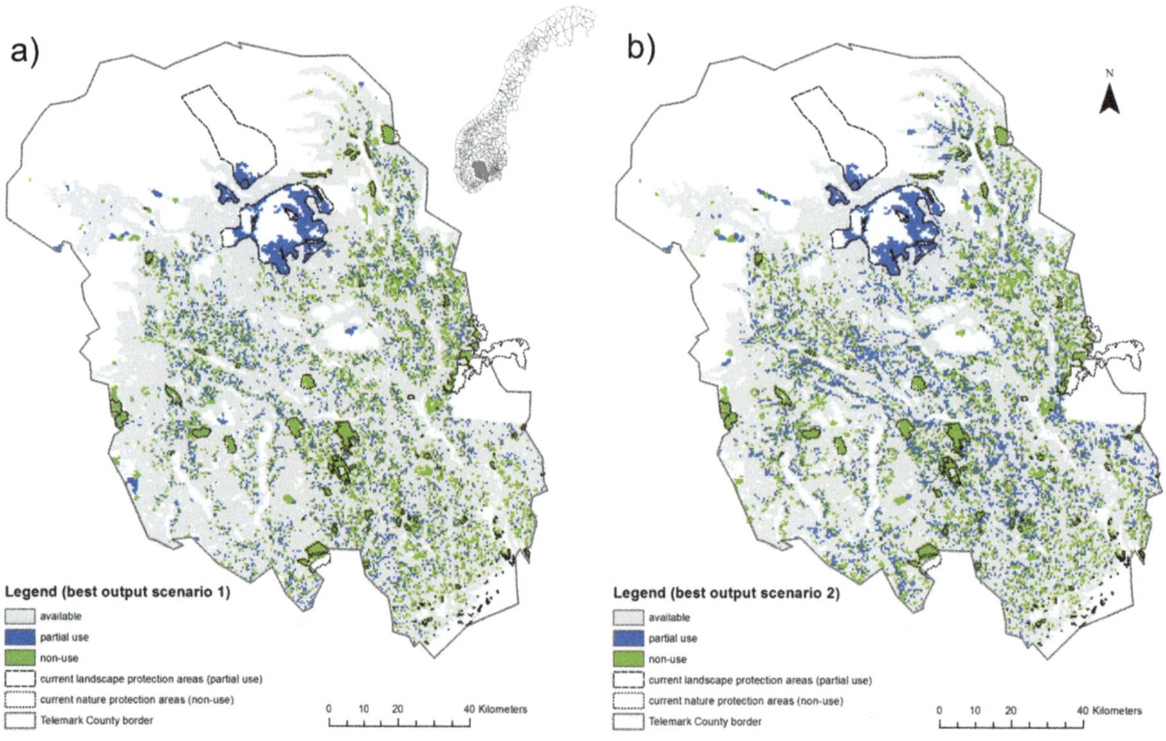

Figure 1. Best solution of the reserve network for scenario 1 (a) and scenario 2 (b). Scenario 1, considers biodiversity conservation criteria only; scenario 2, both biodiversity and ecosystem services criteria. Grey, areas available for forestry; blue, areas in the partial use zone and green, areas in the non-use zone. Current reserves are demarcated in dashed lines. Map inlay shows the location of Telemark within Norway (grey).

important and locally important) were taken from the Norwegian Environmental Agency's database (Naturbase) [51]. In addition, we included ten types of important forest habitats (Table 1) from a Norwegian Forest and Landscape Institute database (MiS) [52].

Marxan with Zones requires setting quantitative conservation feature targets that reflect the proportion of the abundance of each feature to be protected. Targets were based on expert judgments and, wherever possible, on interpretation of policy documents (Table 1, and Text S1 for details). In order to verify targets an expert workshop was organised (Text S1). Written consent to participate in this study was obtained from the participants of the expert workshop.

The policyscape – definition of zones, zone targets, zone contributions. Two types of area protection were included in our analysis, namely a non-use and a partial use zone. Non-use referred to nature reserves, where forestry is completely restricted, i.e. "use" refers to forestry activities. The partial use zone was an 'umbrella' zone covering three different current forms of protection where forestry is partially restricted, namely landscape protection areas, mountain forest ('fjellskog'), and outdoor recreation areas ('friluftsområder') (Text S1). All current nature reserves in Telemark [51] were 'locked-in' as non-use zones and all current landscape protection areas were 'locked-in' as partial use zones, which means that spatial units overlapping with these areas were selected for the respective zone in each run of Marxan.

Marxan with Zones allows for distribution of the targets across zones. Zone targets were defined according to an own expert judgement about how well the non-use and partial use areas were compatible with the persistence of the respective feature. Zone targets (Table 1) were discussed, reviewed and as far as possible confirmed during the expert workshop (Text S1).

Marxan with Zones allows for differentiation of how effective zones are in order to achieve targets (zone contribution). We considered the effectiveness of partial use areas as "the relative contribution of actions to realizing conservation objectives" [53]. We assumed that non-use areas are fully effective to reach the targets of all features (100% contribution). There is growing, but yet inconclusive knowledge on how low impact logging could be compatible with biodiversity conservation [17–21,54,55]. This means that effectiveness of partial use areas is highly uncertain, and may affect features differently. Zone contributions were thus discussed and as far as possible confirmed during the expert workshop. In a sensitivity analysis we further explored the consequences of changing the zone contribution of the partial use zone (File S1).

Planning units. The forest area in Telemark was divided into 43.513 grid planning units of 25 ha size (500 m×500 m). This resolution was suitable in terms of time and computing capacity, and considered relevant for land-use planning. Property sizes in Norwegian forests vary widely from as little as 0.1 ha to several hundred hectares [32] and as such are not a good guide to setting the size of the planning unit.

Opportunity costs of conservation. Foregone timber harvest was selected as an indicator of opportunity costs of conservation since harvest activities are constrained by different forms of protection [25]. We used a net revenue (stumpage value) forest model to determine opportunity costs (Text S1). In non-use areas opportunity costs were set to 100%, while in partial-use areas, we estimated that restrictions would account for 25% of the stumpage value. This estimate was based on different logging restrictions [56] which ranged from 15% (landscape protection area), to 20% (outdoor recreation area) and 30% (mountain forest).

Analyses

Marxan with Zones was run 20 times with the parameters described above (for further parameter adjustments see Table S1 and Table S2). The software was run for both scenarios to determine the best solution and the selection frequency of each planning unit over all runs, which ranged from 0 (never chosen) to the maximum of 20 (chosen in each run) and indicated importance of a particular planning unit to achieve the overall conservation targets [57]. Marxan with Zones input files, including spatial information on all conservation features, can be found in the supporting information for scenario 1 (File S2) and scenario 2 (File S3).

Comparison of scenarios. We used selection frequency of planning units to determine how the policyscapes of both scenarios differed spatially. Selection frequency of each planning unit to each of the two zones in scenario 1 (biodiversity only) was subtracted from selection frequency in scenario 2 (biodiversity and ES) to determine the difference. To compare the spatial configuration of the policyscapes, we calculated Pearson's correlation coefficient between the selection frequency of each scenario for the partial and the non-use zone. We calculated Cohen's Kappa on the selection frequency of each planning unit as a measure of agreement between the scenarios for each zone. To compare the two scenarios in absolute terms we calculated a number of statistics, including total costs, number of planning units without protection, planning units in the partial and non-use zone and average target achievement.

Trade-off between conservation target achievement and timber harvest. The PPF was identified by running a series of cost constraints for scenario 2. Cost constraints are a restricting condition that defines an upper limit of costs when selecting planning units. We started by running the scenario with no cost constraints and close to 100% average target achievement, and recorded the total unconstrained cost. We then introduced cost constraints at different levels (80%, 60%, 40%, 20%, 10%, 5%, 1%) of the total unconstrained cost in consecutive runs (see Table S4 for parameter details). The value of timber production (horizontal axis in the PPF) was determined as the total sum of stumpage value across all planning units in the study area minus the opportunity cost of the best solution of each run. The vertical axis in the PPF was determined as the average percentage of target achievement for all biodiversity and ES features. To assess the opportunity costs of conservation and the conservation target achievement of the current existing reserve network, we used an overlay analysis (r.stats in GRASS GIS).

Conservation burden across Telemark. To determine the conservation burden among the municipalities in Telemark, the expected opportunity cost for each municipality was calculated as the summed expected value of opportunity costs:

$$C_e = \sum \begin{cases} \frac{f_{n_i}}{20} * C_i \text{ if } {}_{n_i} \geq f_{p_i} \\ \frac{f_{p_i}}{20} * 0.25 * C_i \text{ if } f_{n_i} < f_{p_i} \end{cases} \quad \text{for } i = 1, \ldots, 43513 \quad (1)$$

where C_e is the expected opportunity cost, f_{ni} is the selection frequency of non-use areas for planning unit i, f_{pi} is the selection frequency of partial use areas for planning unit i and C_i is the opportunity cost of planning unit i. The denominator 20 stands for the number of runs in our case and the factor 0.25 specifies the harvest restriction in the partial use areas.

This analysis was run on scenario 2 with first, no cost constraint and, second, a medium cost constraint of 60% of the maximum costs needed to achieve close to 100% of the average targets. Opportunity costs per municipality were determined with zonal statistics in ArcMap for both expected opportunity cost layers and for current reserves. Municipalities were ranked according to relative opportunity costs, i.e. opportunity costs divided by municipal forest area. To analyse the spatial shift of the conservation burden across municipalities, Spearman's rank correlation coefficient was calculated between the current situation and the unconstrained scenario, as well as between the 60% cost constraint and the unconstrained scenario.

Results

Incorporating ES in the policyscape for biodiversity conservation

Incorporating ES into the policyscape changed the absolute sum of area in the two zones, the opportunity costs (Table 2) as well as the spatial configuration of the policyscape (Figures 1 and 2). When considering ES, the sum of partial use areas increased by 36.2% and the sum of non-use-areas by 3.2% compared to the scenario that only considered biodiversity. Opportunity costs were 6.6% higher in scenario 2 than in scenario 1. As an illustration of a policyscape, Figure 1 shows the best solution per scenario for scenario 1 (a) and scenario 2 (b). Selection frequencies of planning units for both scenarios can be found in Figure S1. The differences in selection frequencies are shown in Figure 2 for the partial (a) and non-use zone (b). A positive difference means higher selection frequency in the policyscape of scenario 2 than in scenario 1, while a negative difference indicates a lower selection frequency in the policyscape of scenario 2 than in scenario 1. Comparison of the spatial configuration of the policyscapes of both scenarios led to the following results. Pearson's correlation coefficient between selection frequencies of sites in the non-use zone was $r = 0.90$, while for the partial use zone, it was $r = 0.58$. This indicates that relatively larger differences can be expected in the partial use zone than in the non-use zone when ES were considered, which partly rests upon the fact that ES can, in contrast to most of the biodiversity features in this study, partly be protected in this zone. Cohen's Kappa statistics was $K = 0.577$ (sig$\leq$0.0001) for the non-use zone and $K = 0.398$ (sig $\leq$0.0001) for the partial use zone. These results imply 'moderate agreement' in non-use and 'fair agreement' in partial use zone, respectively [58], which supports the observation of a relatively larger agreement between non-use areas in the different spatial configurations of the policyscapes.

Trade-offs between conservation and timber production: Production possibility frontier (PPF)

The PPF shows a concave curve representing the trade-off between timber production and conservation of biodiversity and non-forestry related ES (Figure 3). Creating a reserve network to achieve the conservation targets comes at a cost of timber production. The marginal increase in conservation target achievement is initially high when the current constraint on conservation cost is relaxed (i.e. moving left in Figure 3). This marginal conservation gain decreases more rapidly after having passed a cost constraint of about 40% of the total cost required to achieve 100% of the overall conservation target. The current policyscape (black square) lies under the PPF curve, meaning that more cost-effective policyscape configurations than the current one are possible. This means that higher average target achievement could hypothetically be realised at current levels of timber production, or that the same target could be achieved at lower costs. At the same time, the location of the current policyscape shows a strong preference of decisions towards timber production. Consequently, the conservation targets we set in our scenario are barely met by the current reserve system (average achievement 9.1%).

While Figure 3 shows the average target achievement of all 64 features, Figure 4 shows the development of target achievement along changing opportunity cost constraints for single, exemplary features (for all features see Table S3). Some features meet high targets at low (20%) cost constraints (carbon sequestration and one type of low productive old-growth forest), which means that these features did not constrain the solution to a high degree. Some conservation features decreased at higher rates than the average (e.g., one type of high productive forest and recently burned forest). Such features are more costly to be comprehensively conserved in a compact reserve network.

Distribution of the conservation burden of cost-effective conservation areas

The creation of the policyscape for conservation of biodiversity and ES formed the basis for determining the 'conservation burden' across municipalities of Telemark (Table 3, spatial distribution in Figure S2). Conservation burdens across municipalities were slightly shifted in a (hypothetical) scenario with no cost constraint in which approximately 100% of the average target could be achieved compared to the current situation. For instance, while Porsgrunn ranked 6th in terms of the conservation burden of the current policyscape, it ranked 1st in the policyscape of with no cost constraints. The Spearman's correlation coefficient between the current situation and the scenario with unconstrained costs was $r = 0.67$. The Spearman's correlation coefficient between a 60% cost constraint and the unconstrained scenario was $r = 0.46$, which means that spatial priorities for conservation, and thus conservation burdens, shift with the level of the opportunity cost constraint.

Discussion

A policyscape for conservation of biodiversity and ES

The use of spatial planning tools that simultaneously consider conservation of biodiversity and ES in a cost-effective way is a fairly new approach, facilitated by recent advancement in computational science. This approach provides a range of opportunities [8,10], but still presents challenges in operationalization. Considering ES within biodiversity conservation could be beneficial for incorporating sustainable use of ecosystems [4] when achieving overall conservation goals in land use planning (land sharing), compared to a land use strategy that separates conservation and provision of ES (land sparing). A land sharing principle was included in our study in the partial use zone, which partly allows for the development of synergies between ES, biodiversity and timber production and which complements strict protection zones in policyscapes analysed in this study. In our analysis, we had to rely on expert-backed assumptions when describing the effects of the partial use zone on conservation. This is due to inconclusive knowledge on how restricted logging affects particular elements of biodiversity and ES [17–21,55]. Our study suggests that in forest areas of Telemark the configuration of a policyscape for conservation changes when ES were incorporated (scenario 2) compared to considering only biodiversity conservation criteria (scenario 1). This change was twofold and included a change in total areas assigned to the two protection zones and a change in the spatial configuration of selected sites. Including ES resulted in an increase in the size of the reserve network, a result that is in line with previous studies [8,36] in that when optimizing for cost-effective representation of conservation targets more areas with lower opportunity costs that contribute to target achievements of both biodiversity and ES are selected.

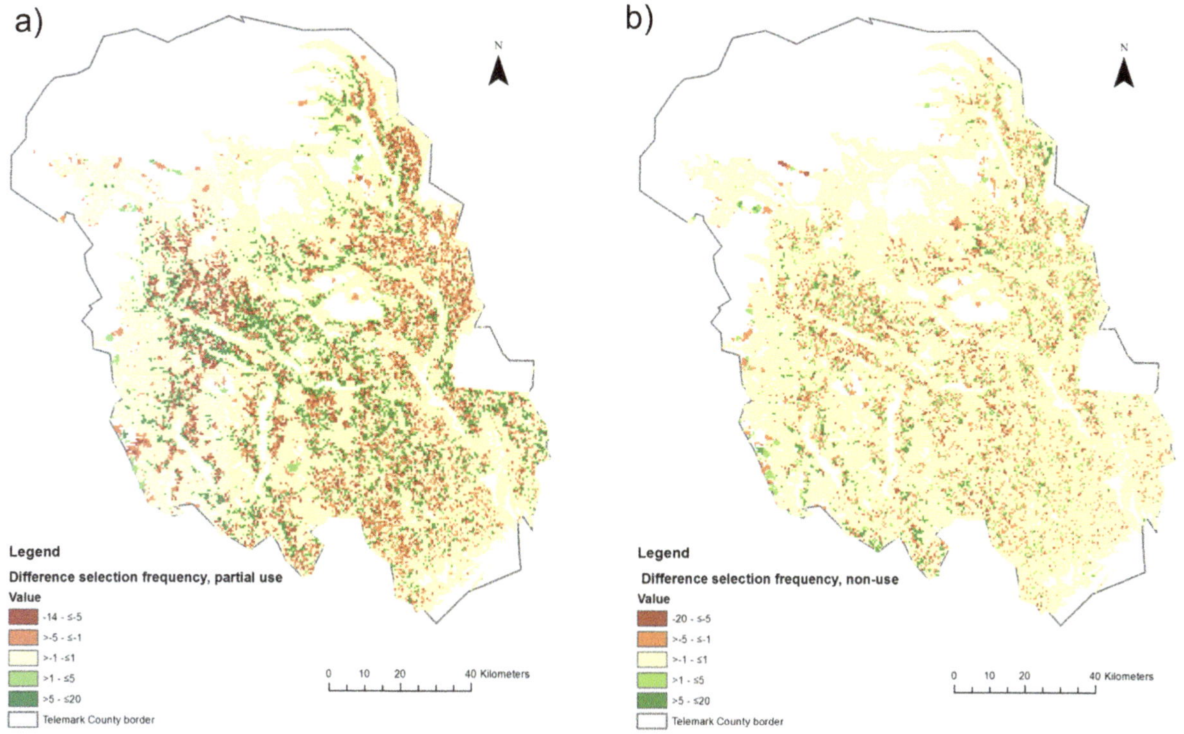

Figure 2. Differences in selection frequency of sites for partial (a) and non-use (b) areas. The maps show the difference of scenario 2 (biodiversity and ES features) versus scenario 1 (biodiversity only). A positive difference means higher selection frequency in scenario 2 than in scenario 1.

In contrast to former studies, we used different levels of protection, which enabled us to also specify the change in the policyscape in terms of the spatial distribution of the different zones. Including ES resulted in a strong increase in partial use areas (+36.2%), which was partly expected due to the fact that ES features were considered to be protected for a relatively larger proportion in partial use zones than biodiversity features (Table 1). The difference in spatial configurations of the policyscapes of the two scenarios can partly be explained by relatively low degrees of pairwise spatial overlaps between some ES and the biodiversity features (File S4). It also depends, for instance, on various combinations of biodiversity and ES features on cost-effective sites and proximity of suitable combinations to existing reserves. The difference in spatial configuration leads to different spatial prioritisations of sites to preserve in both zones and thus would have important implications for regional and local decision making.

Trade-off between commercial timber production and conservation of biodiversity and ES

Including ES next to biodiversity into a conservation scenario reflects different values [4,59] and as such could lead to more informed policy decisions. In our conservation scenario we thus treated ES of public interest representing partly intangible values (regulating and cultural services) as conservation features with an own target. While in the ES discourse, ES are often treated as generally beneficial [4], here we shed light on potential specific trade-offs among ES and between ES and biodiversity conservation priorities. We included timber production in our analysis, a provisioning service that contributes to private economic benefits, and assessed the form of the trade-off curve (PPF) between timber production on the one hand and cultural and regulating services and biodiversity on the other. The existence of a trade-off on a system level was expected based on our assumption that outside

Table 2. Summary statistics describing the difference between scenario 1 (considering biodiversity conservation criteria only) and 2 (considering biodiversity and ecosystem services) in terms of opportunity costs, area in the different zones and average conservation target achievement.

Statistics	Scenario 1	Scenario 2	Difference 2 vs. 1 in %
opportunity costs (billion NOK)	1.912	2.038	+6.6
without protection (no. of planning units of 25 ha)	32,183	30,279	−5.9
partial use area (no. of planning units of 25 ha)	4,661	6,349	+36.2
non-use (no. of planning units of 25 ha)	6,669	6,885	+3.2
average conservation target achievement (%)	99.86	99.23	−0.6

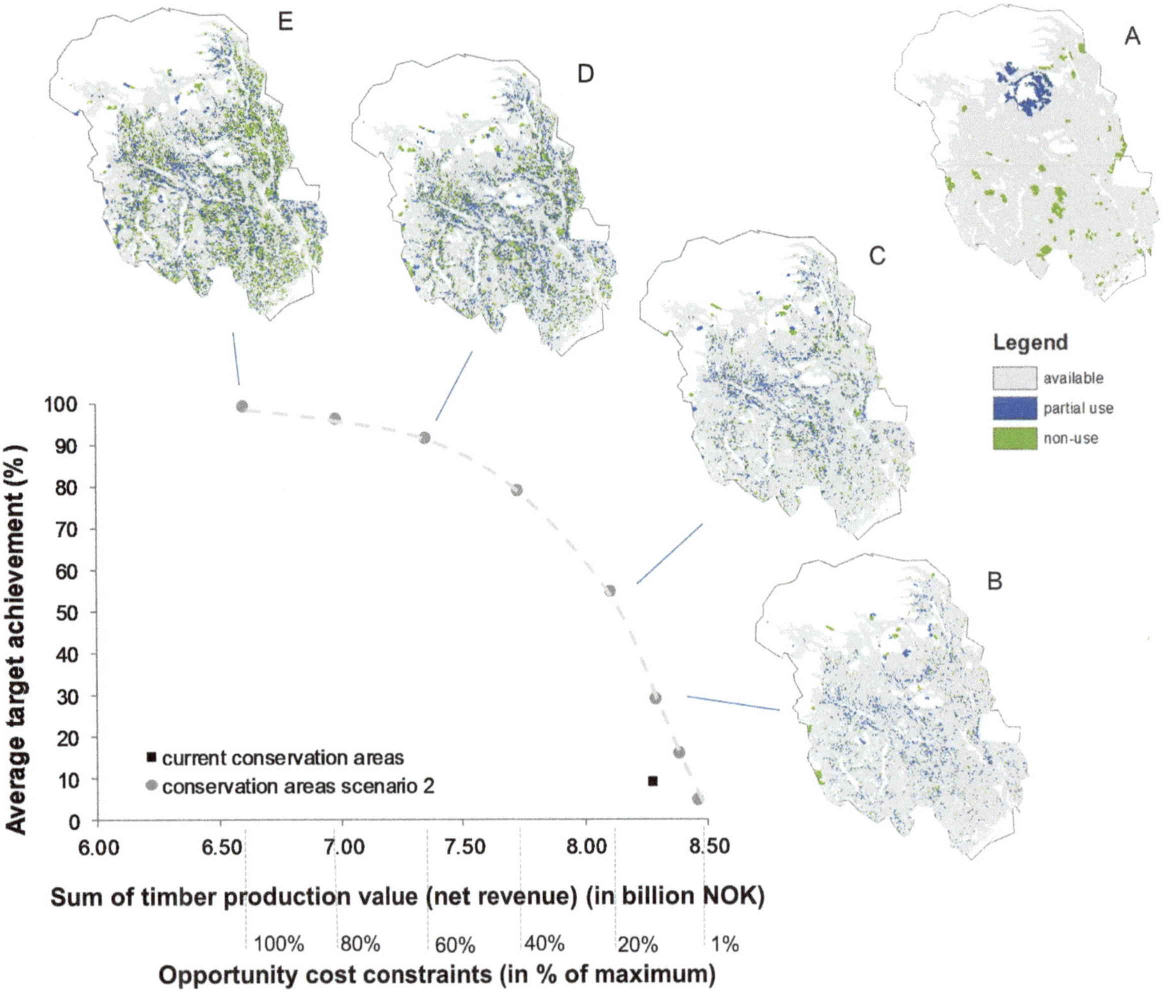

Figure 3. Forest conservation-timber production possibility frontier (PPF). Note that the x-axis (sum of timber production value) starts at 6.00 billion NOK. The maps indicate current reserve network (A) and selected (B–E) available, partial and non-use areas when current reserves are not locked-in. The spatially explicit solutions (policyscapes) are shown as maps on the trade-off between net revenues from timber production and average conservation target achievement, along a range of opportunity costs constraints.

the two conservation zones, elements of biodiversity and ES would not be conserved. This assumption might seem strong, but can be defended by the fact that the dominant form of forest management in Norway is characterised by large-scale clear-cutting [60].

From the PPF, we derive two broad policy conclusions. First, the currently designated nature reserves and landscape protection areas achieved a very low proportion (9.1%) of the conservation targets we set in our scenario. This is partly because the conservation network has not been initially designed to meet the conservation targets we defined in our study. For instance, while attention has been given to rare and threatened forest types [31], we did not assign different conservation targets to the different old-growth forest types, which might in practice be of different importance for forest biodiversity conservation. The result is, however, in agreement with the relatively little forest area that is currently allocated to conservation [31] due to low conservation budgets and conflicts. Further, our findings support the observation of a biased representation of protected areas towards high altitudes and lower opportunity cost areas [61]. This pattern, as well as the under-representation of productive forest in the current conservation network, have also been found for Norway [29,31,62]. Our present scenario was deliberately designed to include high productive forest, which partly explains the low target achievement of the current conservation network.

Second, the PPF analysis also provides insights for policy-makers regarding balancing private and public interests. It is a societal choice to determine the level of production of either timber or biodiversity and regulating and cultural ES. The PPF illustrates the high importance given to timber production at present. At the same time, it shows that the relationship between gains in conservation and opportunity costs is not linear. This means that high marginal improvements in conservation can be obtained with relatively smaller increases in costs when a low opportunity cost constraint is relaxed. Thus, with relatively little investment, e.g. spending 40% of the maximum opportunity costs, on average 79% of the scenario targets could be achieved under the assumptions applied in this study. However, inspection of the PPF curve also reveals that lowering the cost constraint reduces the probability of achieving conservation targets for certain habitats (e.g. recently burned forests, high productive forests) within the reserve network. In contrast, carbon sequestration reaches high proportions of the target at low cost which indicates that carbon sequestration can be seen as a co-benefit of protecting biodiversity and other ES, assessed at the scale of all prioritised full and partial

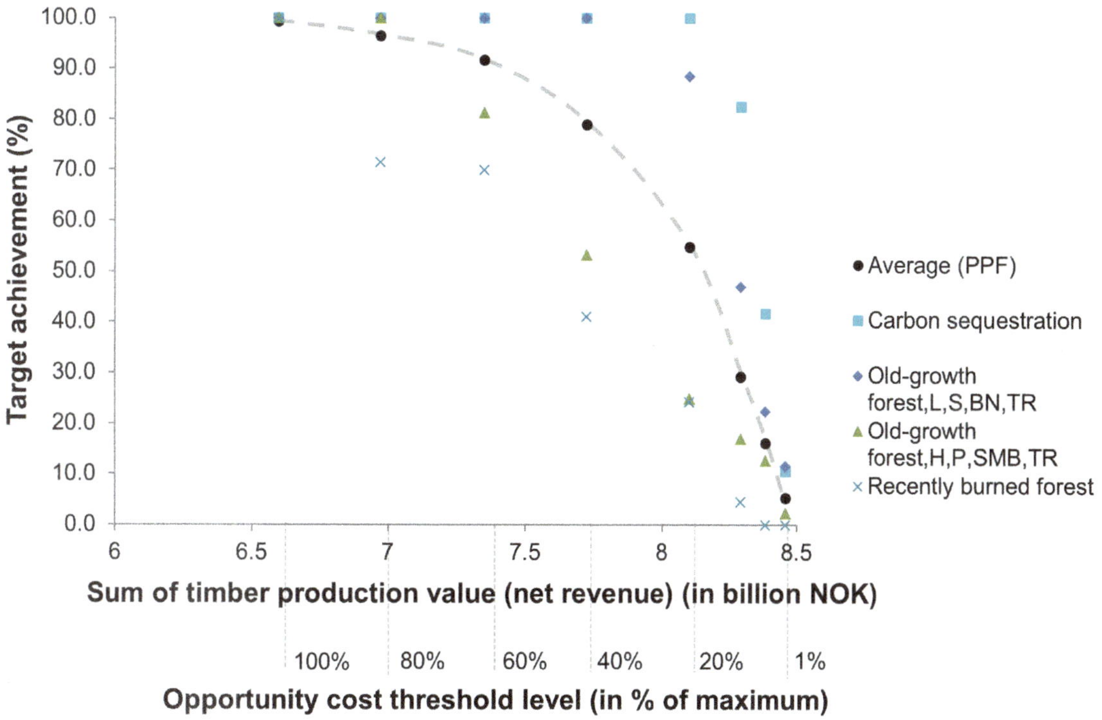

Figure 4. Forest conservation-timber production possibility frontier (PPF) for single, exemplary features. Old-growth forest L, S, BN, TR = impediment and low productivity, spruce dominated, boreonemoral zone, oceanic-inland transition zone. Old-growth forest H, P, SMB, TR = high & very high productivity, pine dominated, South & Mid- boreal zone, oceanic-inland transition zone.

protection areas across the study area. This is the inverse logic of the current international debate (i.e. REDD+), where carbon sequestration is targeted to be protected while (unmeasured) biodiversity is a (hoped for) co-benefit [63], but is in agreement with findings of process-based models in recent studies [17].

Uncertainties in creating the conservation scenario

We encountered several challenges in creating the conservation scenario. The choice of conservation features is a crucial factor that determines the outcome of the site prioritisation. Operationalizing biodiversity conservation requires quantifiable and obtainable indicators [64,65]. Given restrictions on data availability, we believe that our choice of biodiversity surrogates represents a first step for planning the maintenance of biodiversity in Norwegian forest ecosystems.

Despite the "inevitable subjectivity" in setting conservation targets [66], there is some experience in setting targets for biodiversity conservation [65,67]. However, setting explicit targets for ES when determining spatial priorities has seldom been done [68]. Current studies using Marxan for ES conservation have pointed out the need for experimentation, explicitly stated assumptions and expertise in setting targets given the absence of this information [8,11,36,37], particularly because ES targets influence the size of the reserve network [35]. A systematic sensitivity test of target levels was, however, out of scope of this current study. ES targets may vary considerably because alternative means are available for substituting forest ES depending on location. Preferences for recreational hiking can shift outside the forest towards mountainous areas. In some areas, feasible technical substitutes for snow slide prevention by forests are available. Since different interests and values are reflected in ES, a systematic stakeholder involvement could provide more insight on target levels for each conservation feature. In a future study, sensitivity analyses could be run based on integrated consultation of forest owners. Because Marxan is a regional level policy-support tool its suitability to be used for conservation planning at the property level is restricted. For example, once priority areas have been identified in a regional planning exercise, local authorities in collaboration with the local forest association try to reach agreement with several adjacent property owners [32]. The conservation outcome is the result of multiple negotiations to achieve a single voluntary nature reserve, the final spatial configuration of which does not depend on the result of a near-optimal site prioritisation software. However, Marxan with Zones could be run iteratively on different agreement configurations to show how marginal conservation burden and target achievement are shifted to other locations, for instance when particular forest owners have declined to agree with an area which would in the first place have been prioritised. Scenario analyses in Marxan with Zones could help planners evaluate the cost-effectiveness of local level conservation decisions, in light of the portfolio of other options, instead of negotiating about one or a few sites at a time.

Another uncertainty in conservation planning lies in the underlying opportunity costs [69]. While we did not test this uncertainty in our analysis, we point out that the advent of forest harvesting for bioenergy could be a 'game changer' as it would probably change expected returns to forestry and thus change the spatial distribution of opportunity costs.

Partial use areas, where extractive resource exploitation is restricted, can host high levels of biodiversity [17,18,26,55] and integrating such areas in conservation networks may improve overall conservation effectiveness by reducing costs and conflicts between different economic activities [53]. A combination of non-use and partial-use areas may also help to maintain a landscape that enables processes such as colonization and forest succession, particularly if non-use areas are small. The determination of

Table 3. Absolute and relative conservation burden per municipality in the current situation, with a cost constraint of 60% and with no cost constraint.

		Total opportunity costs[1] (million NOK)			Relative opportunity costs (NOK per km² forest area)					Ranks relative opportunity costs (NOK/km²) (largest to smallest)			
Municipality	**Forest area in planning units (km²)**	**Current**	**60% cost con-straint**	**No cost con-straint**	**Current**	**60% cost con-straint**	**No cost con-straint**	**Total addi-tional burden[2] (million NOK)**	**Relative additional burden[2] (NOK/km²)**	**Current**	**60% cost con-straint**	**No cost con-straint**	**Addi-tional burden**
Porsgrunn	175.5	3.2	30.0	60.0	18,457	170,677	341,874	56.8	323,417	6	4	1	1
Bamble	318.8	13.4	110.9	105.0	42,011	347,859	329,518	91.6	287,507	3	3	2	3
Notodden	818.8	14.7	39.0	254.3	17,945	47,655	310,558	239.6	292,613	7	15	3	2
Sauherad	316.5	13.1	52.3	95.3	41,404	165,259	301,208	82.2	259,804	4	5	4	4
Kragerø	341.8	5.8	15.3	88.4	16,979	44,866	258,777	82.6	241,797	8	16	5	5
Nome	412.8	54.2	150.9	105.7	131,320	365,660	256,155	51.5	124,835	1	2	6	10
Drangedal	1050.8	26.2	63.4	265.6	24,970	60,353	252,817	239.4	227,846	5	12	7	7
Bø	239.3	1.5	25.3	56.8	6,122	105,791	237,347	55.3	231,225	14	6	8	6
Skien	582.5	7.0	54.8	138.1	11,996	94,157	237,166	131.2	225,169	11	7	9	8
Siljan	130.5	9.7	57.6	22.2	74,231	441,457	169,989	12.5	95,758	2	1	10	13
Nissedal	855.3	12.5	57.5	110.6	14,630	67,191	129,361	98.1	114,731	9	11	11	11
Tokke	712.0	0.4	57.5	89.6	527	80,752	125,781	89.2	125,255	18	9	12	9
Kviteseid	662.8	0.9	56.6	72.7	1,433	85,374	109,691	71.7	108,258	17	8	13	12
Tinn	880.0	10.4	44.2	91.9	11852	50,223	104,416	81.5	92,564	12	14	14	15
Fyresdal	1147.5	5.0	31.2	113.7	4336	27,190	99,098	108.7	94,762	15	18	15	14
Hjartdal	649.8	5.6	36.2	57.0	8584	55,755	87,702	51.4	79,118	13	13	16	16
Seljord	577.3	1.4	25.2	47.0	2442	43,680	81,465	45.6	79,023	16	17	17	17
Vinje	939.8	12.2	64.4	68.9	13025	68,492	73,306	56.6	60,281	10	10	18	18

[1]Calculated as foregone net stumpage value.
[2]Calculated as the difference between opportunity costs for the case of no cost constraint and the current opportunity costs.

effectiveness of zones to achieve a conservation target has been identified as a major challenge for conservation planning given limited availability of knowledge [34,70]. For the sake of simplicity, we assumed a 100% effectiveness to protect biodiversity and ES for the non-use zone, given that this is the highest level of protection that can be achieved. We acknowledge, however, that considering a lower effectiveness level would most probably have led to a larger network of protected areas. In face of natural dynamics and disturbances, effectiveness of conservation areas should be monitored in terms of representativeness and persistence [66,71]. Because of the uncertainty about the probability of biodiversity persistence in the partial use zone, we explored the consequences of changing the zone contribution for the partial use zone as input in Marxan for 46 biodiversity features (Figure A1 in File S1). With a lower zone contribution, Marxan with Zones tended to select more planning units in the non-use and less in the partial use zone despite considerably lower opportunity costs of the partial use zone; a result that is in line with the findings by Makino et al. [53] in a study of partial protection zones in a marine environment in Fiji.

Assessing regional level implications of site prioritisation for ES and biodiversity: conservation burden

Decision-making about cost-effective area allocation to protect biodiversity and ES takes place at various levels of governance that may justify the design of new policy instruments. Cost-effective selection of priority sites for conservation can guide measures directed to land owners, for instance by consultation with land owners of selected priority sites on whether they would agree to convert forestry land into voluntary nature reserves, as is the current practice in Norway [32]. While land owners voluntarily entering conservation agreements in Norway are generally compensated for their private opportunity cost [32] accumulated loss of forestry activity in a region may, on the one hand, result in unequal public conservation burdens, particularly across different municipalities. Large protected areas may lead to foregone business opportunities, loss of tax income and additional expenses for municipal governments. On the other hand, protected areas can also provide positive externalities to others, through tourism opportunities and protection of biodiversity more generally. Local governments can be compensated for costs of conservation by state-to-municipal "ecological fiscal transfers" [40], an instrument that has been implemented in Brazil and Portugal, and is currently being considered in several European countries [72]. Ecological fiscal transfers have mainly been based on compensation scaled by area. Proposals to scale ecological fiscal transfers using criteria reflecting the effectiveness of conservation in a municipality have generally been limited by the availability of spatially representative data on biodiversity. We have demonstrated how the creation of cost-effective policyscapes could be used to determine distributional effects of additional conservation efforts.

Conclusions

Marxan with Zones provides a spatially explicit way to include different types of ES and biodiversity conservation criteria to study a policyscape for cost-effective conservation. We have shown that, in the case of Telemark, including a number of ES shifts priority sites for conservation and increases the area of both a partial use and a non-use zone, compared to a situation where only biodiversity conservation criteria are considered. Conservation of a number of regulating and cultural ES leads to additional conservation efforts, in terms of higher opportunity costs and a larger area protected. We show how carbon sequestration can be viewed as a side-benefit of the protection of other ES and biodiversity in the context of the current Kyoto-based setting of national targets. This is opposite to current thinking about biodiversity as a hoped-for side-benefit of climate mitigation measures under REDD+. The current conservation situation in Telemark clearly prioritises timber production against the protection of biodiversity and ES, and relatively large marginal increases in conservation target achievement could be reached with modest additional investments in terms of compensation for foregone timber production. Our analysis also shows potential differences in conservation burden among municipalities in Telemark, opening the debate on policy instruments such as ecological fiscal transfers that support county-level cost-effective conservation through stimulation of local conservation efforts.

Although the integration of partial use areas into conservation could provide opportunities to increase cost-effectiveness in conservation, significant work is needed to document effectiveness of different levels of protection on particular conservation features. Despite the high level of uncertainty, a policy mix of conservation measures appears to have the potential to contribute to address the complexity of cost-effective conservation problems.

Conservation targets for many aspects of biodiversity and especially ES are currently absent. Conservation planning could be better operationalised with more knowledge on stakeholder preferences about the importance of ES as well as with more ecological knowledge on area size needed to preserve a biodiversity feature.

Our analysis should not be understood as a concrete regional management plan, but rather as an exploratory analysis to provide insights about the current forest conservation situation, about which conservation outcomes could be achieved at which opportunity costs levels. In practice, selection of protected areas is often based on other criteria and motives than cost-effective, comprehensive site prioritisation [61]. Decision makers could use the results of this study to encourage disproportional conservation efforts at local level that achieve cost-effective, near optimal solutions to a conservation problem of multiple biodiversity and ES features. For this to happen, decision makers have to decide to what extent additional information, such as mapping of ES, could be integrated into land-use planning [73]. We have shown how ES mapping, conservation benchmarking and distributional impact analysis using conservation planning tools could inform decision-making and support compensation of land owners' and local governments' conservation efforts.

Acknowledgments

We thank Lars Hein and Lucie Vermeulen for useful comments on an earlier draft. We thank the advisory board members of the POLICYMIX project as well as all participants of the expert workshop held in Oslo on 7 May 2014. We express our sincere gratitude to two reviewers whose constructive remarks helped to improve the paper.

Author Contributions

Conceived and designed the experiments: MS GMR DNB SB BN. Performed the experiments: MS SB. Analyzed the data: MS GMR DNB SB BN. Contributed reagents/materials/analysis tools: MS GMR DNB SB BN. Contributed to the writing of the manuscript: MS GMR DNB SB BN.

References

1. Haines-Young R, Potschin M (2010) Proposal for a Common International Classification of Ecosystem Goods and Services (CICES) for integrated environmental and economic accounting. New York, USA: European Environment Agency.
2. Carpenter SR, Mooney HA, Agard J, Capistrano D, DeFries RS, et al. (2009) Science for managing ecosystem services: Beyond the Millennium Ecosystem Assessment. Proc Natl Acad Sci U S A 106: 1305–1312.
3. Larigauderie A, Prieur-Richard A-H, Mace GM, Lonsdale M, Mooney HA, et al. (2012) Biodiversity and ecosystem services science for a sustainable planet: the DIVERSITAS vision for 2012–20. Curr Opin Environ Sustain 4: 101–105.
4. Schröter M, van der Zanden EH, van Oudenhoven APE, Remme RP, Serna-Chavez HM, et al. (2014) Ecosystem services as a contested concept: a synthesis of critique and counter-arguments. Conserv Lett: doi:10.1111/conl.12091.
5. Mace GM, Norris K, Fitter AH (2012) Biodiversity and ecosystem services: a multilayered relationship. Trends Ecol Evol 27: 19–26.
6. Faith DP (2012) Common ground for biodiversity and ecosystem services: The "partial protection" challenge [v1; ref status: indexed, http://f1000r.es/QPrmmt]. F1000Research 2012 1.
7. Reyers B, Polasky S, Tallis H, Mooney HA, Larigauderie A (2012) Finding Common Ground for Biodiversity and Ecosystem Services. Bioscience 62: 503–507.
8. Chan KMA, Hoshizaki L, Klinkenberg B (2011) Ecosystem services in conservation planning: Targeted benefits vs. co-benefits or costs? PLoS ONE 6: e24378.
9. Egoh BN, Paracchini ML, Zulian G, Schägner JP, Bidoglio G (2014) Exploring restoration options for habitats, species and ecosystem services in the European Union. J Appl Ecol 51: 899–908.
10. Egoh B, Rouget M, Reyers B, Knight AT, Cowling RM, et al. (2007) Integrating ecosystem services into conservation assessments: A review. Ecol Econ 63: 714–721.
11. Chan KMA, Shaw MR, Cameron DR, Underwood EC, Daily GC (2006) Conservation Planning for Ecosystem Services. PLoS Biol 4: e379.
12. Bennett EM, Peterson GD, Gordon LJ (2009) Understanding relationships among multiple ecosystem services. Ecol Lett 12: 1394–1404.
13. Faith D (in press) Ecosystem services can promote conservation over conversion and protect local biodiversity, but these local win-wins can be a regional disaster. Aust Zool: doi:10.7882/az.2014.7031.
14. Certain G, Skarpaas O, Bjerke J-W, Framstad E, Lindholm M, et al. (2011) The Nature Index: A General Framework for Synthesizing Knowledge on the State of Biodiversity. PLoS ONE 6: e18930.
15. Anderson BJ, Armsworth PR, Eigenbrod F, Thomas CD, Gillings S, et al. (2009) Spatial covariance between biodiversity and other ecosystem service priorities. J Appl Ecol 46: 888–896.
16. Chhatre A, Agrawal A (2009) Trade-offs and synergies between carbon storage and livelihood benefits from forest commons. Proceedings of the National Academy of Sciences 106: 17667–17670.
17. Pichancourt J-B, Firn J, Chadès I, Martin TG (2014) Growing biodiverse carbon-rich forests. Global Change Biology 20: 382–393.
18. Persha L, Agrawal A, Chhatre A (2011) Social and Ecological Synergy: Local Rulemaking, Forest Livelihoods, and Biodiversity Conservation. Science 331: 1606–1608.
19. Götmark F (2013) Habitat management alternatives for conservation forests in the temperate zone: Review, synthesis, and implications. For Ecol Manag 306: 292–307.
20. Lindenmayer DB, Franklin JF, Fischer J (2006) General management principles and a checklist of strategies to guide forest biodiversity conservation. Biol Conserv 131: 433–445.
21. Nordén B, Paltto H, Claesson C, Götmark F (2012) Partial cutting can enhance epiphyte conservation in temperate oak-rich forests. For Ecol Manag 270: 35–44.
22. Bengtsson J, Angelstam P, Elmqvist T, Emanuelsson U, Folke C, et al. (2003) Reserves, Resilience and Dynamic Landscapes. Ambio 32: 389–396.
23. Hanski I (2011) Habitat Loss, the Dynamics of Biodiversity, and a Perspective on Conservation. Ambio 40: 248–255.
24. Daily GC, Ceballos G, Pacheco J, Suzán G, Sánchez-Azofeifa A (2003) Countryside Biogeography of Neotropical Mammals: Conservation Opportunities in Agricultural Landscapes of Costa Rica. Conserv Biol 17: 1814–1826.
25. Hauer G, Cumming S, Schmiegelow F, Adamowicz W, Weber M, et al. (2010) Tradeoffs between forestry resource and conservation values under alternate policy regimes: A spatial analysis of the western Canadian boreal plains. Ecol Model 221: 2590–2603.
26. Eigenbrod F, Anderson BJ, Armsworth PR, Heinemeyer A, Jackson SF, et al. (2009) Ecosystem service benefits of contrasting conservation strategies in a human-dominated region. Proc R Soc Biol Sci Ser B 276: 2903–2911.
27. Nalle DJ, Montgomery CA, Arthur JL, Polasky S, Schumaker NH (2004) Modeling joint production of wildlife and timber. J Environ Econ Manage 48: 997–1017.
28. Murdoch W, Polasky S, Wilson KA, Possingham HP, Kareiva P, et al. (2007) Maximizing return on investment in conservation. Biol Conserv 139: 375–388.
29. Barton DN, Blumentrath S, Rusch GM (2013) Policyscape–A Spatially Explicit Evaluation of Voluntary Conservation in a Policy Mix for Biodiversity Conservation in Norway. Soc Nat Resour 26: 1185–1201.
30. Kålås JA, Viken Å, Henriksen S, Skjelseth S (2010) The 2010 Norwegian Red List for Species. Trondheim: Norwegian Biodiversity Information Centre Norway.
31. Framstad E, Økland B, Bendiksen E, Bakkestuen V, Blom H, et al. (2002) Evaluering av skogvernet i Norge. NINA fagrapport. Oslo: NINA. 146.
32. Skjeggedal T, Gundersen V, Harvold KA, Vistad OI (2010) Frivillig vern av skog - evaluering av arbeidsformen (Norwegian: voluntary forest conservation - an evaluation of the approach). Oslo: NIBR/NINA.
33. Watts ME, Ball IR, Stewart RS, Klein CJ, Wilson K, et al. (2009) Marxan with Zones: Software for optimal conservation based land- and sea-use zoning. Environ Model Softw 24: 1513–1521.
34. Reyers B, O'Farrell P, Nel J, Wilson K (2012) Expanding the conservation toolbox: conservation planning of multifunctional landscapes. Landsc Ecol 27: 1121–1134.
35. Egoh BN, Reyers B, Rouget M, Richardson DM (2011) Identifying priority areas for ecosystem service management in South African grasslands. J Environ Manag 92: 1642–1650.
36. Egoh BN, Reyers B, Carwardine J, Bode M, O'Farrell PJ, et al. (2010) Safeguarding Biodiversity and Ecosystem Services in the Little Karoo, South Africa. Conserv Biol 24: 1021–1030.
37. Izquierdo AE, Clark ML (2012) Spatial Analysis of Conservation Priorities Based on Ecosystem Services in the Atlantic Forest Region of Misiones, Argentina. Forests 3: 764–786.
38. Barton DN, Faith DP, Rusch GM, Acevedo H, Paniagua L, et al. (2009) Environmental service payments: Evaluating biodiversity conservation trade-offs and cost-efficiency in the Osa Conservation Area, Costa Rica. J Environ Manag 90: 901–911.
39. Rusch GM, Barton DN, Bernasconi P, Ramos-Bendaña Z, Pinto R (2013) Best practice guidelines for assessing effectiveness of instruments on biodiversity conservation and ecosystem services provision. POLICYMIX Technical Brief, Issue No 7. Oslo: NINA.
40. Ring I, May P, Loureiro W, Santos R, Antunes P, et al. (2011) Ecological Fiscal Transfers. In: I Ring and C Schröter-Schlaack, editors. Instrument Mixes for Biodiversity Policies POLICYMIX Report, Issue No 2/2011. Leipzig: Helmholtz Centre for Environmental Research - UFZ. 98–118.
41. SSB (2012) Statistisk årbok 2012. Oslo, Kongsvinger: SSB.
42. Meteorological Institute (2012) Monthly normal values. Oslo: Meteorological Institute.
43. Moen A (1999) National Atlas of Norway: Vegetation. Hønefoss: Norwegian Mapping Authority.
44. Schröter M, Barton DN, Remme RP, Hein L (2014) Accounting for capacity and flow of ecosystem services: A conceptual model and a case study for Telemark, Norway. Ecol Indic 36: 539–551.
45. NFLI (2010) Arealressurskart AR5. Ås: National Forest and Landscape Institute (NFLI, Skog og Landskap).
46. Margules CR, Sarkar S (2007) Systematic conservation planning. Cambridge [etc.]: Cambridge University Press. 270 p.

47. Framstad E, Blumentrath S, Erikstad L, Bakkestuen V (2012) Naturfaglig evaluering av norske verneområder: Verneområdenes funksjon som økologisk nettverk og toleranse for klimaendringer. NINA rapport 888. Oslo: NINA.
48. Opdam P, Steingröver E, van Rooij S (2006) Ecological networks: A spatial concept for multi-actor planning of sustainable landscapes. Landsc Urban Plann 75: 322–332.
49. Directorate for Nature Management (2007) Kartlegging av naturtyper - Verdisetting av biologisk mangfold. DN håndbok 13. Trondheim: Directorate for Nature Management.
50. Gjerde I, Baumann C (2002) Miljøregistrering i skog - biologisk mangfold. Ås: Norwegian Institute for Forest Research.
51. Norwegian Environmental Agency (2013) Naturbase. Trondheim: Norwegian Environmental Agency (Miljødirektoratet).
52. NFLI (2013) Miljøregistrering i skog (Register of important forest habitats), Telemark. Ås: National Forest and Landscape Institute (NFLI, Skog og Landskap).
53. Makino A, Klein CJ, Beger M, Jupiter SD, Possingham HP (2013) Incorporating Conservation Zone Effectiveness for Protecting Biodiversity in Marine Planning. PLoS ONE 8: e78986.
54. Faith D (1995) Biodiversity and regional sustainability analysis. Canberra: CSIRO.
55. Fisher B, Edwards DP, Larsen TH, Ansell FA, Hsu WW, et al. (2011) Cost-effective conservation: calculating biodiversity and logging trade-offs in Southeast Asia. Conserv Lett 4: 443–450.
56. Søgaard G, Eriksen R, Astrup R, Øyen B-H (2012) Effekter av ulike miljøhensyn på tilgjengelig skogareal og volum i norske skoger. Rapport fra Skog og Landskap 2/2012. Ås: National Forest and Landscape Institute (NFLI, Skog og Landskap).
57. Wilson KA, Possingham HP, Martin TG, Grantham HS (2010) Key Concepts. In: J. A Ardron, H. P Possingham and C. J Klein, editors. Marxan Good Practices Handbook, Version 2. Victoria, BC, Canada: Pacific Marine Analysis and Research Association. 18–23.
58. Landis JR, Koch GG (1977) The Measurement of Observer Agreement for Categorical Data. Biometrics 33: 159–174.
59. Chan KMA, Satterfield T, Goldstein J (2012) Rethinking ecosystem services to better address and navigate cultural values. Ecol Econ 74: 8–18.
60. Granhus A (2014) Miljøhensyn ved hogst og skogkultur. In: S Tomter and L. S Dalen, editors. Bærekraftig skogbruk i Norge. Ås: Norwegian Forest and Landscape Institute. 90–99.
61. Joppa LN, Pfaff A (2009) High and Far: Biases in the Location of Protected Areas. PLoS ONE 4: e8273.
62. Framstad E, Blindheim T, Erikstad L, Thingstad PG, Sloreid SE (2010) Naturfaglig evaluering av norske verneområder. NINA rapport 535. Oslo: NINA.
63. Venter O, Laurance WF, Iwamura T, Wilson KA, Fuller RA, et al. (2009) Harnessing Carbon Payments to Protect Biodiversity. Science 326: 1368.
64. Sarkar S, Margules C (2002) Operationalizing biodiversity for conservation planning. J Biosci 27: 299–308.
65. Carwardine J, Klein CJ, Wilson KA, Pressey RL, Possingham HP (2009) Hitting the target and missing the point: target-based conservation planning in context. Conserv Lett 2: 4–11.
66. Margules CR, Pressey RL (2000) Systematic conservation planning. Nature 405: 243–253.
67. Margules CR, Pressey RL, Williams PH (2002) Representing biodiversity: Data and procedures for identifying priority areas for conservation. J Biosci 27: 309–326.
68. Luck GW, Chan KMA, Klein CJ (2012) Identifying spatial priorities for protecting ecosystem services [v1; ref status: indexed, http://f1000r.es/T0yHOY].F1000Research 2012.
69. Carwardine J, Wilson KA, Hajkowicz SA, Smith RJ, Klein CJ, et al. (2010) Conservation Planning when Costs Are Uncertain. Conserv Biol 24: 1529–1537.
70. Chape S, Harrison J, Spalding M, Lysenko I (2005) Measuring the extent and effectiveness of protected areas as an indicator for meeting global biodiversity targets. Philos Trans R Soc Lond B Biol Sci 360: 443–455.
71. Gaston KJ, Charman K, Jackson SF, Armsworth PR, Bonn A, et al. (2006) The ecological effectiveness of protected areas: The United Kingdom. Biol Conserv 132: 76–87.
72. Schröter-Schlaack C, Ring I, Koellner T, Santos R, Antunes P, et al. (2014) Intergovernmental fiscal transfers to support local conservation action in Europe. The German Journal of Economic Geography 58: 98–114.
73. European Commission (2014) Mapping and Assessment of Ecosystems and their Services. Technical Report – 2014-080. Brussels.

21

Tree-Ring Stable Isotopes Reveal Twentieth-Century Increases in Water-Use Efficiency of *Fagus sylvatica* and *Nothofagus* spp. in Italian and Chilean Mountains

Roberto Tognetti[1,2]*, Fabio Lombardi[1], Bruno Lasserre[1], Paolo Cherubini[3], Marco Marchetti[1]

1 Dipartimento di Bioscienze e Territorio, Università degli Studi del Molise, Pesche, Italy, **2** The EFI Project Centre on Mountain Forests (MOUNTFOR), Edmund Mach Foundation, San Michele all'Adige, Italy, **3** WSL Swiss Federal Institute for Forest, Snow and Landscape Research, Birmensdorf, Switzerland

Abstract

Changes in intrinsic water use efficiency (iWUE) were investigated in *Fagus sylvatica* and *Nothofagus* spp. over the last century. We combined dendrochronological methods with dual-isotope analysis to investigate whether atmospheric changes enhanced iWUE of *Fagus* and *Nothofagus* and tree growth (basal area increment, BAI) along latitudinal gradients in Italy and Chile. Post-maturation phases of the trees presented different patterns in $\delta^{13}C$, $\Delta^{13}C$, $\delta^{18}O$, Ci (internal CO_2 concentration), iWUE, and BAI. A continuous enhancement in isotope-derived iWUE was observed throughout the twentieth century, which was common to all sites and related to changes in Ca (ambient CO_2 concentration) and secondarily to increases in temperature. In contrast to other studies, we observed a general increasing trend of BAI, with the exception of *F. sylvatica* in Aspromonte. Both iWUE and BAI were uncoupled with the estimated drought index, which is in agreement with the absence of enduring decline in tree growth. In general, $\delta^{13}C$ and $\delta^{18}O$ showed a weak relationship, suggesting the major influence of photosynthetic rate on Ci and $\delta^{13}C$, and the minor contribution of the regulation of stomatal conductance to iWUE. The substantial warming observed during the twentieth century did not result in a clear pattern of increased drought stress along these latitudinal transects, because of the variability in temporal trends of precipitation and in specific responses of populations.

Editor: Sylvain Delzon, INRA - University of Bordeaux, France

Funding: Financial support was provided by the project "Ciclo del carbonio ed altri gas serra in ecosistemi forestali, naturali ed artificiali dell'America Latina: analisi preliminare, studio di fattibilità e comparazione con ecosistemi italiani" (Internationalization 2008, MIUR, Italy) and "Sviluppo di modelli innovativi per il monitoraggio multiscala degli indicatori di servizi ecosistemici nelle foreste Mediterranee" (FIRB 2012, MIUR, Italy) - http://www.istruzione.it/. The funders had no role in study design, data collection and analysis, decision to publish, or preparation of the manuscript.

Competing Interests: The authors have declared that no competing interests exist.

* Email: tognetti@unimol.it

Introduction

In the mid-latitudes, the emerging picture of enhanced evapotranspiration (ET) highlights the possible threat posed by increasing drought frequency to managing water resources in a greenhouse-affected climate [1]. Extreme drought summers in Europe, throughout the twentieth century [2], and recent drying trends are qualitatively consistent with predictions for the coming decades [3]. Extensive tree mortality triggered by dry and hot climatic conditions has been documented for most biomes of both Hemispheres over the past two decades [4]. Multiple mechanisms (hydraulic failure, carbohydrate depletion and insect attack) are involved in drought-induced forest decline [5], which occurs not only in arid regions but also in wet forests not normally considered at drought risk [6] [7] [8]. A decrease in tree growth, in the event of prolonged drought periods might reduce the future economic value of currently productive forests [9], which requires the study of tree physiological adaptations [10] [11].

Variation in intrinsic water-use efficiency (iWUE) estimated as the ratio between photosynthetic rate and stomatal conductance is recorded in the variation of the carbon isotope discrimination ($\Delta^{13}C$) of the annual growth rings that are laid down during each growing season [12]. Elevated atmospheric CO_2 concentration (Ca) is expected to reduce stomatal openings and increase assimilation rates [13] [14], which can alter the Ca to intercellular CO_2 concentration (Ci) gradient and result in increased iWUE in the long-term, which will be recorded in the variation of this $\Delta^{13}C$. However, the $\Delta^{13}C$ in tree rings also responds strongly to other environmental variables, especially climatic ones, such as growing season temperature, relative humidity and precipitation [15]. In either case, along with increases in iWUE (detected through changes in $\Delta^{13}C$) an increasingly faster tree growth (measured by tree-ring width converted into basal area increment, BAI) could also be observed [16]. Nevertheless, an increase in iWUE alone is not directly translated into higher BAI, since other factors, including high temperature, recurrent drought, nutrient limitation and/or plant acclimation, may preclude tree growth. Conversely, if BAI declines while iWUE increases, any photosynthesis advantage conferred by higher CO_2 concentrations should not offset warming-induced stress [17]. The extent that rising Ca has affected long-term iWUE, and whether climate could explain deviations from expected Ca-induced growth enhancement, are still poorly understood [18].

[19] found that climate change towards more arid conditions accounted for higher iWUE in *Fagus sylvatica* L. forests more than the continuous increase in Ca. Drought-induced decline in the productivity of beech forests in recent decades was observed by [20] in the Apennines (Italy). Nevertheless, different bioclimatic regions can have opposite responses to global climatic change, and BAI of old trees showed an increasing trend over time in the Alps [21]. Dendrochronological analysis on beech stands in Switzerland evidenced that trees near their dry distribution limit are adapted to extreme conditions already [22]. Unusual decreases in tree growth over the last 50 years were observed in dry-mesic Patagonia [23]. In Central Andes, *Nothofagus pumilio* (Poepp. & Endl.) Krasser showed reduced radial growth due to high temperatures in spring and summer, which enhanced ET [24]. Contrasting precipitation and temperature patterns have been reported for Northern and Southern Patagonia. [25] observed that, in *N. pumilio*, the relationship between iWUE and climate was more obvious in sites with good water availability, while in drier sites the photosynthetic rate was severely limited by water deficits so that the reduction in radial growth was not compensated by the increase in iWUE.

Tissue $\delta^{13}C$ provides an integrative record of supply and demand for CO_2 [26]. Variation in $\delta^{13}C$ may be driven by changes in stomatal conductance (i.e., supply of CO_2), or in photosynthetic rate (i.e., demand for CO_2), or both. Stomatal closure, because of restrictions in water availability, generally reduces Ci, leading to an increase in $\delta^{13}C$. Because light limitation of photosynthesis increases Ci, $\delta^{13}C$ can also depend on radiation [27]. Therefore, the dependence of $\delta^{13}C$ on the ratio Ci/Ca provides limited information about the strength of stomatal control of Ci and photosynthesis, since a change in Ci (inferred from $\delta^{13}C$) could be the result of a change in either stomatal conductance or photosynthesis. Tissue $\delta^{18}O$ is not strongly influenced by photosynthetic rate, so that combined measurements of $\delta^{13}C$ and $\delta^{18}O$ should allow stomatal and photosynthetic effects on $\delta^{13}C$ to be teased apart [28] [29] [30]. The relative responses of both $\delta^{13}C$ and $\delta^{18}O$ can be related to the sensitivity of a plant to changing evaporative conditions [31]. In a recent study, [32] have assessed soil drought exposure in forest stands by calculating the stand-level increase in $\Delta^{13}C$ of late wood from wet to dry years, which exhibited a negative linear relationship with tree species diversity in two forest types (including temperate beech), but not in others (as mountainous beech).

In this study, we used tree-ring $\delta^{13}C$ and $\delta^{18}O$ to investigate iWUE and tree growth changes during the twentieth century in *F. sylvatica* in Italy and *Nothofagus* spp. [*N. antartica* (Forster) Oerst., *N. betuloides* (Mirb.) Oerst., *N. dombeyi* (Mirb.) Oerst., and *N. pumilio*] in Chile. All the sampled stands were naturally occurring uneven-aged forests in national parks or protected areas that had been left unmanaged for many years, which may serve as reference for tree responses to natural disturbances. Along these two North-to-South transects, there is a substantial geographic gradient, averaged temperature and ET decreasing with increasing latitude. It was hypothesized that these *Fagus* and *Nothofagus* spp. would respond to climate by shifting from a more hydric dependence at lower latitudes to a more temperature dependence at higher latitudes. In particular, we aimed to relate the changes in $\Delta^{13}C$, $\delta^{13}C$ and $\delta^{18}O$, and iWUE with tree growth patterns, and to test whether the beneficial effect of Ca fertilization on tree growth could compensate the negative impact of warming-drought on BAI.

Materials and Methods

Geographic area

Sampling was accomplished along two latitudinal transects: eight forest stands (*F. sylvatica*) were sampled in Italy, and five forest stands (*Nothofagus* spp.) were sampled in Chile (Table 1). The age (years) of trees sampled for stable isotope analyses was the following (mean ± standard errors): North-Central Italy - Pian de Cansiglio 167±17, Sasso Fratino 187±33, Val Cervara 339±15, Montedimezzo 153±4; South Italy - Gargano 193±6, Cilento 158±20, Sila 152±3, Aspromonte 156±11; Chile - Chillan 137±14, Villarrica 230±16, Torres del Paine 144±24, Omora low 203±17, Omora high 178±16. Sampling was conducted in national parks or natural reserves, under the guidance of local authorities (park authority and/or forest service). Sampling did not involve endangered or protected species and no specific permissions were required for these locations/activities.

Italian sites

Pian del Cansiglio is an almost pure mature *F. sylvatica* high stand. The landscape morphology is gently sloping with valleys, and bedrock is mainly limestone and marlstone (Cretaceous).

Sasso Fratino is a stand that has been left to natural evolution for more than seventy years. *Fagus sylvatica* is the prevailing tree species with an average dominant height of about 40 m. Small groups and singles trees of *Abies alba* Mill. eventually occur.

Val Cervara is an old-growth forest located in an amphitheater shaped valley with Cretaceous limestone bedrock and soils that can be referred to the brown group. This almost pure *F. sylvatica* high stand escaped logging because of difficult access [33].

Montedimezzo is a forest that has been un-harvested since 1950. The geology is dominated by Cretacic limestone, with greensands and clay soils prevalent. The structure of this Apennine-Corsican montane beech forest is very diverse [34]. The *F. sylvatica*-dominated stand is associated with *Quercus cerris* L. at the lowest altitudes and sometimes mixed with other species, as *Acer pseudoplatanus* L., *A. campestre* L., *A. obtusatum* W. et K. and *Taxus baccata* L.

Gargano is an almost pure *F. sylvatica* stand with *Ilex aquifolium* L. in the understory, un-harvested since 1954. The geology is dominated by Cretacic limestone, with greensands and clay soils prevalent.

Cilento is a *F. sylvatica* high stand characterized by temperate climatic features (sub-continental and oceanic with mild and cool weather conditions) [35]. The hilly landscape is mainly constituted of stratified flysch formations of Tertiary age, which is complex and chaotic sandstone sedimentation.

Sila is an uneven-aged *F. sylvatica* high stand, mixed with *A. alba*, which has been un-harvested since 1960, and occupies humid locations of the Sila plateau, alternating with forests of *Pinus nigra* laricio (Poir.) Maire. The geology is dominated by Cretacic limestone, with greensands and clay soils prevalent.

Aspromonte is located on a calcareous massif. *Fagus sylvatica* forests cover the summit of the mountains, reaching almost 2000 m a.s.l., representing the Southernmost limit of the European beech forest area. This uneven-aged *F. sylvatica* high stand has been unmanaged since 1900; *A. alba* and *P. laricio* are also present.

Chilean sites

Chillan is a *N. dombeyi* forest stand in the Southern volcanic zone of the Andean Cordillera. This site refers to the phytogeographic transition zone between semi-arid, Mediterranean-type climate to the North and cool temperate rain forest climate to the

Table 1. Location, tree species and main characteristics of the sampling sites (and coordinates of the meteorological CRU TS 3.1 data set), as well as mean annual temperature (T) and total annual precipitation (P) obtained from nearby meteorological station (30-year mean).

Italy

ID	Administrative Region	Site name	Study area	Geographic coordinates		Species	Elevation (m a.s.l.)	Exposure (°N)	Slope (%)	Mean annual T (°C)	Total annual P (mm)	Stand age (years)	Grid point coordinates	
				North	East								North	East
1	Veneto	Pian del Cansiglio	Biogenetic Natural Reserve Campo di Mezzo – Pian Parrocchia	46°03′	12°23′	*F. sylvatica*	1200	120	5	5.6	1660	120	46°25′	12°25′
2	Emilia Romagna	Sasso Fratino	Foreste Casentinesi National Park	43°51′	11°44′	*F. sylvatica*	1550	45	40	9.0	1689	175	43°75′	11°75′
3	Abruzzo	Val Cervara	Abruzzo, Lazio and Molise National Park	41°49′	13°43′	*F. sylvatica*	1780	0	25	7.2	1211	350	41°35′	13°75′
4	Molise	Montedimezzo	Unesco Reserve Collemeluccio-Montedimezzo	41°46′	14°12′	*F. sylvatica*	1100	40	15	8.6	1022	180	41°75′	14°25′
5	Puglia	Gargano	Gargano National Park	41°45′	16°00′	*F. sylvatica*	775	355	40	11.4	800	190	41°75′	16°25′
6	Campania	Cilento	Cilento National Park	40°28′	15°24′	*F. sylvatica*	1290	340	15	7.1	1600	190	40°25′	15°25′
7	Calabria	Sila	Sila National Park	39°08′	16°40′	*F. sylvatica*	1680	225	20	8.9	1550	150	39°25′	16°75′
8	Calabria	Aspromonte	Aspromonte National Park	38°11′	15°52′	*F. sylvatica*	1560	120	15	10.2	1350	180	38°25′	15°75′

Chile

ID	Administrative region	Site name	Study area	Geographic coordinates		Species	Elevation (m a.s.l.)	Exposure (°N)	Slope (%)	Mean annual T (°C)	Total annual P (mm)	Stand age (years)	Grid point coordinates	
				South	West								South	West
9	Magallanes (Provincia de Antártica Chilena)	Omora "low"	Omora Ethnobotanical Park	54°56′	67°38′	*N. pumilio*	120	10	**5**	5.6	467	200	54°75′	67°25′
10	Magallanes (Provincia de Antártica Chilena)	Omora "high"	Omora Ethnobotanical Park	54°55′	67°37′	*N. antarctica*	600	15	40	4.8	540	215	54°75′	67°25′
11	Magallanes	Torres del Paine	Torres del Paine National Park	51°06′	73°08′	*N. pumilio*	600	40	5	7.5	730	130	51°25′	72°75′
12	Araucanía	Villarrica	Villarrica National Park	39°22′	71°58′	*N. betuloides*	1250	340	20	5.8	2200	310	39°25′	71°75′
13	Bio Bio	Chillan	Thermas de Chillan	36°55′	71°26′	*N. dombeyi*	1400	310	10	9.1	1900	340	36°75′	71°25′

Populations are listed from North to South (Northern Hemisphere) or South to North (Southern Hemisphere).

South, a tension zone between warm and cool temperate *Nothofagus* forests.

Villarrica is a coigüe (*N. betuloides*) forest stand on the South side of Villarrica volcano. The region is characterized by a west-coast maritime climate with a mild Mediterranean influence. Throughout the region, most of the soils are derived from recently deposited volcanic ashes that overlie Pleistocene glacial topography.

Torres del Paine is a *N. pumilio* high forest. Much of the geology of the Paine Massif area consists of Cretaceous sedimentary rocks that have been intruded by a Miocene-aged laccolith. Orogenic and erosional processes have shaped the topography, and glacial erosion is mainly responsible for the sculpturing of the massif.

Omora "low" and Omora "high" are old-growth forest stands on the Navarino Island, about 4 km away from each other. These rainy sub-Antarctic forests are the planet's Southern-most forest ecosystems [36]. Omora "low" is characterized by deciduous Magellanic forest, where lenga beech (*N. pumilio*) is the dominant tree species. Omora "high" is a pure ñirre beech (*N. antarctica*) stand. The climate is oceanic, while soils are composed of medium sized gravel, with a sandy-clayey matrix, showing no internal stratification.

Meteorological data

Observed climatic data near these sites cover only a short climatic period, and the record is spatially fragmented. For this reason, we used interpolated climatic data. The meteorological data used was from the CRU TS 3.1 0.5° gridded product [37]. Monthly minimum, mean and maximum temperatures, as well as total monthly precipitation from 1901 to 2009 were used. The coordinates of the closest grid point selected at each site are described in Table 1.

To assess water availability for tree growth, a monthly drought index (DI = P−ETo) based on precipitation sums (P) and the potential evapotranspiration (ETo), following [38], was calculated.

Tree sampling

At each site, 16÷20 dominant trees were sampled, and two increment cores at breast height were collected from each tree, with large-diameter borer. Four dominant trees per site were selected and cores extracted for analyses of stable isotopes. To avoid the effect of any wood alteration and of exogenous disturbances on ring growth, only trees without abrasion scars or other visible evidence of injury were selected.

Ring-width chronologies

Ring widths were measured with a resolution of 0.01 mm on each of the cores, using LINTAB measurement equipment (Frank Rinn, Heidelberg, Germany) fitted with a Leica MS5 stereoscope (Leica Microsystems, Germany) and analysed with the TSAP software package. Crossdating of all the tree-ring data was verified using the Program COFECHA, which assesses the quality of crossdating and measurement accuracy of tree-ring series using the segmented time-series correlation technique [39]. We only used the successfully crossdated cores, having a significant *Gleichläufigkeit*. This is a statistical measure of the year-to-year agreement between the interval trends of the chronologies, based on the sign of agreement [40] and Student's t test, which determines the degree of correlation between curves. With an overlap of 50 years (which is commonly used in tree-ring studies), GLK becomes significant ($p<0.05$) at 62% and highly significant ($p<0.01$) at 67%. In this study, the analysed time series were always longer than 50 years and cross-dating was considered successful if GLK was higher than 62%. The statistical significance of the GLK (GSL) was also computed. The TVBP is a Student's t-value commonly used as a statistical tool for comparing and cross-dating ring widths series. It determines the degree of correlation between curves. This method eliminates low-frequency variations in the time series, as each value is divided by the corresponding 5-year moving average. Each tree's ring-width record was standardized and then averaged with the other trees' records to obtain the mean standard chronology for the study site [41]. The standardization procedure involves fitting the observed ring-width series to a curve or a straight line and computing an index of the observed ring widths divided by the expected value. This reduces the variance between the cores and transforms the ring widths into dimensionless index values, and computes the tree-ring chronologies' average that is not influenced by the fastest growing trees, which have large ring widths. The ARSTAN program was used to process the tree-ring data into final chronologies [42]. To preserve the long-term fluctuations in the series, a conservative method of detrending was used. Only the negative exponential curves and/or linear regressions with negative or zero slopes were used.

Ring width was converted into tree basal area increment (BAI) according to the following standard formula:

$$BAI = \pi(r_n^2 - r_{(n-1)}^2)$$

where r is the radius of the tree and n is the year of tree-ring formation. To examine the mean growth trend of the canopy trees, BAI for each year was averaged over all individuals, to remove variation in radial growth attributable to increasing circumference, and averaged on a 10-year period basis, starting from 1900 (when BAI of the 20-year period was ≥10 cm^2, except for Omora sites). We worked with mean not-standardized BAI values across all trees for each year to preserve long-term cumulative effects of climatic events on tree growth.

Isotope analysis

The pooled rings of four trees per site (10-years blocks) were examined under a microscope to ensure that no false rings were included (one core from each of the four trees with no evident growth disturbance was selected for isotope analysis). The blocks were cut using a thin sharp blade. The samples were ground with a centrifugal mill (ZM 1000, Retsch, Retsch Technology, Haan, Germany) using a mesh size of 0.5 mm to assure homogeneity [43]. Cellulose was extracted from all samples to avoid isotope variations that are purely based on the changes in the relative abundance of individual wood constituents, differing typically in their isotope signatures. The method for cellulose extraction (modified from [44]) was based on a double-step digestion: the first step consisted of the treatment of wood with a solution of 5% NaOH for 2 h at 60°C, repeated twice, to remove lipids, resins, oil, tannins and hemicelluloses. In the second step, samples were washed with a solution of 7% $NaClO_2$ and 3–4 mL of acetic acid (pH 4) for a minimum of 36 h at 60°C. Because the solution is only reacting for 10 h, it was changed daily and refilled as necessary [45]. During this stage lignin was digested. Finally, samples were washed three to four times with boiling distilled water and dried overnight at 50°C. The $\delta^{13}C$ values were measured by combustion and $\delta^{18}O$ by pyrolysis of the cellulose samples in an elemental analyzer (Carlo Erba 1110, Milano, Italy) interfaced via a Conflo II Interface (Thermo Finnigan, Bremen, Germany) to a dual inlet/continuous flow isotope ratio mass spectrometer (Delta S, Thermo Finnigan, Bremen, Germany), in the continuous flow mode. Isotope ratio deviation results are presented in the common δ notation,

$$\delta = \frac{R_{sa}}{R_{ref}} - 1$$

with δ-values expressed in per mill (‰) on the international VPDB scale for carbon-13 and VSMOW scale for oxygen-18; R refers to the number ratio of ^{13}C to ^{12}C or ^{18}O to ^{16}O isotopes in the sample ('*sa*') and the reference ('*ref*'), respectively.

The standard deviation for the repeated analysis of an internal standard (commercial cellulose) was <0.1% for $\delta^{13}C$ and 0.3% for $\delta^{18}O$. The calibration was carried out by measurement of IAEA USGS-24 (graphite), IAEA CH7 (polyethylene) and IAEA CH3 (cellulose).

The raw $\delta^{13}C$ chronology exhibited a decline in $\delta^{13}C$ mainly in the twentieth century, which may be attributed to the lowering of $\delta^{13}C$ of air through anthropogenic-related increases in CO_2 concentration (the "^{13}C Suess effect"). We removed this trend in the carbon isotope chronology using the annual records of past atmospheric $\delta^{13}C$ obtained from ice cores [46].

We, thereafter, calculated $\Delta^{13}C$, Ci and iWUE. The $\Delta^{13}C$ (carbon isotopic discrimination) was calculated as

$$\Delta^{13}C = \frac{\Delta^{13}C_{aair} - \Delta^{13}C_{plant}}{1 + \frac{\Delta^{13}C_{plant}}{1000}}$$

following [26] and using published values for air $\delta^{13}C_{plant}$ from ice core measurements, direct atmospheric measurements and inferred from C4 plants [47]. Since $\Delta^{13}C$ is related to Ci (intercellular CO_2 concentration) and Ca (ambient CO_2 concentration) by the following equation:

$$\Delta^{13}C_{‰} = a + (b - a)\frac{\text{Ci}}{\text{Ca}}$$

where a (≈4.4‰) is the discrimination against $^{13}CO_2$ during CO_2 diffusion through stomata [48], b (≈27‰) is the discrimination associated with carboxylation [49], and given Fick's Law:

$$A = gCO_2(\text{Ca} - \text{Ci})$$

where A is the net photosynthesis, measured as CO_2 uptake, and gCO_2 is the leaf conductance to CO_2, and given that gH_2O, the leaf conductance to water vapour is 1.6 gCO_2, $\Delta^{13}C$ can be finally related to the ratio A/gH_2O (intrinsic water-use efficiency, iWUE) [50] by the following equation:

$$\Delta^{13}C = a + (b - a)(1 - 1.6\frac{A}{\text{Ca}gH_2O})$$

we determined the Ci, using [51]'s equation:

$$\text{Ci} = \text{Ca}\frac{\delta^{13}C_{plant} - \delta^{13}C_{\text{air}} + a}{a - b}$$

We calculated iWUE as:

$$\text{iWUE} = \text{Ca}\frac{b - \Delta^{13}C}{1.6(b - a)}$$

Data treatment and trend analysis

Regression analyses were used to identify significant temporal trends in $\delta^{13}C$, $\Delta^{13}C$, $\delta^{18}O$, Ci, iWUE and BAI of individual trees over the last century, when most trees were in their mature phase. ANCOVA was used to test for differences in past decade trends between mean mature trees $\delta^{13}C$, $\Delta^{13}C$, $\delta^{18}O$, Ci, iWUE and BAI at each site and the influence of Ca and climate. Differences between forest sites were tested by *post hoc* Bonferroni test. For correlation analyses of these variables with climate, expressed as average (10 years), annual Ca and index of aridity were calculated for the period beginning October (Italy) or April (Chile) of the previous year and ending with September (Italy) or March (Chile) of the current year. This procedure avoids relating growth to climate in the months after leaf fall when growth has ceased. Regression, ANCOVA, correlation and all the statistical calculations and analyses were conducted using the SPSS software package (SPSS Inc., Chicago, IL, USA).

Trend analysis of $\delta^{13}C$, $\Delta^{13}C$, $\delta^{18}O$, Ci, iWUE, BAI, and DI was investigated using the non-parametric Man-Kendall test [52] [53], which compares the relative magnitudes of the samples and provides information on whether the null hypothesis of no trend in the data can be rejected or not. The data values are evaluated as an ordered time series. Each data value is compared to all subsequent data values. The initial value of the Mann-Kendall statistic, S, is assumed to be zero (i.e., no trend; null hypothesis). If a data value from a later time period is higher than a data value from an earlier time period (i.e., a trend is detected; alternative hypothesis), S is incremented by one. On the other hand, if the data value from a later time period is lower than a data value sampled earlier, S is reduced by one. The net result of all such increments and decrements yields the final value of S. This test allows for investigating long-term trends of data without assuming any particular distribution. The rank correlation coefficient evaluates the degree of similarity between the ranks of two sets of measured data. The rank correlation provides a distribution free test of the independence and a measure of strength of dependence between two variables. The tau coefficient is defined as:

$$\tau = \frac{n_c - n_d}{\frac{1}{2}n(n-1)}$$

where n_c is the number of concordant pairs (i.e., pairs ranks with moving in the same direction), and n_d is the number of discordant pairs (i.e., pairs with ranks moving in different directions), in the data set. The approach is to count the number of concordant and discordant pairs between two ordered sets. This number gives a distance between sets called symmetric difference distance. The coefficient of correlation is obtained by normalizing the symmetric difference by the total number of pairs such that it will take values between −1 and +1. The tau coefficient has the following properties: 1) If the agreement between the two rankings is perfect (i.e., the two rankings are the same) the coefficient has value 1; and 2) If the disagreement between the two rankings is perfect (i.e., one ranking is the reverse of the other) the coefficient has value −1. For all other arrangements the values lies between −1 and 1, and increasing values imply increasing agreement between the rankings. If the rankings are completely independent, the coefficient has value 0 on average.

Results

Patterns of climate index and $\delta^{13}C$, $\Delta^{13}C$, $\delta^{18}O$, Ci, iWUE, and BAI

A downward trend of climate index at the end of last century, indicating increasingly drier conditions, was found for Val Cervara, Sasso Fratino and Montedimezzo (Fig. 1). However, the climate index was variable from year to year and below −25 mm only in several years (and only for Val Cervara, Gargano, Cilento and Chillan), indicating that trees were rarely subjected to severe and prolonged drought stress over the past century.

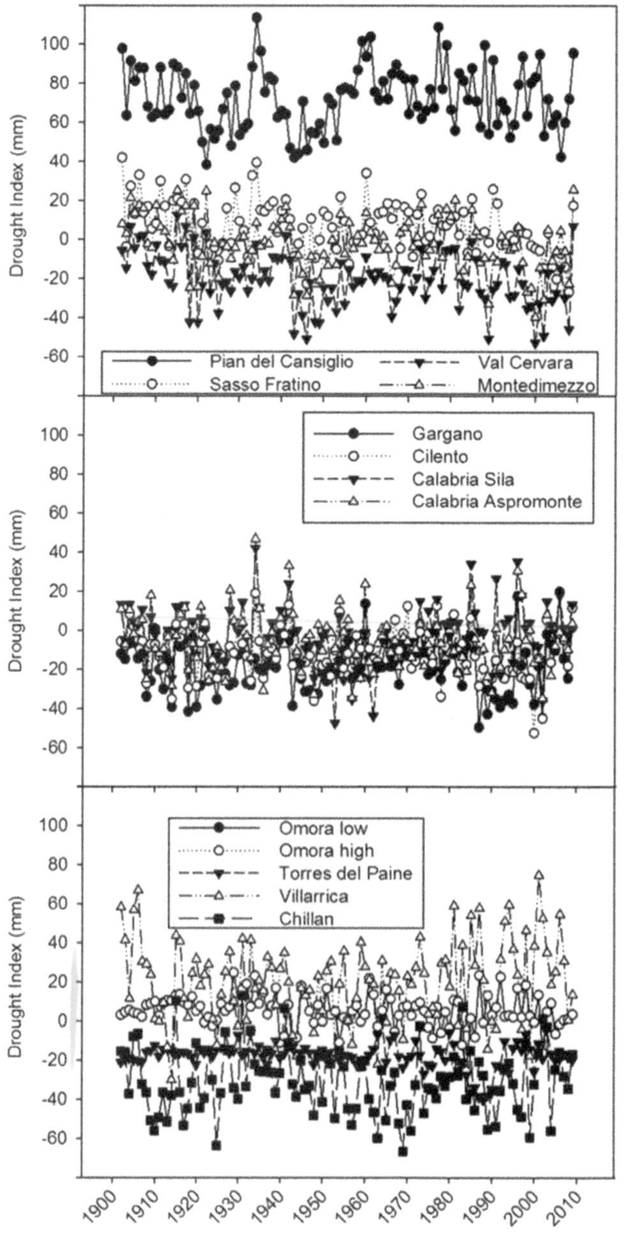

Figure 1. Temporal evolution of drought index (DI = P−ETo) calculated for the period beginning October (Italy) or April (Chile) of the previous year and ending with September (Italy) or March (Chile) of the current year, and used to select climatically contrasting years. It must be pointed out that the great heterogeneity of the local environment in these mountainous areas can contribute errors to estimated values of drought index.

In general terms, similar trends in $\delta^{13}C$, $\Delta^{13}C$, $\delta^{18}O$, Ci and iWUE were found throughout the chronologies corresponding to *F. sylvatica* and *Notophagus* spp. trees from the different forest sites (Fig. 2). Tree rings $\delta^{13}C$ on forest trees showed variable patterns (e.g., increasing at Torres del Paine, while decreasing at Omora high). When the atmospheric effect over plants carbon isotopic ratios is excluded, the discrimination ($\Delta^{13}C$) trends during the past decades showed the opposite (decreasing) trend. The more negative $\delta^{13}C$ values resulted in higher discrimination ($\Delta^{13}C$) and lower iWUE. Yet, $\delta^{18}O$ showed relatively stable patterns throughout the past century, with some increases in the last thirty years in several sites, but not in others.

iWUE increased throughout the 20th century, particularly in the second half, simultaneously with changes in Ca and Ci (Fig. 2). Ci ranged between 220 and 240 ppm in the forest until mid 20th century, reaching recently values as high as 270 ppm. This rise in Ci as well as in iWUE represented a significant trend, which followed the increase in Ca in the past decades.

The BAI curve of *F. sylvatica* and *Notophagus* spp. trees showed a continuous and steep increase in the first years of the 20th century in most of the sites, before levelling off to mature levels of growth, with signals of changes in growth patterns for trees of Cilento (Southern Italy), and Torres del Paine and Villarrica (Chile) (Fig. 2). The analysis was limited to those years (20th century) for which growth suppression (before the release phase) was considered negligible in most stands of these late successional species. There were, however, exceptions to this abrupt increasing trend; more specifically, in the oldest site (Val Cervara) and in Aspromonte in Southern Italy for *F. sylvatica*, and in the Southernmost sites (Omora high and Omora low) for *Notophagus* spp. BAI showed smoother patterns over time, with values between 10 and 20 cm^2 (Italian sites) or barely reaching 15 cm^2 (Chilean sites) The levelling off to low BAI values marks the beginning of late successional stages. Nevertheless, a slight increase in the most recent decade was common to most sites (except for Pian del Cansiglio and Omora low).

Table 2 reports the results of trend analysis for parameters of Fig. 1 and 2, based on the Mann–Kendall test. A significant decreasing trend of DI, indicating drier conditions, was observed for Sasso Fratino, Val Cervara, Montedimezzo, and Omora sites. In general, iWUE showed significant increasing trends with strong correlation (tau values close to 1); BAI behaved similarly, with the exception of Cilento (low tau values) and Aspromonte (negative *S* values). Correspondingly, $\Delta^{13}C$ showed marked decreasing trends. More uncertain were correlation coefficients for the other parameters.

The absolute averaged values for the 20th century evidenced a rather marked trend of increasing BAI with decreasing latitude in Chile, while higher variable occurred in Italy (Fig. 3) (ANCOVA, $P<0.01$); the old-growth forest of Val Cervara showed the lowest averaged values among the Italian sites and the highest BAI was observed in Pian del Cansiglio (Northernmost site) and Montedimezzo (ANCOVA, $P<0.01$). Pian del Cansiglio also showed the lowest averaged $\delta^{18}O$ values for the Italian site, while Omora high had the lowest $\delta^{18}O$ amongst all sites (ANCOVA, $P<0.01$). The latter site, which was the coldest one, showed the highest averaged $\delta^{13}C$ and $\Delta^{13}C$, in absolute terms, and Ci, and the lowest iWUE amongst all sites (ANCOVA, $P<0.01$).

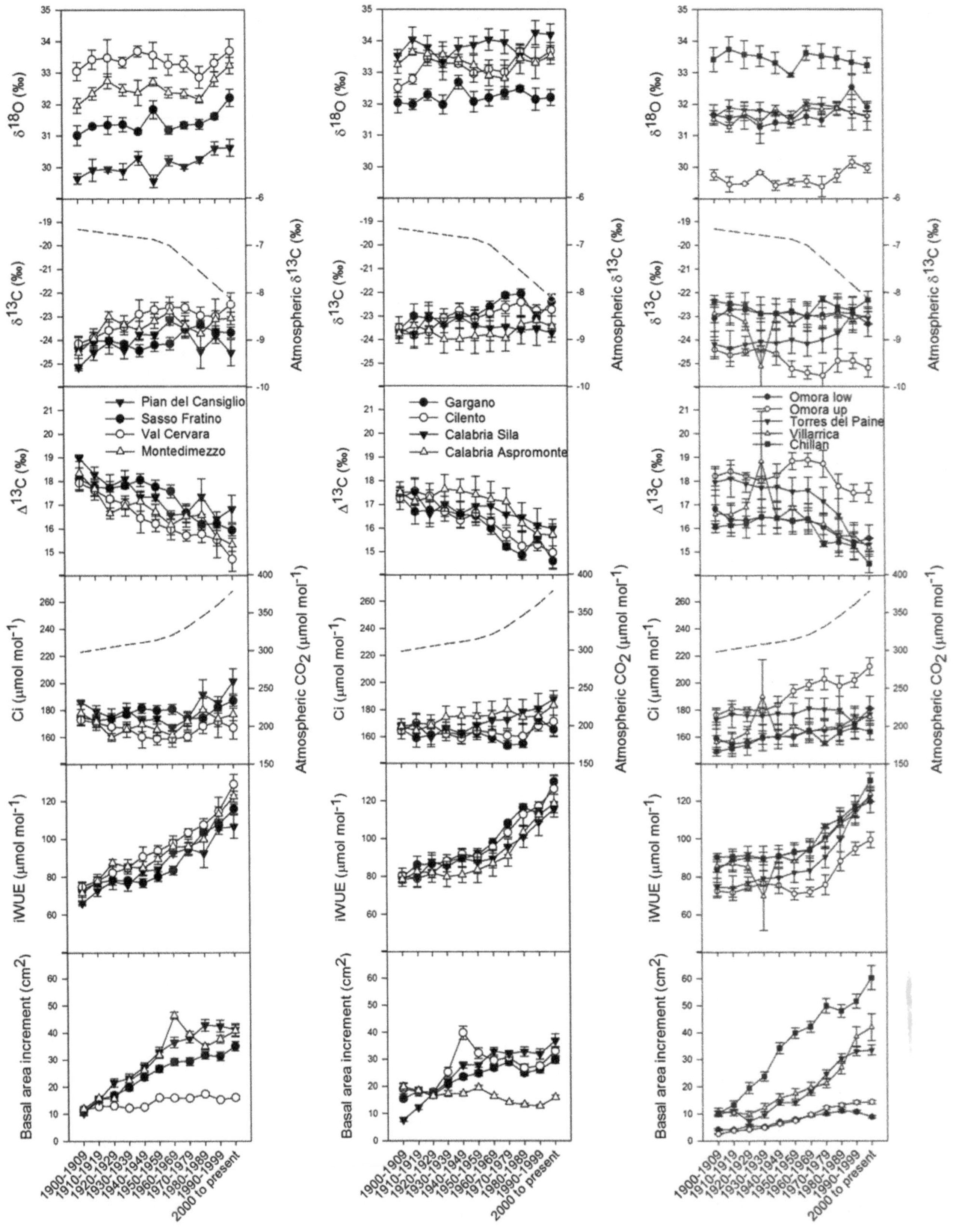

Figure 2. Variation in mean *Fagus sylvatica* and *Nothofagus* spp. $\delta^{13}C$, $\Delta^{13}C$, $\delta^{18}O$, Ci, iWUE, and BAI. Symbols with the error bars (±1 standard error, n = 4) are means of individuals per site in 10-years total values (±1 standard error n = 4). Temporal trends of Ca and atmospheric $\delta^{13}C$ are also reported (dashed lines). Sites are referred to by symbols in the legend: Pian del Cansiglio, Sasso Fratino, Val Cervara, and Montedimezzo (North-Central Italy); Gargano, Cilento, Aspromonte, and Sila (Southern Italy); Omora low, Omora high, Torres del Paine, Villarrica, and Chillan (Chile). Site and species details are reported in Table 1. Populations are listed from North to South (Northern Hemisphere) or South to North (Southern Hemisphere).

Table 2. Trend analysis results (based on the Mann–Kendall test) for parameters reported in Fig. 1 and 2, expressed as negative or positive *S* values, indicating decreasing or increasing trend, respectively.

		Pian del Cansiglio	Sasso Fratino	Val Cervara	Montedimezzo	Gargano	Cilento	Calabria Sila	Calabria Aspromonte	Omora low	Omora high	Torres del Paine	Villarrica	Chillan
DI	tau	0.02	−0.24	−0.21	−0.18	0.01	−0.09	0.01	−0.11	−0.14	−0.14	−0.11	0.06	−0.01
	S	92.00	−1390.00	−1242.00	−1028.00	72.00	−528.00	34.00	−610.00	−818.00	−818.00	−654.00	324.00	−54.00
	Var(S)	141882.00	141882.00	141882.00	141882.00	141882.00	141882.00	141882.00	141882.00	141882.00	141882.00	141882.00	141882.00	141882.00
	p-value (bilateral)	-	≤0.01	≤0.01	≤0.01	-	-	-	-	≤0.05	≤0.05	-	-	-
$\delta^{18}O$	tau	0.60	0.56	0.05	0.31	0.31	0.45	0.38	−0.27	0.35	0.27	−0.02	0.31	−0.35
	S	33.00	31.00	3.00	17.00	17.00	25.00	21.00	−15.00	19.00	15.00	−1.00	17.00	−19.00
	Var(S)	0.00	0.00	0.00	0.00	0.00	0.00	0.00	0.00	0.00	0.00	0.00	0.00	0.00
	p-value (bilateral)	≤0.01	≤0.05	-	-	-	-	-	-	-	-	-	-	-
$\delta^{13}C$	tau	0.24	0.42	0.64	0.35	0.49	0.67	−0.24	0.24	−0.27	−0.49	0.75	−0.05	−0.02
	S	13.00	23.00	35.00	19.00	27.00	37.00	−13.00	13.00	−15.00	−27.00	41.00	−3.00	−1.00
	Var(S)	0.00	0.00	0.00	0.00	0.00	0.00	0.00	0.00	0.00	0.00	0.00	0.00	0.00
	p-value (bilateral)	-	-	≤0.01	-	≤0.05	≤0.01	-	-	-	≤0.05	≤0.01	-	-
$\Delta^{13}C$	tau	−0.67	−0.71	−0.96	−0.75	−0.85	−0.85	−0.82	−0.64	−0.71	−0.20	−0.89	−0.67	−0.38
	S	−37.00	−39.00	−53.00	−41.00	−47.00	−47.00	−45.00	−35.00	−39.00	−11.00	−49.00	−37.00	−21.00
	Var(S)	0.00	0.00	0.00	0.00	0.00	0.00	0.00	0.00	0.00	0.00	0.00	0.00	0.00
	p-value (bilateral)	≤0.01	≤0.01	≤0.01	≤0.01	≤0.01	≤0.01	≤0.01	≤0.01	≤0.01	-	≤0.01	≤0.01	-
Ci	tau	0.09	0.49	−0.20	0.24	0.02	−0.05	0.71	0.64	0.89	0.78	0.20	0.60	0.75
	S	5.00	27.00	−11.00	13.00	1.00	−3.00	39.00	35.00	49.00	43.00	11.00	33.00	41.00
	Var(S)	0.00	0.00	0.00	0.00	0.00	0.00	0.00	0.00	0.00	0.00	0.00	0.00	0.00
	p-value (bilateral)	-	≤0.05	-	-	-	-	≤0.01	≤0.01	≤0.01	≤0.01	-	≤0.01	≤0.01
iWUE	tau	0.89	0.85	1.00	0.85	0.96	0.93	0.85	0.82	0.96	0.56	0.96	0.78	0.85
	S	49.00	47.00	55.00	47.00	53.00	51.00	47.00	45.00	53.00	31.00	53.00	43.00	47.00
	Var(S)	0.00	0.00	0.00	0.00	0.00	0.00	0.00	0.00	0.00	0.00	0.00	0.00	0.00
	p-value (bilateral)	≤0.01	≤0.01	≤0.01	≤0.01	≤0.01	≤0.01	≤0.01	≤0.01	≤0.01	≤0.05	≤0.01	≤0.01	≤0.01
BAI	tau	0.89	0.96	0.56	0.75	0.82	0.38	0.82	−0.67	0.75	1.00	0.82	0.89	0.96
	S	49.00	53.00	31.00	41.00	45.00	21.00	45.00	−37.00	41.00	55.00	45.00	49.00	53.00
	Var(S)	0.00	0.00	0.00	0.00	0.00	0.00	0.00	0.00	0.00	0.00	0.00	0.00	0.00
	p-value (bilateral)	≤0.01	≤0.01	≤0.05	≤0.01	≤0.01	-	≤0.01	≤0.01	≤0.01	≤0.01	≤0.01	≤0.01	≤0.01

The tau coefficient varies between −1 and 1, and values closer to unity imply increasing agreement between the rankings.

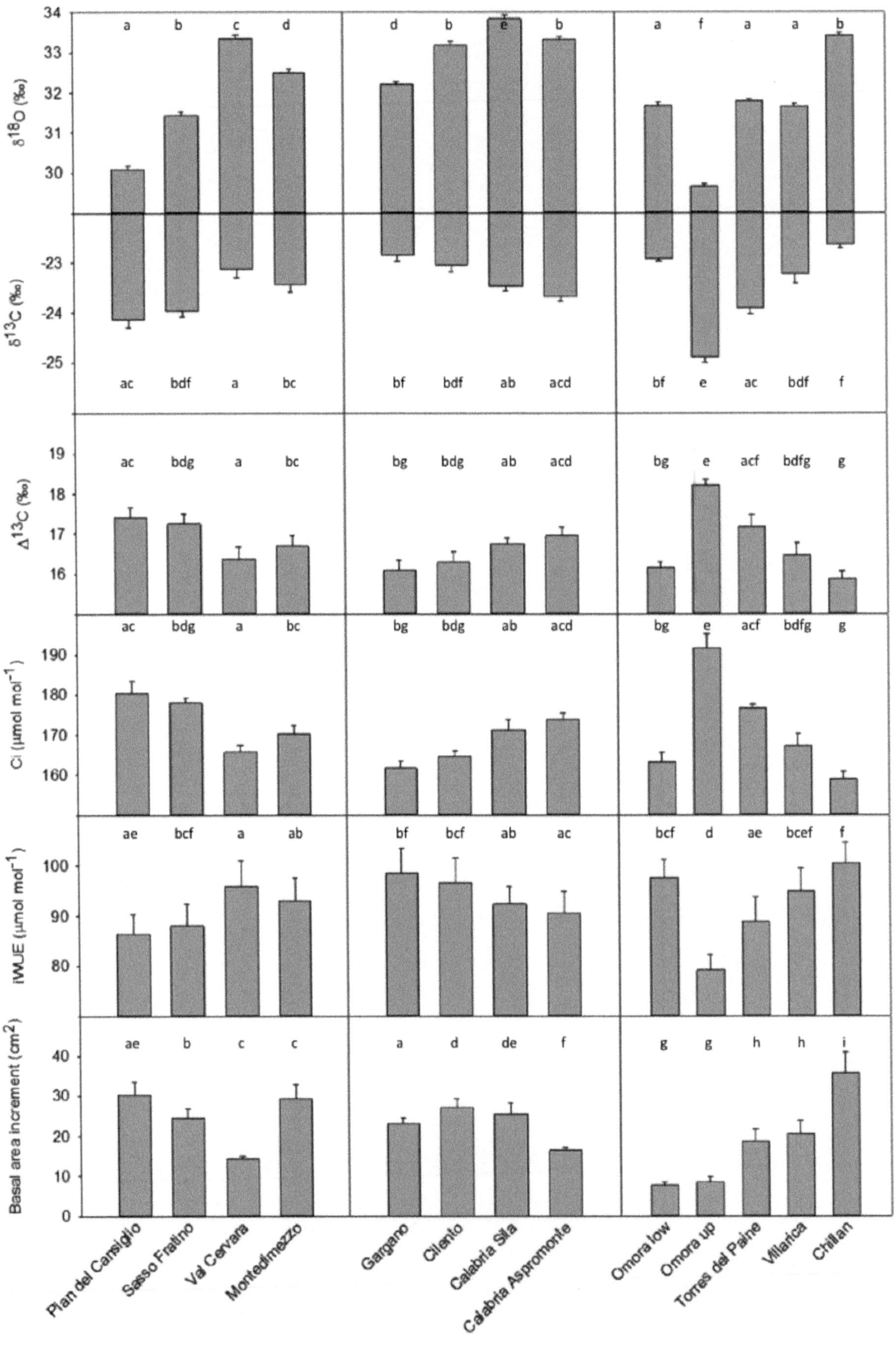
δ18O (‰)
δ13C (‰)
Δ13C (‰)
Ci (µmol mol-1)
iWUE (µmol mol-1)
Basal area increment (cm2)
Pian del Cansiglio
Sasso Fratino
Val Cervara
Montedimezzo
Gargano
Cilento
Calabria Sila
Calabria Aspromonte
Omora low
Omora up
Torres del Paine
Villarica
Chillan

Figure 3. Mean (+1 standard error) $\delta^{13}C$, $\Delta^{13}C$, $\delta^{18}O$, Ci, iWUE, and BAI of *Fagus sylvatica* and *Nothofagus* spp. during the 20th century. Different letters indicate significantly different values between sites ($P<0.05$; post hoc Boferroni test, after ANCOVA analyses). Populations are listed from North to South (Northern Hemisphere) or South to North (Southern Hemisphere).

Relations of iWUE, $\delta^{13}C$, $\delta^{18}O$ and BAI with climate index, atmospheric CO_2 concentration (and temperature)

The estimated iWUE was positively correlated with Ca, while the correlation with the drought index was significant and negative only for Sasso Fratino (Fig. 4). Correlations between iWUE and temperature were clear and positive for *Fagus*, while this was not the case for *Nothofagus* spp. (not shown). The calculated BAI was also positively correlated with Ca, though with varying slopes in the early decades of the 20th century (Fig. 4). Val Cervara exhibited the lowest slope among the populations of North-Central Italy. The Chilean species and populations showed a latitudinal gradient with decreasing slopes with increasing latitude. Even in the case of BAI, the relationship with the drought index was vague, with the sites clustered in some cases (North-Central Italy and Chile) and not in others (South Italy). Correlations between BAI and temperature were again positively clearer for *Fagus* than for *Nothofagus* spp., and particularly for the populations of North-Central Italy (not shown). This clustering pattern was maintained for the relationship between $\delta^{18}O$ and Ca or the drought index (and temperature), with rather shallow slopes (not shown). Similarly, the relationship between $\delta^{13}C$ and the drought index (or temperature) resulted in the separation among sites in the case of North-Central Italy and Chile, while this was not the case for South Italy (not shown). The relationship between $\delta^{13}C$ and Ca was unclear, somewhat following the pattern of iWUE. Populations of North-Central Italy showed an increase of BAI with iWUE, Val Cervara showing a lower slope in comparison with all others (Fig. 4); populations of South Italy also showed a similar increase of BAI with iWUE, though levelling off at higher iWUE values, and with the exception of Aspromonte, showing a negative relationship; also in the case of Chile, the Southernmost site (Omora) showed rather shallow relationships between BAI and iWUE, the other sites exhibiting positive correlations.

The regression coefficients for equations describing the relationship between variables are reported in Table 3.

Correlations between $\delta^{18}O$ and $\delta^{13}C$

Overall, $\delta^{18}O$ and $\delta^{13}C$ were positively correlated when considering pooled Chilean sites. Correlations varied between the two Italian clusters (North-Central and Southern sites), with Southern sites showing a reverse (negative) and weak relationship in comparison with North-Central sites (Fig. 5). At each site, $\delta^{18}O$ and $\delta^{13}C$ were poorly correlated (regressions not shown). Montedimezzo and Cilento had correlation coefficients (R^2) of 0.49 and 0.30 (positive correlation), respectively. Aspromonte, Gargano, Sassofratino and Villarrica had R^2 ranging from 0.21 to 0.14 (positive correlation); other populations showed rather shallow relationships ($R^2<0.1$). The Chilean sites (and species) were fairly distinguishable from each other, with the Northernmost site (Chillan, *N. dombeyi*) showing the highest (more positive) and Omora high (*N. antartica*) the lowest (more negative) $\delta^{18}O$ and $\delta^{13}C$ values, *N. betuloides* and *N. pumilio* showing intermediate values. The Italian sites were also rather clearly discernible.

Discussion

Temporal changes in iWUE and BAI

Over the twentieth century, the iWUE increased 35.3±1.4% in the sites of Southern Italy and 39.8±1.3% in those of North-Central Italy sites (Fig. 2, 3), which is somewhat in agreement with the 44% increase reported by [12] in Northeastern France and with 50% increase reported by [19] in Northeastern Spain. Similar increases in iWUE and decreases in $\Delta^{13}C$ have been found during the twentieth century by using historical records in tree-rings of other (temperate and boreal) Northern Hemisphere species [54] [17]; although this is not always the case [55]. Likewise, the iWUE increased 31.6±2.0% in the Chilean sites, and all these increases in iWUE are in agreement with previous reports for the Southern Hemisphere; e.g., [56] found 34% increase for Brazilian *Cedrela odorata* L. and 52% increase for *Swietenia macrophylla* King, while [57] found 30% increase in iWUE of mature *Araucania angustifolia* (Bertol.) during the second half of the past century.

The large inter-decadal variability of $\delta^{13}C$ may have hidden long-term trends of a regulation of either Ci or Ci/Ca (Fig. 2), although a spatially uneven increase of Ci/Ca during the second half of the twentieth century points to some patchy homoeostasis in iWUE along environmental gradients. Several studies have reported widespread increases in iWUE (and decreases of $\Delta^{13}C$) coinciding with rising Ca over the past century [12] [16]. Inter- and intra-specific differences in ^{13}C discrimination [58] [59] and variation across altitudinal and latitudinal gradients [60] [61] were not pondered in this study, which warrants caution in interpreting trends in iWUE as a coherent global response to rising Ca, because of the correlation with Ca and its use as an independent factor in the iWUE calculation [62]. The increase in iWUE across biomes without univocal effect on tree growth [62], indeed, raises the question of whether iWUE calculations reflect actual physiological responses to elevated Ca; hydraulic adjustment to increasing evaporative demand and nitrogen deposition affecting gas exchange may also have an impact on iWUE.

In contrast with other studies [63] [64], we found evidence of long-term increases in BAI (Fig. 2, 3), associated with increasing iWUE, likely because the ameliorating effects of increasing Ca on water stress. Global estimates of iWUE, inferred from the $\Delta^{13}C$ analyses of comparable mature trees, showed a 20.5% increase and an increase in Ca of over 50 μmol mol^{-1} during the last 40 years, with minor differences across biomes [16]. Several authors [65] [66] suggested that changes in temperature and Ca might have accelerated tree growth over recent decades in temperate climates, though this conclusion is controversial [16]. Although stem growth is the most sensitive component of NPP to drought, because it is low on the carbon allocation hierarchy [67], the documented warming [68] [69] could have exerted a larger control on stem growth than precipitation in these relatively humid forest stands [70]. Prolonged water limitation at the stand level, however, will act as major constraint on the adaptive capacity of drought-sensitive beech forests if warming continues [71] [24] [20]. Therefore, current growth trends may not translate fully into long-term increased tree density for *Fagus* and *Nothofagus* spp. [72] [73] [74]. There is already some evidence that increased variability in climate warming has had little effect on the recruitment of *Nothofagus* close to the tree line in the Southern Hemisphere [75] [76], which contrasts with the shift of beech populations in mountain ranges of the Northern Hemisphere [77].

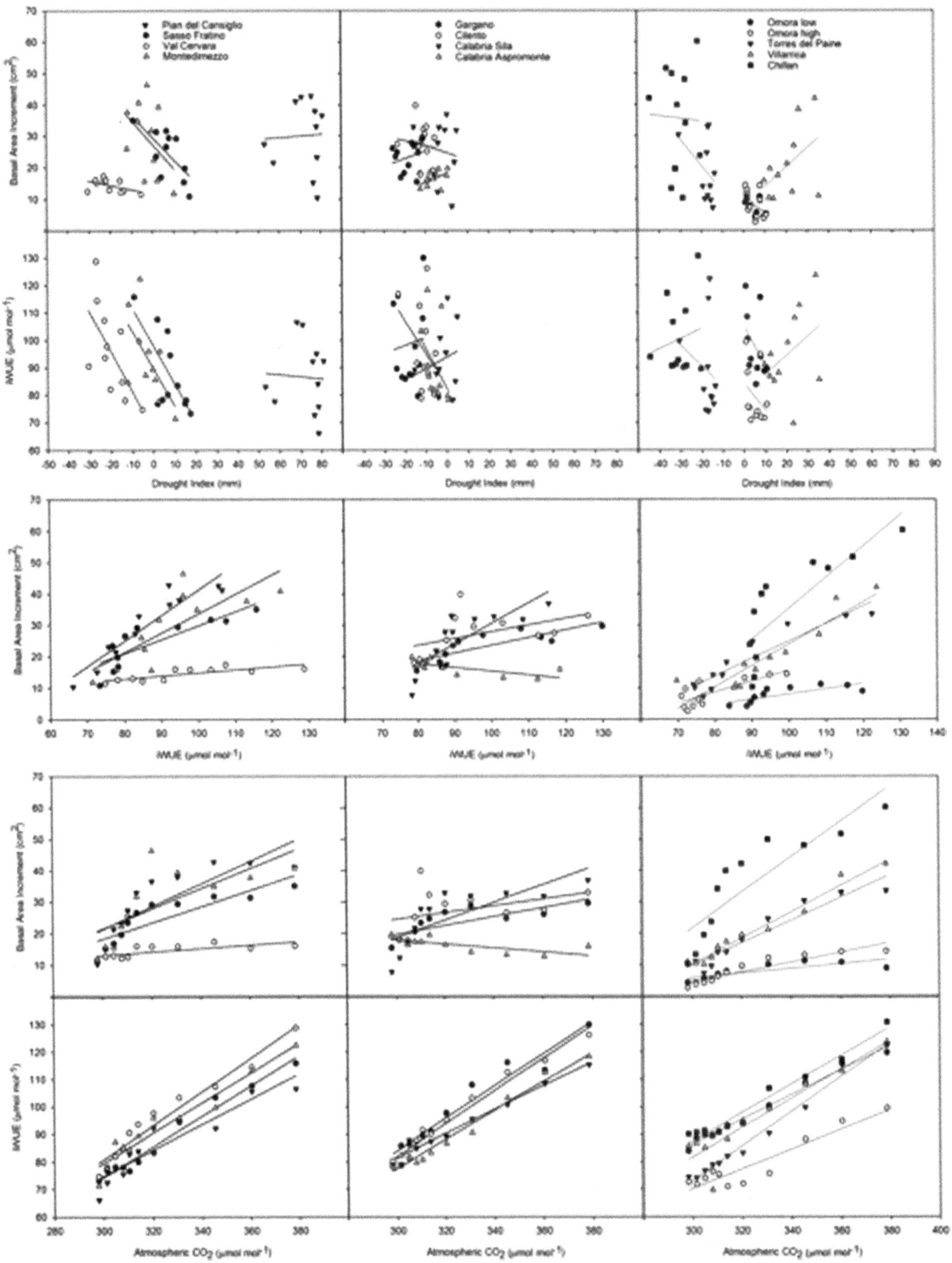

Figure 4. Relationships between BAI and DI, iWUE and DI, BAI and iWUE, BAI and Ca, iWUE and Ca. Sites are referred to by symbols in the legend: Pian del Cansiglio, Sasso Fratino, Val Cervara, and Montedimezzo (North-Central Italy); Gargano, Cilento, Aspromonte, and Sila (Southern Italy); Omora low, Omora high, Torres del Paine, Villarrica, and Chillan (Chile). Site and species details are reported in Table 1. Regression coefficients are reported in Table 2.

Table 3. Regression coefficients of the relationships between variables as shown in Fig. 4.

		Pian del Cansiglio	Sasso Fratino	Val Cervara	Montedimezzo	Gargano	Cilento	Calabria Sila	Calabria Aspromonte	Omora low	Omora high	Torres del Paine	Villarrica	Chillan
BAI/DI	b	26.28	29.64	11.69	27.43	25.58	25.37	24.74	18.14	9.16	11.76	4.11	7.71	33.19
	a	0.05	−0.71	−0.14	−0.77	0.24	−0.16	−0.24	0.29	−0.38	−0.65	−0.82	0.62	−0.08
	R^2	0.01	0.41	0.15	0.10	0.00	0.00	0.00	0.17	0.16	0.16	0.06	0.15	0.00
	p-value	-	≤0.05	-	-	-	-	-	-	-	-	-	-	-
IWUE/DI	b	91.89	98.05	66.73	89.69	103.71	82.94	93.83	82.13	105.27	84.14	75.63	80.67	111.85
	a	−0.08	−1.39	−1.46	−1.36	0.28	−1.20	0.42	−1.47	−1.49	−0.97	−0.74	0.69	0.37
	R^2	0.00	0.45	0.34	0.00	0.00	0.03	0.00	0.09	0.09	0.01	0.00	0.07	0.00
	p-value	-	≤0.05	≤0.05	≤0.05	-	-	-	-	-	-	-	-	-
BAI/IWUE	b	−39.89	−14.53	5.30	−27.84	−1.22	6.35	−34.96	27.14	−8.73	−19.13	−30.84	−43.88	−63.52
	a	0.81	0.44	0.09	0.61	0.25	0.22	0.66	−0.12	0.17	0.35	0.56	0.68	0.99
	R^2	0.88	0.70	0.61	0.61	0.67	0.25	0.65	0.49	0.60	0.62	0.88	0.85	0.68
	p-value	≤0.01	≤0.01	≤0.01	≤0.05	≤0.01	-	≤0.01	≤0.05	≤0.01	≤0.01	≤0.01	≤0.01	≤0.01
BAI/Ca	b	−86.83	−58.54	−3.15	−74.35	−22.71	−7.45	−65.75	36.92	−16.60	−42.88	−98.16	−114.81	−147.65
	a	0.36	0.26	0.05	0.32	0.14	0.11	0.28	−0.06	0.07	0.16	0.36	0.42	0.56
	R^2	0.70	0.74	0.50	0.52	0.59	0.16	0.61	0.49	0.58	0.87	0.91	0.97	0.78
	p-value	≤0.01	≤0.01	≤0.05	≤0.05	≤0.01	-	≤0.01	≤0.05	≤0.01	≤0.01	≤0.01	≤0.01	≤0.01
IWUE/Ca	b	−63.97	−91.83	−103.42	−84.32	−92.96	−97.40	−49.46	−80.51	−48.24	−37.46	−115.27	−77.54	−69.04
	a	0.46	0.55	0.61	0.55	0.59	0.60	0.44	0.53	0.45	0.36	0.63	0.53	0.52
	R^2	0.86	0.98	0.96	0.93	0.95	0.97	0.97	0.97	0.99	0.88	0.98	0.85	0.96
	p-value	≤0.01	≤0.01	≤0.01	≤0.01	≤0.01	≤0.01	≤0.01	≤0.01	≤0.01	≤0.01	≤0.01	≤0.01	≤0.01

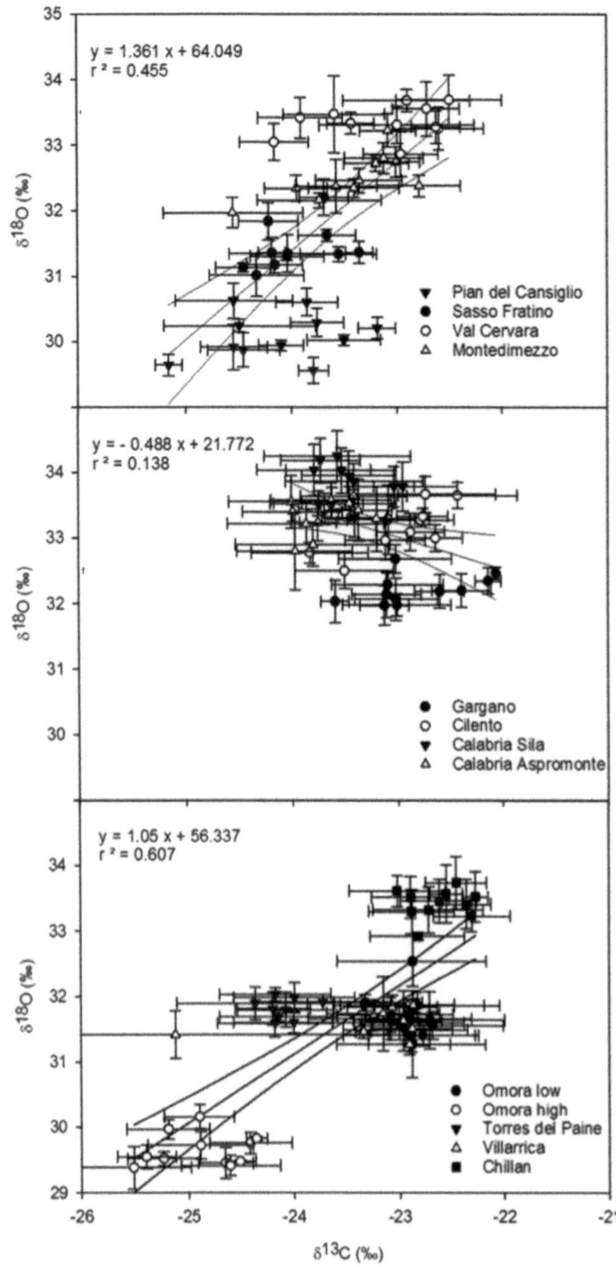

Figure 5. Relationships between $\delta^{18}O$ and $\delta^{13}C$ of cellulose for trees at Pian del Cansiglio, Sasso Fratino, Val Cervara, and Montedimezzo (North-Central Italy); Gargano, Cilento, Aspromonte, and Sila (Southern Italy); Omora low, Omora high, Torres del Paine, Villarrica, and Chillan (Chile). Populations are listed from North to South (Northern Hemisphere) or South to North (Southern Hemisphere). Symbols with the error bars (±1 standard error, n=4) are mean values per each site. The regression equation and coefficient of correlation are also shown ($P \leq 0.01$). Site and species details are reported in Table 1.

The effect of climate

The index used here to describe soil water availability to trees cannot be assumed a priori to be indicative of soil moisture deficit, though providing straight interpretable values for correlation purposes [78] [2] [79]. Long-term trends in BAI and iWUE were uncoupled with the estimated drought index, regardless of the site (Fig. 4). This is consistent with oscillating climate patterns and minor drought cycles at these sites. Differences in soil water availability and stand structural traits among sites could also explain differences in iWUE and stable isotopes and clustering of populations [80]. Since iWUE is function of Ci and Ca, a significant relation (from linear to asymptotic, depending on evaporative demand and water availability) between iWUE and Ca was expected (Fig. 4) [65]. The linear correlation found for *Nothofagus* spp. (and to a lesser extent in populations of Southern Italy) contrasts with asymptotic-like curves as in beech of North-Central Italy, where iWUE may have not increased despite the rise in Ca [19].

Fagus and *Nothofagus* were rather sensitive to increasing Ca (Fig. 4), in the absence of climatic stress. However, our results support only in part the prediction that tree ring cellulose $\delta^{18}O$ will necessarily increase with aridity, because significant correlations between drought index and cellulose $\delta^{18}O$ was evident for sites of North-Central Italy (and for those of Chile, to a lesser extent), taken together, though this was not the case for sites of Southern Italy. Since no influence of marked drought periods was perceived, predictions that follow Ca effects over Ci and iWUE, together with rising temperature and related factors (e.g., length of growing season), would increase or stabilize BAI in healthy trees [81]. The relationships between BAI and iWUE evidenced a saturation effect of tree growth in several populations. Indeed, a decrease in stomatal conductance might induce an increase in iWUE, without changing photosynthetic rate. Regardless of the cause for a reduction in stomatal conductance, an increase in iWUE would not necessarily translate into enhanced BAI [82].

Recently, temperate forests have shown major growth declines, high levels of mortality and delayed multi-year effects from drought and heat spells [83]. While the available evidence is not yet conclusive, if summers become drier, trees growing on mesic sites may undergo growth reductions; whereas at their dry distribution limit, growth of *Fagus* and *Nothofagus* may collapse, inducing dieback and compromising the provision of ecosystem services. *Fagus* and *Nothofagus* have, however, high intra-specific plasticity [84] [85] [86] [87] [88] [89], probably driven by environmental factors, though with weak coordination among functional traits. In a *F. sylvatica* stand of Central Italy, [90] found an increased iWUE with decreasing soil water availability, suggesting that trees were able to adjust carbon–water balance to prevent carbon depletion, maintaining plant growth to some extent during summer.

Physiological implications of correlations between $\delta^{18}O$ and $\delta^{13}C$

In populations with positive correlations between the isotopes, an increased stomatal control of Ci and photosynthesis is expected [91] [92], the slope of the relationship varying with vapour pressure deficit and the sensitivity of plants to evaporative demand [28]. In beech, a positive relationship between $\delta^{18}O$ and $\delta^{13}C$ has been linked to a negative relationship of $\delta^{13}C$ and $\delta^{18}O$ with air moisture and stomatal conductance [93] [94], regulated by leaf transpiration. Yet, an increase in stomatal conductance might decrease evaporative enrichment in ^{18}O by cooling the leaf and by "flushing" it with un-enriched soil-water at higher transpiration rates (Péclet effect) and increase discrimination against ^{13}C (more negative $\delta^{13}C$), because CO_2 supply is greater [95].

In most populations, conversely, $\delta^{13}C$ and $\delta^{18}O$ showed a weak relationship (Fig. 5), suggesting that variation in photosynthetic capacity exerted a controlling influence on Ci and $\delta^{13}C$, and that stomatal conductance was not suppressed during summer. Indeed, if photosynthetic capacity drives change in tree-ring $\delta^{13}C$, then variation in stomatal conductance can be detected from $\delta^{18}O$, and the two isotopes will be either negatively (water is scarce) or not correlated [96], with stomata showing a limited operational range of water availability. On the other hand, positive relationships between $\delta^{13}C$ and $\delta^{18}O$ are mainly related to stomatal limitation and photosynthetic rate is relatively unaffected, which may be expected when there is no need to reduce water loss (water is available), and stomata may operate over a wide range. However, repeated drought may reduce both the sensitivity of stomata to water stress and stomatal control of photosynthesis in *F. sylvatica* [97], metabolic impairment becoming the major limitation to photosynthesis, notwithstanding stomatal closure. Variation in internal conductance of CO_2 may contribute to year-to-year differences in tree-ring $\delta^{13}C$, also potentially dampening the correlation between $\delta^{13}C$ and $\delta^{18}O$, with change in tree-ring $\delta^{18}O$ more strongly related to climate than $\delta^{13}C$ [91].

Differences in vapour pressure deficit can modify leaf evaporative enrichment and modify cellulose $\delta^{18}O$ without necessarily influencing stomatal conductance and its relationship with photosynthetic rate [98]. In the conceptual model used for interpreting $\delta^{13}C$, $\delta^{18}O$ is plotted against $\delta^{13}C$ and the direction of change between conditions is considered diagnostic for interpreting $\delta^{13}C$ signals [96]. However, it requires that environmental influences on evaporative enrichment (i.e., source water, atmospheric vapour $\delta^{18}O$ and vapour pressure deficits) were similar over time or that they also affect stomatal conductance. Post-assimilation fractionations/processes might alter physiological and climatic signals recorded by carbon and oxygen isotopes at the leaf level, which would be decoupled from signals stored in tree rings, adding uncertainties to the dual-isotope conceptual model [99] [100].

Conclusions

The global rise in CO_2 and the changing climate have already influenced the gas exchange of beech forests in both Northern and Southern Hemispheres. A continuous enhancement in iWUE was observed at all the sites during the twentieth century, which was primarily related to changes in Ca. We also observed a common enhancement in BAI that, as for iWUE, was uncoupled with the estimated climate index. The combined analysis of tree rings and stable isotopes yielded insights applicable to long-term responses of these temperate forests to important drivers of global change that might have had negative (drought), or positive (CO_2 fertilization), and even mixed (warming), effects on tree growth and stand ecohydrology [101]. Increasing climatic stress will have negative impacts on the performance of drought-exposed populations.

The control over tree growth by climatic factors within a region differed across latitudinal gradients in environmental conditions, but only to a certain extent. Species- and site-specific responses to long-term environmental changes were unravelled, the dual isotope approach showing how plants were subject to different environmental influences and suggesting that variation in photosynthetic rates still exert a controlling influence on Ci. Population-specific resilience/vulnerability should be taken into account when predicting future forest dynamics under changing climatic conditions. Highly resolved stable isotope analyses would allow for the signals of dry and rainy season to be distinguished more clearly, and improve our understanding of the hydrological cycle over these forest ecosystems.

Acknowledgments

We acknowledge the help of Claudia Cocozza, Emanuele Conte, Andrea Andreoli, and Anibal Pauchard. We thank CONAF (Chile) for assistance in site selection. The research is linked to activities conducted within the COST FP1106 (STReESS - Studying Tree Responses to extreme Events: a SynthesiS) and COST ES1203 (SENSFOR - Enhancing the resilience capacity of SENSitive mountain FORest ecosystems under environmental change). The authors would like to thank the reviewers for their comments that help improve the manuscript.

Author Contributions

Conceived and designed the experiments: RT FL BL MM. Performed the experiments: RT FL BL MM. Analyzed the data: RT FL BL PC. Contributed reagents/materials/analysis tools: RT FL BL PC MM. Wrote the paper: RT FL BL PC MM.

References

1. Douville H, Ribes A, Decharme B, Alkama R, Sheffield J (2012) Anthropogenic influence on multidecadal changes in reconstructed global evapotranspiration. Nature Climate Change 3: 59–62.
2. Kress A, Saurer M, Siegwolf RTW, Frank DC, Esper J, et al. (2010) A 350 year drought reconstruction from Alpine tree ring stable isotopes, Global Biogeochemical Cycles 24: GB2011.
3. Dai A (2011) Characteristics and trends in various forms of the Palmer Drought Severity Index during 1900–2008. Journal of Geophysical Research 116: D12115.
4. Allen CD, Macalady AK, Chenchouni H, Bachelet D. McDowell N, et al. (2010) A global overview of drought and heat induced tree mortality reveals emerging climate change risks for forests. Forest Ecology and Management 259: 660–684.
5. McDowell NG, Beerling DJ, Breshears DD, Fisher RA, Raffa KF, et al. (2011) The interdependence of mechanisms underlying climate-driven vegetation mortality. Trends in Ecology and Evolution 26: 523–532.
6. Suarez ML, Ghermandi L, Kitzberger T (2004) Factors predisposing episodic drought-induced tree mortality in *Nothofagus*: site, climatic sensitivity and growth trends. Journal of Ecology 92: 954–966.
7. Anderegg WRL, Berry JA, Field CB (2012) Linking definitions, mechanisms, and modeling of drought-induced tree death. Trends in Plant Science 17: 693–700.
8. Choat B, Jansen S, Brodribb TJ, Cochard H, Delzon S, et al. (2012) Global convergence in the vulnerability of forests to drought. Nature 491: 752–755.
9. Hanewinkel M, Cullmann DA, Schelhaas M-J, Nabuurs G-J, Zimmermann NE (2012) Climate change may cause severe loss in the economic value of European forest land. Nature Climate Change 3: 203–207.
10. Andreu-Hayles L, Planells O, Gutiérrez E, Muntan E, Helle G, et al. (2011) Long tree-ring chronologies reveal 20th century increases in water-use efficiency but no enhancement of tree growth at five Iberian pine forests. Global Change Biology 17: 2095–2112.
11. Gagen M, Finsinger W, Wagner-Cremer F, McCarroll D, Loader NJ, et al. (2011) Evidence of changing intrinsic water-use efficiency under rising atmospheric CO_2 concentrations in Boreal Fennoscandia from subfossil leaves and tree ring $\delta^{13}C$ ratios. Global Change Biology 17: 1064–1072.
12. Duquesnay A, Bréda N, Stievenard M, Dupouey JL (1998) Changes of tree-ring $\delta^{13}C$ and water-use efficiency of beech (*Fagus sylvatica* L.) in north-eastern France during the past century. Plant, Cell and Environment 21: 565–572.
13. Tognetti R, Longobucco A, Miglietta F, Raschi A (1999) Water relations, stomatal response and transpiration of *Quercus pubescens* trees during summer in a Mediterranean carbon dioxide spring. Tree Physiology 19: 261–270.
14. Tognetti R, Johnson JD, Michelozzi M, Raschi A (1998) Response of foliar metabolism in mature trees of *Quercus pubescens* and *Quercus ilex* to long-term elevated CO_2. Environmental and Experimental Botany 39: 233–245.
15. Switsur R, Waterhouse J (1998) Stable isotopes in tree ring cellulose. In: Griffiths H, editor. Stable Isotopes: Integration of Biological, Ecological and Geochemical Processes. Oxford: BIOS Scientific Publishers Ltd. pp. 303–321.
16. Peñuelas J, Canadell JG, Ogaya R (2011) Increased water-use efficiency during the 20th century did not translate into enhanced tree growth. Global Ecology and Biogeography 20: 597–608.
17. Silva LCR, Anand M, Leithead MD (2010) Recent widespread tree growth decline despite increasing atmospheric CO_2. PLoS ONE 5: e11543.
18. Linares JC, Camarero JJ (2012) From pattern to process: linking intrinsic water-use efficiency to drought-induced forest decline. Global Change Biology 18: 1000–1015.
19. Peñuelas J, Hunt JM, Ogaya R, Jump AS (2008) Twentieth century changes of tree-ring $\delta^{13}C$ at the southern range-edge of *Fagus sylvatica*: increasing water use efficiency does not avoid the growth decline induced by warming at low altitudes. Global Change Biology 14: 1076–1088.
20. Piovesan G, Biondi F, Di Filippo A, Alessandrini A, Maugeri M (2008) Drought-driven growth reduction in old beech (*Fagus sylvatica* L.) forests of the central Apennines, Italy. Global Change Biology 14: 1265–1281.

21. Di Filippo A, Biondi F, Maugeri M, Schirone B, Piovesan G (2012) Bioclimate and growth history affect beech lifespan in the Italian Alps and Apennines. Global Change Biology 18: 960–972.
22. Weber P, Bugmann H, Pluess AR, Walthert L, Rigling A (2013) Drought response and changing mean sensitivity of European beech close to the dry distribution limit. Trees 27: 171–181.
23. Villalba R, Lara A, Masiokas MH, Urrutia R, Luckman BH, et al. (2012) Unusual Southern Hemisphere tree growth patterns induced by changes in the Southern Annular Mode. Nature Geoscience 5: 793–798.
24. Lara A, Aravena JC, Wolodarsky-Franke A, Villalba R, Luckman B, et al. (2001) Dendroclimatology of high-elevation *Nothofagus pumilio* forests in the central Andes of Chile. Canadian Journal of Forestry Research 31: 925–936.
25. Srur AM, Villalba R, Villagra PE, Hertel D (2008) Influences of climatic and CO_2 concentration changes on radial growth of *Nothofagus pumilio* in Patagonia. Revista Chilena de Historia Natural 81: 239–256.
26. Farquhar GD, O'Leary MH, Berry JA (1982) On the relationship between carbon isotope discrimination and the intercellular carbon dioxide concentration in leaves. Australian Journal of Plant Physiology 9: 121–137.
27. Leavitt SW, Long A (1986) Stable-carbon isotope variability in tree foliage and wood. Ecology 67: 1002–1010.
28. Barbour MM, Walcroft AS, Farquhar GD (2002) Seasonal variation in $\delta^{13}C$ and $\delta^{18}O$ of cellulose from growth rings of *Pinus radiata*. Plant, Cell and Environment 25: 1483–1499.
29. Ripullone F, Guerrieri MR, Saurer M, Siegwolf R, Jäggi M, et al. (2009) Testing a dual isotope model to track carbon and water gas exchanges in a Mediterranean forest. iForest 2: 59–66.
30. Barnard HR, Brooks JR, Bond BJ (2012) Applying the dual-isotope conceptual model to interpret physiological trends under controlled conditions. Tree Physiology 32: 1183–1198.
31. Saurer M, Aellen K, Siegwolf R (1997) Correlating $\delta^{13}C$ and $\delta^{18}O$ in cellulose of trees. Plant, Cell and Environment 20: 1543–1550.
32. Grossiord C, Granier A, Ratcliffe S, Bouriaud O, Bruelheide H, et al. (2014) Tree diversity does not always improve resistance of forest ecosystems to drought. Proceedings of the National Academy of Sciences 111: 14812–14815.
33. Lombardi F, Lasserre B, Tognetti R, Marchetti M (2008) Deadwood in relation to stand management and forest type in Central Apennines (Molise, Italy). Ecosystems 11: 882–894.
34. Piovesan G, Biondi F, Bernabei M, Di Filippo A, Schirone B (2005) Spatial and altitudinal bioclimatic zones of the Italian peninsula identified from a beech (*Fagus sylvatica* L.) tree-ring network. Acta Oecologica 27: 197–210.
35. Marchetti M, Tognetti R, Lombardi F, Chiavetta U, Palumbo G, et al. (2010) Ecological portrayal of old-growth forests and persistent woodlands in the Cilento and Vallo di Diano National Park (southern Italy). Plant Biosystems 144: 130–147.
36. Lombardi F, Cocozza C, Lasserre B, Tognetti R, Marchetti M (2011) Dendrochronological assessment of the time since death of dead wood in an old growth Magellan's beech forest, Navarino Island (Chile). Austral Ecology 36: 329–340.
37. Harris I., Jones PD, Osborn TJ, Lister DH (2014) Updated high-resolution grids of monthly climatic observations. International Journal of Climatology 34: 623–642.
38. Hargraeves GH, Samani ZA (1982) Estimating potential evapotranspiration. Journal of Irrigation and Drainage Division 108: 225–230.
39. Holmes RL (1983) Computer-assisted quality control in tree ring dating and measurement. Tree-Ring Bulletin 43: 69–78.
40. Kaennel M, Schweingruber FH (1995) Multilingual Glossary of Dendrochronology. Terms and Definitions in English, Spanish, Italian, Portuguese, and Russian. Bern: Haupt.
41. Fritts H (1976) Tree Rings and Climate. Academic Press: New York.
42. Cook ER, Briffa K, Shiyatov S, Mazepa V (1990) Tree-ring standardization and growth-trend estimation. In: Cook ER, Kairiukstis L, editors. Methods of Dendrochronology. Amsterdam: Kluwer. pp. 104–132.
43. Borella S, Leuenberger M, Saurer M, Siegwolf R (1998) Reducing uncertainties in $\delta^{13}C$ analysis of tree rings: poolings, milling and cellulose extraction. Journal of Geophysical Research 103: 19519–19526.
44. Loader NJ, Robertson I, Barker AC, Switsur VR, Waterhouse JS (1997) An improved technique for the batch processing of small whole wood samples to α-cellulose. Chemical Geology 136: 313–317.
45. Battipaglia G, Jäggi M, Saurer M, Siegwolf RTW, Cotrufo MF (2008) Climatic sensitivity of $\delta^{18}O$ in the wood and cellulose of tree rings: results from a mixed stand of *Acer pseudoplatanus* L. and *Fagus sylvatica* L. Palaeogeography Palaeoclimatology, Palaeoecology 261: 193–202.
46. Francey RJ, Allison CE, Etheridge DM, Trudinger CM, Enting IG, et al. (1999) A 1000-year high precision record of $\delta^{13}C$ in atmospheric CO_2. Tellus Series B 51: 170–193.
47. McCarroll D, Loader NJ (2004) Stable isotopes in tree rings. Quaternary Science Reviews 23: 771–801.
48. O'Leary M (1981) Carbon isotope fractionation in plants. Phytochemistry 20: 553–567.
49. Farquhar GD, Richards RA (1984) Isotopic composition of plant carbon correlates with water-use efficiency of wheat genotypes. Australian Journal of Plant Physiology 11: 539–552.
50. Osmond CB, Björkman O, Anderson DJ (1980) Physiological Processes in Plant Ecology. Toward a Synthesis with Atriplex. New York: Springer.
51. Francey RJ, Farquhar GD (1982) An explanation of $^{13}C/^{12}C$ variations in tree rings. Nature 297: 28–31.
52. Mann H (1945) Nonparametric tests against trend. Econometrica 13: 245–259.
53. Kendall M (1976) Rank Correlation Methods. 4th edn. London: Griffin.
54. Feng X (1999) Trends in intrinsic water-use efficiency of natural trees for the past 100–200 years: a response to atmospheric CO_2 concentrations. Geochimica et Cosmochimica Acta 63: 1891–1903.
55. Marshall JD, Monserud RA (1996) Homeostatic gas-exchange parameters inferred from $^{13}C/^{12}C$ in tree rings of conifers. Oecologia 105: 13–21.
56. Hietz P, Wanek W, Dunisch O (2005) Long-term trends in cellulose delta C-13 and water-use efficiency of tropical *Cedrela* and *Swietenia* from Brazil. Tree Physiology 25: 745–752.
57. Silva LCR, Anand M, Oliveira JM, Pillar VD (2009) Past century changes in *Araucaria angustifolia* (Bertol.) Kuntze water use efficiency and growth in forest and grassland ecosystems of southern Brazil: implications for forest expansion. Global Change Biology 15: 2387–2396.
58. Anderson WT, Sternberg LSL, Pinzon MC, Gann-Troxler T, Childers DL, et al. (2005) Carbon isotopic composition of cypress trees from South Florida and changing hydrologic conditions. Dendrochronologia 23: 1–10.
59. Nock CA, Baker PJ, Wanek W, Leis A, Grabner M, et al. (2011) Long-term increases in intrinsic water-use efficiency do not lead to increased stem growth in a tropical monsoon forest in western Thailand. Global Change Biology 17: 1049–1063.
60. Körner C, Farquhar GD, Wong SC (1991) Carbon isotope discrimination by plants follows latitudinal and altitudinal trends. Oecologia 88: 30–40.
61. Silva LCR, Anand M (2012) Probing for the influence of atmospheric CO_2 and climate change on forest ecosystems across biomes. Global Ecology and Biogeography 22: 83–92.
62. Silva LCR, Horwath WR (2013) Explaining global increases in water use efficiency: why have we overestimated responses to rising atmospheric CO_2 in natural forest ecosystems? PLoS ONE 8: e53089.
63. Barber VA, Juday GP, Finney BP (2000) Reduced growth of Alaskan white spruce in the twentieth century from temperature-induced drought stress. Nature 405: 668–673.
64. Linares JC, Delgado-Huertas A, Camarero JJ, Merino J, Carreira JA (2009) Competition and drought limit the response of water-use efficiency to rising atmospheric carbon dioxide in the Mediterranean fir *Abies pinsapo*. Oecologia 161: 611–624.
65. Salzer MW, Hughes MK, Bunn AG, Kipfmueller KF (2009) Recent unprecedented treering growth in bristlecone pine at the highest elevations and possible causes. Proceedings of the National Academy of Sciences USA 106: 20348–20353.
66. McMahon SM, Parker GG, Miller DR (2010) Evidence for a recent increase in forest growth. Proceedings of the National Academy of Sciences USA 107: 3611–3615.
67. Chapin FS, Schulze ED, Mooney HA (1990) The ecology and economics of storage in plants. Annual Review of Ecology and Systematics 21: 423–447.
68. Giorgi F, Lionello P (2008) Climate change projections for the Mediterranean region. Global Planetary Change 63: 90–104.
69. Villalba R, Lara A, Boninsegna JA, Masiokas M, Delgado S, et al. (2003) Large-scale temperature changes across the southern Andes: 20th-century variations in the context of the past 400 years. Climatic Change 59: 177–232.
70. Lara A, Villalba R, Wolodarsky-Franke A, Aravena JC, Luckman BH, et al. (2005) Spatial and temporal variation in *Nothofagus pumilio* growth at tree line along its latitudinal range (35°40′–55°S) in the Chilean Andes. Journal of Biogeography 32: 879–893.
71. Gutierréz E (1988) Dendroecological study of *Fagus sylvatica* L. in the Montseny Mountains (Spain). Acta Oecologica-Oecologia Plantarum 9: 301–309.
72. Villalba R, Boninsegna JA, Veblen TT, Schmelter A, Rubulis S (1997) Recent trends in tree-ring records from high elevation sites in the Andes of Northern Patagonia. Climatic Change 36: 425–454.
73. Dittmar C, Zech W, Elling W (2003) Growth variations of common beech (*Fagus sylvatica* L.) under different climatic and environmental conditions in Europe - a dendroecological study. Forest Ecology and Management 173: 63–78.
74. Bascietto M, Cherubini P, Scarascia-Mugnozza G (2004) Tree rings from a European beech forest chronosequence are useful for detecting growth trends and carbon sequestration. Canadian Journal of Forest Research 34: 481–492.
75. Cullen LE, Stewart GH, Duncan RP, Palmer JG (2001) Disturbance and climate warming influences in New Zealand *Nothofagus* tree-line population dynamics. Journal of Ecology 89: 1061–1071.
76. Daniels LD, Veblen TT (2004) Spatiotemporal influences of climate on altitudinal treeline in northern Patagonia. Ecology 85: 1284–1296.
77. Holtmeier FK, Broll G (2005) Sensitivity and response of northern hemisphere altitudinal and polar treelines to environmental change at landscape and local scales. Global Ecology and Biogeography 14: 395–410.
78. Eilmann B, Rigling A (2012) Tree-growth analyses to estimate tree species' drought tolerance. Tree Physiology 32: 178–187.
79. Rigling A, Bigler C, Eilmann B, Feldmeyer-Christe E, Gimmi U, et al. (2013) Driving factors of a vegetation shift from Scots pine to pubescent oak in dry Alpine forests. Global Change Biology 19: 229–240.

80. Härdtle W, Niemeyer T, Assmann T, Baiboks S, Fichtner A, et al. (2013) Long-term trends in tree-ring width and isotope signatures ($\delta^{13}C$, $\delta^{15}N$) of *Fagus sylvatica* L. on soils with contrasting water supply. Ecosystems 16: 1413–1428.
81. Li JB, Cook ER, D'Arrigo R, Chen FH, Gou XH, et al. (2008) Common tree growth anomalies over the northeastern Tibetan Plateau during the last six centuries: implications for regional moisture change. Global Change Biology 14: 2096–2107.
82. Granda E, Rossatto DR, Camarero JJ, Voltas J, Valladares F (2014) Growth and carbon isotopes of Mediterranean trees reveal contrasting responses to increased carbon dioxide and drought. Oecologia 174: 307–317.
83. Ciais P, Reichstein M, Viovy N, Granier A, Ogée J, et al. (2005) Europe-wide reduction in primary productivity caused by the heat and drought in 2003. Nature 437: 529–533.
84. Borghetti M, Leonardi S, Raschi A, Snyderman D, Tognetti R (1993) Ecotypic variation of xylem embolism, phenological traits, growth parameters and allozyme characteristics in *Fagus sylvatica*. Functional Ecology 7: 713–720.
85. Tognetti R, Johnson JD, Michelozzi M (1995) The response of European beech (*Fagus sylvatica* L.) seedlings from two Italian populations to drought and recovery. Trees 9: 348–354.
86. Tognetti R, Minotta G, Pinzauti S, Michelozzi M, Borghetti M (1998) Acclimation to changing light conditions of long-term shade-grown beech (*Fagus sylvatica* L.) seedlings of different geographic origins. Trees 12: 326–333.
87. Premoli AC, Raffaele E, Mathiasen P (2007) Morphological and phenological differences in *Nothofagus pumilio* from contrasting elevations: evidence from a common garden. Austral Ecology 32: 515–523.
88. Fajardo A, Piper FI (2011) Intraspecific trait variation and covariation in a widespread tree species (*Nothofagus pumilio*) in southern Chile. New Phytologist 189: 259–271.
89. Richardson SJ, Allen RB, Buxton RP, Easdale TA, Hurst JM, et al. (2013) Intraspecific relationships among wood density, leaf structural traits and environment in four co-occurring species of *Nothofagus* in New Zealand. PLoS ONE 8: e58878.
90. Scartazza A, Moscatello S, Matteucci G, Battistelli A, Brugnoli E (2013) Seasonal and inter-annual dynamics of growth, non-structural carbohydrates and C stable isotopes in a Mediterranean beech forest. Tree Physiology 33: 730–742.
91. Cullen LE, Adams MA, Anderson MJ, Grierson PF (2008) Analyses of $\delta^{13}C$ and $\delta^{18}O$ in tree rings of *Callitris columellaris* provide evidence of a change in stomatal control of photosynthesis in response to regional changes in climate. Tree Physiology 28: 1525–1533.
92. Tognetti R, Lombardi F, Lasserre B, Battipaglia G, Saurer M, et al. (2012) Tree-ring responses in *Araucaria araucana* to two major eruptions of Lonquimay Volcano (Chile). Trees 26: 1805–1819.
93. Keitel C, Matzarakis A, Rennenberg H, Gessler A (2006) Carbon isotopic composition and oxygen isotopic enrichment in phloem and total leaf organic matter of European beech (*Fagus sylvatica* L.) along a climate gradient. Plant, Cell and Environment 29: 1492–1507.
94. Grams TEE, Kozovits AR, Haberle K-H, Matyssek R, Dawson TE (2007) Combining $\delta^{13}C$ and $\delta^{18}O$ analysis to unravel competition, CO_2 and O_3 effects on the physiological performance of different-aged trees. Plant Cell and Environment 30: 1023–1034.
95. Barbour MM (2007) Stable oxygen isotope composition of plant tissue: a review. Functional Plant Biology 34: 83–94.
96. Scheidegger Y, Saurer M, Bahn M, Siegwolf RTW (2000) Linking stable oxygen and carbon isotopes with stomatal conductance and photosynthetic capacity: a conceptual model. Oecologia 125: 350–357.
97. Tognetti R, Michelozzi M, Borghetti M (1994) Response to light of shade-grown beech seedlings subjected to different watering regimes. Tree Physiology 14: 751–758.
98. Roden JS, Farquhar GD (2012) A controlled test of the dual-isotope approach for the interpretation of stable carbon and oxygen isotope ratio variation in tree rings. Tree Physiology 32: 490–503.
99. Gessler A, Ferrio JP, Hommel R, Treydte K, Werner RA, et al. (2014) Stable isotopes in tree rings: towards a mechanistic understanding of isotope fractionation and mixing processes from the leaves to the wood. Tree Physiolgy in press.
100. Song X, Farquhar GD, Gessler A, Barbour MM (2014) Turnover time of the non-structural carbohydrate pool influences $\delta^{18}O$ of leaf cellulose. Plant, Cell and Environment in press.
101. Schollaen K, Heinrich I, Neuwirth B, Krusic PJ, D'Arrigo RD, et al. (2013) Multiple tree-ring chronologies (ring width, $\delta^{13}C$ and $\delta^{18}O$) reveal dry and rainy season signals of rainfall in Indonesia. Quaternary Science Reviews 73: 170–181.

Permissions

All chapters in this book were first published in PLOS ONE, by The Public Library of Science; hereby published with permission under the Creative Commons Attribution License or equivalent. Every chapter published in this book has been scrutinized by our experts. Their significance has been extensively debated. The topics covered herein carry significant findings which will fuel the growth of the discipline. They may even be implemented as practical applications or may be referred to as a beginning point for another development.

The contributors of this book come from diverse backgrounds, making this book a truly international effort. This book will bring forth new frontiers with its revolutionizing research information and detailed analysis of the nascent developments around the world.

We would like to thank all the contributing authors for lending their expertise to make the book truly unique. They have played a crucial role in the development of this book. Without their invaluable contributions this book wouldn't have been possible. They have made vital efforts to compile up to date information on the varied aspects of this subject to make this book a valuable addition to the collection of many professionals and students.

This book was conceptualized with the vision of imparting up-to-date information and advanced data in this field. To ensure the same, a matchless editorial board was set up. Every individual on the board went through rigorous rounds of assessment to prove their worth. After which they invested a large part of their time researching and compiling the most relevant data for our readers.

The editorial board has been involved in producing this book since its inception. They have spent rigorous hours researching and exploring the diverse topics which have resulted in the successful publishing of this book. They have passed on their knowledge of decades through this book. To expedite this challenging task, the publisher supported the team at every step. A small team of assistant editors was also appointed to further simplify the editing procedure and attain best results for the readers.

Apart from the editorial board, the designing team has also invested a significant amount of their time in understanding the subject and creating the most relevant covers. They scrutinized every image to scout for the most suitable representation of the subject and create an appropriate cover for the book.

The publishing team has been an ardent support to the editorial, designing and production team. Their endless efforts to recruit the best for this project, has resulted in the accomplishment of this book. They are a veteran in the field of academics and their pool of knowledge is as vast as their experience in printing. Their expertise and guidance has proved useful at every step. Their uncompromising quality standards have made this book an exceptional effort. Their encouragement from time to time has been an inspiration for everyone.

The publisher and the editorial board hope that this book will prove to be a valuable piece of knowledge for researchers, students, practitioners and scholars across the globe.

List of Contributors

Xiaojun Kou
State Key Laboratory of Earth Surface Processes and Resource Ecology, Beijing Normal University, Beijing, PR China
College of Life Sciences, Beijing Normal University, Beijing, PR China

Qin Li
Department of Botany, University of British Columbia, Vancouver, B.C., Canada

Carl Beierkuhnlein
Department of Biogeography, University of Bayreuth, Bayreuth, Germany

Yiheng Zhao
College of Life Sciences, Beijing Normal University, Beijing, PR China

Shirong Liu
Institute of Forest Environment and Ecology, Chinese Academy of Forestry, Beijing, PR China

Hai Ren, Linjun Li, Shuguang Jian, Jun Wang, Hongfang Lu, Guoyi Zhou, Xuli Tang, Qianmei Zhang, Lianlian Yuan and Xubing Chen
Key Laboratory of Vegetation Restoration and Management of Degraded Ecosystems, South China Botanical Garden, Chinese Academy of Sciences, Guangzhou, China

Qiang Liu
College of Life Science, Hainan Normal University, Haikou, China

Xu Wang
College of Environment and Plant Protection, Hainan University, Haikou, China

Yide Li and Huai Yang
Research Institute of Tropical Forestry, Chinese Academy of Forestry, Guangzhou, China

Dafeng Hui
Department of Biological Sciences, Tennessee State University, Nashville, TN, United States of America

Dong Wang
School of Life Sciences, Central China Normal University, Wuhan, China

Rebecca Ostertag
Department of Biology, University of Hawai'i at Hilo, Hilo, Hawai'i, United States of America

Faith Inman-Narahari
Department of Natural Resources and Environmental Management, University of Hawai'i at Mānoa, Honolulu, Hawai'i, United States of America

Susan Cordell and Christian P. Giardina
Institute of Pacific Islands Forestry, Pacific Southwest Research Station, USDA Forest Service, Hilo, Hawai'i, United States of America

Lawren Sack
Department of Ecology and Evolutionary Biology, University of California Los Angeles, Los Angeles, California, United States of America

António M. Lopes
Institute of Mechanical Engineering, Faculty of Engineering, University of Porto, Porto, Portugal

J. A. Tenreiro Machado
Institute of Engineering, Polytechnic of Porto, Porto, Portugal

Heather Keith, David B. Lindenmayer, David Blair, Lauren Carter, Lachlan McBurney and Sachiko Okada
The Fenner School of Environment and Society, Australian National University, Building 48, Canberra, ACT, Australia

Brendan G. Mackey
Griffith Climate Change Response Program, Griffith University, Queensland, Australia

Tomoko Konishi-Nagano
Fujitsu Laboratories Ltd., Kawasaki, Japan

Massimo Ponti, Rossella Angela Perlini and Vincenzo Ventra
Dipartimento di Scienze Biologiche, Geologiche e Ambientali, University of Bologna, UO CoNISMa, Ravenna, Italy

Daniele Grech and Carlo Cerrano
Dipartimento di Scienze della Vita e dell'Ambiente, Polytechnic University of Marche, UO CoNISMa, Ancona, Italy

Marco Abbiati
Dipartimento di Scienze Biologiche, Geologiche e Ambientali, University of Bologna, UO CoNISMa, Ravenna, Italy

ISMAR, Consiglio Nazionale delle Ricerche - Istituto di Scienze Marine, Bologna, Italy

Mioko Ataka and Makoto Tani
Laboratory of Forest Hydrology, Division of Environmental Science and Technology, Graduate School of Agriculture, Kyoto University, Kyoto, Japan

Yuji Kominami, Kenichi Yoshimura and Takafumi Miyama
Kansai Research Center, Forestry and Forest Products Research Institute (FFPRI), Kyoto, Japan

Mayuko Jomura
College of Bioresource Sciences, Nihon University, Fujisawa, Kanagawa, Japan

Rosemary L. Sherriff
Department of Geography, Humboldt State University, Arcata, California, United States of America

Rutherford V. Platt
Department of Environmental Studies, Gettysburg College, Gettysburg, Pennsylvania, United States of America

Thomas T. Veblen, Tania L. Schoennagel and Meredith H. Gartner
Department of Geography, University of Colorado, Boulder, Colorado, United States of America

Tomoharu Inoue, Shin Nagai, Hadi Fadaei, Reiichiro Ishii and Rikie Suzuki
Department of Environmental Geochemical Cycle Research, Japan Agency for Marine-Earth Science and Technology (JAMSTEC), Yokohama, Japan

Satoshi Yamashita, Kimiko Okabe and Hisatomo Taki
Department of Forest Entomology, Forestry and Forest Products Research Institute (FFPRI), Tsukuba, Ibaraki, Japan

Yoshiaki Honda and Koji Kajiwara
Center of Environmental Remote Sensing, Chiba University, Chiba, Japan

Peter O. Alele
Key Laboratory of Tropical Forest Ecology, Xishuangbanna Tropical Botanical Garden (XTBG), Chinese Academy of Sciences, Kunming, Yunnan, P. R. China
University of the Chinese Academy of Sciences, Beijing, P. R. China
Great Nile Conservation Centre (GNCC), Lira, Uganda
Institute of Tropical Forest Conservation (ITFC), Mbarara University of Science and Technology (MUST), Kabale, Uganda

Douglas Sheil
Department of Ecology and Natural Resource Management, Norwegian University of Life Sciences, Ås, Norway
Center for International Forestry Research (CIFOR), Bogor, Indonesia
Department of Ecology and Natural Resource Management, School of Environment, Science and Engineering, Southern Cross University, Lismore, New South Wales, Australia
Institute of Tropical Forest Conservation (ITFC), Mbarara University of Science and Technology (MUST), Kabale, Uganda

Yann Surget-Groba
Key Laboratory of Tropical Forest Ecology, Xishuangbanna Tropical Botanical Garden (XTBG), Chinese Academy of Sciences, Kunming, Yunnan, P. R. China

Shi Lingling
Key Laboratory of Tropical Forest Ecology, Xishuangbanna Tropical Botanical Garden (XTBG), Chinese Academy of Sciences, Kunming, Yunnan, P. R. China
University of the Chinese Academy of Sciences, Beijing, P. R. China

Charles H. Cannon
Key Laboratory of Tropical Forest Ecology, Xishuangbanna Tropical Botanical Garden (XTBG), Chinese Academy of Sciences, Kunming, Yunnan, P. R. China
Texas Tech University, Lubbock, Texas, United States of America

Simon Thorn and Claus Bässler
Sachgebiet Forschung und Dokumentation, Nationalparkverwaltung Bayerischer Wald, Grafenau, Germany

Thomas Gottschalk
Hochschule für Forstwirtschaft Rottenburg, Rottenburg am Neckar, Germany

Torsten Hothorn
Abteilung Biostatistik, Universität Zürich, Zürich, Switzerland

Heinz Bussler
Bavarian State Institute for Forestry, Freising, Germany

Kenneth Raffa
Department of Entomology, University of Wisconsin-Madison, Madison, United States of America

Jörg Müller
Sachgebiet Forschung und Dokumentation, Nationalparkverwaltung Bayerischer Wald, Grafenau, Germany
Chair for Terrestrial Ecology, Department of Ecology and Ecosystem Management, Technische Universität München, Freising, Germany

Kun Huang
Key Laboratory of Ecosystem Network Observation and Modeling, Institute of Geographic Sciences and Natural Resources Research, Chinese Academy of Sciences, Beijing 100101, China
University of Chinese Academy of Sciences, Beijing 100049, China

Shaoqiang Wang, Lei Zhou, Huimin Wang and Peili Shi
Key Laboratory of Ecosystem Network Observation and Modeling, Institute of Geographic Sciences and Natural Resources Research, Chinese Academy of Sciences, Beijing 100101, China

Junhui Zhang
Institute of Applied Ecology, Chinese Academy of Sciences, Shenyang 110016, China

Junhua Yan
South China Botanical Garden, Chinese Academy of Sciences, Guangzhou 510650, China

Liang Zhao
Northwest Plateau Institute of Biology, Chinese Academy of Sciences, Xining 810001, China

Yanfen Wang
University of Chinese Academy of Sciences, Beijing 100049, China

Arnaud Goessens
Laboratory of Systems Ecology and Resource Management, Universite´ Libre de Bruxelles - ULB, CPI 264/1, Brussels, Belgium
Mangrove Research Unit (MARU), Institute of Oceanography and Environment (INOS), Universiti Malaysia Terengganu - UMT, Kuala Terengganu, Malaysia

Behara Satyanarayana
Laboratory of Systems Ecology and Resource Management, Universite´ Libre de Bruxelles - ULB, CPI 264/1, Brussels, Belgium
Mangrove Research Unit (MARU), Institute of Oceanography and Environment (INOS), Universiti Malaysia Terengganu - UMT, Kuala Terengganu, Malaysia
Laboratory of Plant Biology and Nature Management, Vrije Universiteit Brussel - VUB, Brussels, Belgium

Tom Van der Stocken, Melissa Quispe Zuniga and Farid Dahdouh-Guebas
Laboratory of Systems Ecology and Resource Management, Universite´ Libre de Bruxelles - ULB, CPI 264/1, Brussels, Belgium
Laboratory of Plant Biology and Nature Management, Vrije Universiteit Brussel - VUB, Brussels, Belgium

Husain Mohd-Lokman and Ibrahim Sulong
Mangrove Research Unit (MARU), Institute of Oceanography and Environment (INOS), Universiti Malaysia Terengganu - UMT, Kuala Terengganu, Malaysia

Claire M. P. Ozanne and Peter J. Shaw
Centre for Research in Ecology, Department of Life Sciences, University of Roehampton, London, United Kingdom

Christie Cabral
School for Social & Community Medicine, University of Bristol, Bristol, United Kingdom

Xiang Li, Jianzhi Niu and Baoyuan Xie
Key Laboratory of Soil and Water Conservation and Desertification Combating of Education Ministry, Beijing Forestry University, Beijing, China

Qiongyu Huang, Anu Swatantran and Ralph Dubayah
Department of Geographical Sciences, University of Maryland, College Park, Maryland, United States of America

Scott J. Goetz
Woods Hole Research Center, Falmouth, Massachusetts, United States of America

Pekka E. Kauppi
Department of Environmental Sciences, University of Helsinki, Helsinki, Finland
Finnish Forest Research Institute (METLA), Vantaa, Finland

Maximilian Posch
Coordination Centre for Effects (CCE), National Institute for Public Health and the Environment (RIVM), Bilthoven, The Netherlands

Pentti Pirinen
Climate Service Centre, Finnish Meteorological Institute, Helsinki, Finland

Robert C. Morrissey
Department of Forest Ecology, Faculty of Forestry and Wood Sciences, Czech University of Life Sciences, Prague, Czech Republic
Hardwood Tree Improvement and Regeneration Center, Department of Forestry and Natural Resources, Purdue University, West Lafayette, Indiana, United States of America

Michael A. Jenkins and Michael R. Saunders
Hardwood Tree Improvement and Regeneration Center, Department of Forestry and Natural Resources, Purdue University, West Lafayette, Indiana, United States of America

Lin Li and Shi Guang Wei
Guilin University of Electronic Technology, Guilin, China

Ju Yu Lian, Zhong Liang Huang and Wan Hui Ye
South China Botanical Garden, Chinese Academy of Science, Guangzhou, China

Matthias Schröter
Environmental Systems Analysis Group, Wageningen University, Wageningen, The Netherlands
Norwegian Institute for Nature Research (NINA), Trondheim/Oslo, Norway

Graciela M. Rusch, David N. Barton, Stefan Blumentrath and Björn Nordén
Norwegian Institute for Nature Research (NINA), Trondheim/Oslo, Norway

Roberto Tognetti
Dipartimento di Bioscienze e Territorio, Università degli Studi del Molise, Pesche, Italy
The EFI Project Centre on Mountain Forests (MOUNTFOR), Edmund Mach Foundation, San Michele all'Adige, Italy

Fabio Lombardi, Bruno Lasserre and Marco Marchetti
Dipartimento di Bioscienze e Territorio, Università degli Studi del Molise, Pesche, Italy

Paolo Cherubini
WSL Swiss Federal Institute for Forest, Snow and Landscape Research, Birmensdorf, Switzerland

Index